Springer Optimization and Its Applications

Volume 23

Aims and Scope

Optimization has continued to expand in all directions at an astonishing rate. New algorithmic and theoretical techniques are continually developing and the diffusion into other disciplines is proceeding at a rapid pace, with a spot light on machine learning, artificial intelligence, and quantum computing. Our knowledge of all aspects of the field has grown even more profound. At the same time, one of the most striking trends in optimization is the constantly increasing emphasis on the interdisciplinary nature of the field. Optimization has been a basic tool in areas not limited to applied mathematics, engineering, medicine, economics, computer science, operations research, and other sciences.

The series **Springer Optimization and Its Applications (SOIA)** aims to publish state-of-the-art expository works (monographs, contributed volumes, textbooks, handbooks) that focus on theory, methods, and applications of optimization. Topics covered include, but are not limited to, nonlinear optimization, combinatorial optimization, continuous optimization, stochastic optimization, Bayesian optimization, optimal control, discrete optimization, multi-objective optimization, and more. New to the series portfolio include Works at the intersection of optimization and machine learning, artificial intelligence, and quantum computing.

Volumes from this series are indexed by Web of Science, zbMATH, Mathematical Reviews, and SCOPUS.

Zdeněk Dostál

Optimal Quadratic Programming and QCQP Algorithms with Case Studies

Second Edition

Springer

Zdeněk Dostál
National Supercomputer Center and
Department of Applied Mathematics
VŠB - Technical University Ostrava
Ostrava, Czech Republic

ISSN 1931-6828 ISSN 1931-6836 (electronic)
Springer Optimization and Its Applications
ISBN 978-3-031-95166-4 ISBN 978-3-031-95167-1 (eBook)
https://doi.org/10.1007/978-3-031-95167-1

Originally published with the title: "Optimal Quadratic Programming Algorithms: with Applications to Variational Inequalities"

This Springer imprint is published by the registered company Springer Nature Switzerland AG
The registered company address is: Gewerbestrasse 11, 6330 Cham, Switzerland

To Maruška, for her support and patience

Preface

The primary purpose of this book is to present some recently developed practical algorithms for solving large bound and/or equality-constrained quadratic programming (QP) problems. The unique feature of these algorithms is the rate of convergence in terms of the bounds on the spectrum of the cost function's Hessian matrix. If we apply the algorithms to the class of QP problems with the cost functions whose Hessian has the spectrum confined to a given positive interval, they can find approximate solutions in a uniformly bounded number of simple iterations, such as the matrix-vector multiplications. Moreover, if the problems admit a sparse representation of the Hessian, it simply follows that the cost of the solution is proportional to the number of unknowns.

Notice also that the cost of duplicating the solution is proportional to the number of variables. The only difference is a constant coefficient. But the constants are essential; people are interested in their salaries, as Professor Babuška used to point out. We therefore tried hard to present a *quantitative* theory of convergence of our algorithms. In particular, we tried to avoid unspecified constants and give realistic bounds on the convergence rate, usually in terms of the extreme nonzero eigenvalues of the matrices involved in defining the problem. The theory also admits dependent constraints.

The presentation of each new algorithm is complete in the sense that it starts from its classical predecessors and describes their drawbacks, introduce modifications that improve their performance, and documents the improvements through numerical experiments. Since the exposition is self-contained, the book can serve as an introductory text for anybody interested in QP. Moreover, since we can reduce the solution of more general nonlinear problems to the solution of a sequence of QP problems, the book can also serve as a convenient introduction to effective nonlinear programming. Such presentation has also a considerable methodological appeal as it enables us to separate the simple geometrical ideas which are behind many theoretical results and algorithms, from the technical difficulties arising in the analysis of more sophisticated optimization algorithms.

Our algorithms are based on modifications of the active set strategy that are optionally combined with variants of the augmented Lagrangian method. Simple observations and careful analysis resulted in their qualitatively improved performance. Surprisingly, these methods can solve some large QP problems with less effort than a single step of the popular interior point methods. The reason is that the standard implementation of the interior point methods can hardly use a favorable distribution of the spectrum of the Hessian due to the barrier function. On the other hand, the standard implementations of interior point methods do not rely on the conditioning of the Hessian. They can efficiently exploit the Hessian's sparsity pattern to simplify LU decomposition. Hence, interior point methods can solve many problems more efficiently, and our approach may complement them.

Motivation: Contact Problems and Scalable Algorithms

The development of the algorithms presented in this book was motivated by an effort to solve the large and sparse problems arising from the discretization of elliptic variational inequalities, such as those describing the equilibrium of elastic bodies in mutual contact. A simple academic example is the contact problem of elasticity to describe the deformation and contact pressure due to volume forces of the cantilever cube over the obstacle in Fig. 1.

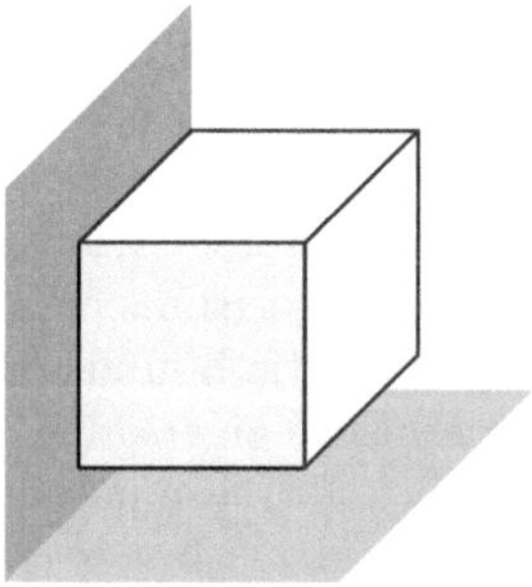

Fig. 1. Cantilever cube over the obstacle

The class of problems arising from various discretizations of a given variational inequality by the finite element or boundary element method can be reduced to the class of QP problems with a uniformly bounded spectrum by an application of the FETI/BETI (Finite/Boundary Element Tearing and Interconnecting)-based domain decomposition methods. Let us recall that the basic idea of these methods is to decompose the domain into subdomains, as in Fig. 2, and then "interconnect" them by the Lagrange multipliers found by an iterative procedure.

Combination of the results on scalability of variants of the FETI methods for unconstrained problems with the algorithms presented in this book resulted in the development of *scalable* algorithms for elliptic boundary variational inequalities. Let us recall that an algorithm is numerically scalable if the cost of the solution is nearly proportional to the number of unknowns, and it enjoys the parallel scalability if the time required for the solution can be reduced nearly proportionally to the number of available processors. For example, the solution to our toy problem needed from 111 to 133 sparse matrix-vector multiplications for varying discretizations, using from 417 to 163,275 nodes on the surface (Fig. 3).

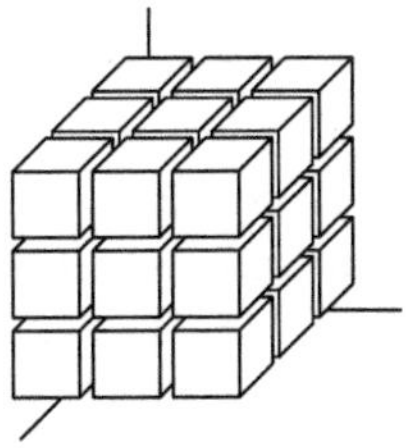

Fig. 2. Decomposed cube

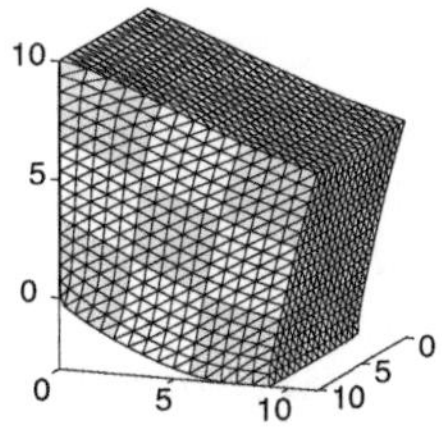

Fig. 3. Solution

As a more realistic example, let us consider the problem to describe the deformation and contact pressure in the ball bearings in Fig. 4. We can easily recognize that it comprises several bodies—balls, rings, and cages. The balls can move in their cages, so their stiffness matrices are necessarily singular and the discretized nonpenetration conditions can be described naturally by dependent constraints. Though the displacements and forces are typically given on parts of the surfaces of some bodies, exact places where the deformed balls come into contact with the cages or the rings are not known till the problem is solved.

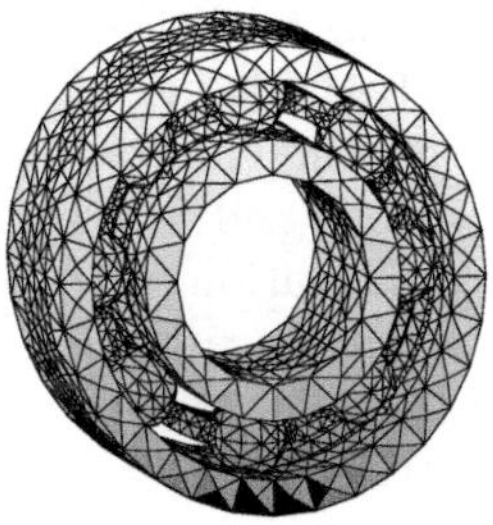

Fig. 4. Ball bearings

It should not be surprising that duality-based methods can solve constrained problems more efficiently than primal ones, as the duality typically turns the general inequality constraints into simpler constraints. The first fully scalable algorithm for numerical solution of linear problems, FETI, was introduced only in the early 1990s. It took a lot of work to get similar results for variational inequalities. Since the cost of the solution of a linear problem is proportional to the number of variables, a scalable algorithm must identify the active constraints in a sense for free!

Preface to the Second Edition

Since the first edition of this book appeared 15 years ago, numerous researchers and engineers have successfully applied the algorithms presented to solve problems of both practical and academic interest. The algorithms were implemented in ESPRESO [121] and PERMON [227] packages that can exploit the massively parallel facilities of modern supercomputers. The algorithms demonstrated their efficiency by solving many problems, including a contact problem discretized by billions of nodal variables solved in less than a minute.

The algorithms also found applications in the new fields, such as optimal control [243], image registration [238], or classification by support vector machines [43]. The solution of some problems required a modification of the algorithms to enhance their specific features, such as positive semidefinite Hessian of the cost function for solving the particle dynamics problems [232].

The optimal algorithms were also adapted to solve QCQP (quadratic constraints quadratic programming) problems with separable quadratic convex constraints [106]. Our research of algorithms for QCQP problems was motivated by contact problems with Tresca and Coulomb friction.

To enhance the experience overviewed above, we reorganized the book into four parts. We split the exposition of the algorithms into two middle parts, presenting the classical algorithms with their ingredients and enriched but mostly standard staff separately, and four chapters with the updated presentation of the optimal algorithms. The new presentation reflects improved insight into the convergence theory and improvements of the algorithms, including the adaptive reorthogonalization of equality constraints in the semimonotonic augmented Lagrangian algorithm for problems with separable and equality constraints. The last four new chapters comprise case studies reflecting increasing fields of applications.

Synopsis of the Book

We organized the book into four parts. We start the introductory part by reviewing some well-known facts on linear algebra in the form that is useful in the following analysis. The review includes less standard estimates, stable matrix decompositions of semidefinite matrices with known kernels, and spectral theory. Then, we review the standard theory of convex and quadratic programming. Though many results concerning the existence and uniqueness of QP problems are particular cases of a more general theory of nonlinear programming, it is often possible to develop more straightforward proofs that exploit a specific structure of the QP problems, mainly the three-term Taylor's expansion, and sometimes to get stronger results. We paid particular attention to the results for dependent constraints and/or positive semidefinite Hessian, including the sensitivity analysis and the duality theory. Two new Sects. 2.8 and 2.8.2, discuss the separable quadratic constraints.

The second part overviews the classical algorithms and their ingredients. It starts with a new Chap. 3, which overviews the results on *gradient* methods, including the discussion of the spectral (Barzilai–Borwain) gradient method and a tight bound on the Euclidean error of the Polyak's Heavy ball method. The following Chap. 4 summarizes the basic facts on the *conjugate gradient* (CG) method for unconstrained QP problems viewed as a direct method. The new Chap. 5 on the *gradient projection* comprises both the classical and more recent original results on decreasing the cost function along the projected gradient path. The following Chap. 6 discusses the *augmented Lagrangians* for the equality-constrained problems with the exact solution of auxiliary unconstrained problems. We pay special attention to the analysis of the penalty method. The second part's final Chap. 7 discusses variants of the *Polyak algorithm* for solving bound-constrained problems. The modified algorithm preserves the finite termination property of the original algorithm but accepts the inexact solutions of auxiliary problems.

The third part is the core of the book and comprises four chapters on the algorithms for specific types of constraints. It starts with Chap. 8 describing the convergence analysis of *conjugate gradient (CG) algorithm* that supports solving of large *unconstrained* problems. Special attention is paid to the preconditioning by a conjugate projector (deflation) in Sect. 8.3. A new Sect. 8.4 describes CG for solving *linear least square problems* with the same convergence rate in the regular condition number as standard CG.

Chapter 9 reviews in detail the semi-monotonic augmented Lagrangian algorithm *SMALE* for solving large QP problems with *equality constraints*. The algorithm uses an inexact solution of auxiliary unconstrained problems with efficient precision control that guarantees the increase of Lagrangian. We prove optimality for two variants of SMALE. The bounds on the parameters that guarantee linear convergence are also presented. We present a variant of SMALE that converges with an arbitrary positive regularization parameter.

In Chap. 10, we describe the MPRGP algorithm for solving QP problems with the *bound constraints.* MPRGP enjoys the linear rate of convergence of both the iterates and their projected gradients in terms of the extreme eigenvalues of the Hessian of the cost function. We show that preconditioning by the conjugate projectors can improve the convergence rate of MPRGP. The new Sect. 10.7 describes the algorithm MPRGPS for bound-constrained QP problems with an SPS Hessian of the cost function.

The brand new Chap. 11 introduces the optimal MPGP and PBBF algorithms for solving QCQP problems with *separable quadratic constraints.* MPGP has similar convergence properties as MPRGP. We paid particular attention to the reliable treatment of the constraints with strong curvature, which is specific for nonlinear constraints. PBBF is based on the Barzilai–Borwein spectral gradient method adapted for separable constraints.

The last Chap. 12 of the third part combines the results of our previous investigations to develop efficient algorithms for the minimization of strictly convex quadratic functions subject to *convex separable inequality and linear equality* constraints. An essential part of the last chapter is *adaptive reorthogonalization*, which improves the convergence rate of the outer loop. All the chapters of the third part provide results of numerical experiments.

The last four chapters of the book provide four *case studies*: solving a huge model variational inequality, a contact problem with friction, an optimal control problem, and an SVM (support vector machines) problem. The brief last chapter of the book provides information on the software implementing the presented algorithms for solving large problems, which was used in most of the benchmarks.

Acknowledgments

Most of the nonstandard results presented in the first edition of this book have been found by the author and his colleagues over the last 25 years, often in cooperation with other colleagues. I would like to acknowledge here my thanks especially to Ana Friedlander and Mario Martínez for sharing their insight into augmented Lagrangian methods, to Sandra A. Santos and F.A.M. Gomes for their share in early development of our algorithms for the solution of variational inequalities, to Joachim Schöberl for sharing his original insight into the gradient projection method, and to Charbel Farhat for drawing attention to the practical aspects of our algorithms and an inspiration for thinking twice about simple topics.

We presented the first results on optimal algorithms at the summer schools organized by Ivo Marek, whose encouragement was essential in the decision to prepare the first edition of this book.

My thanks go also to my colleagues and students from the Faculty of Electrical Engineering and Computer Science of VŠB–Technical University of Ostrava. Vít Vondrák implemented the first versions of the algorithms to the solution of contact problems of mechanics and shape optimization, David Horák first implemented many variants of the algorithms that appear in this book, and Dalibor Lukáš adapted the algorithms of Chap. 4 to the solution of the Stokes problem. My thanks also go to Marta Domorádová for her share in the research of conjugate projectors for bound-constrained problems, and assistance with figures, to Radek Kučera who adopted the algorithms for bound-constrained QP to the solution of more general problems with separable convex constraints and carried out a lot of joint work, and to Petr Beremlijski, Tomáš Kozubek, and Oldřich Vlach, who applied at least some of these algorithms to the solution of engineering benchmarks. Marta Domorádová, Marie Sadowská, Dalibor Lukáš, and David Horák kindly assisted with numerical experiments. It was a pleasure to work on the book with the publication staff at Springer. I am very grateful to Elizabeth Loew and Frank Ganz for their share in the final refinements of the book.

Acknowledgments for the Second Edition

Many of the colleagues mentioned above contributed to the second edition of this book. I especially thank Tomáš Brzobohatý and David Horák, who implemented the most relevant algorithms into their packages ESPRESO and PERMON. They worked with Luboš Říha, Ondřej Meca, Jakub Kružík, and Marek Pecha. Oldřich Vlach contributed to the development of adaptive reorthogonalization, figures, and many other issues; Jiří Bouchala and Petr Vodstrčil were instrumental in generalizing the results related to bound constraints to separable convex constraints. The research related to the spectral gradients and the algorithms for SPS problems would hardly have started without Lukáš Pospíšil, who connected these activities with the applications in particle dynamics. I am grateful to Marta Domorádová – Jarošová for her part in optimal control numerical experiments and to Marie Sadowská for reading most of the new staff and her comments. I am happy that I could share the inspiring environment of the Department of Applied Mathematics and National Supercomputing Center VŠB Technical University of Ostrava.

The author is grateful to Vladimír Havlena and Ondřej Šantin from ČVUT Praha for the optimal control benchmark (Chap. 15) and Serena Crisci, David Horák, Jakub Kružík, and Marek Pecha for the SVM applications (Chap. 16). The author thanks the Springer staff, particularly Kritheka Elango and Mahalakshmi Saravanan, for their expertise and patience in completing the book.

This work was supported by the Ministry of Education, Youth and Sports of the Czech Republic through the e-INFRA CZ (ID:90254).

Ostrava, Czech Republic
August 2024

Zdeněk Dostál

Competing Interests The author has no competing interests to declare that are relevant to the content of this manuscript.

Contents

Part I
Background

1
Linear Algebra

The purpose of this chapter is to briefly review definitions, notations, and results of linear algebra used in the rest of the book. We also include a few results specially developed for analyzing our algorithms. There is no claim of completeness as the reader is assumed to be familiar with the basic concepts of college linear algebra. More systematic exposition and additional material can be found, e.g., in the books by Demmel [49] or Golub and Van Loan [147].

1.1 Vectors

In this book we work with *n-dimensional arithmetic vectors* $\mathbf{v} \in \mathbb{R}^n$, where $\mathbb{R}$ denotes the set of real numbers. The only exception is Sect. 1.10, where vectors with complex entries are considered. We denote the ith component of an arithmetic vector $\mathbf{v} \in \mathbb{R}^n$ by $[\mathbf{v}]_i$. Thus $[\mathbf{v}]_i = v_i$ if $\mathbf{v} = [v_i]$ is defined by its components v_i. We assume that all the arithmetic vectors are column vectors. The relations between vectors $\mathbf{u}, \mathbf{v} \in \mathbb{R}^n$ are defined componentwise. Thus $\mathbf{u} = \mathbf{v}$ is equivalent to $[\mathbf{u}]_i = [\mathbf{v}]_i, i = 1, \ldots, n$, and $\mathbf{u} \le \mathbf{v}$ is equivalent to $[\mathbf{u}]_i \le [\mathbf{v}]_i, i = 1, \ldots, n$. We sometimes call the elements of $\mathbb{R}^n$ *points* to indicate that the concepts of length and direction are unimportant.

Having arithmetic vectors $\mathbf{u}, \mathbf{v} \in \mathbb{R}^n$ and a scalar $\alpha \in \mathbb{R}$, we define the *addition* and *multiplication by scalar* componentwise by

$$[\mathbf{u} + \mathbf{v}]_i = [\mathbf{u}]_i + [\mathbf{v}]_i \quad \text{and} \quad [\alpha \mathbf{v}]_i = \alpha [\mathbf{v}]_i, \quad i = 1, \ldots, n.$$

The rules that govern these operations, such as associativity, may be easily deduced from the related rules for computations with real numbers.

The vector analog of $0 \in \mathbb{R}$ is the *zero vector* $\mathbf{o}_n \in \mathbb{R}^n$ with all the entries equal to zero. When the dimension can be deduced from the context, possibly

Z. Dostál, *Optimal Quadratic Programming and QCQP Algorithms with Case Studies*, Springer Optimization and Its Applications 23,
https://doi.org/10.1007/978-3-031-95167-1_1

using the assumption that all the expressions in our book are well defined, we often drop the subscript and write simply $\mathbf{o}$.

A nonempty set $\mathcal{V} \subseteq \mathbb{R}^n$ with the operations defined above is a *vector space* if $\alpha \in \mathbb{R}$ and $\mathbf{u}, \mathbf{v} \in \mathcal{V}$ imply $\mathbf{u}+\mathbf{v} \in \mathcal{V}$ and $\alpha\mathbf{u} \in \mathcal{V}$. In particular, both $\mathbb{R}^n$ and $\{\mathbf{o}\}$ are vector spaces. Given vectors $\mathbf{v}_1, \ldots, \mathbf{v}_k \in \mathbb{R}^n$, the set

$$\operatorname{Span}\{\mathbf{v}_1, \ \ldots, \ \mathbf{v}_k\} = \{\mathbf{v} \in \mathbb{R}^n : \mathbf{v} = \alpha_1\mathbf{v}_1 + \cdots + \alpha_k\mathbf{v}_k, \ \alpha_i \in \mathbb{R}\}$$

is a vector space called the *linear span* of $\mathbf{v}_1, \ldots, \mathbf{v}_k$. If $\mathcal{U}$ and $\mathcal{V}$ are vector spaces, then the sets $\mathcal{U} \cap \mathcal{V}$ and

$$\mathcal{U} + \mathcal{V} = \{\mathbf{x} + \mathbf{y} : \mathbf{x} \in \mathcal{U} \quad \text{and} \quad \mathbf{y} \in \mathcal{V}\}$$

are also vector spaces. If $\mathcal{W} = \mathcal{U} + \mathcal{V}$ and $\mathcal{U} \cap \mathcal{V} = \{\mathbf{o}\}$, then $\mathcal{W}$ is said to be the *direct sum* of $\mathcal{U}$ and $\mathcal{V}$. We denote it by

$$\mathcal{W} = \mathcal{U} \oplus \mathcal{V}.$$

If $\mathcal{U}, \mathcal{V} \subseteq \mathbb{R}^n$ are vector spaces and $\mathcal{U} \subseteq \mathcal{V}$, then $\mathcal{U}$ is a *subspace* of $\mathcal{V}$.

A vector space $\mathcal{V} \subseteq \mathbb{R}^n$ can be spanned by different sets of vectors. A finite set of vectors $\mathcal{E} \subset \mathbb{R}^n$ that spans a given vector space $\mathcal{V} \neq \{\mathbf{o}\}$ is called a *basis* of $\mathcal{V}$ if no proper subset of $\mathcal{E}$ spans $\mathcal{V}$. For example, the set of vectors

$$\mathcal{S} = \{\mathbf{s}_1, \ \ldots, \ \mathbf{s}_n\}, \ [\mathbf{s}_i]_j = \delta_{ij}, \ i, \ j = 1, \ \ldots, \ n,$$

where δ_{ij} denotes the *Kronecker symbol* defined by $\delta_{ij} = 1$ for $i = j$ and $\delta_{ij} = 0$ for $i \neq j$, is the *standard basis* of $\mathbb{R}^n$. If $\mathcal{E} = \{\mathbf{e}_1, \ \ldots, \ \mathbf{e}_d\}$ is a basis of a vector space $\mathcal{V}$, then $\mathcal{E}$ is *independent*, that is,

$$\alpha_1\mathbf{e}_1 + \cdots + \alpha_d\mathbf{e}_d = \mathbf{o}$$

implies

$$\alpha_1 = \cdots = \alpha_d = 0.$$

Any two bases of a vector space $\mathcal{V}$ have the same number of vectors. We call it the *dimension* of $\mathcal{V}$ and denote it $\dim \mathcal{V}$. Obviously $\dim \mathbb{R}^n = n$ and $\dim \mathcal{V} \leq n$ for any subspace $\mathcal{V} \subseteq \mathbb{R}^n$. For convenience, we define $\dim\{\mathbf{o}\} = 0$.

We sometimes use the componentwise extensions of scalar functions to vectors. Thus if $\mathbf{v} \in \mathbb{R}^n$, then $\mathbf{v}^+$ and $\mathbf{v}^-$ are the vectors whose ith components are $\max\{[\mathbf{v}]_i, \ 0\}$ and $\min\{[\mathbf{v}]_i, \ 0\}$, respectively. Similarly, if $\mathbf{u}, \mathbf{v} \in \mathbb{R}^n$, then $\max\{\mathbf{u}, \mathbf{v}\}$ and $\min\{\mathbf{u}, \mathbf{v}\}$ denote the vectors whose ith components are $\max\{[\mathbf{u}]_i, \ [\mathbf{v}]_i\}$ and $\min\{[\mathbf{u}]_i, \ [\mathbf{v}]_i\}$, respectively.

If $\mathcal{I}$ is a nonempty subset of $\{1, \ \ldots, n\}$ and $\mathbf{v} \in \mathbb{R}^n$, then we denote by $[\mathbf{v}]_\mathcal{I}$ or simply $\mathbf{v}_\mathcal{I}$ the subvector of $\mathbf{v}$ with components $[\mathbf{v}]_i, i \in \mathcal{I}$. Thus if $\mathcal{I}$ has m elements, then $\mathbf{v}_\mathcal{I} \in \mathbb{R}^m$, so that we can refer to the components of $\mathbf{v}_\mathcal{I}$ either by the *global indices* $i \in \mathcal{I}$ or by the *local indices* $j \in \{1, \ \ldots, \ m\}$. We usually rely on the reader's judgment to recognize the appropriate type of indexing.

1.2 Matrices and Matrix Operations

Throughout the whole book, all the matrices are assumed to be real except Sect. 1.10, where also complex matrices are considered. Similarly to the related convention for vectors, the $(i,\ j)$th component of a matrix $\mathsf{A} \in \mathbb{R}^{m\times n}$ is denoted by $[\mathsf{A}]_{ij}$, so that $[\mathsf{A}]_{ij} = a_{ij}$ for $\mathsf{A} = [a_{ij}]$ which is defined by its entries a_{ij}. A matrix $\mathsf{A} \in \mathbb{R}^{m\times n}$ is called an $(m,\ n)$-*matrix*; a matrix $\mathsf{A} \in \mathbb{R}^{n\times n}$ is called a *square matrix* of the order n.

Having $(m,\ n)$-matrices A, B and a scalar $\alpha \in \mathbb{R}$, we define *addition* and *multiplication by a scalar* by

$$[\mathsf{A}+\mathsf{B}]_{ij} = [\mathsf{A}]_{ij} + [\mathsf{B}]_{ij} \quad \text{and} \quad [\alpha\mathsf{A}]_{ij} = \alpha[\mathsf{A}]_{ij}.$$

The rules that govern the addition of matrices and their multiplication by scalars are the same as those for corresponding vector operations.

The matrix analog of 0 is the *zero matrix* $\mathsf{O}_{mn} \in \mathbb{R}^{m\times n}$ with all the entries equal to zero. When the dimension is clear from the context, we often drop the subscripts and write simply O.

Having matrices $\mathsf{A} \in \mathbb{R}^{m\times k}$ and $\mathsf{B} \in \mathbb{R}^{k\times n}$, we define their *product* $\mathsf{AB} \in \mathbb{R}^{m\times n}$ by

$$[\mathsf{AB}]_{ij} = \sum_{l=1}^{k} [\mathsf{A}]_{il}[\mathsf{B}]_{lj}.$$

Matrix multiplication is associative, therefore we do not need to use brackets to specify the order of multiplication. In particular, given a positive integer k and a square matrix A, we can define the kth *power of a square matrix* A by

$$\mathsf{A}^k = \underbrace{\mathsf{AA}\ldots\mathsf{A}}_{k\text{-times}}.$$

Matrix multiplication is not commutative.

The matrix counterpart of $1 \in \mathbb{R}$ in $\mathbb{R}^{n\times n}$ is the *identity matrix* $\mathsf{I}_n = [\delta_{ij}]$ of the order n. When the dimension may be deduced from the context, we often drop the subscripts and write simply I. Thus we can write

$$\mathsf{A} = \mathsf{IA} = \mathsf{AI}$$

for any matrix A, having in mind that the order of I on the left may be different from that on the right.

Given $\mathsf{A} \in \mathbb{R}^{m\times n}$, we define the *transposed matrix* $\mathsf{A}^T \in \mathbb{R}^{n\times m}$ to A by $[\mathsf{A}^T]_{ij} = [\mathsf{A}]_{ji}$. Having matrices $\mathsf{A} \in \mathbb{R}^{m\times k}$ and $\mathsf{B} \in \mathbb{R}^{k\times n}$, it may be checked that

$$(\mathsf{AB})^T = \mathsf{B}^T\mathsf{A}^T. \tag{1.1}$$

A square matrix A is *symmetric* if $\mathsf{A} = \mathsf{A}^T$.

A matrix A is *positive definite* if $\mathbf{x}^T\ \mathsf{A}\mathbf{x} > 0$ for any $\mathbf{x} \neq \mathbf{o}$, *positive semidefinite* if $\mathbf{x}^T\mathsf{A}\mathbf{x} \geq 0$ for any $\mathbf{x}$, and *indefinite* if neither A nor $-\mathsf{A}$ is positive definite or semidefinite. We are especially interested in symmetric positive definite (SPD) and semidefinite (SPS) matrices.

If $\mathsf{A} \in \mathbb{R}^{m\times n}, \mathcal{I} \subseteq \{1, \dots, m\}$, and $\mathcal{J} \subseteq \{1, \dots, n\}, \mathcal{I}$ and $\mathcal{J}$ nonempty, we denote by $\mathsf{A}_{\mathcal{I}\mathcal{J}}$ the submatrix of A with the components $[\mathsf{A}]_{ij}, i \in \mathcal{I}, j \in \mathcal{J}$. The local indexing of the entries of $\mathsf{A}_{\mathcal{I}\mathcal{J}}$ is used whenever it is convenient in a similar way as the local indexing of subvectors which was introduced in Sect. 1.1. The full set of indices may be replaced by $*$ so that $\mathsf{A} = \mathsf{A}_{**}$ and $\mathsf{A}_{\mathcal{I}*}$ denotes the submatrix of A with the row indices belonging to $\mathcal{I}$.

Sometimes it is useful to rearrange the matrix operations into manipulations with submatrices called blocks. A *block matrix* $\mathsf{A} \in \mathbb{R}^{m\times n}$ is defined by its blocks $\mathsf{A}_{ij} = \mathsf{A}_{\mathcal{I}_i\mathcal{J}_j}$, where $\mathcal{I}_i$ and $\mathcal{J}_j$ denote nonempty contiguous sets of indices decomposing $\{1, \dots, m\}$ and $\{1, \dots, n\}$, respectively.

Very large matrices are often *sparse* in the sense that they have a small number of nonzero entries distributed in a pattern that can be exploited to the reduction of storage requirements or to the effective solution of standard linear algebra problems. Such matrices arise, e.g., from the discretization of problems described by differential operators. The matrices with a large number of nonzero entries are often called *full* or *dense matrices.*

1.3 Matrices and Mappings

Each matrix $\mathsf{A} \in \mathbb{R}^{m\times n}$ defines the mapping which assigns to each $\mathbf{x} \in \mathbb{R}^n$ the vector $\mathsf{A}\mathbf{x} \in \mathbb{R}^m$. Two subspaces associated with this mapping are *range* or *image space* $\mathrm{Im}\mathsf{A}$ and its *kernel* or *null space* $\mathrm{Ker}\mathsf{A}$; they are defined by

$$\mathrm{Im}\mathsf{A} = \{\mathsf{A}\mathbf{x} : \ \mathbf{x} \in \mathbb{R}^n\} \quad \text{and} \quad \mathrm{Ker}\mathsf{A} = \{\mathbf{x} \in \mathbb{R}^n : \ \mathsf{A}\mathbf{x} = \mathbf{o}\}.$$

The range of A is the span of its columns.

If f is a mapping defined on $\mathcal{D} \subseteq \mathbb{R}^n$ and $\Omega \subseteq \mathcal{D}$, then $f|\Omega$ denotes the *restriction* of f to Ω; it assigns to each $\mathbf{x} \in \Omega$ the value $f(\mathbf{x})$. If $\mathsf{A} \in \mathbb{R}^{n\times n}$ and $\mathcal{V} = \mathrm{Im}\mathsf{V}$, $\mathsf{V} \in \mathbb{R}^{n\times m}$, is a subspace of $\mathbb{R}^n$, we define $\mathsf{A}|\mathcal{V}$ as a restriction of the mapping associated with A to $\mathcal{V}$. The restriction $\mathsf{A}|\mathcal{V}$ is said to be positive definite or SPD if $\mathbf{x}^T\mathsf{A}\mathbf{x} > 0$ for $\mathbf{x} \in \mathcal{V}, \mathbf{x} \neq \mathbf{o}$, and positive semidefinite or SPS if $\mathbf{x}^T\mathsf{A}\mathbf{x} \geq 0$ for $\mathbf{x} \in \mathcal{V}$. The abbreviations refer to the matrix $\mathsf{V}^T\mathsf{A}\mathsf{V}$ of $\mathsf{A}|\mathcal{V}$ in the basis defined by the columns of V.

The mapping associated with A is *injective* if $\mathsf{A}\mathbf{x} = \mathsf{A}\mathbf{y}$ implies $\mathbf{x} = \mathbf{y}$. It is easy to check that the mapping associated with A is injective if and only if $\mathrm{Ker}\mathsf{A} = \{\mathbf{o}\}$. More generally, it may be proved that

$$\dim \mathrm{Im}\mathsf{A} + \dim\, \mathrm{Ker}\mathsf{A} = n \tag{1.2}$$

for any $\mathsf{A} \in \mathbb{R}^{m\times n}$. If $m = n$, then A is injective if and only if $\mathrm{Im}\mathsf{A} = \mathbb{R}^n$.

The *rank* or *column rank* of a matrix A is equal to the dimension of the range of A. The *column rank* is known to be equal to the *row rank*, the number of linearly independent rows. A matrix is of *full row rank* or *full column rank* when its rank is equal to the number of its rows or columns, respectively. A matrix $\mathsf{A} \in \mathbb{R}^{m\times n}$ is of *full rank* when its rank is $\min\{m, n\}$.

A subspace $\mathcal{V} \subseteq \mathbb{R}^n$ which satisfies

$$\mathsf{A}\mathcal{V} = \{\mathsf{A}\mathbf{x} : \mathbf{x} \in \mathcal{V}\} \subseteq \mathcal{V}$$

is an *invariant subspace* of A. Obviously

$$\mathsf{A}(\mathrm{Im}\mathsf{A}) \subseteq \mathrm{Im}\mathsf{A},$$

so that $\mathrm{Im}\mathsf{A}$ is an invariant subspace of A.

A *projector* is a square matrix P that satisfies

$$\mathsf{P}^2 = \mathsf{P}.$$

Such a matrix is also said to be *idempotent.* A vector $\mathbf{x} \in \mathrm{Im}\mathsf{P}$ if and only if there is $\mathbf{y} \in \mathbb{R}^n$ such that $\mathbf{x} = \mathsf{P}\mathbf{y}$, so that

$$\mathsf{P}\mathbf{x} = \mathsf{P}(\mathsf{P}\mathbf{y}) = \mathsf{P}\mathbf{y} = \mathbf{x}.$$

If P is a projector, then $\mathsf{Q} = \mathsf{I} - \mathsf{P}$ and P^T are also projectors as

$$(\mathsf{I} - \mathsf{P})^2 = \mathsf{I} - 2\mathsf{P} + \mathsf{P}^2 = \mathsf{I} - \mathsf{P} \quad \text{and} \quad (\mathsf{P}^T)^2 = (\mathsf{P}^2)^T = \mathsf{P}^T.$$

Since for any $\mathbf{x} \in \mathbb{R}^n$

$$\mathbf{x} = \mathsf{P}\mathbf{x} + (\mathsf{I} - \mathsf{P})\mathbf{x},$$

it simply follows that $\mathrm{Im}\mathsf{Q} = \mathrm{Ker}\mathsf{P}$,

$$\mathbb{R}^n = \mathrm{Im}\mathsf{P} + \mathrm{Ker}\mathsf{P}, \quad \text{and} \quad \mathrm{Ker}\mathsf{P} \cap \mathrm{Im}\mathsf{P} = \{\mathbf{o}\}.$$

We say that P is a projector onto $\mathcal{U} = \mathrm{Im}\mathsf{P}$ along $\mathcal{V} = \mathrm{Ker}\mathsf{P}$ and Q is a complementary projector onto $\mathcal{V}$ along $\mathcal{U}$. The above relations may also be rewritten as

$$\mathrm{Im}\mathsf{P} \oplus \mathrm{Ker}\mathsf{P} = \mathbb{R}^n. \tag{1.3}$$

Let $(\pi(1), \ldots, \pi(n))$ be a permutation of numbers $1, \ldots, n$. Then the mapping which assigns to each $\mathbf{v} = [v_i] \in \mathbb{R}^n$ a vector $[v_{\pi(1)}, \ldots, v_{\pi(n)}]^T$ is associated with the *permutation matrix*

$$\mathsf{P} = [\mathbf{s}_{\pi(1)}, \ldots, \mathbf{s}_{\pi(n)}],$$

where $\mathbf{s}_i$ denotes the ith column of the identity matrix I_n. If P is a permutation matrix, then

$$\mathsf{P}\mathsf{P}^T = \mathsf{P}^T\mathsf{P} = \mathsf{I}.$$

A by reordering of the rows of A, then there is

1.4 Inverse and Generalized Inverse Matrices

If A is a square full rank matrix, then there is the unique *inverse matrix* A^{-1} such that

$$\mathsf{A}\mathsf{A}^{-1} = \mathsf{A}^{-1}\mathsf{A} = \mathsf{I}. \tag{1.4}$$

The mapping associated with A^{-1} is inverse to that associated with A.

If A^{-1} exists, we say that A is *nonsingular*. A square matrix is *singular* if its inverse matrix does not exist. Any positive definite matrix is nonsingular. If P is a permutation matrix, then P is nonsingular and

$$\mathsf{P}^{-1} = \mathsf{P}^T.$$

If A is a nonsingular matrix, then $\mathsf{A}^{-1}\mathbf{b}$ is the unique solution of the system of linear equations $\mathsf{A}\mathbf{x} = \mathbf{b}$.

If A is a nonsingular matrix, then we can transpose (1.4) and use (1.1) to get

$$\left(\mathsf{A}^{-1}\right)^T \mathsf{A}^T = \mathsf{A}^T \left(\mathsf{A}^{-1}\right)^T = \mathsf{I},$$

so that

$$(\mathsf{A}^T)^{-1} = (\mathsf{A}^{-1})^T. \tag{1.5}$$

It follows that if A is symmetric, then A^{-1} is symmetric.

If $\mathsf{A} \in \mathbb{R}^{n\times n}$ is positive definite, then A^{-1} is also positive definite, as any vector $\mathbf{x} \neq \mathbf{o}$ can be expressed as $\mathbf{x} = \mathsf{A}\mathbf{y}, \mathbf{y} \neq \mathbf{o}$, and

$$\mathbf{x}^T\mathsf{A}^{-1}\mathbf{x} = (\mathsf{A}\mathbf{y})^T\mathsf{A}^{-1}\mathsf{A}\mathbf{y} = \mathbf{y}^T\mathsf{A}^T\mathbf{y} = \mathbf{y}^T\mathsf{A}\mathbf{y} > 0.$$

If A and B are nonsingular matrices, then it is easy to check that AB is also nonsingular and

$$(\mathsf{AB})^{-1} = \mathsf{B}^{-1}\mathsf{A}^{-1}.$$

If $\mathsf{U}, \mathsf{V} \in \mathbb{R}^{m\times n}, m < n$, and $\mathsf{A}, \mathsf{A} + \mathsf{U}^T\mathsf{V}$ are nonsingular, then it can be verified directly that

$$(\mathsf{A} + \mathsf{U}^T\mathsf{V})^{-1} = \mathsf{A}^{-1} - \mathsf{A}^{-1}\mathsf{U}^T(\mathsf{I} + \mathsf{V}\mathsf{A}^{-1}\mathsf{U}^T)^{-1}\mathsf{V}\mathsf{A}^{-1}. \tag{1.6}$$

The formula (1.6) is *Sherman-Morrison-Woodbury's formula* (see [147]).

If $\mathsf{A} \in \mathbb{R}^{m\times n}$ and $\mathbf{b} \in \mathrm{Im}\mathsf{A}$, then we can express a solution of the system of linear equations $\mathsf{A}\mathbf{x} = \mathbf{b}$ by means of a *left generalized inverse matrix* $\mathsf{A}^+ \in \mathbb{R}^{n\times m}$ which satisfies $\mathsf{A}\mathsf{A}^+\mathsf{A} = \mathsf{A}$. Indeed, if $\mathbf{b} \in \mathrm{Im}\mathsf{A}$, then there is $\mathbf{y}$ such that $\mathbf{b} = \mathsf{A}\mathbf{y}$ and $\overline{\mathbf{x}} = \mathsf{A}^+\mathbf{b}$ satisfies

$$\mathsf{A}\overline{\mathbf{x}} = \mathsf{A}\mathsf{A}^+\mathbf{b} = \mathsf{A}\mathsf{A}^+\mathsf{A}\mathbf{y} = \mathsf{A}\mathbf{y} = \mathbf{b}.$$

Thus A^+ acts on $\mathrm{Im}\mathsf{A}$ like the inverse matrix. Moreover, if $\mathsf{S} \in \mathbb{R}^{n\times p}$ satisfies $\mathsf{AS} = \mathsf{O}$ and $\mathsf{N} \in \mathbb{R}^{n\times p}$, then $(\mathsf{A}^+) + \mathsf{S}\mathsf{N}^T$ is also a left generalized inverse as

$$\mathsf{A}\left((\mathsf{A}^+) + \mathsf{SN}^T\right)\mathsf{A} = \mathsf{AA}^+\mathsf{A} + \mathsf{ASN}^T\mathsf{A} = \mathsf{A}.$$

If A is an SPS matrix, then there is a permutation matrix P such that

$$\mathsf{A} = \mathsf{P}^T \begin{bmatrix} \mathsf{B} & \mathsf{C}^T \\ \mathsf{C} & \mathsf{CB}^{-1}\mathsf{C}^T \end{bmatrix} \mathsf{P},$$

where B is a nonsingular matrix whose dimension is equal to the rank of A. It may be verified directly that the matrix

$$\mathsf{A}^\# = \mathsf{P}^T \begin{bmatrix} \mathsf{B}^{-1} & \mathsf{O}^T \\ \mathsf{O} & \mathsf{O} \end{bmatrix} \mathsf{P} \tag{1.7}$$

is a left generalized inverse of A. If A is SPS, then $\mathsf{A}^\#$ is also SPS. Notice that if $\mathsf{AS} = \mathsf{O}$, then $\mathsf{A}^+ = \mathsf{A}^\# + \mathsf{SS}^T$ is also an SPS generalized inverse.

1.5 Direct Methods for Solving Linear Equations

The inverse matrix is a useful tool for theoretical developments, but not for computations, especially when sparse matrices are involved. It is often much more efficient to implement the multiplication of a vector by the inverse matrix by solving the related system of linear equations. We recall here briefly the *direct methods*, which reduce solving of the original system of linear equations to solving of systems of linear equations with triangular matrices.

A matrix $\mathsf{L} = [l_{ij}]$ is *lower triangular* if $l_{ij} = 0$ for $i < j$. It is easy to solve a system $\mathsf{L}\mathbf{x} = \mathbf{b}$ with the nonsingular lower triangular matrix $\mathsf{L} \in \mathbb{R}^{n\times n}$. As there is only one unknown in the first equation, we can find it and then substitute it into the remaining equations to obtain a system with the same structure, but with only $n-1$ remaining unknowns. We can repeat the procedure until we find all the components of $\mathbf{x}$. A similar procedure, but starting from the last equation, can be applied to the system with a nonsingular *upper triangular matrix* $\mathsf{U} = [u_{ij}]$ with $u_{ij} = 0$ for $i > j$. The solution costs of a system with triangular matrices is proportional to the number of its nonzero entries.

If we are to solve the system of linear equations with a nonsingular matrix, we can use systematically *equivalent transformations* that do not change the solution in order to modify the original system to that with an upper triangular matrix. It is well-known that the solutions of a system of linear equations are the same as the solutions of a system of linear equations obtained from the original system by interchanging two equations, replacing an equation by its nonzero multiple, or adding a multiple of one equation to another equation.

The *Gauss elimination* for the solution of a system of linear equations with a nonsingular matrix thus consists of two steps: the *forward reduction*, which exploits equivalent transformations to reduce the original system to the system with an upper triangular matrix, and the *backward substitution*, which solves the resulting system with the upper triangular matrix.

Alternatively, we can use suitable matrix factorizations. For example, it is well-known that any positive definite matrix A can be decomposed into the product

$$\mathsf{A} = \mathsf{L}\mathsf{L}^T, \tag{1.8}$$

where L is a nonsingular lower triangular matrix with positive diagonal entries. Having the decomposition, we can evaluate $\mathbf{z} = \mathsf{A}^{-1}\mathbf{x}$ by solving the systems

$$\mathsf{L}\mathbf{y} = \mathbf{x} \quad \text{and} \quad \mathsf{L}^T\mathbf{z} = \mathbf{y}.$$

The factorization-based solvers may be especially useful when we are to solve several systems of equations with the same coefficients but different right-hand sides coming one after another.

The method of evaluation of the factor L is known as the *Cholesky factorization*. The Cholesky factor L can be computed in a number of equivalent ways. For example, we may compute it column by column. Suppose that

$$\mathsf{A} = \begin{bmatrix} a_{11} & \mathbf{a}_1^T \\ \mathbf{a}_1 & \mathsf{A}_{22} \end{bmatrix} \quad \text{and} \quad \mathsf{L} = \begin{bmatrix} l_{11} & \mathbf{o} \\ \mathbf{l}_1 & \mathsf{L}_{22} \end{bmatrix}.$$

Substituting for A and L into (1.8) and comparing the corresponding terms immediately reveals that

$$l_{11} = \sqrt{a_{11}}, \quad \mathbf{l}_1 = l_{11}^{-1}\mathbf{a}_1, \quad \mathsf{L}_{22}\mathsf{L}_{22}^T = \mathsf{A}_{22} - \mathbf{l}_1\mathbf{l}_1^T. \tag{1.9}$$

This gives us the first column of L, and the remaining factor L_{22} is simply the Cholesky factor of the Schur complement $\mathsf{A}_{22} - \mathbf{l}_1\mathbf{l}_1^T$ which is known to be positive definite, so we can find its first column by the above procedure. The algorithm can be implemented to exploit a sparsity pattern of A, e.g., when $\mathsf{A} = [a_{ij}] \in \mathbb{R}^{n \times n}$ is a *band matrix* with $a_{ij} = 0$ for $|i - j| > b, b \ll n$.

If $\mathsf{A} \in \mathbb{R}^{n \times n}$ is only positive semidefinite, it can happen that $a_{11} = 0$. Then

$$0 \leq \mathbf{x}^T\mathsf{A}\mathbf{x} = \mathbf{y}^T\mathsf{A}_{22}\mathbf{y} + 2x_1\mathbf{a}_1^T\mathbf{y}$$

for any vector $\mathbf{x} = [x_1, \mathbf{y}^T]^T$. The inequality implies that $\mathbf{a}_1 = \mathbf{o}$, as otherwise we could take $\mathbf{y} = -\mathbf{a}_1$ and large x_1 to get

$$\mathbf{y}^T\mathsf{A}_{22}\mathbf{y} + 2x_1\mathbf{a}_1^T\mathbf{y} = \mathbf{a}_1^T\mathsf{A}_{22}\mathbf{a}_1 - 2x_1\|\mathbf{a}_1\|^2 < 0.$$

Thus for A symmetric positive semidefinite and $a_{11} = 0$, (1.9) reduces to

$$l_{11} = 0, \quad \mathbf{l}_1 = \mathbf{o}, \quad \mathsf{L}_{22}\mathsf{L}_{22}^T = \mathsf{A}_{22}. \tag{1.10}$$

This simple modification assumes exact arithmetics. In computer arithmetics, the decision whether a_{11} is to be treated as zero depends on a small

$\varepsilon > 0$. The following lemma shows that we can use any basis of Ker$\mathsf{A}$ to identify the zero rows (and columns) of a Cholesky factor.

Lemma 1.1. *Let* $\mathsf{A} = \mathsf{L}\mathsf{L}^T$ *denote a triangular decomposition of a symmetric positive semidefinite matrix* A*, let* $\mathsf{A}\mathbf{e} = \mathbf{o}$*, and let* $l(\mathbf{e})$ *denote the largest index of a nonzero entry of* $\mathbf{e} \in \mathrm{Ker}\mathsf{A}$*, so that*

$$[\mathbf{e}]_{l(\mathbf{e})} \neq 0 \quad \textit{and} \quad [\mathbf{e}]_j = 0 \quad \textit{for} \quad j > l(\mathbf{e}).$$

Then

$$[\mathsf{L}]_{l(\mathbf{e})l(\mathbf{e})} = 0.$$

Proof. If $\mathsf{A}\mathbf{e} = \mathbf{o}$, then

$$\mathbf{e}^T\mathsf{A}\mathbf{e} = \mathbf{e}^T\mathsf{L}\mathsf{L}^T\mathbf{e} = (\mathsf{L}^T\mathbf{e})^T(\mathsf{L}^T\mathbf{e}) = 0.$$

Thus $\mathsf{L}^T\mathbf{e} = \mathbf{o}$ and in particular

$$[\mathsf{L}^T\mathbf{e}]_{l(\mathbf{e})} = [\mathsf{L}]_{l(\mathbf{e})l(\mathbf{e})}[\mathbf{e}]_{l(\mathbf{e})} = 0.$$

Since $[\mathbf{e}]_{l(\mathbf{e})} \neq 0$, we have new $[\mathsf{L}]_{l(\mathbf{e})l(\mathbf{e})} = 0$. □

Let $\mathsf{A} \in \mathbb{R}^{n\times n}$ be SPS and let $\mathsf{R} \in \mathbb{R}^{n\times d}$ denote a full column rank matrix such that Ker$\mathsf{A}$ = ImR. Observing that application of equivalent transformations to the columns of R preserves the image space and the rank of R, we can modify the forward reduction to find R which satisfies

$$l(\mathsf{R}_{*1}) < \cdots < l(\mathsf{R}_{*d}).$$

It follows by Lemma 1.1 that $l(\mathsf{R}_{*i})$, $i = 1,\dots d$, are the indices of the Cholesky factors' zero columns.

Using the following refinement of Lemma 1.1, we can use any orthonormal basis of Ker$\mathsf{A}$ to control the conditioning of the regular diagonal block of A.

Lemma 1.2. *Let* $\mathsf{A} \in \mathbb{R}^{n\times n}$ *denote an SPS matrix with the kernel spanned by the matrix* $\mathsf{R} \in \mathbb{R}^{n\times d}$ *with orthonormal columns. Let*

$$\mathcal{I} = \{i_1, \dots, i_d\}, \quad 1 \leq i_1 < i_2 < \cdots < i_d \leq n,$$

denote a set of indices, and let $\mathcal{J} = \mathcal{N} - \mathcal{I}$, $\quad \mathcal{N} = \{1, 2, \dots, n\}$.

Then

$$\lambda_{\min}(\mathsf{A}_{\mathcal{J}\mathcal{J}}) \geq \overline{\lambda}_{\min}(\mathsf{A})\, \sigma^4_{\min}(\mathsf{R}_{\mathcal{I}*}), \tag{1.11}$$

where $\overline{\lambda}_{\min}(\mathsf{A})$ *and* $\sigma_{\min}(\mathsf{R}_{\mathcal{I}*})$ *denote the least nonzero eigenvalue of* A *and the least singular value of* $\mathsf{R}_{\mathcal{I}*}$*, respectively.*

Proof. See Dostál et al. [105]. □

1.6 Schur Complement

Let A and $\mathbf{b}$ denote a square matrix and a vector, respectively,

$$\mathsf{A} = \begin{bmatrix} \mathsf{A}_{\mathcal{I}\mathcal{I}} & \mathsf{A}_{\mathcal{I}\mathcal{J}} \\ \mathsf{A}_{\mathcal{J}\mathcal{I}} & \mathsf{A}_{\mathcal{J}\mathcal{J}} \end{bmatrix}, \quad \mathbf{b} = \begin{bmatrix} \mathbf{b}_{\mathcal{I}} \\ \mathbf{b}_{\mathcal{J}} \end{bmatrix}.$$

If the block $\mathsf{A}_{\mathcal{I}\mathcal{I}}$ of A is nonsingular, then we can solve the system of linear equations

$$\begin{aligned} \mathsf{A}_{\mathcal{I}\mathcal{I}}\mathbf{x}_{\mathcal{I}} + \mathsf{A}_{\mathcal{I}\mathcal{J}}\mathbf{x}_{\mathcal{J}} &= \mathbf{b}_{\mathcal{I}}, \\ \mathsf{A}_{\mathcal{J}\mathcal{I}}\mathbf{x}_{\mathcal{I}} + \mathsf{A}_{\mathcal{J}\mathcal{J}}\mathbf{x}_{\mathcal{J}} &= \mathbf{b}_{\mathcal{J}} \end{aligned}$$

in two steps. First evaluate $\mathbf{x}_{\mathcal{I}}$ from the first equation to get

$$\mathbf{x}_{\mathcal{I}} = \mathsf{A}_{\mathcal{I}\mathcal{I}}^{-1}\mathbf{b}_{\mathcal{I}} - \mathsf{A}_{\mathcal{I}\mathcal{I}}^{-1}\mathsf{A}_{\mathcal{I}\mathcal{J}}\mathbf{x}_{\mathcal{J}}$$

and then substitute the result into the second equation to get

$$(\mathsf{A}_{\mathcal{J}\mathcal{J}} - \mathsf{A}_{\mathcal{J}\mathcal{I}}\mathsf{A}_{\mathcal{I}\mathcal{I}}^{-1}\mathsf{A}_{\mathcal{I}\mathcal{J}})\mathbf{x}_{\mathcal{J}} = \mathbf{b}_{\mathcal{J}}.$$

The matrix

$$\mathsf{S} = \mathsf{A}_{\mathcal{J}\mathcal{J}} - \mathsf{A}_{\mathcal{J}\mathcal{I}}\mathsf{A}_{\mathcal{I}\mathcal{I}}^{-1}\mathsf{A}_{\mathcal{I}\mathcal{J}}$$

is called the *Schur complement of the block* $\mathsf{A}_{\mathcal{I}\mathcal{I}}$.

Let us assume that A is symmetric, so that $\mathsf{A}_{\mathcal{I}\mathcal{J}}^T = \mathsf{A}_{\mathcal{J}\mathcal{I}}$, let $\mathsf{A}_{\mathcal{I}\mathcal{I}}$ be nonsingular, and let us consider all $\mathbf{x}$ and $\mathbf{y}$ such that

$$\mathsf{A}_{\mathcal{I}\mathcal{I}}\mathbf{x} + \mathsf{A}_{\mathcal{I}\mathcal{J}}\mathbf{y} = \mathbf{o}.$$

Denoting

$$\mathbf{x} = \mathbf{x}(\mathbf{y}) = -\mathsf{A}_{\mathcal{I}\mathcal{I}}^{-1}\mathsf{A}_{\mathcal{I}\mathcal{J}}\mathbf{y}, \quad \mathbf{z} = \begin{bmatrix} \mathbf{x} \\ \mathbf{y} \end{bmatrix},$$

we get

$$\begin{aligned} \mathbf{z}^T\mathsf{A}\mathbf{z} &= \begin{bmatrix} \mathbf{x}^T & \mathbf{y}^T \end{bmatrix} \begin{bmatrix} \mathsf{A}_{\mathcal{I}\mathcal{I}} & \mathsf{A}_{\mathcal{I}\mathcal{J}} \\ \mathsf{A}_{\mathcal{J}\mathcal{I}} & \mathsf{A}_{\mathcal{J}\mathcal{J}} \end{bmatrix} \begin{bmatrix} \mathbf{x} \\ \mathbf{y} \end{bmatrix} \\ &= \mathbf{y}^T\mathsf{A}_{\mathcal{I}\mathcal{J}}^T\mathsf{A}_{\mathcal{I}\mathcal{I}}^{-1}\mathsf{A}_{\mathcal{I}\mathcal{J}}\mathbf{y} - \mathbf{y}^T\mathsf{A}_{\mathcal{I}\mathcal{J}}^T\mathsf{A}_{\mathcal{I}\mathcal{I}}^{-1}\mathsf{A}_{\mathcal{I}\mathcal{J}}\mathbf{y} \\ &\quad + \mathbf{y}^T\mathsf{A}_{\mathcal{I}\mathcal{J}}^T\mathsf{A}_{\mathcal{I}\mathcal{I}}^{-1}\mathsf{A}_{\mathcal{I}\mathcal{J}}\mathbf{y} + \mathbf{y}^T\mathsf{A}_{\mathcal{J}\mathcal{J}}\mathbf{y} \\ &= \mathbf{y}^T\mathsf{S}\mathbf{y}. \end{aligned}$$

It follows that if A is an SPS or SPD matrix, then the Schur complement of A with respect to an arbitrary nonsingular diagonal submatrix of A is also SPS or SPD, respectively. Moreover, we have

$$\min_{\|\mathbf{z}\|=1} \mathbf{z}^T\mathsf{A}\mathbf{z} \le \min_{\|\mathbf{x}\|=1} \mathbf{x}^T\mathsf{S}\mathbf{x} \le \max_{\|\mathbf{x}\|=1} \mathbf{x}^T\mathsf{S}\mathbf{x} \le \max_{\|\mathbf{z}\|=1} \mathbf{z}^T\mathsf{A}\mathbf{z}. \tag{1.12}$$

If A is an SPS matrix, then it is easy to check that S "inherits" its kernel from A. Indeed, if $\mathsf{A}\mathbf{e} = \mathbf{o}$, then

$$\mathsf{A}_{\mathcal{I}\mathcal{I}}\mathbf{e}_{\mathcal{I}} + \mathsf{A}_{\mathcal{I}\mathcal{J}}\mathbf{e}_{\mathcal{J}} = \mathbf{o},$$
$$\mathsf{A}_{\mathcal{J}\mathcal{I}}\mathbf{e}_{\mathcal{I}} + \mathsf{A}_{\mathcal{J}\mathcal{J}}\mathbf{e}_{\mathcal{J}} = \mathbf{o},$$

and

$$\begin{aligned}\mathsf{S}\mathbf{e}_{\mathcal{J}} =& (\mathsf{A}_{\mathcal{J}\mathcal{J}} - \mathsf{A}_{\mathcal{J}\mathcal{I}}\mathsf{A}_{\mathcal{I}\mathcal{I}}^{-1}\mathsf{A}_{\mathcal{I}\mathcal{J}})\mathbf{e}_{\mathcal{J}}\\ =& \mathsf{A}_{\mathcal{J}\mathcal{J}}\mathbf{e}_{\mathcal{J}} - \mathsf{A}_{\mathcal{J}\mathcal{I}}\mathsf{A}_{\mathcal{I}\mathcal{I}}^{-1}(-\mathsf{A}_{\mathcal{I}\mathcal{I}}\mathbf{e}_{\mathcal{I}})\\ =& \mathsf{A}_{\mathcal{J}\mathcal{J}}\mathbf{e}_{\mathcal{J}} + \mathsf{A}_{\mathcal{J}\mathcal{I}}\mathbf{e}_{\mathcal{I}} = \mathbf{o},\end{aligned}$$

i.e., $\mathbf{e}_{\mathcal{J}} \in \mathrm{Ker}\mathsf{S}$.

The Schur complement is a valuable tool for reducing the problems of manipulating singular matrices to those with smaller matrices. Alternatively, we can use the Cholesky decomposition of the nonsingular block $\mathsf{A}_{\mathcal{I}\mathcal{I}}$ to decompose an SPS matrix A into the product of triangular and nearly triangular factors. To specify the procedure in more detail, let

$$\mathsf{A}_{\mathcal{I}\mathcal{I}} = \mathsf{L}_{\mathcal{I}\mathcal{I}}\mathsf{L}_{\mathcal{I}\mathcal{I}}^{T}.$$

Denoting

$$\mathsf{L}_{\mathcal{J}\mathcal{I}} = \mathsf{A}_{\mathcal{J}\mathcal{I}}\mathsf{L}_{\mathcal{I}\mathcal{I}}^{-T}, \quad \mathsf{S} = \mathsf{A}_{\mathcal{J}\mathcal{J}} - \mathsf{A}_{\mathcal{J}\mathcal{I}}\mathsf{A}_{\mathcal{I}\mathcal{I}}^{-1}\mathsf{A}_{\mathcal{I}\mathcal{J}},$$

we can check that

$$\mathsf{A} = \begin{bmatrix}\mathsf{A}_{\mathcal{I}\mathcal{I}} & \mathsf{A}_{\mathcal{I}\mathcal{J}}\\ \mathsf{A}_{\mathcal{I}\mathcal{I}} & \mathsf{A}_{\mathcal{J}\mathcal{J}}\end{bmatrix} = \begin{bmatrix}\mathsf{L}_{\mathcal{I}\mathcal{I}} & \mathsf{O}\\ \mathsf{L}_{\mathcal{J}\mathcal{I}} & \mathsf{I}\end{bmatrix}\begin{bmatrix}\mathsf{L}_{\mathcal{I}\mathcal{I}}^{T} & \mathsf{L}_{\mathcal{J}\mathcal{I}}^{T}\\ \mathsf{O} & \mathsf{S}\end{bmatrix}. \tag{1.13}$$

If S is a small matrix, we can use the decomposition (1.13) for the effective construction of a generalized inverse. We shall summarize the main observations in the following lemma.

Lemma 1.3. *Let A denote an SPS block matrix as in (1.13), let $\mathbf{e} \in \mathrm{Ker}\mathsf{A}$, and let S denote the Schur complement matrix of the block $\mathsf{A}_{\mathcal{I}\mathcal{I}}$. Then*

$$\mathbf{e}_{\mathcal{J}} \in \mathrm{Ker}\mathsf{S}.$$

If S^{+} denote any generalized inverse of S, then the matrix A^{+} defined by

$$\mathsf{A}^{+} = \begin{bmatrix}\mathsf{L}_{\mathcal{J}\mathcal{J}}^{-T} & -\mathsf{L}_{\mathcal{J}\mathcal{J}}^{-T}\mathsf{L}_{\mathcal{I}\mathcal{J}}^{T}\mathsf{S}^{+}\\ \mathsf{O} & \mathsf{S}^{+}\end{bmatrix}\begin{bmatrix}\mathsf{L}_{\mathcal{J}\mathcal{J}}^{-1} & \mathsf{O}\\ -\mathsf{L}_{\mathcal{I}\mathcal{J}}\mathsf{L}_{\mathcal{J}\mathcal{J}}^{-1} & \mathsf{I}\end{bmatrix} \tag{1.14}$$

is a generalized inverse of A which satisfies

$$[\mathsf{A}^{+}]_{\mathcal{J}\mathcal{J}} = \mathsf{S}^{+}. \tag{1.15}$$

1.7 Norms

General concepts of size and distance in a vector space are expressed by norms. A *norm* on $\mathbb{R}^n$ is a function which assigns to each $\mathbf{x} \in \mathbb{R}^n$ a number $\|\mathbf{x}\| \in \mathbb{R}$ in such a way that for any vectors $\mathbf{x}, \mathbf{y} \in \mathbb{R}^n$ and any scalar $\alpha \in \mathbb{R}$, the following three conditions are satisfied:

(i) $\|\mathbf{x}\| \geq 0$, and $\|\mathbf{x}\| = 0$ if and only if $\mathbf{x} = \mathbf{o}$.
(ii) $\|\mathbf{x} + \mathbf{y}\| \leq \|\mathbf{x}\| + \|\mathbf{y}\|$.
(iii) $\|\alpha\mathbf{x}\| = |\alpha|\|\mathbf{x}\|$.

It is easy to check that the functions

$$\|\mathbf{x}\|_1 = |\mathbf{x}_1| + \cdots + |\mathbf{x}_n| \quad \text{and} \quad \|\mathbf{x}\|_\infty = \max\{|\mathbf{x}_1|, \ \ldots, \ |\mathbf{x}_n|\}$$

are norms. They are called ℓ_1 and ℓ_∞ norms, respectively. We often use the Euclidean norm defined by

$$\|\mathbf{x}\|_2 = \sqrt{x_1^2 + \cdots + x_n^2}.$$

The norms on $\mathbb{R}^n$ introduced above satisfy the inequalities

$$\|\mathbf{x}\|_\infty \leq \|\mathbf{x}\|_2 \leq \|\mathbf{x}\|_1 \leq \sqrt{n}\|\mathbf{x}\|_2 \leq n\|\mathbf{x}\|_\infty.$$

Given a norm defined on the domain and the range of a matrix A, we can define the *induced norm* $\|\mathsf{A}\|$ of A by

$$\|\mathsf{A}\| = \sup_{\|\mathbf{x}\|=1} \|\mathsf{A}\mathbf{x}\| = \sup_{\mathbf{x}\neq\mathbf{o}} \frac{\|\mathsf{A}\mathbf{x}\|}{\|\mathbf{x}\|}.$$

If $\mathsf{B} \neq \mathsf{O}$, then

$$\|\mathsf{AB}\| = \sup_{\mathbf{x}\neq\mathbf{o}} \frac{\|\mathsf{AB}\mathbf{x}\|}{\|\mathbf{x}\|} = \sup_{\mathsf{B}\mathbf{x}\neq\mathbf{o}} \frac{\|\mathsf{AB}\mathbf{x}\|}{\|\mathsf{B}\mathbf{x}\|}\frac{\|\mathsf{B}\mathbf{x}\|}{\|\mathbf{x}\|} \leq \sup_{\substack{\mathbf{y}\in\mathrm{Im}\mathsf{B},\\ \mathbf{y}\neq\mathbf{o}}} \frac{\|\mathsf{A}\mathbf{y}\|}{\|\mathbf{y}\|} \sup_{\mathbf{x}\neq\mathbf{o}} \frac{\|\mathsf{B}\mathbf{x}\|}{\|\mathbf{x}\|}.$$

It follows easily that the induced norm is *submultiplicative*, i.e.,

$$\|\mathsf{AB}\| \leq \|\mathsf{A}|\mathrm{Im}\mathsf{B}\| \, \|\mathsf{B}\| \leq \|\mathsf{A}\| \, \|\mathsf{B}\|. \tag{1.16}$$

If $\mathsf{A} = [a_{ij}] \in \mathbb{R}^{m\times n}$ and $\mathbf{x} = [x_i] \in \mathbb{R}^n$, then

$$\|\mathsf{A}\mathbf{x}\|_\infty = \max_{i=1,\ldots,m} |\sum_{j=1}^{n} a_{ij}x_j| \leq \max_{i=1,\ldots,m} \sum_{j=1}^{n} |a_{ij}||x_j| \leq \|\mathbf{x}\|_\infty \max_{i=1,\ldots,m} \sum_{j=1}^{n} |a_{ij}|.$$

1.8 Scalar Products

General concepts of length and angle in a vector space are introduced by means of a *scalar product*; it is the mapping which assigns to each couple $\mathbf{x}, \mathbf{y} \in \mathbb{R}^n$ a number $\langle \mathbf{x}, \mathbf{y} \rangle \in \mathbb{R}$ in such a way that for any vectors $\mathbf{x}, \mathbf{y}, \mathbf{z} \in \mathbb{R}^n$ and any scalar $\alpha \in \mathbb{R}$, the following four conditions are satisfied:

(i) $\langle \mathbf{x}, \mathbf{y} + \mathbf{z} \rangle = \langle \mathbf{x}, \mathbf{y} \rangle + \langle \mathbf{x}, \mathbf{z} \rangle$.
(ii) $\langle \alpha\mathbf{x}, \mathbf{y} \rangle = \alpha \langle \mathbf{x}, \mathbf{y} \rangle$.
(iii) $\langle \mathbf{x}, \mathbf{y} \rangle = \langle \mathbf{y}, \mathbf{x} \rangle$.
(iv) $\langle \mathbf{x}, \mathbf{x} \rangle > 0$ for $\mathbf{x} \neq \mathbf{o}$.

We often use the *Euclidean scalar product* or *Euclidean inner product* which assigns to each couple of vectors $\mathbf{x}, \mathbf{y} \in \mathbb{R}^n$ a number defined by

$$(\mathbf{x},\ \mathbf{y}) = \mathbf{x}^T \mathbf{y}.$$

If A is a symmetric positive definite matrix, then we can define the more general A-*scalar product* on $\mathbb{R}^n$ by

$$(\mathbf{x},\ \mathbf{y})_{\mathsf{A}} = \mathbf{x}^T \mathsf{A} \mathbf{y}.$$

Using a scalar product, we can define the norm $\|\mathbf{x}\|$ of $\mathbf{x}$ and the angle α between $\mathbf{x}$ and $\mathbf{y}$ by

$$\|\mathbf{x}\|^2 = \langle \mathbf{x},\ \mathbf{x} \rangle, \quad \cos\alpha = \frac{\langle \mathbf{x}, \mathbf{y} \rangle}{\|\mathbf{x}\|\|\mathbf{y}\|}.$$

We denote for any $\mathbf{x} \in \mathbb{R}^n$ its *Euclidean norm* and A-*norm* by

$$\|\mathbf{x}\| = (\mathbf{x},\ \mathbf{x})^{1/2}, \quad \|\mathbf{x}\|_{\mathsf{A}} = (\mathbf{x},\ \mathbf{x})_{\mathsf{A}}^{1/2}.$$

It is easy to see that any norm induced by a scalar product satisfies the properties (i) and (iii) of the norm. The property (ii) follows from the *Cauchy–Schwarz inequality*

$$\langle \mathbf{x},\ \mathbf{y} \rangle^2 \leq \|\mathbf{x}\|^2\ \|\mathbf{y}\|^2, \tag{1.17}$$

which is valid for any $\mathbf{x}, \mathbf{y} \in \mathbb{R}^n$ and any scalar product. The bound is tight in the sense that the inequality becomes the equality when $\mathbf{x}, \mathbf{y}$ are dependent. The property (ii) of the norm then follows by

$$\|\mathbf{x} + \mathbf{y}\|^2 = \|\mathbf{x}\|^2 + 2\langle \mathbf{x}, \mathbf{y} \rangle + \|\mathbf{y}\|^2 \leq \|\mathbf{x}\|^2 + 2\|\mathbf{x}\|\|\mathbf{y}\| + \|\mathbf{y}\|^2 = (\|\mathbf{x}\| + \|\mathbf{y}\|)^2.$$

A pair of vectors $\mathbf{x}$ and $\mathbf{y}$ is *orthogonal* (with respect to a given scalar product) if

$$\langle \mathbf{x},\ \mathbf{y} \rangle = 0.$$

If the scalar product is not specified, then we assume by default the Euclidean scalar product. The vectors $\mathbf{x}$ and $\mathbf{y}$ that are orthogonal in A-scalar product are also called A-*conjugate* or briefly *conjugate.*

Two *sets of vectors $\mathcal{E}$ and $\mathcal{F}$ are orthogonal* if every $\mathbf{x} \in \mathcal{E}$ is orthogonal to any $\mathbf{y} \in \mathcal{F}$. The set $\mathcal{E}^{\perp}$ of all the vectors of $\mathbb{R}^n$ that are orthogonal to $\mathcal{E} \subseteq \mathbb{R}^n$ is a vector space called the *orthogonal complement* of $\mathcal{E}$. If $\mathcal{E} \subseteq \mathbb{R}^n$, then

$$\mathbb{R}^n = \text{Span}\, \mathcal{E} \oplus \mathcal{E}^{\perp}.$$

A *set of vectors $\mathcal{E}$ is orthogonal* if its elements are pairwise orthogonal, i.e., any $\mathbf{x} \in \mathcal{E}$ is orthogonal to any $\mathbf{y} \in \mathcal{E}, \mathbf{y} \neq \mathbf{x}$. A set of vectors $\mathcal{E}$ is *orthonormal* if it is orthogonal and $\langle \mathbf{x}, \mathbf{x} \rangle = 1$ for any $\mathbf{x} \in \mathcal{E}$.

Any orthogonal set $\mathcal{E} = \{\mathbf{e}_1, \ldots, \mathbf{e}_n\}$ of nonzero vectors $\mathbf{e}_i$ is independent. Indeed, if

$$\alpha_1 \mathbf{e}_1 + \cdots + \alpha_n \mathbf{e}_n = \mathbf{o},$$

then we can take the scalar product of both sides of the equation with $\mathbf{e}_i$ and use the assumption on orthogonality of $\mathcal{E}$ to get that

$$\alpha_i \langle \mathbf{e}_i, \mathbf{e}_i \rangle = 0, .$$

If $\mathcal{E}$ is an orthonormal basis of a vector space $\mathcal{V} \subseteq \mathbb{R}^n$, then the same procedure as above may be used to get conveniently the coordinates ξ_i of any $\mathbf{x} \in \mathcal{V}$. For example, if $\mathcal{E}$ is orthonormal with respect to the Euclidean scalar product, it is enough to multiply

$$\mathbf{x} = \xi_1 \mathbf{e}_1 + \cdots + \xi_n \mathbf{e}_n$$

on the left by $\mathbf{e}_i^T$ to get $\xi_i = \mathbf{e}_i^T \mathbf{x}$.

A square matrix U is orthogonal if $\mathsf{U}^T \mathsf{U} = \mathsf{I}$, that is, $\mathsf{U}^{-1} = \mathsf{U}^T$. Multiplication by an orthogonal matrix U preserves both the angles between any two vectors and the Euclidean norm of any vector as

$$(\mathsf{U}\mathbf{x})^T \mathsf{U}\mathbf{y} = \mathbf{x}^T \mathsf{U}^T \mathsf{U}\mathbf{y} = \mathbf{x}^T \mathbf{y}.$$

A matrix $\mathsf{P} \in \mathbb{R}^{n \times n}$ is an *orthogonal projector* if P is a projector, i.e., $\mathsf{P}^2 = \mathsf{P}$, and $\text{Im}\mathsf{P}$ is orthogonal to $\text{Ker}\mathsf{P}$. The latter condition reads

$$\mathsf{P}^T (\mathsf{I} - \mathsf{P}) = \mathsf{O}.$$

It simply follows that

$$\mathsf{P}^T = \mathsf{P}^T \mathsf{P} = \mathsf{P},$$

so that orthogonal projectors are symmetric matrices and symmetric projectors are orthogonal projectors. If P is an orthogonal projector, then $\mathsf{I} - \mathsf{P}$ is also an orthogonal projector as

$$(\mathsf{I} - \mathsf{P})^2 = I - 2\mathsf{P} + \mathsf{P}^2 = \mathsf{I} - \mathsf{P} \quad \text{and} \quad (\mathsf{I} - \mathsf{P})^T \mathsf{P} = (\mathsf{I} - \mathsf{P})\mathsf{P} = \mathsf{O}.$$

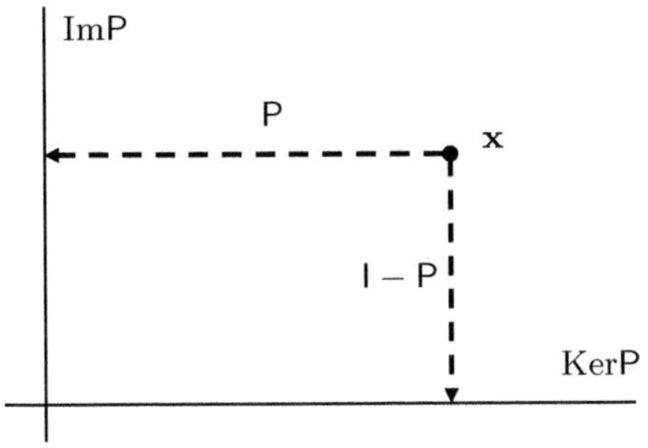

Fig. 1.1. Orthogonal projector

See Fig. 1.1 for a geometric interpretation.

If $\mathcal{U} \subseteq \mathbb{R}^n$ is the subspace spanned by the columns of a full column rank matrix $\mathsf{U} \in \mathbb{R}^{m \times n}$, then

$$\mathsf{P} = \mathsf{U}(\mathsf{U}^T\mathsf{U})^{-1}\mathsf{U}^T$$

is an orthogonal projector as

$$\mathsf{P}^2 = \mathsf{U}(\mathsf{U}^T\mathsf{U})^{-1}\mathsf{U}^T\mathsf{U}(\mathsf{U}^T\mathsf{U})^{-1}\mathsf{U}^T = \mathsf{P} \quad \text{and} \quad \mathsf{P}^T = \mathsf{P}.$$

Since any vector $\mathbf{x} \in \mathcal{U}$ may be written in the form $\mathbf{x} = \mathsf{U}\mathbf{y}$ and

$$\mathsf{P}\mathbf{x} = \mathsf{U}(\mathsf{U}^T\mathsf{U})^{-1}\mathsf{U}^T\mathsf{U}\mathbf{y} = \mathsf{U}\mathbf{y} = \mathbf{x},$$

it follows that

$$\mathcal{U} = \mathrm{Im}\mathsf{P}.$$

Observe that $\mathsf{U}^T\mathsf{U}$ is nonsingular; since $\mathsf{U}^T\mathsf{U}\mathbf{x} = \mathbf{o}$ implies

$$\|\mathsf{U}\mathbf{x}\|^2 = \mathbf{x}^T(\mathsf{U}^T\mathsf{U}\mathbf{x}) = 0,$$

it follows that $\mathbf{x} = \mathbf{o}$ by the assumption on the full column rank of U.

Let $\mathsf{B} \in \mathbb{R}^{m \times n}$ and $\mathbf{x} \in \mathrm{Im}\mathsf{B}^T$, so that there is $\mathbf{y}$ such that $\mathbf{x} = \mathsf{B}^T\mathbf{y}$. Then for any $\mathbf{z} \in \mathrm{Ker}\mathsf{B}$

$$\mathbf{x}^T\mathbf{z} = (\mathsf{B}^T\mathbf{y})^T\mathbf{z} = \mathbf{y}^T(\mathsf{B}\mathbf{z}) = 0,$$

so that $\mathrm{Ker}\mathsf{B}$ is orthogonal to $\mathrm{Im}\mathsf{B}^T$. An important result of linear algebra is that

$$(\mathrm{Ker}\mathsf{B})^\perp = \mathrm{Im}\mathsf{B}^T, \tag{1.18}$$

thus

$$\mathbb{R}^n = \mathrm{Ker}\mathsf{B} \oplus \mathrm{Im}\mathsf{B}^T. \tag{1.19}$$

The orthogonal projectors and their generalization, the conjugate projectors that we introduce in Sect. 8.3.1, are useful computational tools for manipulations with subspaces.

1.9 Orthogonalization

If $\mathsf{A} = [\mathbf{a}_1, \ldots, \mathbf{a}_m] \in \mathbb{R}^{n\times m}$ is a full column rank matrix, then it is often useful to find an orthonormal basis $\mathbf{q}_1, \ldots, \mathbf{q}_m$ of ImA.

The *classical Gram–Schmidt* orthogonalization generates an orthogonal basis $\mathbf{u}_1, \ldots, \mathbf{u}_m$ and orthonormal basis $\mathbf{q}_1, \ldots, \mathbf{q}_m$ of ImA starting from

$$\mathbf{q}_1 = \mathbf{a}_1, \quad \mathbf{q}_1 = \|\mathbf{u}_1\|^{-1}\mathbf{u}_1, \quad r_{11} = (\mathbf{q}_1, \mathbf{a}_1) = \|\mathbf{u}_1\|, \quad \mathbf{a}_1 = r_{11}\mathbf{q}_1. \quad (1.20)$$

Having $\mathbf{q}_1, \ldots, \mathbf{q}_{k-1}$, $k \le m$, the algorithm looks for $\mathbf{q}_k$ in the form

$$\mathbf{q}_k = \|\mathbf{u}_k\|^{-1}\mathbf{u}_k, \quad \mathbf{u}_k = \mathbf{a}_k - r_{1k}\mathbf{q}_1 - \cdots - r_{k-1,k}\mathbf{q}_{k-1}.$$

Taking the scalar product of the right equation of (1.20) with $\mathbf{q}_i$, we get

$$r_{ik} = (\mathbf{q}_i, \mathbf{a}_k), \quad r_{kk} = \|\mathbf{u}_k\| = \sqrt{\|\mathbf{a}_k\|^2 - \sum_{i=1}^{k-1} r_{ik}^2}, \quad i = 1, \ldots k-1. \quad (1.21)$$

If we denote by $\mathsf{R} \in \mathbb{R}^{m\times m}$ the upper triangular matrix with the nonzero entries defined by (1.21), we get the *QR decomposition*

$$\mathsf{A} = \mathsf{QR}, \quad \mathsf{QQ}^T = \mathsf{I}, \quad \mathsf{Q} = [\mathbf{q}_1, \ldots, \mathbf{q}_m].$$

It follows that we can obtain R alternatively by the Cholesky factorization

$$\mathsf{A}^T\mathsf{A} = \mathsf{LL}^T, \quad \mathsf{L} = \mathsf{R}^T.$$

If the columns of A are not strongly independent, then the application of the classical Gram-Schmidt orthogonalization can be seriously flawed by rounding errors. This effect can be reduced without increasing the cost by replacing (1.21) by

$$r_{ik} = (\mathbf{q}_i, \mathbf{a}_k - \sum_{j=1}^{i-1} r_{jk}\mathbf{q}_j), \quad i = 1, \ldots k-1. \quad (1.22)$$

Though (1.21) and (1.22) are identical in exact arithmetic, it has been proved that the application of *modified Gram–Schmidt* orthogonalization using (1.22) generates better Q, i.e., the bound on the violation of orthogonality given by

$$e = \|\mathsf{Q}^T\mathsf{Q} - \mathsf{I}\|$$

is smaller for the modified Gram–Schmidt algorithm (see Giraud et al. [142]). Still, better orthogonality can be achieved by executing the Gram-Schmidt twice.

1.10 Eigenvalues and Eigenvectors

If $\mathsf{A} \in \mathbb{R}^{n \times n}$ is a square matrix, then it may happen that there is a vector $\mathbf{e} \in \mathbb{R}^n$ such that $\mathsf{A}\mathbf{e}$ is just a scalar multiple of $\mathbf{e}$. Such vectors turned out to be useful for analysis of problems described by matrices. Since the theory has been developed for complex matrices, we consider in this section the vectors and matrices with the entries belonging to the set of complex numbers $\mathbb{C}$ in order to simplify our exposition.

Let $\mathsf{A} \in \mathbb{C}^{n \times n}$ denote a square matrix with complex entries. If a vector $\mathbf{e} \in \mathbb{C}^n$ and a scalar $\lambda \in \mathbb{C}$ satisfy

$$\mathsf{A}\mathbf{e} = \lambda \mathbf{e},$$

then $\mathbf{e}$ is said to be an *eigenvector* of A associated with an *eigenvalue* λ. A vector $\mathbf{e}$ is an eigenvector of A if and only if $\mathrm{Span}\{\mathbf{e}\}$ is an invariant subspace of A; the restriction $\mathsf{A}|\mathrm{Span}\{\mathbf{e}\}$ reduces to the multiplication by λ. If $\{\mathbf{e}_1, \ldots, \mathbf{e}_k\}$ are eigenvectors of a symmetric matrix A, then it is easy to check that $\mathrm{Span}\{\mathbf{e}_1, \ldots, \mathbf{e}_k\}$ and $\mathrm{Span}\{\mathbf{e}_1, \ldots, \mathbf{e}_k\}^\perp$ are invariant subspaces.

The set of all eigenvalues of A is called the *spectrum* of A; we denote it by $\sigma(\mathsf{A})$. Obviously $\lambda \in \sigma(\mathsf{A})$ if and only if $\mathsf{A} - \lambda\mathsf{I}$ is singular, and $0 \in \sigma(\mathsf{A})$ if and only if A is singular. Nonconvex set is in Fig. 1.2.

$$\begin{bmatrix} 2 & 2 & 2 \\ 2 & 2 & 2 \\ 2 & 2 & 2 \end{bmatrix}$$

Fig. 1.2. The spectrum of a matrix

Many important relations in this book are proved by analysis of the spectrum of a given matrix. The information about the spectrum is typically obtained indirectly, in a similar way as information about a ghost in a spirit session, when the participants are not assumed to see the ghost, but to observe that the table is moving.

The eigenvalues can be characterized algebraically by means of the *determinant.* Since it is well-known that a matrix is singular if and only if its determinant is equal to zero, it follows that the eigenvalues of A are the roots of the *characteristic equation*

$$\det(\mathsf{A} - \lambda\mathsf{I}) = 0. \tag{1.23}$$

The *characteristic polynomial* $p_{\mathsf{A}}(\lambda) = \det(\mathsf{A} - \lambda\mathsf{I})$ is of the degree n. Thus there are at most n distinct eigenvalues and $\sigma(\mathsf{A})$ is not the empty set.

In what follows, we associate with each matrix $\mathsf{A} \in \mathbb{C}^{n\times n}$ a sequence $\lambda_1, \ldots, \lambda_n$ of the eigenvalues of A labeled as the roots of the characteristic polynomial. Each distinct eigenvalue appears in this sequence as many times as corresponds to its *algebraic multiplicity* as a root of the characteristic polynomial. Using the factorization

$$p_{\mathsf{A}}(\lambda) = (\lambda_1 - \lambda)\cdots(\lambda_n - \lambda)\ ,$$

we get

$$\det(\mathsf{A}) = p_{\mathsf{A}}(0) = \lambda_1 \cdots \lambda_n. \tag{1.24}$$

Comparing the coefficients in the factorization of $p_{\mathsf{A}}(\lambda)$ with those arising from evaluation of $\det(\mathsf{A} - \lambda\mathsf{I})$, we get

$$\operatorname{trace}(\mathsf{A}) = \lambda_1 + \cdots + \lambda_n, \tag{1.25}$$

where $\operatorname{trace}(\mathsf{A})$ denotes the sum of all diagonal entries of A.

The dimension of $\operatorname{Ker}(\mathsf{A} - \lambda\mathsf{I})$ is called the *geometric multiplicity* of λ. If the algebraic multiplicity of an eigenvalue λ is k, then the number of the independent eigenvectors corresponding to λ is an integer between 1 and k.

Even though it is in general difficult to evaluate the eigenvalues of a given matrix A, it is still possible to get nontrivial information about $\sigma(\mathsf{A})$ without heavy computations. Useful information about the location of eigenvalues can be obtained by *Gershgorin's theorem*, which guarantees that every eigenvalue of $\mathsf{A} = [a_{ij}] \in \mathbb{C}^{n\times n}$ is located in at least one of the n circular *Gershgorin disks* in the complex plane with the centers a_{ii} and radii $r_i = \sum_{j\neq i} |a_{ij}|$.

The *eigenvalues of a real symmetric matrix are real.* Since it is easy to check whether a matrix is symmetric, this gives us useful information about the location of eigenvalues.

Let $\mathsf{A} \in \mathbb{R}^{n\times n}$ denote a real symmetric matrix, let $\mathcal{I} = \{1,\ \ldots,\ n-1\}$, and let $\mathsf{A}^1 = \mathsf{A}_{\mathcal{I}\mathcal{I}}$. Let $\lambda_1 \geq \cdots \geq \lambda_n$ and $\lambda_1^1 \geq \cdots \geq \lambda_{n-1}^1$ denote the eigenvalues of A and A^1, respectively. Then by the *Cauchy interlacing theorem*

$$\lambda_1 \geq \lambda_1^1 \geq \lambda_2 \geq \lambda_2^1 \geq \cdots \geq \lambda_{n-1}^1 \geq \lambda_n. \tag{1.26}$$

There is a beautiful result by Smith [248] on the interlacing of the eigenvalues of A and its Schur complements. Let $\mathcal{I} = \{1, \ldots, m\}$, $\mathcal{J} = \{m+1, \ldots, n\}$, and let $\lambda_i(\mathsf{A})$, $i = 1, \ldots, n$, and $\lambda_j(\mathsf{S})$, $j = 1, \ldots, n-m$, denote the entries of the decreasing series of the eigenvalues of A and of the Schur complement

$$\mathsf{S} = \mathsf{A}_{\mathcal{J}\mathcal{J}} - \mathsf{A}_{\mathcal{J}\mathcal{I}}\mathsf{A}_{\mathcal{I}\mathcal{I}}^{-1}\mathsf{A}_{\mathcal{J}\mathcal{I}}.$$

Then

$$\|\mathsf{A}_{\mathcal{I}\mathcal{I}}\| \geq \lambda_i(\mathsf{S}) \geq \lambda_{i+m}(\mathsf{A}),\ \ i = 1, \ldots, n-m. \tag{1.27}$$

1.11 Matrix Decompositions

If $\mathsf{A} \in \mathbb{R}^{n\times n}$ is a symmetric matrix, then it is possible to find n orthonormal eigenvectors $\mathbf{e}_1, \ldots, \mathbf{e}_n$ that form the basis of $\mathbb{R}^n$. Moreover, the corresponding eigenvalues are real. Denoting by $\mathsf{U} = [\mathbf{e}_1, \ldots, \mathbf{e}_n] \in \mathbb{R}^{n\times n}$ an orthogonal matrix whose columns are the eigenvectors, we may write the *spectral decomposition* of A as

$$\mathsf{A} = \mathsf{U}\mathsf{D}\mathsf{U}^T, \tag{1.28}$$

where $\mathsf{D} = \mathrm{diag}(\lambda_1, \ldots, \lambda_n) \in \mathbb{R}^{n\times n}$ is the diagonal matrix whose diagonal entries are the eigenvalues corresponding to the eigenvectors $\mathbf{e}_1, \ldots, \mathbf{e}_n$. Reordering the columns of U, we can achieve that $\lambda_1 \geq \cdots \geq \lambda_n$.

The spectral decomposition reveals close relations between the properties of a symmetric matrix and its eigenvalues. Thus a symmetric matrix is SPD if and only if all its eigenvalues are positive, and it is SPS if and only if they are nonnegative. The rank of a symmetric matrix is equal to the number of nonzero entries of D.

If A is symmetric, then we can use (1.28) to check that for any nonzero $\mathbf{x}$

$$\lambda_1 = \lambda_{\max} \geq \|\mathbf{x}\|^{-2}\mathbf{x}^T\mathsf{A}\mathbf{x} \geq \lambda_{\min} = \lambda_n. \tag{1.29}$$

Thus for any symmetric positive definite matrix A

$$\|\mathsf{A}\| = \lambda_{\max},\, \|\mathsf{A}^{-1}\| = \lambda_{\min}^{-1},\, \|\mathbf{x}\|_{\mathsf{A}} \leq \lambda_{\max}\|\mathbf{x}\|,\, \|\mathbf{x}\|_{\mathsf{A}^{-1}} \leq \lambda_{\min}^{-1}\|\mathbf{x}\|. \tag{1.30}$$

The *spectral condition number* $\kappa(\mathsf{A}) = \|\mathsf{A}\|\|\mathsf{A}^{-1}\|$ can be expressed for real symmetric matrix by

$$\kappa(\mathsf{A}) = \lambda_{\max}/\lambda_{\min}.$$

Another consequence of the spectral decomposition theorem is the *Courant–Fischer minimax principle*, (see, e.g., [147]) which states that if $\lambda_1 \geq \cdots \geq \lambda_n$ are the eigenvalues of a real symmetric matrix $\mathsf{A} \in \mathbb{R}^{n\times n}$, then

$$\lambda_k = \max_{\substack{\mathcal{V} \subseteq \mathbb{R}^n,\\ \dim \mathcal{V} = k}} \min_{\substack{\mathbf{x} \in \mathcal{V},\\ \|\mathbf{x}\| = 1}} \mathbf{x}^T\mathsf{A}\mathbf{x}, \quad k = 1, \ldots, n. \tag{1.31}$$

If A is a real symmetric matrix and f is a real function defined on $\sigma(\mathsf{A})$, we can use the spectral decomposition to define the *scalar function* by

$$f(\mathsf{A}) = \mathsf{U}f(\mathsf{D})\mathsf{U}^T,$$

where $f(\mathsf{D}) = \mathrm{diag}\,(f(\lambda_1), \ldots, f(\lambda_n))$. It is easy to check that if a is the identity function on $\mathbb{R}$ defined by $a(x) = x$, then

$$a(\mathsf{A}) = \mathsf{A},$$

and if f and g are real functions defined on $\sigma(\mathsf{A})$, then

$$(f+g)(\mathsf{A}) = f(\mathsf{A}) + g(\mathsf{A}) \quad \text{and} \quad (f \cdot g)(\mathsf{A}) = f(\mathsf{A})g(\mathsf{A}).$$

Moreover, if $f(x) \geq 0$ for $x \in \sigma(\mathsf{A})$, then $f(\mathsf{A})$ is SPS, and if $f(x) > 0$ for $x \in \sigma(\mathsf{A})$, then $f(\mathsf{A})$ is SPD. Obviously

$$\sigma(f(\mathsf{A})) = f(\sigma(\mathsf{A})), \tag{1.32}$$

and if $\mathbf{e}_i$ is an eigenvector corresponding to $\lambda_i \in \sigma(\mathsf{A})$, then it is also an eigenvector of $f(\mathsf{A})$ corresponding to $f(\lambda_i)$. It follows easily that for any symmetric positive semidefinite matrix

$$\mathrm{Im}\mathsf{A} = \mathrm{Im}\mathsf{A}^{1/2} \quad \text{and} \quad \mathrm{Ker}\mathsf{A} = \mathrm{Ker}\mathsf{A}^{1/2}. \tag{1.33}$$

For any $\mathsf{B} \in \mathbb{R}^{n \times n}$, there is an invertible matrix U such that

$$\mathsf{B} = \mathsf{U}\,\mathrm{diag}(\mathsf{J}_1, \dots, \mathsf{J}_k)\,\mathsf{U}^{-1}, \quad \mathsf{J}_i = \lambda_i \mathsf{I} + \mathsf{N}_i, \quad i = 1, \dots, k, \tag{1.34}$$

where N_i has zero entries except $[\mathsf{N}_i]_{j,j+1} = 1$. Matrices J_i define *Jordan decomposition*; they are called *Jordan blocks*.

A key to understanding nonsymmetric matrices is the *singular value decomposition* (SVD). If $\mathsf{B} \in \mathbb{R}^{m \times n}$, then SVD of B is given by

$$\mathsf{B} = \mathsf{U}\mathsf{S}\mathsf{V}^T, \tag{1.35}$$

where $\mathsf{U} \in \mathbb{R}^{m \times m}$ and $\mathsf{V} \in \mathbb{R}^{n \times n}$ are orthogonal, and $\mathsf{S} \in \mathbb{R}^{m \times n}$ is a diagonal matrix with nonnegative diagonal entries $\sigma_1 \geq \dots \geq \sigma_{\min\{m,n\}} = \sigma_{\min}$ called *singular values* of B. If $\mathsf{A} \neq \mathsf{O}$, it is often more convenient to use the *reduced (slim) singular value decomposition* (RSVD)

$$\mathsf{B} = \widehat{\mathsf{U}}\widehat{\mathsf{S}}\widehat{\mathsf{V}}^T, \tag{1.36}$$

where $\widehat{\mathsf{U}} \in \mathbb{R}^{m \times r}$ and $\widehat{\mathsf{V}} \in \mathbb{R}^{n \times r}$ are matrices with orthonormal columns, $\widehat{\mathsf{S}} \in \mathbb{R}^{r \times r}$ is a nonsingular diagonal matrix with positive diagonal entries $\sigma_1 \geq \dots \geq \sigma_r = \overline{\sigma}_{\min}$, and $r \leq \min\{m, n\}$ is the rank of B. The matrices $\widehat{\mathsf{U}}$ and $\widehat{\mathsf{V}}$ are formed by the first r columns of U and V. If $\mathbf{x} \in \mathbb{R}^m$, then

$$\mathsf{B}\mathbf{x} = \widehat{\mathsf{U}}\widehat{\mathsf{S}}\widehat{\mathsf{V}}^T\mathbf{x} = (\widehat{\mathsf{U}}\widehat{\mathsf{S}}\widehat{\mathsf{V}}^T)(\widehat{\mathsf{V}}\widehat{\mathsf{S}}\widehat{\mathsf{U}}^T)(\widehat{\mathsf{U}}\widehat{\mathsf{S}}^{-1}\mathbf{x}) = \mathsf{B}\mathsf{B}^T\mathbf{y},$$

so that

$$\mathrm{Im}\mathsf{B} = \mathrm{Im}\mathsf{B}\mathsf{B}^T. \tag{1.37}$$

The singular value decomposition reveals close relations between the properties of a matrix and its singular values. Thus the rank of $\mathsf{B} \in \mathbb{R}^{m \times n}$ is equal to the number of its nonzero singular values,

$$\|\mathsf{B}\| = \|\mathsf{B}^T\| = \sigma_1, \tag{1.38}$$

and for any vector $\mathbf{x} \in \mathbb{R}^n$

$$\sigma_{\min}\|\mathbf{x}\| \le \|\mathsf{B}\mathbf{x}\| \le \|\mathsf{B}\|\|\mathbf{x}\|. \tag{1.39}$$

Let $\overline{\sigma}_{\min}$ denote the least nonzero singular value of $\mathsf{B} \in \mathbb{R}^{m\times n}$, let $\mathbf{x} \in \mathrm{Im}\mathsf{B}^T$, and consider RSVD decomposition $\mathsf{B} = \widehat{\mathsf{U}}\widehat{\mathsf{S}}\widehat{\mathsf{V}}^T$ with $\widehat{\mathsf{U}} \in \mathbb{R}^{m\times r}$, $\widehat{\mathsf{V}} \in \mathbb{R}^{n\times r}$, and $\widehat{\mathsf{S}} \in \mathbb{R}^{r\times r}$. Then there is $\mathbf{y} \in \mathbb{R}^r$ such that $\mathbf{x} = \widehat{\mathsf{V}}\mathbf{y}$ and

$$\|\mathsf{B}\mathbf{x}\| = \|\widehat{\mathsf{U}}\widehat{\mathsf{S}}\widehat{\mathsf{V}}^T\widehat{\mathsf{V}}\mathbf{y}\| = \|\widehat{\mathsf{U}}\widehat{\mathsf{S}}\mathbf{y}\| = \|\widehat{\mathsf{S}}\mathbf{y}\| \ge \overline{\sigma}_{\min}\|\mathbf{y}\|.$$

Since

$$\|\mathbf{x}\| = \|\widehat{\mathsf{V}}\mathbf{y}\| = \|\mathbf{y}\|,$$

we conclude that

$$\overline{\sigma}_{\min}\|\mathbf{x}\| \le \|\mathsf{B}\mathbf{x}\| \le \|\mathsf{B}\|\|\mathbf{x}\| \quad \text{for any} \quad \mathbf{x} \in \mathrm{Im}\mathsf{B}^T. \tag{1.40}$$

The singular value decomposition (1.35) can be used to introduce the *Moore–Penrose generalized inverse* of an $m \times n$ matrix B by

$$\mathsf{B}^\dagger = \mathsf{V}\mathsf{S}^\dagger\mathsf{U}^T = \widehat{\mathsf{V}}\widehat{\mathsf{S}}^\dagger\widehat{\mathsf{U}}^T,$$

where $\mathsf{S}^\dagger$ is the diagonal matrix with the entries $[\mathsf{S}^\dagger]_{ii} = 0$ if $\sigma_i = 0$ and $[\mathsf{S}^\dagger]_{ii} = \sigma_i^{-1}$ otherwise. It is easy to check that

$$\mathsf{B}\mathsf{B}^\dagger\mathsf{B} = \widehat{\mathsf{U}}\widehat{\mathsf{S}}\widehat{\mathsf{V}}^T\widehat{\mathsf{V}}\widehat{\mathsf{S}}^\dagger\widehat{\mathsf{U}}^T\widehat{\mathsf{U}}\widehat{\mathsf{S}}\widehat{\mathsf{V}}^T = \widehat{\mathsf{U}}\widehat{\mathsf{S}}\widehat{\mathsf{V}}^T = \mathsf{B}, \tag{1.41}$$

so that the Moore–Penrose generalized inverse is a generalized inverse. If B is a full row rank matrix, then it may be checked directly that

$$\mathsf{B}^\dagger = \mathsf{B}^T(\mathsf{B}\mathsf{B}^T)^{-1}.$$

If B is a singular matrix and $\mathbf{c} \in \mathrm{Im}\mathsf{B}$, then $\mathbf{x}_{\mathrm{LS}} = \mathsf{B}^\dagger\mathbf{c}$ is a solution of the system of linear equations $\mathsf{B}\mathbf{x} = \mathbf{c}$, i.e.,

$$\mathsf{B}\mathbf{x}_{\mathrm{LS}} = \mathbf{c}.$$

Notice that $\mathbf{x}_{\mathrm{LS}} \in \mathrm{Im}\mathsf{B}^T$, so that if $\overline{\mathbf{x}}$ is any other solution, then $\overline{\mathbf{x}} = \mathbf{x}_{\mathrm{LS}} + \mathbf{d}$, where $\mathbf{d} \in \mathrm{Ker}\mathsf{B}$, $\mathbf{x}_{\mathrm{LS}}^T\mathbf{d} = 0$ by (1.18), and

$$\|\mathbf{x}_{\mathrm{LS}}\|^2 \le \|\overline{\mathbf{x}}_{\mathrm{LS}}\|^2 + \|\mathbf{d}\|^2 = \|\overline{\mathbf{x}}\|^2. \tag{1.42}$$

The vector $\mathbf{x}_{\mathrm{LS}}$ is called the *least square solution* of $\mathsf{B}\mathbf{x} = \mathbf{c}$. Obviously

$$\|\mathsf{B}^\dagger\| = \overline{\sigma}_{\min}^{-1}, \tag{1.43}$$

where $\overline{\sigma}_{\min}$ denotes the least nonzero singular value of B, so that

$$\|\mathbf{x}_{\mathrm{LS}}\| = \|\mathsf{B}^\dagger\mathbf{c}\| \le \overline{\sigma}_{\min}^{-1}\|\mathbf{c}\|. \tag{1.44}$$

1.12 Angles of Subspaces

In Sect. 1.8, we introduced the concept of the angle between vectors in terms of their scalar product. A similar definition can be formulated for subspaces $\mathcal{X}, \mathcal{Y} \subset \mathbb{R}^n$, but in this case, we shall use a collection of angles starting from the angle between the nearest vectors $\mathbf{x} \in \mathcal{X}$, $\mathbf{y} \in \mathcal{Y}$. More formally, the *canonical angle* γ_1 between the subspaces $\mathcal{X}$ and $\mathcal{Y}$ of the dimensions p and q, $p \leq q$, is defined recurrently by

$$\cos(\gamma_1) = \mathbf{x}_1^T \mathbf{y}_1 = \max\{\mathbf{x}^T \mathbf{y} : \mathbf{x} \in \mathcal{X}, \mathbf{y} \in \mathcal{Y}, \|\mathbf{x}\| = \|\mathbf{y}\| = 1\}$$

and

$$\cos(\gamma_k) = \mathbf{x}_k^T \mathbf{y}_k = \max\{\mathbf{x}^T \mathbf{y} : \mathbf{x} \in \mathcal{X}_{k-1}, \mathbf{y} \in \mathcal{Y}_{k-1}, \|\mathbf{x}\| = \|\mathbf{y}\| = 1\},$$

where $k = 2, \ldots, p$ and

$$\mathcal{X}_k = \left\{\mathbf{x} \in \operatorname{Span}\{\mathbf{x}_1, \ldots, \mathbf{x}_k\}^{\perp}, \quad \mathcal{Y}_k = \{\mathbf{y} \in \operatorname{Span}\{\mathbf{y}_1, \ldots, \mathbf{y}_k\}^{\perp}\right\}.$$

To see the relation to SVD, let us denote by $\mathsf{X} \in \mathbb{R}^{n \times p}$ and $\mathsf{Y} \in \mathbb{R}^{n \times q}$ the full rank matrices with orthonormal columns such that

$$\mathcal{X} = \operatorname{Im} \mathsf{X} \quad \text{and} \quad \mathcal{Y} = \operatorname{Im} \mathsf{Y}.$$

Then

$$\begin{aligned}\cos(\gamma_1) = \mathbf{x}_1^T \mathbf{y}_1 &= \max\{\boldsymbol{\xi}^T \mathsf{X}^T \mathsf{Y} \boldsymbol{\eta} : \boldsymbol{\xi} \in \mathbb{R}^p, \boldsymbol{\eta} \in \mathbb{R}^q, \|\boldsymbol{\xi}\| = \|\boldsymbol{\eta}\| = 1\} \\ &= \boldsymbol{\xi}_1^T \mathsf{X}^T \mathsf{Y} \boldsymbol{\eta}_1,\end{aligned}$$

i.e., we can take $\mathbf{x}_1 = \mathsf{X}\boldsymbol{\xi}_1, \quad \mathbf{y}_1 = \mathsf{Y}\boldsymbol{\eta}_1$. Introducing SVD

$$\mathsf{X}^T \mathsf{Y} = \mathsf{U}\mathsf{S}\mathsf{V}^T, \quad \mathsf{S} = \operatorname{diag}(\sigma_1, \ldots, \sigma_p) \in \mathbb{R}^{p \times q},$$

and taking into account the extremal properties of singular values, we get

$$\cos(\gamma_1) = \boldsymbol{\xi}_1^T \mathsf{X}^T \mathsf{Y} \boldsymbol{\eta}_1 = \max\{\boldsymbol{\xi}^T \mathsf{U}\mathsf{S}\mathsf{V}^T \boldsymbol{\eta} : \boldsymbol{\xi} \in \mathbb{R}^p, \boldsymbol{\eta} \in \mathbb{R}^q, \|\boldsymbol{\xi}\| = \|\boldsymbol{\eta}\| = 1\} = \sigma_1.$$

Using similar arguments, we get

$$\cos(\gamma_k) = \mathbf{x}_k^T \mathbf{y}_k = \boldsymbol{\xi}_k^T \mathsf{X}^T \mathsf{Y} \boldsymbol{\eta}_k = \boldsymbol{\xi}_k^T \mathsf{U}\mathsf{S}\mathsf{V}^T \boldsymbol{\eta}_k = \sigma_k, \quad k = 2, \ldots, p,$$

and

$$[\mathbf{x}_1, \ldots, \mathbf{x}_p] = \mathsf{X}\,[\boldsymbol{\xi}_1, \ldots, \boldsymbol{\xi}_p] \quad [\mathbf{y}_1, \ldots, \mathbf{y}_p] = \mathsf{Y}\,[\boldsymbol{\eta}_1, \ldots, \boldsymbol{\eta}_p].$$

The largest canonical angle γ_p between two subspaces, the cosine of which corresponds to the smallest singular value σ_p, incorporates the information about their distance. If applied to one or two-dimensional subspaces associated with lines or planes in $\mathbb{R}^3$, γ_p coincides with the notion of the angle between a vector and a plane or between two planes.

If $p = q$, we can characterize the distance between $\mathcal{X}$ and $\mathcal{Y}$ by the *gap*

$$\text{gap}(\mathcal{X}, \mathcal{Y}) = \|\mathsf{P}_{\mathcal{X}} - \mathsf{P}_{\mathcal{Y}}\|,$$

where $\mathsf{P}_{\mathcal{X}}$ denotes the orthogonal projector onto $\mathcal{X}$. To see the relation between the gap and canonical angles, denote $\mathcal{X} = \text{Im}\mathsf{X}$, $\mathcal{Y} = \text{Im}\mathsf{Y}$ and let

$$\mathsf{W} = [\mathsf{X} \ \ \mathsf{X}_\perp], \quad \mathsf{Z} = [\mathsf{Y} \ \ \mathsf{Y}_\perp]$$

denote orthogonal matrices. Then

$$\begin{aligned}\text{gap}(\mathcal{X}, \mathcal{Y}) &= \|\mathsf{X}\mathsf{X}^T - \mathsf{Y}\mathsf{Y}^T\| = \left\|\mathsf{W}^T\left(\mathsf{X}\mathsf{X}^T - \mathsf{Y}\mathsf{Y}^T\right)\mathsf{Z}\right\| \\ &= \left\|\begin{bmatrix}\mathsf{X}^T \\ \mathsf{X}_\perp^T\end{bmatrix}\left(\mathsf{X}\mathsf{X}^T - \mathsf{Y}\mathsf{Y}^T\right)[\mathsf{Y} \ \ \mathsf{Y}_\perp]\right\| = \left\|\begin{bmatrix}\mathsf{O} & \mathsf{X}^T\mathsf{Y}_\perp \\ \mathsf{X}_\perp^T\mathsf{Y} & \mathsf{O}\end{bmatrix}\right\|.\end{aligned}$$

Since $\mathsf{W}^T\mathsf{Z}$ is orthogonal, it follows that there are orthogonal matrices

$$\begin{aligned}&\mathsf{U}_1, \mathsf{V}_1 \in \mathbb{R}^{p\times p}, \ \mathsf{U}_2, \mathsf{V}_2 \in \mathbb{R}^{(n-p)\times(n-p)}, \text{ and} \\ &\mathsf{U} = \text{diag}(\mathsf{U}_1, \mathsf{U}_2), \ \mathsf{V} = \text{diag}(\mathsf{V}_1, \mathsf{V}_2),\end{aligned}$$

such that the CS decomposition of $\mathsf{W}^T\mathsf{Z}$ has the form

$$\mathsf{U}^T\mathsf{W}^T\mathsf{Z}\mathsf{V} = \left[\begin{array}{ccc|ccc}\mathsf{I} & \mathsf{O} & \mathsf{O} & \mathsf{O} & \mathsf{O} & \mathsf{O} \\ \mathsf{O} & \mathsf{C} & \mathsf{O} & \mathsf{O} & \mathsf{S} & \mathsf{O} \\ \mathsf{O} & \mathsf{O} & \mathsf{O} & \mathsf{O} & \mathsf{O} & \mathsf{I} \\ \hline \mathsf{O} & \mathsf{O} & \mathsf{O} & \mathsf{I} & \mathsf{O} & \mathsf{O} \\ \mathsf{O} & -\mathsf{S} & \mathsf{O} & \mathsf{O} & \mathsf{C} & \mathsf{O} \\ \mathsf{O} & \mathsf{O} & \mathsf{I} & \mathsf{O} & \mathsf{O} & \mathsf{O}\end{array}\right] = \begin{bmatrix}\mathsf{U}_1^T & \mathsf{O} \\ \mathsf{O} & \mathsf{U}_2^T\end{bmatrix}\left[\begin{array}{c|c}\mathsf{X}^T\mathsf{Y} & \mathsf{X}^T\mathsf{Y}_\perp \\ \hline \mathsf{X}_\perp^T\mathsf{Y} & \mathsf{X}_\perp^T\mathsf{Y}_\perp\end{array}\right]\begin{bmatrix}\mathsf{V}_1 & \mathsf{O} \\ \mathsf{O} & \mathsf{V}_2\end{bmatrix}. \tag{1.45}$$

Comparing the blocks of the right matrix with its CS decomposition, we get that the *nonzero canonical cosines of* $\mathsf{X}^T\mathsf{Y}$ *and* $\mathsf{X}_\perp^T\mathsf{Y}_\perp$ are the same and

$$\|\mathsf{X}^T\mathsf{Y}_\perp\| = \|\mathsf{U}_1^T\mathsf{X}^T\mathsf{Y}_\perp\mathsf{V}_2\| = \|\text{diag}(\mathsf{O}, \mathsf{S}, \mathsf{I})\| = \|\mathsf{U}_2^T\mathsf{X}_\perp^T\mathsf{Y}\mathsf{V}_2\| = \|\mathsf{X}_\perp^T\mathsf{Y}\|.$$

Moreover, if $\mathbf{x} \in \mathbb{R}^{p\times p}$, $\|\mathbf{x}\| = 1$, then

$$1 = \left\|\begin{bmatrix}\mathsf{X}^T \\ \mathsf{X}_\perp^T\end{bmatrix}[\mathsf{Y}, \mathsf{Y}_\perp]\begin{bmatrix}\mathbf{x} \\ \mathbf{o}\end{bmatrix}\right\| = \|[\mathsf{X}^T\mathsf{Y}\mathbf{x}]\|^2 + \|[\mathsf{X}_\perp^T\mathsf{Y}\mathbf{x}]\|^2.$$

It follows that

$$\begin{aligned}\|\mathsf{X}^T\mathsf{Y}_\perp\| &= \max_{\|\mathbf{x}\|=1}\|[\mathsf{X}^T\mathsf{Y}_\perp\mathbf{x}]\| = \sqrt{1 - \min_{\|\mathbf{x}\|=1}\|[\mathsf{X}^T\mathsf{Y}\mathbf{x}]\|^2} \\ \text{gap}(\mathcal{X}, \mathcal{Y}) &= \|\mathsf{X}^T\mathsf{Y}_\perp\| = \sqrt{1 - \sigma_p^2} = \sin(\gamma_p).\end{aligned} \tag{1.46}$$

See also the Golub and van Loan [147, Chap. 2] or Ipsen and Mayer [175].

1.13 Augmented Matrices

We often use the matrices

$$\mathsf{A}_\varrho = \mathsf{A} + \varrho \mathsf{B}^T \mathsf{B},$$

where $\mathsf{A} \in \mathbb{R}^{n\times n}$ is a symmetric matrix, $\mathsf{B} \in \mathbb{R}^{m\times n}$, and $\varrho \geq 0$. The matrix A_ϱ is called the *augmented* or *penalized matrix* as it is closely related to the penalty method described in Sect. 6.2. Let us first give a simple sufficient condition that enforces A_ϱ to be positive definite.

Lemma 1.4. *Let* $\mathsf{A} \in \mathbb{R}^{n\times n}$ *be a symmetric positive semidefinite matrix, let* $\mathsf{B} \in \mathbb{R}^{m\times n}$, $\varrho > 0$, *and let* $\mathrm{Ker}\mathsf{A} \cap \mathrm{Ker}\mathsf{B} = \{\mathbf{o}\}$. *Then* A_ϱ *is positive definite.*

Proof. If $\mathbf{x} \neq \mathbf{o}$ and $\mathrm{Ker}\mathsf{A} \cap \mathrm{Ker}\mathsf{B} = \{\mathbf{o}\}$, then either $\mathsf{A}\mathbf{x} \neq \mathbf{o}$ or $\mathsf{B}\mathbf{x} \neq \mathbf{o}$. Since $\mathsf{A}\mathbf{x} \neq \mathbf{o}$ is by (1.33) equivalent to $\mathsf{A}^{1/2}\mathbf{x} \neq \mathbf{o}$, we get for $\varrho > 0$

$$\mathbf{x}^T \mathsf{A}_\varrho \mathbf{x} = \mathbf{x}^T \mathsf{A}\mathbf{x} + \varrho\|\mathsf{B}\mathbf{x}\|^2 = \|\mathsf{A}^{1/2}\mathbf{x}\|^2 + \varrho\|\mathsf{B}\mathbf{x}\|^2 > 0.$$

Thus A_ϱ is positive definite. □

If A is positive definite, then it can be verified either directly or by Lemma 1.4, using $\mathrm{Ker}\mathsf{A} = \{\mathbf{o}\}$, that A_ϱ is also positive definite. It follows that we can use the Sherman–Morrison–Woodbury formula (1.6) to get

$$\mathsf{A}_\varrho^{-1} = \mathsf{A}^{-1} - \mathsf{A}^{-1}\mathsf{B}^T(\varrho^{-1}\mathsf{I} + \mathsf{B}\mathsf{A}^{-1}\mathsf{B}^T)^{-1}\mathsf{B}\mathsf{A}^{-1}. \tag{1.47}$$

The following lemma shows that A_ϱ can be positive definite even when A is indefinite.

Lemma 1.5. *Let* $\mathsf{A} \in \mathbb{R}^{n\times n}$ *denote a symmetric matrix, let* $\mathsf{B} \in \mathbb{R}^{m\times n}$, *and let there be* $\mu > 0$ *such that*

$$\mathbf{x}^T \mathsf{A}\mathbf{x} \geq \mu\|\mathbf{x}\|^2, \quad \mathbf{x} \in \mathrm{Ker}\mathsf{B}.$$

Then A_ϱ *is positive definite for sufficiently large* ϱ.

Proof. Let $\lambda_{\min}$ and $\overline{\sigma}_{\min}$ denote the least eigenvalue of A and the least positive singular value of B, respectively, and recall that by (1.19) any $\mathbf{x} \in \mathbb{R}^n$ can be written in the form

$$\mathbf{x} = \mathbf{y} + \mathbf{z}, \quad \mathbf{y} \in \mathrm{Ker}\mathsf{B}, \quad \mathbf{z} \in \mathrm{Im}\mathsf{B}^T.$$

Using the definition of A_ϱ, the assumptions, and (1.40), we get

$$\begin{aligned}\mathbf{x}^T \mathsf{A}_\varrho \mathbf{x} &= \mathbf{y}^T \mathsf{A}\mathbf{y} + 2\mathbf{y}^T \mathsf{A}\mathbf{z} + \mathbf{z}^T \mathsf{A}\mathbf{z} + \varrho\|\mathsf{B}\mathbf{z}\|^2 \\ &\geq \mu\|\mathbf{y}\|^2 - 2\|\mathsf{A}\|\|\mathbf{y}\|\|\mathbf{z}\| + \left(\lambda_{\min} + \varrho\overline{\sigma}_{\min}^2\right)\|\mathbf{z}\|^2 \qquad (1.48) \\ &= [\|\mathbf{y}\|, \|\mathbf{z}\|] \begin{bmatrix} \mu, & -\|\mathsf{A}\| \\ -\|\mathsf{A}\|, \lambda_{\min} & +\varrho\overline{\sigma}_{\min}^2 \end{bmatrix} \begin{bmatrix} \|\mathbf{y}\| \\ \|\mathbf{z}\| \end{bmatrix}.\end{aligned}$$

We shall complete the proof by showing that the matrix

$$\mathrm{H}_\varrho = \begin{bmatrix} \mu, & -\|\mathsf{A}\| \\ -\|\mathsf{A}\|, \lambda_{\min} & +\varrho\overline{\sigma}_{\min}^2 \end{bmatrix}$$

is positive definite for sufficiently large values of ϱ. To this end, it is enough to show that the eigenvalues λ_1, λ_2 of H_ϱ are positive for sufficiently large ϱ.

First observe that H_ϱ is symmetric, so that λ_1, λ_2 are real. Moreover, both the determinant and the trace of H_ϱ are obviously positive for a sufficiently large ϱ, so that by (1.24) and (1.25) both $\lambda_1\lambda_2 > 0$ and $\lambda_1 + \lambda_2 > 0$ for sufficiently large values of ϱ. Since the latter implies that at least one of the eigenvalues of H_ϱ is positive for sufficiently large ϱ, it follows from $\lambda_1\lambda_2 > 0$ that $\lambda_1 > 0$ and $\lambda_2 > 0$ provided ϱ is sufficiently large. Thus H_ϱ is positive definite for sufficiently large ϱ and the statement of our lemma follows by (1.48). □

We often use bounds on the spectrum of some matrix expressions with penalized matrices that are based on the following lemma.

Lemma 1.6. *Let $m < n$ be given positive integers, let $\mathsf{A} \in \mathbb{R}^{n\times n}$ denote a symmetric positive definite matrix, and let $\mathsf{B} \in \mathbb{R}^{m\times n}$.*

Then

$$\mathsf{BA}_\varrho^{-1} = (\mathsf{I} + \varrho\mathsf{BA}^{-1}\mathsf{B}^T)^{-1}\mathsf{BA}^{-1} \qquad (1.49)$$

and the eigenvalues μ_i of $\mathsf{BA}_\varrho^{-1}\mathsf{B}^T$ are related to the eigenvalues β_i of $\mathsf{BA}^{-1}\mathsf{B}^T$ by

$$\mu_i = \beta_i/(1 + \varrho\beta_i), \ i = 1, \ \ldots, \ n. \qquad (1.50)$$

Proof. Following [90], we can use a special form (1.47) of the Sherman–Morrison–Woodbury formula to get

$$\begin{aligned}\mathsf{B}(\mathsf{A} + \varrho\mathsf{B}^T\mathsf{B})^{-1} &= \mathsf{BA}^{-1} - \mathsf{BA}^{-1}\mathsf{B}^T(\varrho^{-1}\mathsf{I} + \mathsf{BA}^{-1}\mathsf{B}^T)^{-1}\mathsf{BA}^{-1} \\ &= (\mathsf{I} - \varrho\mathsf{BA}^{-1}\mathsf{B}^T(\mathsf{I} + \varrho\mathsf{BA}^{-1}\mathsf{B}^T)^{-1})\mathsf{BA}^{-1} \\ &= (\mathsf{I} - ((\mathsf{I} + \varrho\mathsf{BA}^{-1}\mathsf{B}^T) - \mathsf{I})(\mathsf{I} + \varrho\mathsf{BA}^{-1}\mathsf{B}^T)^{-1})\mathsf{BA}^{-1} \\ &= (\mathsf{I} + \varrho\mathsf{BA}^{-1}\mathsf{B}^T)^{-1}\mathsf{BA}^{-1}.\end{aligned}$$

To prove (1.50), notice that by (1.49)

$$\mathsf{BA}_\varrho^{-1}\mathsf{B}^T = (\mathsf{I} + \varrho\mathsf{BA}^{-1}\mathsf{B}^T)^{-1}\mathsf{BA}^{-1}\mathsf{B}^T$$

and apply (1.32) with $f(x) = x/(1 + \varrho x)$.

□

The eigenvalues of the restriction of $\mathsf{BA}^{-1}\mathsf{B}^T$ to its invariant subspace $\mathrm{Im}\mathsf{B}$ are given by the following lemma.

Lemma 1.7. *Let $m < n$ be positive integers, let $\mathsf{A} \in \mathbb{R}^{n\times n}$ denote a symmetric positive definite matrix, let $\mathsf{B} \in \mathbb{R}^{m\times n}$, and let r denote the rank of the matrix B.*

Then

$$\mathrm{Im}\mathsf{B} = \mathrm{Im}\mathsf{BA}^{-1} = \mathrm{Im}\mathsf{BA}^{-1}\mathsf{B}^T \tag{1.51}$$

and the eigenvalues $\overline{\beta}_i$ of $\mathsf{BA}_\varrho^{-1}\mathsf{B}^T|\mathrm{Im}\mathsf{B}$ of the restriction of $\mathsf{BA}_\varrho^{-1}\mathsf{B}^T$ to $\mathrm{Im}\mathsf{B}$ are related to the positive eigenvalues

$$\beta_1 \geq \beta_2 \geq \cdots \geq \beta_r$$

of $\mathsf{BA}^{-1}\mathsf{B}^T$ by

$$\overline{\beta}_i = \beta_i/(1 + \varrho\beta_i), \ i = 1, \ \ldots, \ r. \tag{1.52}$$

Proof. First observe that if $\mathsf{C} \in \mathbb{R}^{n\times n}$ is nonsingular and $\mathsf{B} \in \mathbb{R}^{m\times n}$, then $\mathsf{B}\mathbf{x} = \mathsf{BC}(\mathsf{C}^{-1}\mathbf{x})$, so that

$$\mathrm{Im}\mathsf{B} = \mathrm{Im}\mathsf{BC}.$$

It follows that if $\mathsf{A} \in \mathbb{R}^{n\times n}$ is positive definite, so that $\mathsf{A}^{1/2}$ is a well-defined nonsingular matrix, then we can use (1.33) and (1.37) to get

$$\mathrm{Im}\mathsf{B} = \mathrm{Im}\mathsf{BA}^{-1} = \mathrm{Im}\mathsf{BA}^{-1/2} = \mathrm{Im}\mathsf{BA}^{-1/2}(\mathsf{BA}^{-1/2})^T = \mathrm{Im}\mathsf{BA}^{-1}\mathsf{B}^T.$$

We have thus proved (1.51).

To prove (1.52), notice that we can use (1.51) with $\mathsf{A} = \mathsf{A}_\varrho$ to get $\mathrm{Im}\mathsf{B} = \mathrm{Im}\mathsf{BA}_\varrho^{-1}\mathsf{B}^T$. Since $\mathrm{Im}\mathsf{BA}_\varrho^{-1}\mathsf{B}^T$ is spanned by the eigenvectors of $\mathsf{BA}_\varrho^{-1}\mathsf{B}^T$ which correspond to the positive eigenvalues of $\mathsf{BA}_\varrho^{-1}\mathsf{B}^T$, we can use (1.50) to finish the proof. □

The following lemma gives the estimates that are useful in the analysis of the Uzawa-type algorithms.

Lemma 1.8. *Let $m < n$ be given positive integers, let $\mathsf{A} \in \mathbb{R}^{n\times n}$ denote a symmetric positive definite matrix, and let $\mathsf{B} \in \mathbb{R}^{m\times n}$. Let $\overline{\beta}_{\min} > 0$ and $\beta_{\max} \geq \overline{\beta}_{\min}$ denote the least nonzero eigenvalue and the largest eigenvalue of the matrix $\mathsf{BA}^{-1}\mathsf{B}^T$, respectively.*

Then for any $\varrho > 0$

$$\|\mathsf{BA}_\varrho^{-1}\| \le \|\mathsf{BA}^{-1}\|/(1+\varrho\overline{\beta}_{\min}), \tag{1.53}$$

$$\|(\mathsf{I}+\varrho\mathsf{BA}^{-1}\mathsf{B}^T)^{-1}|\mathrm{Im}\mathsf{B}\| = 1/(1+\varrho\overline{\beta}_{\min}), \tag{1.54}$$

and

$$\|\mathsf{BA}_\varrho^{-1}\mathsf{B}^T\| = \beta_{\max}/(1+\varrho\beta_{\max}) < \varrho^{-1}. \tag{1.55}$$

Moreover

$$\lim_{\varrho\to\infty} \kappa(\mathsf{BA}_\varrho^{-1}\mathsf{B}^T|\mathrm{Im}\mathsf{B}) = 1. \tag{1.56}$$

Proof. Applying submultiplicativity of the matrix norms (1.16) to (1.49), we get that

$$\|\mathsf{BA}_\varrho^{-1}\| \le \|(\mathsf{I}+\varrho\mathsf{BA}^{-1}\mathsf{B}^T)^{-1}|\mathrm{Im}\mathsf{BA}^{-1}\|\|\mathsf{BA}^{-1}\|.$$

To evaluate the first factor, notice that by (1.51) $\mathrm{Im}\mathsf{B} = \mathrm{Im}\mathsf{BA}^{-1} = \mathrm{Im}\mathsf{BA}^{-1}\mathsf{B}^T$, so that our task reduces to the evaluation of

$$\|(\mathsf{I}+\varrho\mathsf{BA}^{-1}\mathsf{B}^T)^{-1}|\mathrm{Im}\mathsf{BA}^{-1}\mathsf{B}^T\|.$$

Since $\mathrm{Im}\mathsf{BA}^{-1}\mathsf{B}^T$ is spanned by the eigenvectors of $\mathsf{BA}^{-1}\mathsf{B}^T$ which correspond to the positive eigenvalues, and the eigenvectors of $\mathsf{I}+\varrho\mathsf{BA}^{-1}\mathsf{B}^T$ are just the eigenvectors of $\mathsf{BA}^{-1}\mathsf{B}^T$, it follows by (1.32) with $f(x) = 1/(1+\varrho x)$ that

$$\|(\mathsf{I}+\varrho\mathsf{BA}^{-1}\mathsf{B}^T)^{-1}|\mathrm{Im}\mathsf{BA}^{-1}\mathsf{B}^T\| = \max_{\substack{i=1,\dots,m\\ \beta_i>0}} 1/(1+\varrho\beta_i) = 1/(1+\varrho\overline{\beta}_{\min}).$$

This completes the proof of (1.53) and (1.54).

To prove (1.55), recall that the eigenvalues μ_i of $\mathsf{BA}_\varrho^{-1}\mathsf{B}^T$ are related to the eigenvalues β_i of $\mathsf{BA}^{-1}\mathsf{B}^T$ by (1.50), so that

$$\|\mathsf{BA}_\varrho^{-1}\mathsf{B}^T\| = \max_{i=1,\dots,m} \mu_i = \max_{i=1,\dots,m} \beta_i(1+\varrho\beta_i) = \beta_{\max}/(1+\varrho\beta_{\max}). \tag{1.57}$$

Finally, using Lemma 1.7, we get that

$$\min_{\substack{\|\mathbf{x}\|=1\\ \mathbf{x}\in\mathrm{Im}\mathsf{B}}} \|\mathsf{BA}^{-1}\mathsf{B}^T\mathbf{x}\| = \min_{\substack{i=1,\dots,m\\ \beta_i>0}} \beta_i/(1+\varrho\beta_i) = \overline{\beta}_{\min}/(1+\varrho\overline{\beta}_{\min}),$$

so that

$$\lim_{\varrho\to\infty} \kappa(\mathsf{BA}_\varrho^{-1}\mathsf{B}^T|\mathrm{Im}\mathsf{B}) = \lim_{\varrho\to\infty} \frac{\beta_{\max}}{1+\varrho\beta_{\max}} \frac{1+\varrho\overline{\beta}_{\min}}{\overline{\beta}_{\min}} = 1.$$

□

The last result to be presented here concerns the distribution of the eigenvalues of the penalized matrices. To simplify its formulation, let us denote for each symmetric matrix M of the order n by $\lambda_i(\mathsf{M})$ the ith eigenvalue of M in the decreasing order, so that

$$\lambda_1(\mathsf{M}) \geq \lambda_2(\mathsf{M}) \geq \cdots \geq \lambda_n(\mathsf{M}) \ .$$

Lemma 1.9. *Let* $\mathsf{A} \in \mathbb{R}^{n\times n}$ *denote a symmetric matrix, let* $\mathsf{B} \in \mathbb{R}^{m\times n}$ *denote a matrix of the rank* $r, 0 < r \leq m < n$, *and* $\varrho > 0$.

Then $\lambda_r(\mathsf{B}^T\mathsf{B}) > 0$ *and*

$$\lambda_r(\mathsf{A}_\varrho) \geq \lambda_n(\mathsf{A}) + \varrho\lambda_r(\mathsf{B}^T\mathsf{B}), \tag{1.58}$$

$$\lambda_{r+1}(\mathsf{A}_\varrho) \leq \lambda_1(\mathsf{A}). \tag{1.59}$$

Proof. Using the spectral decomposition (1.28), we can find an orthogonal matrix U such that

$$\mathsf{U}^T\mathsf{B}^T\mathsf{B}\mathsf{U} = \mathrm{diag}(\gamma_1, \ \ldots, \ \gamma_n)$$

with

$$\gamma_i = \lambda_i(\mathsf{B}^T\mathsf{B}) \quad \text{and} \quad \gamma_1 \geq \cdots \geq \gamma_r > \gamma_{r+1} = \gamma_n = 0.$$

Thus

$$\mathsf{U}^T(\mathsf{A} + \varrho\mathsf{B}^T\mathsf{B})\mathsf{U} = \mathsf{U}^T\mathsf{A}\mathsf{U} + \varrho\mathsf{U}^T\mathsf{B}^T\mathsf{B}\mathsf{U} = \begin{bmatrix} \mathsf{E} + \varrho\mathsf{G} & \mathsf{F} \\ \mathsf{F}^T & \mathsf{H} \end{bmatrix}$$

with

$$\mathsf{G} = \mathrm{diag}(\gamma_1, \ \ldots, \ \gamma_r) \quad \text{and} \quad \mathsf{U}^T\mathsf{A}\mathsf{U} = \begin{bmatrix} \mathsf{E} & \mathsf{F} \\ \mathsf{F}^T & \mathsf{H} \end{bmatrix},$$

where H is a square matrix of the order $n - r$. The eigenvalues of $\mathsf{U}^T\mathsf{A}\mathsf{U}$ and A are identical.

To prove (1.58), notice that the elementary properties of the spectrum of symmetric matrices and the *Cauchy interlacing property* of bordering matrices (1.26) imply that

$$\begin{aligned}\lambda_r(\mathsf{A}_\varrho) &\geq \lambda_r(\mathsf{E} + \varrho\mathsf{G}) \geq \lambda_r(\lambda_n(\mathsf{A})\mathsf{I} + \varrho\mathsf{G}) = \lambda_n(\mathsf{A}) + \varrho\lambda_r(\mathsf{G}) \\ &= \lambda_n(\mathsf{A}) + \varrho\gamma_r = \lambda_n(\mathsf{A}) + \varrho\lambda_r(\mathsf{B}^T\mathsf{B}) \ .\end{aligned}$$

To prove (1.59), observe that by the Courant–Fischer min-max principle (1.31)

$$
\begin{aligned}
\lambda_{r+1}\left(\mathsf{A}_{\varrho}\right) &= \max_{\substack{\mathcal{V} \subseteq \mathbb{R}^n \\ \dim \mathcal{V} = r+1}} \; \min_{\substack{\mathbf{x} \in \mathcal{V} \\ \|\mathbf{x}\| = 1}} \mathbf{x}^T \left(\mathsf{A} + \varrho \mathsf{B}^T \mathsf{B}\right) \mathbf{x} \\
&\leq \max_{\substack{\mathcal{V} \subseteq \mathbb{R}^n \\ \dim \mathcal{V} = r+1}} \; \min_{\substack{\mathbf{x} \in \mathcal{V} \cap \operatorname{Ker} \mathsf{B} \\ \|\mathbf{x}\| = 1}} \mathbf{x}^T \left(\mathsf{A} + \varrho \mathsf{B}^T \mathsf{B}\right) \mathbf{x} \\
&= \max_{\substack{\mathcal{V} \subseteq \mathbb{R}^n \\ \dim \mathcal{V} = r+1}} \; \min_{\substack{\mathbf{x} \in \mathcal{V} \cap \operatorname{Ker} \mathsf{B} \\ \|\mathbf{x}\| = 1}} \mathbf{x}^T \mathsf{A} \mathbf{x} \leq \lambda_1(\mathsf{A}), \qquad (1.60)
\end{aligned}
$$

so that the inequality (1.59) is proved. We used the fact that $\dim \mathcal{V} = r+1$ implies

$$\mathcal{V} \cap \operatorname{Ker}\mathsf{B} \neq \{\mathbf{o}\}.$$

□

2
Optimization

In this chapter, we briefly review the optimization results in the extent sufficient for understanding the rest of our book. Since we are interested mainly in quadratic programming (QP) and quadratically constrained quadratic programs (QCQP), we present most results with specialized arguments exploiting the specific structure of these problems. Since QP problems can locally approximate more general optimization problems, this chapter can serve as a simple introduction to nonlinear optimization. A systematic exposition of the optimization theory in the framework of nonlinear optimization can be found in the books by Bertsekas [16], Nocedal and Wright [220], Conn et al. [41], or Bazaraa et al. [12].

2.1 Optimization Problems and Solutions

Optimization problems are described by a *cost* (*objective*) *function* f defined on a subset $\mathcal{D} \subseteq \mathbb{R}^n$ and by a *constraint set* $\Omega \subseteq \mathcal{D}$. The elements of Ω are called *feasible vectors*. The main topic of this book is the development of efficient algorithms for solving *QP* (quadratic programming) problems with quadratic cost and linear constraints and special *QCQP* (quadratically constrained quadratic program) problems with quadratic cost and $\Omega \subseteq \mathbb{R}^n$ described by separable quadratic and optional linear constraints.

We look either for a solution $\overline{\mathbf{x}} \in \mathbb{R}^n$ of the *unconstrained minimization problem* which satisfies

$$f(\overline{\mathbf{x}}) \leq f(\mathbf{x}), \ \mathbf{x} \in \mathbb{R}^n, \tag{2.1}$$

or for a solution $\overline{\mathbf{x}} \in \Omega$ of the *constrained minimization problem*

$$f(\overline{\mathbf{x}}) \leq f(\mathbf{x}), \ \mathbf{x} \in \Omega. \tag{2.2}$$

Z. Dostál, *Optimal Quadratic Programming and QCQP Algorithms with Case Studies*, Springer Optimization and Its Applications 23,
https://doi.org/10.1007/978-3-031-95167-1_2

Such an approach simplifies the analysis and sometimes enables us to obtain stronger results. A solution to the minimization problem is called its *minimizer* or *global minimizer*. The value of f corresponding to a minimizer is the *minimum*. If (2.1) holds only for $\|\mathbf{x}-\overline{\mathbf{x}}\| \leq \delta$ with some $\delta > 0$, we call $\overline{\mathbf{x}}$ a *local minimizer*.

A nonzero vector $\mathbf{d} \in \mathbb{R}^n$ is a *feasible direction* of Ω at a feasible point $\mathbf{x}$ if $\mathbf{x} + \varepsilon\mathbf{d} \in \Omega$ for all sufficiently small $\varepsilon > 0$. Feasible directions are useful in the analysis of local minimizers.

A nonzero vector $\mathbf{d} \in \mathbb{R}^n$ is a *recession direction*, or simply a *direction*, of Ω if for each $\mathbf{x} \in \Omega$, $\mathbf{x} + \alpha\mathbf{d} \in \Omega$ for all $\alpha > 0$.

2.2 Unconstrained Quadratic Programming

Let us first recall some simple results concerning unconstrained QP.

2.2.1 Quadratic Cost Functions

We consider the cost functions in the form

$$f(\mathbf{x}) = \frac{1}{2}\mathbf{x}^T\mathsf{A}\mathbf{x} - \mathbf{b}^T\mathbf{x}, \tag{2.3}$$

where $\mathsf{A} \in \mathbb{R}^{n\times n}$ denotes a given symmetric matrix of order n and $\mathbf{b} \in \mathbb{R}^n$.

If $\mathbf{x}, \mathbf{d} \in \mathbb{R}^n$, then using elementary computations and $\mathsf{A} = \mathsf{A}^T$, we get

$$f(\mathbf{x}+\mathbf{d}) = f(\mathbf{x}) + (\mathsf{A}\mathbf{x}-\mathbf{b})^T\mathbf{d} + \frac{1}{2}\mathbf{d}^T\mathsf{A}\mathbf{d}. \tag{2.4}$$

The formula (2.4) is *Taylor's expansion* of f at $\mathbf{x}$, so that the *gradient* and *Hessian* of f at $\mathbf{x}$ are given by

$$\nabla f(\mathbf{x}) = \mathsf{A}\mathbf{x} - \mathbf{b} \quad \text{and} \quad \nabla^2 f(\mathbf{x}) = \mathsf{A}, \tag{2.5}$$

respectively. Taylor's expansion will be our useful tool in what follows.

A vector $\mathbf{d}$ is a *decrease direction* of f at $\mathbf{x}$ if

$$f(\mathbf{x}+\varepsilon\mathbf{d}) < f(\mathbf{x})$$

for all sufficiently small values of $\varepsilon > 0$. Using Taylor's expansion (2.4)

$$f(\mathbf{x}+\varepsilon\mathbf{d}) = f(\mathbf{x}) + \varepsilon(\mathsf{A}\mathbf{x}-\mathbf{b})^T\mathbf{d} + \frac{\varepsilon^2}{2}\mathbf{d}^T\mathsf{A}\mathbf{d},$$

we get that $\mathbf{d}$ is a decrease direction if and only if

$$(\mathsf{A}\mathbf{x}-\mathbf{b})^T\mathbf{d} < 0.$$

2.2.2 Unconstrained Minimization of Quadratic Functions

The following proposition gives algebraic conditions that are satisfied by the solutions of the unconstrained QP problem to find

$$\min_{\mathbf{x}\in\mathbb{R}^n} f(\mathbf{x}), \tag{2.6}$$

where f is a quadratic function defined by (2.3).

Proposition 2.1. *Let the quadratic function f be defined by a symmetric matrix $\mathsf{A} \in \mathbb{R}^{n\times n}$ and $\mathbf{b} \in \mathbb{R}^n$. Then the following statements hold:*
(i) A vector $\overline{\mathbf{x}}$ solves (2.1) if and only if A is SPS and

$$\nabla f(\overline{\mathbf{x}}) = \mathsf{A}\overline{\mathbf{x}} - \mathbf{b} = \mathbf{o}. \tag{2.7}$$

(ii) The problem (2.1) has a unique solution if and only if A is SPD.

Proof. (i) If $\overline{\mathbf{x}}$ and $\mathbf{d}$ denote arbitrary n-vectors and $\alpha \in \mathbb{R}$, then we can use Taylor's expansion (2.4) to get

$$f(\overline{\mathbf{x}} + \alpha\mathbf{d}) - f(\overline{\mathbf{x}}) = \alpha(\mathsf{A}\overline{\mathbf{x}} - \mathbf{b})^T\mathbf{d} + \frac{\alpha^2}{2}\mathbf{d}^T\mathsf{A}\mathbf{d}.$$

Let us first assume that $\overline{\mathbf{x}}$ solves (2.1), so that the right-hand side of the above equation is nonnegative for any α and $\mathbf{d}$. For α sufficiently large and $\mathbf{d} \in \mathbb{R}^n$ fixed, the nonnegativity of the right-hand side implies that $\mathbf{d}^T\mathsf{A}\mathbf{d} \geq 0$; thus A is SPS. On the other hand, for α sufficiently small, the sign of the linear term determines the sign of the right-hand side, so the nonnegativity of the right-hand side implies that $(\mathsf{A}\overline{\mathbf{x}} - \mathbf{b})^T\mathbf{d} = 0$ for any $\mathbf{d} \in \mathbb{R}^n$. Thus $\mathsf{A}\overline{\mathbf{x}} - \mathbf{b} = \mathbf{o}$.

If A is SPS and $\overline{\mathbf{x}}$ satisfies (2.7), then for any $\mathbf{d} \in \mathbb{R}^n$

$$f(\overline{\mathbf{x}} + \mathbf{d}) - f(\overline{\mathbf{x}}) = \frac{1}{2}\mathbf{d}^T\mathsf{A}\mathbf{d} \geq 0;$$

therefore $\overline{\mathbf{x}}$ is a solution of (2.1).

(ii) If $\widehat{\mathbf{x}}$ is the unique solution of (2.1), then by (i) A is SPS and $\widehat{\mathbf{x}}$ is the only vector satisfying $\mathsf{A}\widehat{\mathbf{x}} = \mathbf{b}$. Thus A is SPD. On the other hand, if A is SPD, then the gradient condition (2.7) has the unique solution. □

Examining the gradient condition (2.7), we get that problem (2.1) has a solution if and only if A is positive semidefinite and

$$\mathbf{b} \in \mathrm{Im}\mathsf{A}. \tag{2.8}$$

Denoting by R a matrix whose columns span $\text{Ker}\mathsf{A}$, we can write

$$\mathbf{b}^T\mathsf{R} = \mathbf{o}.$$

This condition has a simple mechanical interpretation: if a mechanical system is in equilibrium, the external forces must be orthogonal to the rigid body motions.

If $\mathbf{b} \in \text{Im}\mathsf{A}$, a solution of (2.6) is given by

$$\overline{\mathbf{x}} = \mathsf{A}^+\mathbf{b},$$

where A^+ is a left generalized inverse introduced in Sect. 1.4. After substituting into f and simple manipulations, we get

$$\min_{\mathbf{x}\in\mathbb{R}^n} f(\mathbf{x}) = -\frac{1}{2}\mathbf{b}^T\mathsf{A}^+\mathbf{b}. \tag{2.9}$$

In particular, if A is positive definite, then

$$\min_{\mathbf{x}\in\mathbb{R}^n} f(\mathbf{x}) = -\frac{1}{2}\mathbf{b}^T\mathsf{A}^{-1}\mathbf{b}. \tag{2.10}$$

The formulae for the minimum of the unconstrained minimization problems can be used to develop useful estimates. Indeed, if (2.8) holds and $\mathbf{x} \in \mathbb{R}^n$, we can use (2.9), properties of generalized inverses, and (1.43) to get

$$f(\mathbf{x}) \geq -\frac{1}{2}\mathbf{b}^T\mathsf{A}^+\mathbf{b} = -\frac{1}{2}\mathbf{b}^T\mathsf{A}^\dagger\mathbf{b} \geq -\frac{1}{2}\|\mathsf{A}^\dagger\|\|\mathbf{b}\|^2 = -\|\mathbf{b}\|^2/(2\overline{\lambda}_{\min}),$$

where $\mathsf{A}^\dagger$ denotes the Moore–Penrose generalized inverse and $\overline{\lambda}_{\min}$ denotes the least nonzero eigenvalue of A. In particular, it follows that if A is positive definite and $\lambda_{\min}$ denotes the least eigenvalue of A, then for any $\mathbf{x} \in \mathbb{R}^n$

$$f(\mathbf{x}) \geq -\frac{1}{2}\mathbf{b}^T\mathsf{A}^{-1}\mathbf{b} \geq -\frac{1}{2}\|\mathsf{A}^{-1}\|\|\mathbf{b}\|^2 = -\|\mathbf{b}\|^2/(2\lambda_{\min}). \tag{2.11}$$

If the dimension of the unconstrained minimization problem (2.6) is large, then it can be too ambitious to look for a solution that satisfies the gradient condition (2.7) exactly. The natural idea is to consider the weaker condition

$$\|\nabla f(\mathbf{x})\| \leq \varepsilon \tag{2.12}$$

with a small epsilon. If $\mathbf{x}$ satisfies the latter condition with ε sufficiently small and A nonsingular, then $\mathbf{x}$ is near the unique solution $\widehat{\mathbf{x}}$ as

$$\|\mathbf{x} - \widehat{\mathbf{x}}\| = \|\mathsf{A}^{-1}\mathsf{A}(\mathbf{x} - \widehat{\mathbf{x}})\| = \|\mathsf{A}^{-1}(\mathsf{A}\mathbf{x} - \mathbf{b})\| \leq \|\mathsf{A}^{-1}\|\|\nabla f(\mathbf{x})\|. \tag{2.13}$$

The typical "solution" returned by an iterative solver is just $\mathbf{x}$ that satisfies the condition (2.12).

2.3 Convexity

Many strong results can be proved when the problem obeys convexity assumptions. Intuitively, convexity is a property of the sets containing the joining segment of any couple of its points, as in Fig. 2.1. More formally, a subset Ω of $\mathbb{R}^n$ is *convex* if for any $\mathbf{x}$ and $\mathbf{y}$ in Ω and $\alpha \in (0,1)$, the vector $\mathbf{s} = \alpha\mathbf{x} + (1-\alpha)\mathbf{y}$ is also in Ω. Nonconvex set is in Fig. 2.2.

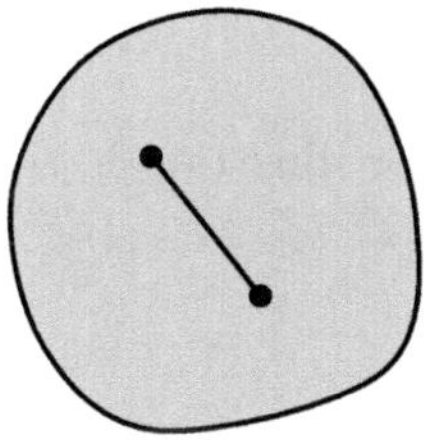

Fig. 2.1. Convex set

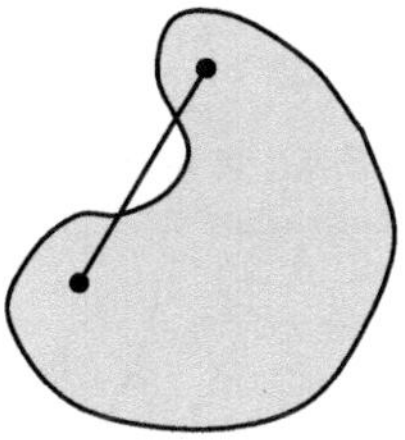

Fig. 2.2. Nonconvex set

Let $\mathbf{x}_1, \ldots, \mathbf{x}_k$ be vectors of $\mathbb{R}^n$. If $\alpha_1, \ldots, \alpha_k$ are scalars such that

$$\alpha_i \geq 0, \quad i = 1, \ldots, k, \quad \sum_{i=1}^{k} \alpha_i = 1,$$

then the vector

$$\mathbf{v} = \sum_{i=1}^{k} \alpha_i \mathbf{x}_i$$

is said to be a *convex combination* of vectors $\mathbf{x}_1, \ldots, \mathbf{x}_k$.

The *convex hull* of $\mathbf{x}_1, \ldots, \mathbf{x}_k$, denoted $\mathrm{Conv}\{\mathbf{x}_1, \ldots, \mathbf{x}_k\}$, is the set of all convex combinations of $\mathbf{x}_1, \ldots, \mathbf{x}_k$. The convex hull of $\mathbf{x}_1, \ldots, \mathbf{x}_k$ is the smallest convex set to which $\mathbf{x}_1, \ldots, \mathbf{x}_k$ belong.

Caratheodory's theorem guarantees that we can represent any point of $\mathsf{Conv}\{\mathbf{x}_1, \ldots, \mathbf{x}_k\}$ as a convex combination of at most $n+1$ elements of $\{\mathbf{x}_1, \ldots, \mathbf{x}_k\}$.

The *convex boundary* of a convex set Ω is a set of vectors that cannot be expressed as a convex combination of any other vectors of Ω. Thus the convex boundary of a closed square comprises its four corners, while the convex boundary of a closed circle is identical to its boundary.

The intersection of convex sets is also convex. In this book, we consider minimization over the convex sets defined by linear constraints and/or separable quadratic inequalities.

2.3.1 Convex Quadratic Functions

Given a convex set $\Omega \in \mathbb{R}^n$, a mapping $h : \Omega \to \mathbb{R}$ is said to be a *convex function* if its epigraph is convex, that is, if

$$h(\alpha\mathbf{x} + (1-\alpha)\mathbf{y}) \le \alpha h(\mathbf{x}) + (1-\alpha)h(\mathbf{y})$$

for all $\mathbf{x}, \mathbf{y} \in \Omega$ and $\alpha \in (0,1)$, and it is *strictly convex* if

$$h(\alpha\mathbf{x} + (1-\alpha)\mathbf{y}) < \alpha h(\mathbf{x}) + (1-\alpha)h(\mathbf{y})$$

for all $\mathbf{x}, \mathbf{y} \in \Omega$, $\mathbf{x} \ne \mathbf{y}$, and $\alpha \in (0,1)$. For illustration, see Fig. 2.3.

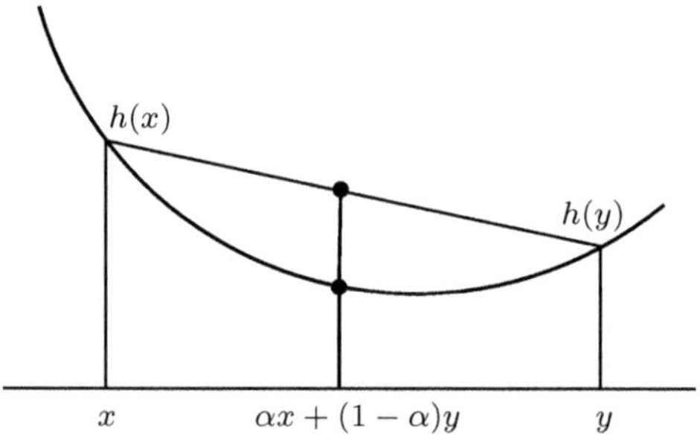

Fig. 2.3. Strictly convex function

The following proposition offers an algebraic characterization of convex functions.

Proposition 2.2. *Let $\mathcal{V}$ be a subspace of $\mathbb{R}^n$. The restriction $f|\mathcal{V}$ of a quadratic function f with the Hessian matrix A to $\mathcal{V}$ is convex if and only if $\mathsf{A}|\mathcal{V}$ is SPS, and $f|\mathcal{V}$ is strictly convex if and only if $\mathsf{A}|\mathcal{V}$ is SPD.*

Proof. Let $\mathcal{V}$ be a subspace, let $\mathbf{x}, \mathbf{y} \in \mathcal{V}$, $\alpha \in (0,1)$, and $\mathbf{s} = \alpha\mathbf{x} + (1-\alpha)\mathbf{y}$. Then by Taylor's expansion (2.4) of f at $\mathbf{s}$

$$f(\mathbf{s}) + \nabla f(\mathbf{s})^T(\mathbf{x}-\mathbf{s}) + \frac{1}{2}(\mathbf{x}-\mathbf{s})^T\mathsf{A}(\mathbf{x}-\mathbf{s}) = f(\mathbf{x}),$$
$$f(\mathbf{s}) + \nabla f(\mathbf{s})^T(\mathbf{y}-\mathbf{s}) + \frac{1}{2}(\mathbf{y}-\mathbf{s})^T\mathsf{A}(\mathbf{y}-\mathbf{s}) = f(\mathbf{y}).$$

Multiplying the first equation by α, the second equation by $1-\alpha$, and summing up, we get

$$f(\mathbf{s}) + \frac{\alpha}{2}(\mathbf{x}-\mathbf{s})^T\mathsf{A}(\mathbf{x}-\mathbf{s}) + \frac{1-\alpha}{2}(\mathbf{y}-\mathbf{s})^T\mathsf{A}(\mathbf{y}-\mathbf{s}) = \alpha f(\mathbf{x}) + (1-\alpha)f(\mathbf{y}). \quad (2.14)$$

It follows that if $\mathsf{A}|\mathcal{V}$ is positive semidefinite, then $f|\mathcal{V}$ is convex. Moreover, since $\mathbf{x} = \mathbf{y}$ is equivalent to $\mathbf{x} = \mathbf{s}$ and $\mathbf{y} = \mathbf{s}$, it follows that if $\mathsf{A}|\mathcal{V}$ is positive definite, then $f|\mathcal{V}$ is strictly convex.

Let us now assume that $f|\mathcal{V}$ is convex, let $\mathbf{z} \in \mathcal{V}$, set $\alpha = \frac{1}{2}$, and denote $\mathbf{x} = 2\mathbf{z}$, $\mathbf{y} = \mathbf{o}$. Then $\mathbf{s} = \mathbf{z}$, $\mathbf{x} - \mathbf{s} = \mathbf{z}$, $\mathbf{y} - \mathbf{s} = -\mathbf{z}$, and substituting into (2.14) results in

$$f(\mathbf{s}) + \frac{1}{2}\mathbf{z}^T\mathsf{A}\mathbf{z} = \alpha f(\mathbf{x}) + (1-\alpha)f(\mathbf{y}).$$

Since $\mathbf{z} \in \mathcal{V}$ is arbitrary and $f|\mathcal{V}$ is assumed to be convex, it follows that

$$\frac{1}{2}\mathbf{z}^T\mathsf{A}\mathbf{z} = \alpha f(\mathbf{x}) + (1-\alpha)f(\mathbf{y}) - f(\alpha\mathbf{x} + (1-\alpha)\mathbf{y}) \geq 0.$$

Thus $\mathsf{A}|\mathcal{V}$ is positive semidefinite. Moreover, if $f|\mathcal{V}$ is strictly convex, then $\mathsf{A}|\mathcal{V}$ is positive definite. □

The following simple corollary is useful in the analysis of equality constrained problems.

Corollary 2.3. *Let f denote a quadratic function with the Hessian $\mathsf{A} \in \mathbb{R}^{n\times n}$, let $\mathsf{B} \in \mathbb{R}^{m\times n}$, $\mathbf{c} \in \mathbb{R}^m$, and $\Omega = \{\mathbf{x} \in \mathbb{R}^n : \mathsf{B}\mathbf{x} = \mathbf{c}\}$.*

Then $f|\Omega$ is convex if and only if $f|\mathrm{Ker}\mathsf{B}$ is convex, and $f|\Omega$ is strictly convex if and only if $f|\mathrm{Ker}\mathsf{B}$ is strictly convex.

Proof. Observe that Ω is convex. If $\overline{\mathbf{x}} \in \Omega$, then $\Omega = \{\overline{\mathbf{x}} + \mathbf{d} : \mathbf{d} \in \mathrm{Ker}\mathsf{B}\}$. It follows that the restriction of $f(\mathbf{x})$ to Ω has the same graph as the restriction of $f(\overline{\mathbf{x}} + \mathbf{d})$ to $\mathrm{Ker}\mathsf{B}$. The statement then follows by Proposition 2.2 and

$$\nabla^2 f(\mathbf{x}) = \nabla^2_{\mathbf{dd}} f(\overline{\mathbf{x}} + \mathbf{d}) = \mathsf{A}.$$

□

The strictly convex functions have a nice property that $f(\mathbf{x}) \to \infty$ when $\|\mathbf{x}\| \to \infty$. The functions with this property are called *coercive functions.* More generally, a function $f : \mathbb{R}^n \mapsto \mathbb{R}$ is said to be *coercive on* $\Omega \subseteq \mathbb{R}^n$ if

$$f(\mathbf{x}) \to \infty \quad \text{for} \quad \|\mathbf{x}\| \to \infty,\ \mathbf{x} \in \Omega.$$

A quadratic function need not be strictly convex to be coercive on a given set, as in the case of $f(x, y) = x^2 - y$, which is coercive on $\Omega = \mathbb{R} \times (-\infty, 1]$. More generally, a quadratic function f with a semidefinite Hessian A is coercive on a convex set Ω if $\mathbf{d}^T\mathbf{b} < 0$ for any recession direction $\mathbf{d}$ of Ω which belongs to $\mathrm{Ker}\mathsf{A}$. The coercive quadratic function with a semidefinite Hessian matrix is also called a *semicoercive function.* For example, the function $f(x, y) = x^2 - y$ is semicoercive on $\Omega = \mathbb{R} \times (-\infty, 1]$.

2.3.2 Minimizers of Convex Function

Under the convexity assumptions, each local minimizer is a global minimizer.

Proposition 2.4. *Let f and $\Omega \subseteq \mathbb{R}^n$ be a quadratic function defined by (2.3) and a closed convex set, respectively. Then the following statements hold:*
(i) If f is convex, then each local minimizer of f subject to $\mathbf{x} \in \Omega$ is a global minimizer of f subject to $\mathbf{x} \in \Omega$.
(ii) If f is convex and $\overline{\mathbf{x}}, \overline{\mathbf{y}}$ are two minimizers of f subject to $\mathbf{x} \in \Omega$, then

$$\overline{\mathbf{x}} - \overline{\mathbf{y}} \in \mathrm{Ker}\mathsf{A} \cap \mathrm{Span}\{\mathbf{b}\}^{\perp}.$$

(iii) If f is strictly convex on Ω and $\overline{\mathbf{x}}$, $\overline{\mathbf{y}}$ are two minimizers of f subject to $\mathbf{x} \in \Omega$, then $\overline{\mathbf{x}} = \overline{\mathbf{y}}$.

Proof. (i) Let $\overline{\mathbf{x}}$, $\overline{\mathbf{y}}$ be local minimizers of f subject to $\mathbf{x} \in \Omega$, $f(\overline{\mathbf{x}}) < f(\overline{\mathbf{y}})$. Denoting $\mathbf{y}_\alpha = \alpha\overline{\mathbf{x}} + (1-\alpha)\overline{\mathbf{y}}$ and using the convexity of f, we get

$$f(\mathbf{y}_\alpha) = f(\alpha\overline{\mathbf{x}} + (1-\alpha)\overline{\mathbf{y}}) \leq \alpha f(\overline{\mathbf{x}}) + (1-\alpha) f(\overline{\mathbf{y}}) < f(\overline{\mathbf{y}})$$

for every $\alpha \in (0,1)$. Since $\|\overline{\mathbf{y}} - \mathbf{y}_\alpha\| = \alpha\|\overline{\mathbf{y}} - \overline{\mathbf{x}}\|$, the inequality contradicts the assumption that $\overline{\mathbf{y}}$ is a local minimizer.
(ii) Let $\overline{\mathbf{x}}$ and $\overline{\mathbf{y}}$ be global minimizers of f on Ω. Then for any $\alpha \in [0,1]$

$$\overline{\mathbf{x}} + \alpha(\overline{\mathbf{y}} - \overline{\mathbf{x}}) = (1-\alpha)\overline{\mathbf{x}} + \alpha\overline{\mathbf{y}} \in \Omega, \quad \overline{\mathbf{y}} + \alpha(\overline{\mathbf{x}} - \overline{\mathbf{y}}) = (1-\alpha)\overline{\mathbf{y}} + \alpha\overline{\mathbf{x}} \in \Omega.$$

Moreover, using Taylor's formula, we get

$$0 \leq f(\overline{\mathbf{x}} + \alpha(\overline{\mathbf{y}} - \overline{\mathbf{x}})) - f(\overline{\mathbf{x}}) = \alpha(\mathsf{A}\overline{\mathbf{x}} - \mathbf{b})^T(\overline{\mathbf{y}} - \overline{\mathbf{x}}) + \frac{\alpha^2}{2}(\overline{\mathbf{y}} - \overline{\mathbf{x}})^T\mathsf{A}(\overline{\mathbf{y}} - \overline{\mathbf{x}}),$$

$$0 \leq f(\overline{\mathbf{y}} + \alpha(\overline{\mathbf{x}} - \overline{\mathbf{y}})) - f(\overline{\mathbf{y}}) = \alpha(\mathsf{A}\overline{\mathbf{y}} - \mathbf{b})^T(\overline{\mathbf{x}} - \overline{\mathbf{y}}) + \frac{\alpha^2}{2}(\overline{\mathbf{x}} - \overline{\mathbf{y}})^T\mathsf{A}(\overline{\mathbf{x}} - \overline{\mathbf{y}}).$$

Since the latter inequalities hold for arbitrarily small α, it follows that

$$(\mathsf{A}\overline{\mathbf{x}} - \mathbf{b})^T(\overline{\mathbf{y}} - \overline{\mathbf{x}}) \geq 0 \quad \text{and} \quad (\mathsf{A}\overline{\mathbf{y}} - \mathbf{b})^T(\overline{\mathbf{x}} - \overline{\mathbf{y}}) \geq 0.$$

After summing up the latter inequalities and simple manipulations, we have

$$-(\overline{\mathbf{x}} - \overline{\mathbf{y}})^T\mathsf{A}(\overline{\mathbf{x}} - \overline{\mathbf{y}}) \geq 0.$$

Since the convexity of f implies by Proposition 2.2 that A is SPS, it follows that $\overline{\mathbf{x}} - \overline{\mathbf{y}} \in \mathrm{Ker}\mathsf{A}$. To prove that $(\overline{\mathbf{x}} - \overline{\mathbf{y}})^T\mathbf{b} = 0$, recall that for $\mathbf{d} = \overline{\mathbf{x}} - \overline{\mathbf{y}}$

$$0 = f(\overline{\mathbf{x}}) - f(\overline{\mathbf{y}}) = f(\overline{\mathbf{y}} + \mathbf{d}) - f(\overline{\mathbf{y}}) = \frac{1}{2}\mathbf{d}^T\mathsf{A}\mathbf{d} + (\mathsf{A}\mathbf{d} - \mathbf{b})^T\mathbf{d} = -\mathbf{b}^T\mathbf{d}.$$

(iii) Let f be strictly convex and let $\overline{\mathbf{x}} \in \Omega$ and $\overline{\mathbf{y}} \in \Omega$ be different global minimizers of f on Ω. Then $\mathrm{Ker}\mathsf{A} = \{\mathbf{o}\}$ and by (ii) $\overline{\mathbf{x}} - \overline{\mathbf{y}} = \mathbf{o}$. □

2.3.3 Existence of Minimizers

Since quadratic functions are continuous, the existence of at least one minimizer is guaranteed by the Weierstrass theorem provided Ω is compact, that is, closed and bounded. We can also use the following standard results, which do not assume that Ω is bounded.

Proposition 2.5. *Let f be a quadratic function defined on a nonempty closed convex set $\Omega \subseteq \mathbb{R}^n$. Then the following statements hold:*
(i) If f is strictly convex, then a global minimizer of f subject to $\mathbf{x} \in \Omega$ exists and is necessarily unique.
(ii) If f is coercive on Ω, then a global minimizer of f subject to $\mathbf{x} \in \Omega$ exists.
(iii) A global minimizer of f subject to $\mathbf{x} \in \Omega$ exists if and only if f is bounded from below on Ω.

Proof. (i) If f is strictly convex, it follows by Proposition 2.2 that its Hessian A is positive definite, and $\mathbf{z} = \mathsf{A}^{-1}\mathbf{b}$ is by Proposition 2.1 the unique minimizer of f on $\mathbb{R}^n$. Thus for any $\mathbf{x} \in \mathbb{R}^n$

$$f(\mathbf{x}) \geq f(\mathbf{z}).$$

It follows that the infimum of $f(\mathbf{x})$ subject to $\mathbf{x} \in \Omega$ exists, and there is a sequence of vectors $\mathbf{x}^k \in \Omega$ such that

$$\lim_{k \to \infty} f(\mathbf{x}^k) = \inf_{\mathbf{x} \in \Omega} f(\mathbf{x}).$$

The sequence $\{\mathbf{x}^k\}$ is bounded as

$$f(\mathbf{x}^k) - f(\mathbf{z}) = \frac{1}{2}(\mathbf{x}^k - \mathbf{z})^T \mathsf{A}(\mathbf{x}^k - \mathbf{z}) \geq \frac{\lambda_{\min}}{2}\|\mathbf{x}^k - \mathbf{z}\|^2,$$

where $\lambda_{\min}$ denotes the least eigenvalue of A. It follows that $\{\mathbf{x}^k\}$ has at least one cluster point $\overline{\mathbf{x}} \in \Omega$. Since f is continuous, we get

$$f(\overline{\mathbf{x}}) = \inf_{\mathbf{x} \in \Omega} f(\mathbf{x}).$$

The uniqueness follows by Proposition 2.4.
(ii) The proof is similar to that of (i). See Bertsekas [16, Proposition A.8].
(iii) The statement is the well-known Frank–Wolfe theorem [136]. See also Eaves [118] or Blum and Oettli [21]. □

The special structure of the feasible set enables us to get stronger existence results. For example, it is known that the quadratic function with a positive semidefinite Hessian attains its minimum on a polyhedral cone if and only if its linear term satisfies $\mathbf{b}^T\mathbf{d} \leq 0$ for any recession direction $\mathbf{d} \in \mathrm{Ker}\mathsf{A}$ (see Zeidler [269, pp. 553–556]). See also Sect. 2.5.4.

2.3.4 Projections to Convex Sets

Having the results of the previous subsection, we can naturally define the *projection* P_Ω *to the (closed) convex set* $\Omega \subset \mathbb{R}^n$ as a mapping which assigns to each $\mathbf{x} \in \mathbb{R}^n$ its nearest vector $\widehat{\mathbf{x}} \in \Omega$ as in Fig. 2.4. The distance can be measured by the norm induced by any scalar product. The following proposition concerns the projection induced by the Euclidean scalar product.

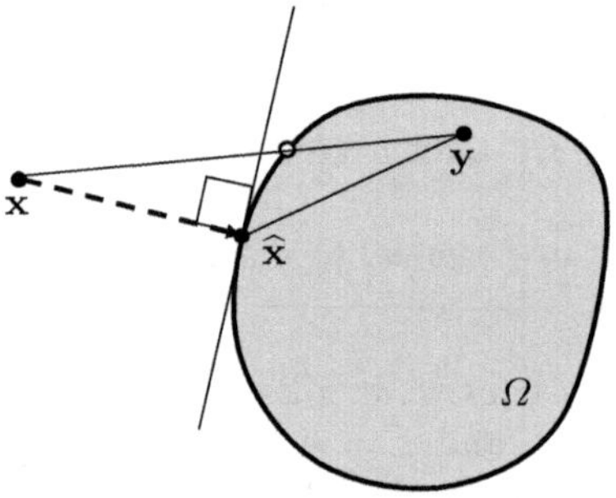

Fig. 2.4. Projection to the convex set

Proposition 2.6. *Let* $\Omega \subseteq \mathbb{R}^n$ *be a nonempty closed convex set and* $\mathbf{x} \in \mathbb{R}^n$*. Then there is a unique point* $\widehat{\mathbf{x}} \in \Omega$ *with the minimum Euclidean distance from* $\mathbf{x}$*, and for any* $\mathbf{y} \in \Omega$

$$(\mathbf{x} - \widehat{\mathbf{x}})^T(\mathbf{y} - \widehat{\mathbf{x}}) \leq 0. \tag{2.15}$$

Proof. Since the proof is trivial for $\mathbf{x} \in \Omega$, let us assume that $\mathbf{x} \notin \Omega$ is arbitrary but fixed and observe that the function f defined on $\mathbb{R}^n$ by

$$f(\mathbf{y}) = \|\mathbf{x} - \mathbf{y}\|^2 = \mathbf{y}^T\mathbf{y} - 2\mathbf{y}^T\mathbf{x} + \|\mathbf{x}\|^2$$

has the Hessian

$$\nabla^2 f(\mathbf{y}) = 2\mathsf{I}.$$

The identity matrix being positive definite, it follows by Proposition 2.2 that f is strictly convex, so that the unique minimizer $\widehat{\mathbf{x}} \in \Omega$ of $f(\mathbf{y})$ with respect to $\mathbf{y} \in \Omega$ exists by Proposition 2.5(i).

If $\mathbf{y} \in \Omega$ and $\alpha \in (0, 1)$, then by convexity of Ω

$$(1 - \alpha)\widehat{\mathbf{x}} + \alpha\mathbf{y} = \widehat{\mathbf{x}} + \alpha(\mathbf{y} - \widehat{\mathbf{x}}) \in \Omega,$$

so that for any $\mathbf{x} \in \mathbb{R}^n$

$$\|\mathbf{x} - \widehat{\mathbf{x}}\|^2 \leq \|\mathbf{x} - \widehat{\mathbf{x}} - \alpha(\mathbf{y} - \widehat{\mathbf{x}})\|^2.$$

Using simple manipulations and the latter inequality, we get

$$\begin{aligned}\|\mathbf{x}-\widehat{\mathbf{x}}-\alpha(\mathbf{y}-\widehat{\mathbf{x}})\|^2 &= \|\widehat{\mathbf{x}}-\mathbf{x}\|^2+\alpha^2\|\mathbf{y}-\widehat{\mathbf{x}}\|^2-2\alpha(\mathbf{x}-\widehat{\mathbf{x}})^T(\mathbf{y}-\widehat{\mathbf{x}})\\ &\le \|\mathbf{x}-\widehat{\mathbf{x}}-\alpha(\mathbf{y}-\widehat{\mathbf{x}})\|^2\\ &\quad+\alpha^2\|\mathbf{y}-\widehat{\mathbf{x}}\|^2-2\alpha(\mathbf{x}-\widehat{\mathbf{x}})^T(\mathbf{y}-\widehat{\mathbf{x}}).\end{aligned}$$

Thus

$$2\alpha(\mathbf{x}-\widehat{\mathbf{x}})^T(\mathbf{y}-\widehat{\mathbf{x}})\le\alpha^2\|\mathbf{y}-\widehat{\mathbf{x}}\|^2$$

for any $\alpha\in(0,1)$. To obtain (2.15), just divide the last inequality by $\alpha>0$ and observe that α may be arbitrarily small. □

Using Proposition 2.6, it is not difficult to show that the mapping P_Ω which assigns to each $\mathbf{x}\in\mathbb{R}^n$ its projection to Ω is *nonexpansive* as in Fig. 2.5.

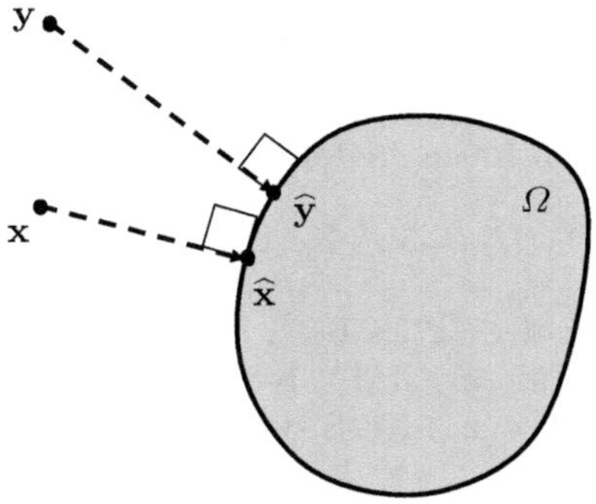

Fig. 2.5. Projection P_Ω is nonexpansive

Corollary 2.7. *Let $\Omega\subseteq\mathbb{R}^n$ be a nonempty closed convex set, and for any $\mathbf{x}\in\mathbb{R}^n$, let $\widehat{\mathbf{x}}\in\Omega$ denote the projection of $\mathbf{x}$ to Ω. Then for any $\mathbf{x},\mathbf{y}\in\Omega$*

$$\|P_\Omega(\mathbf{x})-P_\Omega(\mathbf{y})\|\le\|\mathbf{x}-\mathbf{y}\|. \tag{2.16}$$

Proof. If $\mathbf{x},\mathbf{y}\in\mathbb{R}$, then by Proposition 2.6 their projections $\widehat{\mathbf{x}},\widehat{\mathbf{y}}$ to Ω satisfy

$$(\mathbf{x}-\widehat{\mathbf{x}})^T(\mathbf{z}-\widehat{\mathbf{x}})\le 0 \quad\text{and}\quad (\mathbf{y}-\widehat{\mathbf{y}})^T(\mathbf{z}-\widehat{\mathbf{y}})\le 0$$

for any $\mathbf{z}\in\Omega$. Substituting $\mathbf{z}=\widehat{\mathbf{y}}$ into the first inequality, $\mathbf{z}=\widehat{\mathbf{x}}$ into the second inequality, and summing up, we get

$$(\mathbf{x}-\widehat{\mathbf{x}}-\mathbf{y}+\widehat{\mathbf{y}})^T(\widehat{\mathbf{y}}-\widehat{\mathbf{x}})\le 0.$$

After rearranging the entries and using the Schwarz inequality, we get

$$\|\widehat{\mathbf{x}} - \widehat{\mathbf{y}}\|^2 \leq (\mathbf{x} - \mathbf{y})^T(\widehat{\mathbf{x}} - \widehat{\mathbf{y}}) \leq \|\mathbf{x} - \mathbf{y}\|\|\widehat{\mathbf{x}} - \widehat{\mathbf{y}}\|. \qquad \square$$

2.4 Equality Constrained QP Problems

We shall now consider the problems with the constraint set described by a set of linear equations. More formally, we shall look for

$$\min_{\mathbf{x}\in\Omega_E} f(\mathbf{x}), \quad \Omega_E = \{\mathbf{x} \in \mathbb{R}^n : \mathsf{B}\mathbf{x} = \mathbf{c}\}, \tag{2.17}$$

where f is a quadratic function defined by (2.3), $\mathsf{B} \in \mathbb{R}^{m\times n}$, and $\mathbf{c} \in \mathrm{Im}\mathsf{B}$. We assume that $\mathsf{B} \neq \mathsf{O}$ is not a full column rank matrix so that $\mathrm{Ker}\mathsf{B} \neq \{\mathbf{o}\}$, but we admit dependent rows of B. It is easy to check that Ω_E is a nonempty closed convex set.

A feasible set Ω_E is a *linear manifold* of the form

$$\Omega_E = \overline{\mathbf{x}} + \mathrm{Ker}\mathsf{B}, \quad \mathsf{B}\overline{\mathbf{x}} = \mathbf{c}.$$

Thus a nonzero vector $\mathbf{d} \in \mathbb{R}^n$ is both a feasible and recession direction of Ω_E at any $\mathbf{x} \in \Omega_E$ if and only if $\mathbf{d} \in \mathrm{Ker}\mathsf{B}$.

Substituting $\mathbf{x} = \overline{\mathbf{x}} + \mathbf{z}$, $\mathbf{z} \in \mathrm{Ker}\mathsf{B}$, we can reduce (2.17) to minimizing

$$f_{\overline{x}}(\mathbf{z}) = \frac{1}{2}\mathbf{z}^T\mathsf{A}\mathbf{z} - (\mathbf{b} - \mathsf{A}\overline{\mathbf{x}})^T\mathbf{z} \tag{2.18}$$

over the subspace $\mathrm{Ker}\mathsf{B}$. Thus we can assume, without loss of generality, that $\mathbf{c} = \mathbf{o}$ in the definition of Ω_E. We shall occasionally use this assumption to simplify our exposition.

A useful tool for the analysis of equality constrained problems is the *Lagrangian function* $L_0 : \mathbb{R}^{n+m} \to \mathbb{R}$ defined by

$$L_0(\mathbf{x}, \boldsymbol{\lambda}) = f(\mathbf{x}) + \boldsymbol{\lambda}^T(B\mathbf{x} - \mathbf{c}) = \frac{1}{2}\mathbf{x}^T\mathsf{A}\mathbf{x} - \mathbf{b}^T\mathbf{x} + (\mathsf{B}\mathbf{x} - \mathbf{c})^T\boldsymbol{\lambda}. \tag{2.19}$$

Obviously

$$\nabla^2_{\mathbf{xx}} L_0(\mathbf{x}, \boldsymbol{\lambda}) = \nabla^2 f(\mathbf{x}) = \mathsf{A}, \tag{2.20}$$

$$\nabla_{\mathbf{x}} L_0(\mathbf{x}, \boldsymbol{\lambda}) = \nabla f(\mathbf{x}) + \mathsf{B}^T\boldsymbol{\lambda} = \mathsf{A}\mathbf{x} - \mathbf{b} + \mathsf{B}^T\boldsymbol{\lambda}, \tag{2.21}$$

$$L_0(\mathbf{x} + \mathbf{d}, \boldsymbol{\lambda}) = L_0(\mathbf{x}, \boldsymbol{\lambda}) + (\mathsf{A}\mathbf{x} - \mathbf{b} + \mathsf{B}^T\boldsymbol{\lambda})^T\mathbf{d} + \frac{1}{2}\mathbf{d}^T\mathsf{A}\mathbf{d}. \tag{2.22}$$

The Lagrangian function is defined in such a way that if considered as a function of $\mathbf{x}$, then its Hessian and its restriction to Ω_E are exactly those of

f, but its gradient $\nabla_{\mathbf{x}} L_0(\mathbf{x}, \boldsymbol{\lambda})$ varies depending on the choice of $\boldsymbol{\lambda}$. It simply follows that if f is convex, then L_0 is convex for any fixed $\boldsymbol{\lambda}$, and the global minimizer of L_0 with respect to $\mathbf{x}$ also varies with $\boldsymbol{\lambda}$. We shall see that it is possible to give conditions on A, B, and $\mathbf{b}$ such that with a suitable choice $\boldsymbol{\lambda} = \widehat{\boldsymbol{\lambda}}$, the solution of the constrained minimization problem (2.17) reduces to the unconstrained minimization of L_0 as in Fig. 2.6.

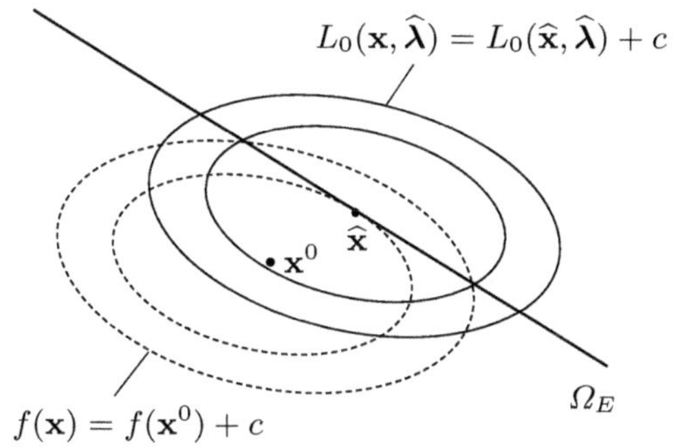

Fig. 2.6. Geometric illustration of the Lagrangian function

2.4.1 Optimality Conditions

The following proposition answers the main questions concerning the optimality and solvability conditions of (2.17).

Proposition 2.8. *Let the equality constrained problem (2.17) be defined by a symmetric matrix* $\mathsf{A} \in \mathbb{R}^{n \times n}$*, a constraint matrix* $\mathsf{B} \in \mathbb{R}^{m \times n}$ *whose rank is less than* n*, and vectors* $\mathbf{b} \in \mathbb{R}^n$, $\mathbf{c} \in \operatorname{Im}\mathsf{B}$. *Then the following statements hold:*
(i) A vector $\overline{\mathbf{x}} \in \Omega_E$ *is a solution of (2.17) if and only if* $\mathsf{A}|\operatorname{Ker}\mathsf{B}$ *is positive semidefinite and*

$$\nabla f(\widehat{\mathbf{x}})^T \mathbf{d} = (\mathsf{A}\overline{\mathbf{x}} - \mathbf{b})^T \mathbf{d} = 0 \tag{2.23}$$

for any $\mathbf{d} \in \operatorname{Ker}\mathsf{B}$.
(ii) A vector $\overline{\mathbf{x}} \in \Omega_E$ *is a solution of (2.17) if and only if* $\mathsf{A}|\operatorname{Ker}\mathsf{B}$ *is positive semidefinite and there is a vector* $\overline{\boldsymbol{\lambda}} \in \mathbb{R}^m$ *such that*

$$\nabla f(\widehat{\mathbf{x}}) + \mathsf{B}^T \overline{\boldsymbol{\lambda}} = \mathsf{A}\overline{\mathbf{x}} - \mathbf{b} + \mathsf{B}^T \overline{\boldsymbol{\lambda}} = \mathbf{o}. \tag{2.24}$$

Proof. (i) Let $\overline{\mathbf{x}}$ be a solution of the equality constrained minimization problem (2.17), so that for any $\mathbf{d} \in \operatorname{Ker}\mathsf{B}$ and $\alpha \in \mathbb{R}$

$$0 \le f(\overline{\mathbf{x}} + \alpha \mathbf{d}) - f(\overline{\mathbf{x}}) = \alpha (\mathsf{A}\overline{\mathbf{x}} - \mathbf{b})^T \mathbf{d} + \frac{\alpha^2}{2} \mathbf{d}^T \mathsf{A} \mathbf{d}. \tag{2.25}$$

Fixing $\mathbf{d} \in \text{Ker}\mathsf{B}$ and taking α sufficiently large, we get that the nonnegativity of the right-hand side of (2.25) implies $\mathbf{d}^T\mathsf{A}\mathbf{d} \geq 0$. Thus $\mathsf{A}|\text{Ker}\mathsf{B}$ must be positive semidefinite. On the other hand, for sufficiently small values of α and $(\mathsf{A}\overline{\mathbf{x}} - \mathbf{b})^T\mathbf{d} \neq 0$, the sign of the right-hand side of (2.25) is determined by the sign of $\alpha(\mathsf{A}\overline{\mathbf{x}} - \mathbf{b})^T\mathbf{d}$. Since we can choose the **sign** of α arbitrarily and the right-hand side of (2.25) is nonnegative, we conclude that (2.23) holds for any $\mathbf{d} \in \text{Ker}\mathsf{B}$.

Let us now assume that (2.23) holds for a vector $\overline{\mathbf{x}} \in \Omega_E$ and $\mathsf{A}|\text{Ker}\mathsf{B}$ is positive semidefinite. Then

$$f(\overline{\mathbf{x}} + \mathbf{d}) - f(\overline{\mathbf{x}}) = \frac{1}{2}\mathbf{d}^T\mathsf{A}\mathbf{d} \geq 0$$

for any $\mathbf{d} \in \text{Ker}\mathsf{B}$, so that $\overline{\mathbf{x}}$ is a solution of (2.17).

(ii) Let $\overline{\mathbf{x}}$ be a solution of (2.17), so that by (i) $\mathsf{A}|\text{Ker}\mathsf{B}$ is positive semidefinite and $\overline{\mathbf{x}}$ satisfies (2.23) for any $\mathbf{d} \in \text{Ker}\mathsf{B}$. The latter condition is by (1.18) equivalent to $\mathsf{A}\overline{\mathbf{x}} - \mathbf{b} \in \text{Im}\mathsf{B}^\mathsf{T}$, so that there is $\overline{\boldsymbol{\lambda}} \in \mathbb{R}^m$ such that (2.24) holds.

Let $\mathsf{A}|\text{Ker}\mathsf{B}$ be positive semidefinite. If there are $\overline{\boldsymbol{\lambda}}$ and $\overline{\mathbf{x}} \in \Omega_E$ such that (2.24) holds, then by Taylor's expansion (2.22)

$$f(\overline{\mathbf{x}} + \mathbf{d}) - f(\overline{\mathbf{x}}) = L_0(\overline{\mathbf{x}} + \mathbf{d}, \overline{\boldsymbol{\lambda}}) - L_0(\overline{\mathbf{x}}, \overline{\boldsymbol{\lambda}}) = \frac{1}{2}\mathbf{d}^T\mathsf{A}\mathbf{d} \geq 0$$

for any $\mathbf{d} \in \text{Ker}\mathsf{B}$, so that $\overline{\mathbf{x}}$ is a solution of the equality constrained problem (2.17). □

We can modify Proposition 2.8(i) to characterize the solutions of the minimization problem whose feasible set is a manifold defined by a vector and a subspace; this modification is helpful in what follows.

Corollary 2.9. *Let f be a convex quadratic function on $\mathbb{R}^n$, let $\mathcal{S}$ be a subspace of $\mathbb{R}^n$, and let $\mathbf{x}^0 \in \mathbb{R}^n$. Then $\overline{\mathbf{x}}$ is a solution of*

$$\min_{\mathbf{x}\in\Omega_S} f(\mathbf{x}), \quad \Omega_S = \mathbf{x}^0 + \mathcal{S} \tag{2.26}$$

if and only if

$$\nabla f(\overline{\mathbf{x}})^T\mathbf{d} = 0 \quad \text{for any} \quad \mathbf{d} \in S.$$

Proof. Let $\mathcal{S}$ be a subspace of $\mathbb{R}^n$, $\mathcal{S} = \text{Im}\mathsf{S}$, where $\mathsf{S} \in \mathbb{R}^{n\times m}$ is a full column rank matrix, $\mathbf{x}^0 \in \mathbb{R}^n$, and let $\mathsf{B} = \mathsf{I} - \mathsf{S}(\mathsf{S}^T\mathsf{S})^{-1}\mathsf{S}^T$, so that

$$\mathcal{S} = \text{Ker}\mathsf{B} \quad \text{and} \quad \Omega_S = \{\mathbf{x} \in \mathbb{R}^n : \mathsf{B}\mathbf{x} = \mathsf{B}\mathbf{x}^0\}.$$

Using Proposition 2.8, we get that $\overline{\mathbf{x}} \in \Omega_S$ is the minimizer of a convex quadratic function f on Ω_S if and only if $\nabla f(\overline{\mathbf{x}})$ is orthogonal to $\mathcal{S}$. □

The conditions (ii) of Proposition 2.8 are known as the *Karush–Kuhn–Tucker (KKT) conditions* for the solution of the equality constrained problem

(2.17). If $\overline{\mathbf{x}} \in \Omega_E$ and $\overline{\boldsymbol{\lambda}} \in \mathbb{R}^m$ satisfy (2.24), then $(\overline{\mathbf{x}}, \overline{\boldsymbol{\lambda}})$ is called a *KKT pair* of problem (2.17). Its second component $\overline{\boldsymbol{\lambda}}$ is called a vector of *Lagrange multipliers* or simply a *multiplier*. We shall often use the notation $\widehat{\mathbf{x}}$ or $\widehat{\boldsymbol{\lambda}}$ to denote the components of a KKT pair that are uniquely determined.

Proposition 2.8 has a simple geometrical interpretation. The condition (2.23) requires that the gradient of f at a solution $\overline{\mathbf{x}}$ is orthogonal to the set of feasible directions $\mathrm{Ker}\mathsf{B}$ of Ω_E so that there is no feasible decrease direction as illustrated in Fig. 2.7.

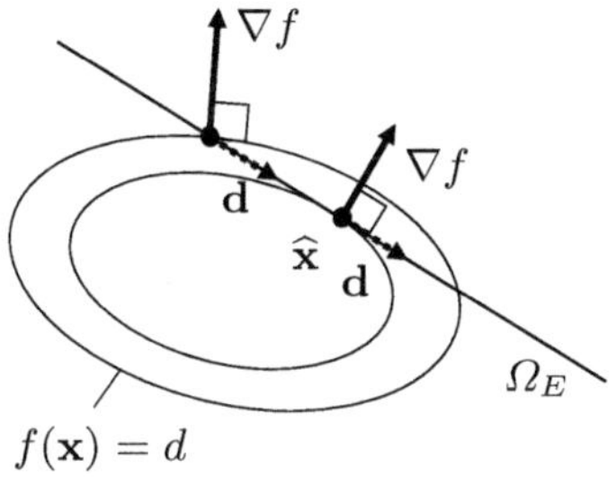

Fig. 2.7. Solvability condition (i)

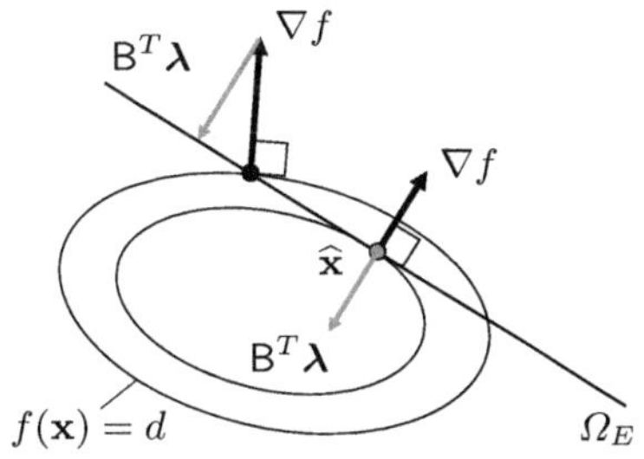

Fig. 2.8. Solvability condition (ii)

Since $\mathbf{d}$ is by (1.18) orthogonal to $\mathrm{Ker}\mathsf{B}$ if and only if $\mathbf{d} \in \mathrm{Im}\mathsf{B}^T$, it follows that (2.23) is equivalent to the possibility to choose $\overline{\boldsymbol{\lambda}}$ so that $\nabla_{\mathbf{x}} L_0(\overline{\mathbf{x}}, \overline{\boldsymbol{\lambda}}) = \mathbf{o}$. If f is convex, then the latter condition is equivalent to the condition for the unconstrained minimizer of L_0 with respect to $\mathbf{x}$ as illustrated in Fig. 2.8.

Notice that if f is convex, then the vector of Lagrange multipliers modifies the linear term of the original problem in such a way that the solution of the unconstrained modified problem is the same as the solution of the original constrained problem. In terms of mechanics, if the original problem describes the equilibrium of an elastic body subject to traction, then the constraints of the modified problem can be replaced by the reaction forces.

2.4.2 Existence and Uniqueness

Using the optimality conditions of Sect. 2.4.1, we can formulate the conditions that guarantee the existence or uniqueness of a solution of (2.17).

Proposition 2.10. *Let the problem (2.17) be defined by a symmetric matrix* $\mathsf{A} \in \mathbb{R}^{n\times n}$, *a constraint matrix* $\mathsf{B} \in \mathbb{R}^{m\times n}$ *with the column rank less than* n, *and vectors* $\mathbf{b} \in \mathbb{R}^n$, $\mathbf{c} \in \mathrm{Im}\mathsf{B}$. *Let* R *denote a matrix whose columns span* $\mathrm{Ker}\mathsf{A}$ *and let* $\mathsf{A}|\mathrm{Ker}\mathsf{B}$ *be SPS. Then the following statements hold:*
(i) Problem (2.17) has a solution if and only if

$$\mathsf{R}^T\mathbf{b} \in \mathrm{Im}(\mathsf{R}^T\mathsf{B}^T). \tag{2.27}$$

(ii) If $\mathsf{A}|\mathrm{Ker}\mathsf{B}$ *is SPD, then problem (2.17) has a unique solution.*
(iii) If $(\overline{\mathbf{x}}, \overline{\boldsymbol{\lambda}})$ *and* $(\overline{\mathbf{y}}, \overline{\boldsymbol{\mu}})$ *are KKT couples for problem (2.17), then*

$$\overline{\mathbf{x}} - \overline{\mathbf{y}} \in \mathrm{Ker}\mathsf{A} \cap \{\mathbf{b}\}^{\perp} \quad \text{and} \quad \overline{\boldsymbol{\lambda}} - \overline{\boldsymbol{\mu}} \in \mathrm{Ker}\mathsf{B}^T.$$

In particular, if problem (2.17) has a solution and $\mathrm{Ker}\mathsf{B}^T = \{\mathbf{o}\}$, *then there is a unique Lagrange multiplier* $\widehat{\boldsymbol{\lambda}}$.

Proof. (i) Using Proposition 2.8(ii), we have that problem (2.17) has a solution if and only if there is $\boldsymbol{\lambda}$ such that $\mathbf{b} - \mathsf{B}^T\boldsymbol{\lambda} \in \mathrm{Im}\mathsf{A}$, or, equivalently, that $\mathbf{b} - \mathsf{B}^T\boldsymbol{\lambda}$ is orthogonal to $\mathrm{Ker}\mathsf{A}$. The latter condition reads

$$\mathsf{R}^T\left(\mathbf{b} - \mathsf{B}^T\boldsymbol{\lambda}\right) = \mathbf{o}$$

and can be rewritten as (2.27).
(ii) First observe that if $\mathsf{A}|\mathrm{Ker}\mathsf{B}$ is positive definite, then $f|\mathrm{Ker}\mathsf{B}$ is strictly convex by Proposition 2.2 and $f|\Omega_E$ is strictly convex by Corollary 2.3. Since Ω_E is closed, convex, and nonempty, it follows by Proposition 2.5(i) that the equality constrained problem (2.17) has a unique solution.
(iii) First observe that $\mathrm{Ker}\mathsf{B} = \{\mathbf{x} - \mathbf{y} : \mathbf{x}, \mathbf{y} \in \Omega_E\}$ and that f is convex on $\mathrm{Ker}\mathsf{B}$ by the assumption and Proposition 2.2. The rest follows by Proposition 2.4(ii) and a simple analysis of the KKT conditions (2.24). □

If B is not a full row rank matrix and $\overline{\boldsymbol{\lambda}}$ is a Lagrange multiplier for (2.17), then by Proposition 2.10(iii) any Lagrange multiplier $\boldsymbol{\lambda}$ can be expressed as

$$\boldsymbol{\lambda} = \overline{\boldsymbol{\lambda}} + \boldsymbol{\delta}, \quad \boldsymbol{\delta} \in \mathrm{Ker}\mathsf{B}^T. \tag{2.28}$$

The Lagrange multiplier $\boldsymbol{\lambda}_{\mathrm{LS}}$ which minimizes the Euclidean norm is called the *least square Lagrange multiplier*; it is a unique multiplier which belongs to $\mathrm{Im}\mathsf{B}$. If $\overline{\boldsymbol{\lambda}}$ denotes Lagrange multipliers, then $\boldsymbol{\lambda}_{\mathrm{LS}}$ can be evaluated by

$$\boldsymbol{\lambda}_{\mathrm{LS}} = (\mathsf{B}^{\dagger})^T\mathsf{B}^T\overline{\boldsymbol{\lambda}}, \quad \overline{\boldsymbol{\lambda}} = \boldsymbol{\lambda}_{\mathrm{LS}} + \boldsymbol{\delta}, \quad \text{and } \boldsymbol{\delta} \in \mathrm{Ker}\mathsf{B}^T. \tag{2.29}$$

2.4.3 KKT Systems

If A is positive definite, then the unique solution $\widehat{\mathbf{x}}$ of (2.17) is by Proposition 2.8 fully determined by the matrix equation

$$\begin{bmatrix} \mathsf{A} & \mathsf{B}^T \\ \mathsf{B} & \mathsf{O} \end{bmatrix} \begin{bmatrix} \mathbf{x} \\ \boldsymbol{\lambda} \end{bmatrix} = \begin{bmatrix} \mathbf{b} \\ \mathbf{c} \end{bmatrix}, \tag{2.30}$$

which is known as the *Karush–Kuhn–Tucker system*, briefly *KKT conditions* (2.17). Proposition 2.8 does not require that the KKT system is nonsingular, in agreement with the observation that the solution of the equality constrained problem should not depend on the description of Ω_E.

An alternative proof of the uniqueness of the component $\widehat{\mathbf{x}}$ of the solution of the KKT system (2.30) for A positive definite and B with dependent rows can be obtained by analysis of the solutions of the homogeneous system

$$\begin{bmatrix} \mathsf{A} & \mathsf{B}^T \\ \mathsf{B} & \mathsf{O} \end{bmatrix} \begin{bmatrix} \mathbf{d} \\ \boldsymbol{\mu} \end{bmatrix} = \begin{bmatrix} \mathbf{o} \\ \mathbf{o} \end{bmatrix}. \tag{2.31}$$

Indeed, after multiplying the first block row of (2.31) by $\mathbf{d}^T$ on the left, $\mathbf{d} \in \mathrm{Ker}\mathsf{B}$, we get

$$\mathbf{d}^T\mathsf{A}\mathbf{d} + \mathbf{d}^T\mathsf{B}^T\mu = 0.$$

Since $\mathsf{B}\mathbf{d} = \mathbf{o}$, it follows that $\mathbf{d}^T\mathsf{B}^T\boldsymbol{\mu} = (\mathsf{B}\mathbf{d})^T\boldsymbol{\mu} = 0$, so that $\mathbf{d}^T\mathsf{A}\mathbf{d} = 0$ and, due to the positive definiteness of A, also $\mathbf{d} = \mathbf{o}$. The same argument is valid for A positive semidefinite provided

$$\mathrm{Ker}\mathsf{A} \cap \mathrm{Ker}\mathsf{B} = \{\mathbf{o}\}.$$

If A and B are respectively positive definite and full row rank matrices, then we can directly evaluate the inverse of the matrix of the KKT system (2.30) to get

$$\begin{bmatrix} \mathsf{A} & \mathsf{B}^T \\ \mathsf{B} & \mathsf{O} \end{bmatrix}^{-1} = \begin{bmatrix} \mathsf{A}^{-1} + \mathsf{A}^{-1}\mathsf{B}^T\mathsf{S}^{-1}\mathsf{B}\mathsf{A}^{-1}, & -\mathsf{A}^{-1}\mathsf{B}^T\mathsf{S}^{-1} \\ -\mathsf{S}^{-1}\mathsf{B}\mathsf{A}^{-1}, & \mathsf{S}^{-1} \end{bmatrix}, \tag{2.32}$$

where $\mathsf{S} = -\mathsf{B}\mathsf{A}^{-1}\mathsf{B}^T$ denotes the *Schur complement matrix.*

Even though not very useful computationally, the inverse matrix is helpful in analysis. For example, in the following lemma, we use it to get information on the distribution of the eigenvalues of the spectrum of a matrix of the KKT system.

Lemma 2.11. *Let $\mathcal{A} \in \mathbb{R}^{(n+m)\times(n+m)}$ denote the matrix of the KKT system (2.30) with a positive definite matrix A and a full rank matrix B. Then $\mathcal{A}$ is nonsingular and its eigenvalues $\alpha_1 \geq \cdots \geq \alpha_{n+m}$ satisfy*

$$\alpha_1 \geq \cdots \geq \alpha_n \geq \lambda_{\min}(\mathsf{A}) > 0 > \alpha_{n+1} \geq \cdots \geq \alpha_{n+m}, \quad \lambda_{\min}(\mathsf{A}) = \|\mathsf{A}^{-1}\|^{-1}.$$

Proof. Using repeatedly the Cauchy interlacing inequalities (1.26) to $\mathcal{A}$, we get

$$\alpha_1 \geq \lambda_1(\mathsf{A}),\ \alpha_2 \geq \lambda_2(\mathsf{A}), \dots, \alpha_n \geq \lambda_n(\mathsf{A}) = \lambda_{\min}(\mathsf{A}),$$

where

$$\lambda_1(\mathsf{A}) \geq \cdots \geq \lambda_n(\mathsf{A})$$

denote the eigenvalues of A.

Now observe that if $\sigma_1 \leq \cdots \leq \sigma_m < 0$ denote the eigenvalues of $\mathsf{S} = -\mathsf{B}\mathsf{A}^{-1}\mathsf{B}^T$, then by (1.32) $0 > \sigma_1^{-1} \geq \cdots \geq \sigma_m^{-1}$ are the eigenvalues of S^{-1}. Thus we can apply the Cauchy interlacing inequalities (1.26) to the formula (2.32) for $\mathcal{A}^{-1}$ to get the inequalities

$$\sigma_1^{-1} \geq \mu_{n+1},\ \sigma_2^{-1} \geq \mu_{n+2}, \dots, \sigma_m^{-1} \geq \mu_{n+m}$$

for the m smallest eigenvalues $\mu_{n+1} \geq \cdots \geq \mu_{n+m}$ of $\mathcal{A}^{-1}$. It follows that $\mathcal{A}$ has at least m negative eigenvalues $\mu_{n+1}^{-1}, \dots, \mu_{n+m}^{-1}$. As $\mathcal{A}$ has altogether $n+m$ eigenvalues counting multiplicity and includes at least n positive ones, we conclude that $0 > \alpha_{n+1}$. □

Using the extreme singular values of B, Rusten and Winther [239] established stronger bounds on the eigenvalues of $\mathcal{A}$ including

$$\frac{1}{2}\left(\lambda_1(\mathsf{A}) - \sqrt{\lambda_1(\mathsf{A})^2 + 4\sigma_{\min}(\mathsf{B})^2}\right) \geq \alpha_{n+1},$$

where $\sigma_{\min}(\mathsf{B})$ denotes the smallest singular value of B. More results concerning the spectrum of $\mathcal{A}$ may be found in Benzi et al. [14].

2.4.4 Min-Max, Dual, and Saddle Point Problems

To simplify our exposition, we shall assume in this subsection that A and B are positive definite and full row rank matrices, respectively, postponing the analysis of more general convex problems to Sects. 2.6.4 and 2.7. The assumptions imply that the related KKT system

$$\nabla_{\mathbf{x}} L_0(\mathbf{x}, \boldsymbol{\lambda}) = \mathsf{A}\mathbf{x} - \mathbf{b} + \mathsf{B}^T\boldsymbol{\lambda} = \mathbf{o}, \tag{2.33}$$

$$\nabla_{\boldsymbol{\lambda}} L_0(\mathbf{x}, \boldsymbol{\lambda}) = \mathsf{B}\mathbf{x} - \mathbf{c} = \mathbf{o} \tag{2.34}$$

has a unique solution $(\widehat{\mathbf{x}}, \widehat{\boldsymbol{\lambda}})$, which can be found by first solving (2.33) with respect to $\mathbf{x}$, and then substituting for $\mathbf{x}$ into (2.34) to get an equation for $\widehat{\boldsymbol{\lambda}}$. We shall now associate these two steps with optimization problems.

First, observe that by the gradient argument of Proposition 2.1, Eq. (2.33) is just the condition for $\mathbf{x}$ to be the unconstrained minimizer of L_0 with respect to $\mathbf{x}$. Thus for a given $\boldsymbol{\lambda} \in \mathbb{R}^m$, the first step is equivalent to evaluating the minimizer

$$\mathbf{x} = \mathbf{x}(\boldsymbol{\lambda}) = \mathsf{A}^{-1}(\mathbf{b} - \mathsf{B}^T\boldsymbol{\lambda})$$

of $L_0(\mathbf{x}, \boldsymbol{\lambda})$ with respect to $\mathbf{x}$. We can express the *dual function* explicitly

$$\begin{aligned} \Theta(\boldsymbol{\lambda}) &= \inf_{\mathbf{x}\in\mathbb{R}^n} L_0(\mathbf{x}, \boldsymbol{\lambda}) = \min_{\mathbf{x}\in\mathbb{R}^n} L_0(\mathbf{x}, \boldsymbol{\lambda}) = L_0(\mathbf{x}(\boldsymbol{\lambda}), \boldsymbol{\lambda}) \qquad (2.35)\\ &= -\frac{1}{2}\boldsymbol{\lambda}^T\mathsf{B}\mathsf{A}^{-1}\mathsf{B}^T\boldsymbol{\lambda} + (\mathsf{B}\mathsf{A}^{-1}\mathbf{b} - \mathbf{c})^T\boldsymbol{\lambda} - \frac{1}{2}\mathbf{b}^T\mathsf{A}^{-1}\mathbf{b} \\ \nabla\Theta(\boldsymbol{\lambda}) &= -\mathsf{B}\mathsf{A}^{-1}\mathsf{B}^T\boldsymbol{\lambda} + (\mathsf{B}\mathsf{A}^{-1}\mathbf{b} - \mathbf{c}). \qquad (2.36) \end{aligned}$$

We can also substitute for $\mathbf{x}$ into (2.34) to get

$$-\mathsf{B}\mathsf{A}^{-1}\mathsf{B}^T\boldsymbol{\lambda} + (\mathsf{B}\mathsf{A}^{-1}\mathbf{b} - \mathbf{c}) = \mathbf{o}.$$

Comparing the left-hand side of the last equation with the explicit expression (2.36) for $\nabla\Theta(\boldsymbol{\lambda})$, we get that the last equation can be written in the form

$$\nabla\Theta(\boldsymbol{\lambda}) = \mathbf{o}.$$

As $\mathsf{BA}^{-1}\mathsf{B}^T$, the Hessian of $-\Theta$, is positive definite, we conclude that the latter is equivalent to the condition (2.7) for the minimizer of $-\Theta$ or, equivalently, for the maximizer of Θ. Therefore the KKT couple $(\widehat{\mathbf{x}}, \widehat{\boldsymbol{\lambda}})$ for the equality constrained problem (2.17) solves the *min-max problem*

$$L_0(\widehat{\mathbf{x}}, \widehat{\boldsymbol{\lambda}}) = \max_{\boldsymbol{\lambda}\in\mathbb{R}^m} \min_{\mathbf{x}\in\mathbb{R}^n} L_0(\mathbf{x}, \boldsymbol{\lambda}), \tag{2.37}$$

$\widehat{\boldsymbol{\lambda}}$ solves the *dual problem*

$$\Theta(\widehat{\boldsymbol{\lambda}}) = \max_{\boldsymbol{\lambda}\in\mathbb{R}^m} \Theta(\boldsymbol{\lambda}), \tag{2.38}$$

and, since $\widehat{\mathbf{x}}$ is feasible, we get

$$f(\widehat{\mathbf{x}}) = L_0(\widehat{\mathbf{x}}, \boldsymbol{\lambda}) = L_0(\widehat{\mathbf{x}}, \widehat{\boldsymbol{\lambda}}) = \Theta(\widehat{\boldsymbol{\lambda}}). \tag{2.39}$$

Moreover, it follows that

$$f(\widehat{\mathbf{x}}) = L_0(\widehat{\mathbf{x}}, \widehat{\boldsymbol{\lambda}}) = \min_{\mathbf{x}\in\mathbb{R}^n} L_0(\mathbf{x}, \widehat{\boldsymbol{\lambda}}) \le L_0(\mathbf{x}, \widehat{\boldsymbol{\lambda}}), \quad \mathbf{x} \in \mathbb{R}^n. \tag{2.40}$$

There is another equivalent problem related to the penalty method. Since

$$\sup_{\boldsymbol{\lambda}\in\mathbb{R}^m} L_0(\mathbf{x}, \boldsymbol{\lambda}) = \infty \text{ for } \mathbf{x} \notin \Omega_E \quad \text{and} \quad \sup_{\boldsymbol{\lambda}\in\mathbb{R}^m} L_0(\mathbf{x}, \boldsymbol{\lambda}) = f(\mathbf{x}) \text{ for } \mathbf{x} \in \Omega_E,$$

it follows that the solution $\widehat{\mathbf{x}}$ of the KKT system (2.31) satisfies

$$f(\widehat{\mathbf{x}}) = \min_{\mathbf{x}\in\mathbb{R}^n} \sup_{\boldsymbol{\lambda}\in\mathbb{R}^m} L_0(\mathbf{x}, \boldsymbol{\lambda}). \tag{2.41}$$

Comparing (2.41) with (2.37) and (2.39), we get the *duality relation*

$$\max_{\boldsymbol{\lambda}\in\mathbb{R}^m} \min_{\mathbf{x}\in\mathbb{R}^n} L_0(\mathbf{x}, \boldsymbol{\lambda}) = \min_{\mathbf{x}\in\mathbb{R}^n} \sup_{\boldsymbol{\lambda}\in\mathbb{R}^m} L_0(\mathbf{x}, \boldsymbol{\lambda}). \tag{2.42}$$

Using (2.39) and (2.40), we get that $(\widehat{\mathbf{x}}, \widehat{\boldsymbol{\lambda}})$ solves the *saddle point problem* to find $(\widehat{\mathbf{x}}, \widehat{\boldsymbol{\lambda}})$ so that for any $\mathbf{x} \in \mathbb{R}^n$ and $\boldsymbol{\lambda} \in \mathbb{R}^m$

$$L_0(\widehat{\mathbf{x}}, \boldsymbol{\lambda}) \le L_0(\widehat{\mathbf{x}}, \widehat{\boldsymbol{\lambda}}) \le L_0(\mathbf{x}, \widehat{\boldsymbol{\lambda}}). \tag{2.43}$$

We have thus obtained two *unconstrained* problems which are equivalent to the original equality constrained problem (2.17). The saddle point formulation explicitly enhances the Lagrange multipliers. The saddle point problem is unconstrained at the cost of two sets of variables, while the dual formulation may enjoy a small dimension at the expense of dealing with more complex matrices. The dual problem may also be better conditioned. Notice that the left inequality in (2.43) can be replaced by equality.

2.4.5 Sensitivity

The Lagrange multipliers emerged in Proposition 2.8 as auxiliary variables that nobody had asked for but were useful in alternative formulations of the optimality conditions. However, it turns out that the Lagrange multipliers frequently have an interesting interpretation in specific practical contexts, as we have mentioned at the end of Sect. 2.4.1, where we briefly described their mechanical interpretation. Here we show that if they are uniquely determined by the KKT conditions (2.30), then they are related to the optimal cost change rates due to the violation of constraints.

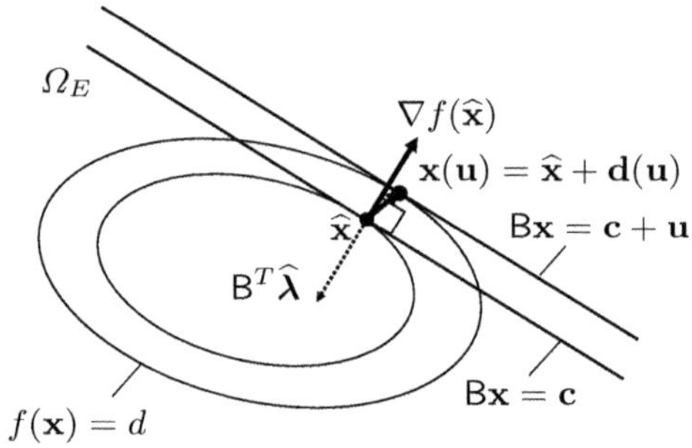

Fig. 2.9. Minimization with perturbed constraints

Let us assume that A and B are positive definite and full rank matrices, respectively, so that there is a unique KKT couple $(\widehat{\mathbf{x}}, \widehat{\boldsymbol{\lambda}})$ of the equality constrained problem (2.17). For $\mathbf{u} \in \mathbb{R}^m$, let us consider also the perturbed problem

$$\min_{\mathsf{B}\mathbf{x}=\mathbf{c}+\mathbf{u}} f(\mathbf{x})$$

as in Fig. 2.9. Its solution $\mathbf{x}(\mathbf{u})$ and the corresponding vector of Lagrange multipliers $\boldsymbol{\lambda}(\mathbf{u})$ are fully determined by the KKT conditions

$$\begin{bmatrix} \mathsf{A} & \mathsf{B}^T \\ \mathsf{B} & \mathsf{O} \end{bmatrix} \begin{bmatrix} \mathbf{x}(\mathbf{u}) \\ \boldsymbol{\lambda}(\mathbf{u}) \end{bmatrix} = \begin{bmatrix} \mathbf{b} \\ \mathbf{c} + \mathbf{u} \end{bmatrix},$$

so that

$$\begin{bmatrix} \mathbf{x}(\mathbf{u}) \\ \boldsymbol{\lambda}(\mathbf{u}) \end{bmatrix} = \begin{bmatrix} \mathsf{A} & \mathsf{B}^T \\ \mathsf{B} & \mathsf{O} \end{bmatrix}^{-1} \begin{bmatrix} \mathbf{b} \\ \mathbf{c} + \mathbf{u} \end{bmatrix} = \begin{bmatrix} \mathsf{A} & \mathsf{B}^T \\ \mathsf{B} & \mathsf{O} \end{bmatrix}^{-1} \begin{bmatrix} \mathbf{b} \\ \mathbf{c} \end{bmatrix} + \begin{bmatrix} \mathsf{A} & \mathsf{B}^T \\ \mathsf{B} & \mathsf{O} \end{bmatrix}^{-1} \begin{bmatrix} \mathbf{o} \\ \mathbf{u} \end{bmatrix}.$$

First observe that $\mathbf{d}(\mathbf{u}) = \mathbf{x}(\mathbf{u}) - \widehat{\mathbf{x}}$ satisfies

$$\mathsf{B}\mathbf{d}(\mathbf{u}) = \mathsf{B}\mathbf{x}(\mathbf{u}) - \mathsf{B}\widehat{\mathbf{x}} = \mathbf{u},$$

so that we can use $\nabla f(\widehat{\mathbf{x}}) = -\mathsf{B}^T\widehat{\boldsymbol{\lambda}}$ to approximate the change of optimal cost by

$$\nabla f(\widehat{\mathbf{x}})^T\mathbf{d}(\mathbf{u}) = -(\mathsf{B}^T\widehat{\boldsymbol{\lambda}})^T\mathbf{d}(\mathbf{u}) = -\widehat{\boldsymbol{\lambda}}^T\mathsf{B}\mathbf{d}(\mathbf{u}) = -\widehat{\boldsymbol{\lambda}}^T\mathbf{u}.$$

It follows that $-[\widehat{\boldsymbol{\lambda}}]_i$ can be used to approximate the change of the optimal cost due to the violation of the ith constraint by $[\mathbf{u}]_i$.

To give a more detailed analysis of the sensitivity of the optimal cost with respect to the violation of constraints, let us define for each $\mathbf{u} \in \mathbb{R}^m$ the *primal function*

$$p(\mathbf{u}) = f(\mathbf{x}(\mathbf{u})).$$

Observing that $\widehat{\mathbf{x}} = \mathbf{x}(\mathbf{o})$ and using the explicit formula (2.32) to evaluate the inverse of the KKT system, we get

$$\mathbf{x}(\mathbf{u}) = \widehat{\mathbf{x}} + \mathsf{A}^{-1}\mathsf{B}^T\mathsf{S}^{-1}\mathbf{u},$$

where $\mathsf{S} = \mathsf{B}\mathsf{A}^{-1}\mathsf{B}^T$ denotes the Schur complement matrix. Thus

$$\mathbf{x}(\mathbf{u}) - \widehat{\mathbf{x}} = \mathsf{A}^{-1}\mathsf{B}^T\mathsf{S}^{-1}\mathbf{u},$$

so that

$$\begin{aligned} p(\mathbf{u}) - p(\mathbf{o}) &= f(\mathbf{x}(\mathbf{u})) - f(\widehat{\mathbf{x}}) \\ &= \nabla f(\widehat{\mathbf{x}})^T(\mathbf{x}(\mathbf{u}) - \widehat{\mathbf{x}}) + \frac{1}{2}(\mathbf{x}(\mathbf{u}) - \widehat{\mathbf{x}})^T\mathsf{A}(\mathbf{x}(\mathbf{u}) - \widehat{\mathbf{x}}) \\ &= \nabla f(\widehat{\mathbf{x}})^T\mathsf{A}^{-1}\mathsf{B}^T\mathsf{S}^{-1}\mathbf{u} + \frac{1}{2}\mathbf{u}^T\mathsf{S}^{-1}\mathsf{B}\mathsf{A}^{-1}\mathsf{B}^T\mathsf{S}^{-1}\mathbf{u}. \end{aligned}$$

It follows that the gradient of the primal function p at $\mathbf{o}$ is given by

$$\nabla p(\mathbf{o}) = (\nabla f(\widehat{\mathbf{x}})^T\mathsf{A}^{-1}\mathsf{B}^T\mathsf{S}^{-1})^T = \mathsf{S}^{-1}\mathsf{B}\mathsf{A}^{-1}\nabla f(\widehat{\mathbf{x}}).$$

Recalling that $\nabla f(\widehat{\mathbf{x}}) = -\mathsf{B}^T\widehat{\boldsymbol{\lambda}}$, we get

$$\nabla p(\mathbf{o}) = -\mathsf{S}^{-1}\mathsf{B}\mathsf{A}^{-1}\mathsf{B}^T\widehat{\boldsymbol{\lambda}} = -\widehat{\boldsymbol{\lambda}}. \tag{2.44}$$

Our analysis shows that if the total differential of f at $\widehat{\mathbf{x}}$ decreases outside Ω_E, then this decrease is compensated by the increase of $\widehat{\boldsymbol{\lambda}}^T(\mathsf{B}\mathbf{x} - \mathbf{c})$. See also Fig. 2.6. The components of $\widehat{\boldsymbol{\lambda}}$ are also called the *shadow prices* after their meaning in the applications in economics.

For the sensitivity analysis of the solution of more general equality constrained problems, we refer to the book by Bertsekas [16].

2.4.6 Error Analysis

We shall now give the bounds on the error of the solution of the KKT system (2.30) in terms of perturbation of the right-hand side. As we do not assume here that the constraints are necessarily defined by a full rank matrix B, we shall use bounds on the nonzero singular values of the constraint matrix.

Proposition 2.12. *Let matrices* A, B *and vectors* $\mathbf{b}, \mathbf{c}$ *be those from the definition of problem (2.17) with* A *SPD and* $\mathsf{B} \in \mathbb{R}^{m\times n}$ *not necessarily a full rank matrix of the column rank less than* n. *Let* $\lambda_{\min}(\mathsf{A})$ *denote the least eigenvalue of* A, *let* $\overline{\sigma}_{\min}(\mathsf{B})$ *denote the least nonzero singular value of* B, *let* $(\widehat{\mathbf{x}}, \overline{\boldsymbol{\lambda}})$ *denote any KKT pair for the equality constrained problem (2.17), let* $\mathbf{g} \in \mathbb{R}^n$, $\mathbf{e} \in \mathbb{R}^m$, *and let* $(\mathbf{x}, \boldsymbol{\lambda})$ *denote an approximate KKT pair which satisfies*

$$\begin{aligned} \mathsf{A}\mathbf{x} + \mathsf{B}^T\boldsymbol{\lambda} &= \mathbf{b} + \mathbf{g}, \\ \mathsf{B}\mathbf{x} \qquad\quad &= \mathbf{c} + \mathbf{e}. \end{aligned} \tag{2.45}$$

Then

$$\|\mathsf{B}^T(\boldsymbol{\lambda} - \overline{\boldsymbol{\lambda}})\| \le \kappa(\mathsf{A})\|\mathbf{g}\| + \frac{\|\mathsf{A}\|}{\overline{\sigma}_{\min}(\mathsf{B})}\|\mathbf{e}\| \tag{2.46}$$

and

$$\|\mathbf{x} - \widehat{\mathbf{x}}\| \le \frac{\kappa(\mathsf{A}) + 1}{\lambda_{\min}(\mathsf{A})}\|\mathbf{g}\| + \frac{\kappa(\mathsf{A})}{\overline{\sigma}_{\min}(\mathsf{B})}\|\mathbf{e}\|. \tag{2.47}$$

Moreover, if $\boldsymbol{\lambda}_{\mathrm{LS}}$ *denotes the least square Lagrange multiplier for (2.17) and* $\boldsymbol{\lambda} \in \mathrm{Im}\mathsf{B}$, *then*

$$\|\boldsymbol{\lambda} - \boldsymbol{\lambda}_{\mathrm{LS}}\| \le \frac{1}{\overline{\sigma}_{\min}(\mathsf{B})}\left(\kappa(\mathsf{A})\|\mathbf{g}\| + \frac{\|\mathsf{A}\|}{\overline{\sigma}_{\min}(\mathsf{B})}\|\mathbf{e}\|\right). \tag{2.48}$$

Proof. Let us recall that we assume $\mathbf{c} \in \mathrm{Im}\mathsf{B}$, so that also $\mathbf{e} \in \mathrm{Im}\mathsf{B}$. If $\mathsf{B}^\dagger$ denotes the Moore–Penrose pseudoinverse of B, it follows that $\boldsymbol{\delta} = \mathsf{B}^\dagger\mathbf{e}$ satisfies $\mathsf{B}\boldsymbol{\delta} = \mathbf{o}$ and $\|\boldsymbol{\delta}\| \le \overline{\sigma}_{\min}(\mathsf{B})^{-1}\|\mathbf{e}\|$ (see (1.44)). Moreover, $\mathbf{y} = \mathbf{x} - \boldsymbol{\delta}$ satisfies

$$\begin{aligned} \mathsf{A}(\mathbf{y} - \widehat{\mathbf{x}}) + \mathsf{B}^T(\boldsymbol{\lambda} - \overline{\boldsymbol{\lambda}}) &= \mathbf{g} + \mathsf{A}\boldsymbol{\delta}, \\ \mathsf{B}(\mathbf{y} - \widehat{\mathbf{x}}) \qquad\qquad\qquad &= \mathbf{o}. \end{aligned}$$

After eliminating $\mathbf{y} - \widehat{\mathbf{x}}$ from the first equation, we get

$$\mathsf{B}\mathsf{A}^{-1}\mathsf{B}^T(\boldsymbol{\lambda} - \overline{\boldsymbol{\lambda}}) = \mathsf{B}(\mathsf{A}^{-1}\mathbf{g} + \boldsymbol{\delta}),$$

so that, after multiplication on the left by $(\boldsymbol{\lambda}-\overline{\boldsymbol{\lambda}})^T$ and taking norms, we get

$$\|\mathsf{A}\|^{-1}\|\mathsf{B}^T(\boldsymbol{\lambda}-\overline{\boldsymbol{\lambda}})\|^2 \leq \|\mathsf{B}^T(\boldsymbol{\lambda}-\overline{\boldsymbol{\lambda}})\|\|\mathsf{A}^{-1}\mathbf{g}+\boldsymbol{\delta}\|.$$

Thus

$$\|\mathsf{B}^T(\boldsymbol{\lambda}-\overline{\boldsymbol{\lambda}})\| \leq \kappa(\mathsf{A})\|\mathbf{g}\| + \|\mathsf{A}\|\|\boldsymbol{\delta}\| \leq \kappa(\mathsf{A})\|\mathbf{g}\| + \frac{\|\mathsf{A}\|}{\overline{\sigma}_{\min}(\mathsf{B})}\|\mathbf{e}\|.$$

After subtracting $\mathsf{A}\widehat{\mathbf{x}}+\mathsf{B}^T\overline{\boldsymbol{\lambda}} = \mathbf{b}$ from the first equation of (2.45), multiplying the result on the left by A^{-1}, and taking the norms, we get

$$\|\mathbf{x}-\widehat{\mathbf{x}}\| = \|\mathsf{A}^{-1}(\mathbf{g}-\mathsf{B}^T(\boldsymbol{\lambda}-\overline{\boldsymbol{\lambda}}))\| \leq \frac{\kappa(\mathsf{A})+1}{\lambda_{\min}(\mathsf{A})}\|\mathbf{g}\| + \frac{\kappa(\mathsf{A})}{\overline{\sigma}_{\min}(\mathsf{B})}\|\mathbf{e}\|.$$

If $\boldsymbol{\lambda} \in \mathrm{Im}\mathsf{B}$, then $\boldsymbol{\lambda}-\boldsymbol{\lambda}_{\mathrm{LS}} \in \mathrm{Im}\mathsf{B}$ and

$$\boldsymbol{\lambda}-\boldsymbol{\lambda}_{\mathrm{LS}} = (\mathsf{B}^\dagger)^T\mathsf{B}^T(\boldsymbol{\lambda}-\boldsymbol{\lambda}_{\mathrm{LS}}).$$

The last inequality then follows by (1.43) and (2.46). □

2.5 Inequality Constrained QP Problems

Let us consider the problems with feasible sets described by linear inequalities. Such sets are called *polyhedral sets.* More formally, we look for

$$\min_{\mathbf{x}\in\Omega_I} f(\mathbf{x}), \tag{2.49}$$

where f is a quadratic function defined by (2.3), $\Omega_I = \{\mathbf{x}\in\mathbb{R}^n : \mathsf{B}\mathbf{x}\leq\mathbf{c}\}$, $\mathsf{B} = [\mathbf{b}_1,\ldots,\mathbf{b}_m]^T \in \mathbb{R}^{m\times n}$, and $\mathbf{c} = [c_i] \in \mathbb{R}^m$. We assume that $\Omega_I \neq \emptyset$.

At any feasible point $\mathbf{x}$, we define the *active set*

$$\mathcal{A}(\mathbf{x}) = \{i \in \{1,\ldots,m\} : \mathbf{b}_i^T\mathbf{x} = c_i\}.$$

In particular, if $\overline{\mathbf{x}}$ is a local solution of (2.49), then each feasible direction of $\overline{\Omega}_E = \{\mathbf{x}\in\mathbb{R}^n : [\mathsf{B}\mathbf{x}]_{\mathcal{A}(\overline{\mathbf{x}})} = \mathbf{c}_{\mathcal{A}(\overline{\mathbf{x}})}\}$ at $\overline{\mathbf{x}}$ is a feasible direction of Ω_I at $\overline{\mathbf{x}}$. Using the arguments of Sect. 2.4.1, we get that $\overline{\mathbf{x}}$ is also a local solution of the equality constrained problem

$$\min_{\mathbf{x}\in\overline{\Omega}_E} f(\mathbf{x}), \quad \overline{\Omega}_E = \{\mathbf{x}\in\mathbb{R}^n : [\mathsf{B}\mathbf{x}]_{\mathcal{A}(\overline{\mathbf{x}})} = \mathbf{c}_{\mathcal{A}(\overline{\mathbf{x}})}\}. \tag{2.50}$$

Thus (2.49) is a more difficult problem than the equality constrained problem (2.17) as its solution necessarily enhances the identification of $\mathcal{A}(\overline{\mathbf{x}})$.

2.5.1 Polyhedral Sets

To understand the conditions of solvability of the inequality constrained problem (2.49), it is useful to get some insight into the geometry of polyhedral sets. We shall need a few new concepts.

A set $\mathcal{C} \subseteq \mathbb{R}^n$ is a (convex) *cone* (with its vertex at the origin) if $\mathbf{x}+\mathbf{y} \in \mathcal{C}$ and $\alpha\mathbf{x} \in \mathcal{C}$ for all $\alpha \geq 0$, $\mathbf{x} \in \mathcal{C}$, and $\mathbf{y} \in \mathcal{C}$. We are interested in *polyhedral cones* which are defined by

$$\mathcal{C} = \{\mathbf{x} \in \mathbb{R}^n : \mathsf{B}\mathbf{x} \leq \mathbf{o}\},$$

where $\mathsf{B} \in \mathbb{R}^{m\times n}$ is a given matrix. The *Minkowski–Weyl Theorem* (see, e.g., [16]) says that polyhedral cones are *finitely generated*, i.e., there is a matrix C such that

$$\mathcal{C} = \{\mathbf{x} \in \mathbb{R}^n : \mathbf{x} = \mathsf{C}\mathbf{y},\ \mathbf{y} \geq \mathbf{o}\}.$$

A polyhedral cone is a closed convex set.

A polyhedral set Ω can be represented as the sum of the convex hull of a finite set of points and the *recession cone* formed by the recession directions of Ω. Let us formulate this nontrivial statement more formally.

Proposition 2.13. *A set $\Omega \subseteq \mathbb{R}^n$ is polyhedral if and only if there is a nonempty set of vectors $\{\mathbf{x}_1, \ldots, \mathbf{x}_k\}$ and a polyhedral cone $\mathcal{C} \subseteq \mathbb{R}^n$ such that*

$$\Omega = \mathcal{C} + \mathrm{Conv}\{\mathbf{x}_1, \ldots, \mathbf{x}_k\}.$$

Proof. See, e.g., [16, Proposition B.17]. □

2.5.2 Farkas's Lemma

Farkas's lemma is used in the proof of the KKT conditions for inequality constrained QP problems in a similar way as the statement that $\mathrm{Im}\mathsf{B}^T$ is the orthogonal complement of $\mathrm{Ker}\mathsf{B}$ in the analysis of equality constrained problems. A geometric illustration of Farkas's lemma is in Fig. 2.10.

Lemma 2.14. *Let $\mathsf{B} \in \mathbb{R}^{m\times n}$ and $\mathbf{h} \in \mathbb{R}^n$. Then exactly one of the following problems has a solution:*

$$\begin{array}{llll} (I) & \textit{Find}\ \ \mathbf{d} \in \mathbb{R}^n & \textit{such that}\ \ \mathsf{B}\mathbf{d} \leq \mathbf{o} & \textit{and}\ \ \mathbf{h}^T\mathbf{d} > 0. \\ (II) & \textit{Find}\ \ \mathbf{y} \in \mathbb{R}^m & \textit{such that}\ \ \mathsf{B}^T\mathbf{y} = \mathbf{h} & \textit{and}\ \ \mathbf{y} \geq \mathbf{o}. \end{array}$$

Proof. Suppose first that (II) has a solution, so that there is $\mathbf{y} \geq \mathbf{o}$ such that $\mathsf{B}^T\mathbf{y} = \mathbf{h}$. Let $\mathbf{d} \in \mathbb{R}^n$ be such that $\mathsf{B}\mathbf{d} \leq \mathbf{o}$. Then

$$\mathbf{h}^T\mathbf{d} = (\mathsf{B}^T\mathbf{y})^T\mathbf{d} = \mathbf{y}^T\mathsf{B}\mathbf{d} \leq 0,$$

so that the problem (I) has no solution.

Now suppose that the problem (II) has no solution and denote

$$\Omega = \{\mathbf{x} \in \mathbb{R}^n : \mathbf{x} = \mathsf{B}^T\mathbf{y}, \mathbf{y} \geq \mathbf{o}\},$$

so that our assumption amounts to $\mathbf{h} \notin \Omega$. Denoting by $\widehat{\mathbf{h}} \in \Omega$ the projection of $\mathbf{h}$ to Ω and $\mathbf{d} = \mathbf{h} - \widehat{\mathbf{h}}$, we get by (2.15) that for any $\mathbf{x} \in \Omega$

$$\mathbf{d}^T(\mathbf{x} - \widehat{\mathbf{h}}) = (\mathbf{h} - \widehat{\mathbf{h}})^T(\mathbf{x} - \widehat{\mathbf{h}}) \leq 0,$$

or alternatively $\mathbf{d}^T\mathbf{x} \leq \mathbf{d}^T\widehat{\mathbf{h}} = \alpha$. Since $\mathbf{o} \in \Omega$, we get $\alpha \geq 0$. Moreover, substituting $\mathbf{x} = \mathsf{B}^T\mathbf{y}$, we get for $\mathbf{y} \geq \mathbf{o}$

$$\mathbf{y}^T\mathsf{B}\mathbf{d} = \mathbf{d}^T\mathsf{B}^T\mathbf{y} \leq \alpha.$$

Since the components of $\mathbf{y}$ can be arbitrarily large, $\mathsf{B}\mathbf{d} \leq \mathbf{o}$. To finish, recall that by assumption $\mathbf{h} \notin \Omega$, so that $\mathbf{d} \neq \mathbf{o}$, so that $\mathbf{d}^T(\mathbf{h} - \widehat{\mathbf{h}}) = \|\mathbf{d}\|^2 > 0$ and

$$\mathbf{d}^T\mathbf{h} > \mathbf{d}^T\widehat{\mathbf{h}} = \alpha \geq 0. \qquad \square$$

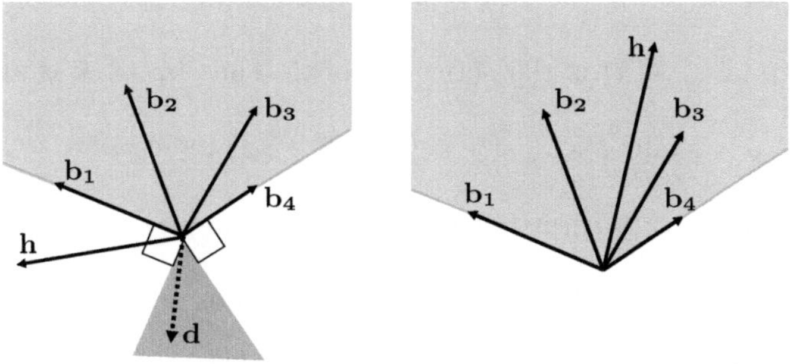

Fig. 2.10. Farkas's lemma: solution of (I) (left) and (II) (right)

2.5.3 Necessary Conditions

Let us start our exposition with the necessary optimality conditions.

Proposition 2.15. *Let the inequality constrained problem (2.49) be defined by a symmetric matrix* $\mathsf{A} \in \mathbb{R}^{n \times n}$*, the constraint matrix* $\mathsf{B} \in \mathbb{R}^{m \times n}$ *whose column rank is less than* n*, and the vectors* $\mathbf{b}, \mathbf{c}$*. Let* $\mathcal{C}$ *denote the cone of directions of the feasible set* Ω_I*. Then the following statements hold:*
(i) If $\overline{\mathbf{x}} \in \Omega_I$ *is a local solution of (2.49), then*

$$(\mathsf{A}\overline{\mathbf{x}} - \mathbf{b})^T\mathbf{d} \geq 0 \tag{2.51}$$

for any feasible direction $\mathbf{d}$ *of* Ω_I *at* $\overline{\mathbf{x}}$.
(ii) If $\overline{\mathbf{x}} \in \Omega_I$ *is a local solution of (2.49), then there is* $\overline{\boldsymbol{\lambda}} \in \mathbb{R}^m$ *such that*

$$\overline{\boldsymbol{\lambda}} \geq \mathbf{o}, \quad \mathsf{A}\overline{\mathbf{x}} - \mathbf{b} + \mathsf{B}^T\overline{\boldsymbol{\lambda}} = \mathbf{o}, \quad \textit{and} \quad \overline{\boldsymbol{\lambda}}^T(\mathsf{B}\overline{\mathbf{x}} - \mathbf{c}) = 0. \tag{2.52}$$

Proof. (i) Let $\overline{\mathbf{x}}$ and $\mathbf{d}$ be a local solution of the problem (2.49) and a feasible direction of Ω_I at $\overline{\mathbf{x}}$, respectively, so that the right-hand side of

$$f(\overline{\mathbf{x}} + \alpha\mathbf{d}) - f(\overline{\mathbf{x}}) = \alpha(\mathsf{A}\overline{\mathbf{x}} - \mathbf{b})^T\mathbf{d} + \frac{\alpha^2}{2}\mathbf{d}^T\mathsf{A}\mathbf{d} \tag{2.53}$$

is nonnegative for all sufficiently small $\alpha > 0$. To prove (2.51), it is enough to take $\alpha > 0$ so small that the nonnegativity of the right-hand side of (2.53) implies that

$$\alpha(\mathsf{A}\overline{\mathbf{x}} - \mathbf{b})^T\mathbf{d} \geq 0.$$

(ii) First observe that if $\overline{\mathbf{x}}$ is a local solution of (2.49), then $\mathbf{d}$ is a feasible direction of Ω_I at $\overline{\mathbf{x}}$ if and only if $\mathbf{d}$ is a feasible direction of

$$\overline{\Omega}_I = \{\mathbf{x} \in \mathbb{R}^n : \mathsf{B}_{\mathcal{A}*}\mathbf{x} \leq \mathbf{c}_{\mathcal{A}}\}$$

at $\overline{\mathbf{x}}$, where $\mathcal{A} = \mathcal{A}(\overline{\mathbf{x}})$ is the active set of $\overline{\mathbf{x}}$. Thus by (i) $\overline{\mathbf{x}}$ is also a local solution of

$$\min_{\mathbf{x} \in \overline{\Omega}_I} f(\mathbf{x}), \quad \overline{\Omega}_I = \{\mathbf{x} \in \mathbb{R}^n : \mathsf{B}_{\mathcal{A}*}\mathbf{x} \leq \mathbf{c}_{\mathcal{A}}\}.$$

Denoting $\mathbf{h} = -(\mathsf{A}\overline{\mathbf{x}} - \mathbf{b})$, it follows by (i) that the problem to find $\mathbf{d} \in \mathbb{R}^n$ such that

$$\mathsf{B}_{\mathcal{A}*}\mathbf{d} \leq \mathbf{o} \quad \text{and} \quad \mathbf{h}^T\mathbf{d} > 0$$

has no solution. Using Farkas's Lemma 2.14, we get $\mathbf{y} \in \mathbb{R}^m$ such that

$$(\mathsf{B}_{\overline{\mathcal{A}}*})^T\mathbf{y} = \mathbf{h}_A \quad \text{and} \quad \mathbf{y} \geq \mathbf{o}.$$

Denoting by $\overline{\boldsymbol{\lambda}} \in \mathbb{R}^m$ the vector obtained by padding $\mathbf{y}$ with zeros, so that $[\overline{\boldsymbol{\lambda}}]_i = 0$ for $i \notin \mathcal{A}$ and $\overline{\boldsymbol{\lambda}}_{\mathcal{A}} = \mathbf{y}$, it is easy to check that $\overline{\boldsymbol{\lambda}}$ satisfies (2.52). □

The conditions (2.52) are called the *KKT conditions for inequality constraints.* The last of these conditions, the equation $\overline{\boldsymbol{\lambda}}^T(\mathsf{B}\overline{\mathbf{x}} - \mathbf{c}) = 0$, is called the *condition of complementarity.* Notice that (ii) can be proved without any reference to Farkas's lemma as any solution $\overline{\mathbf{x}}$ of (2.49) solves (2.50), so that by Proposition 2.8(ii) there is $\mathbf{y}$ such that

$$\mathsf{A}\overline{\mathbf{x}} - \mathbf{b} + \mathsf{B}^T_{\mathcal{A}(\overline{\mathbf{x}})}\mathbf{y} = \mathbf{c}_{\mathcal{A}(\overline{\mathbf{x}})},$$

and $\mathbf{y} \geq \mathbf{o}$ by the arguments based on the sensitivity in Sect. 2.4.5.

2.5.4 Existence and Uniqueness

In our discussion of the existence and uniqueness results for the inequality constrained QP problem (2.49), we restrict our attention to the following results that are useful in our applications.

Proposition 2.16. *Let the inequality constrained problem (2.49) be defined by a symmetric matrix* $\mathsf{A} \in \mathbb{R}^{n\times n}$*, a constraint matrix* $\mathsf{B} \in \mathbb{R}^{m\times n}$*, and vectors* $\mathbf{b}, \mathbf{c}$*. Let* $\mathcal{C}$ *denote the cone of recession directions of the nonempty feasible set* Ω_I*. Then the following statements hold:*
(i) If problem (2.49) has a solution, then $\mathbf{d}^T\mathsf{A}\mathbf{d} \geq 0$ *for* $\mathbf{d} \in \mathcal{C}$ *and*

$$\mathbf{d}^T\mathbf{b} \leq 0 \quad \mathit{for} \quad \mathbf{d} \in \mathcal{C} \cap \mathrm{Ker}\mathsf{A}. \tag{2.54}$$

(ii) If (2.54) holds and f *is convex, then problem (2.49) has a solution.*
(iii) If f *is convex and* $(\overline{\mathbf{x}}, \overline{\boldsymbol{\lambda}})$ *and* $(\overline{\mathbf{y}}, \overline{\boldsymbol{\mu}})$ *are KKT couples for (2.49), then*

$$\overline{\mathbf{x}} - \overline{\mathbf{y}} \in \mathrm{Ker}\mathsf{A} \cap \{\mathbf{b}\}^{\perp} \quad \mathit{and} \quad \overline{\boldsymbol{\lambda}} - \overline{\boldsymbol{\mu}} \in \mathrm{Ker}\mathsf{B}^T. \tag{2.55}$$

(iv) If A *is positive definite, then the inequality constrained minimization problem (2.49) has the unique solution.*

Proof. (i) Let $\overline{\mathbf{x}}$ be a global solution of the inequality constrained minimization problem (2.49), and recall that

$$f(\overline{\mathbf{x}} + \alpha\mathbf{d}) - f(\overline{\mathbf{x}}) = \alpha(\mathsf{A}\overline{\mathbf{x}} - \mathbf{b})^T\mathbf{d} + \frac{\alpha^2}{2}\mathbf{d}^T\mathsf{A}\mathbf{d} \tag{2.56}$$

for any $\mathbf{d} \in \mathbb{R}^n$ and $\alpha \in \mathbb{R}$. Taking $\mathbf{d} \in \mathcal{C}$ arbitrary but fixed and α sufficiently large, we get that the nonnegativity of the right-hand side requires $\mathbf{d}^T\mathsf{A}\mathbf{d} \geq 0$. Moreover, if $\mathbf{d} \in \mathcal{C} \cap \mathrm{Ker}\mathsf{A}$, then (2.56) reduces to

$$f(\overline{\mathbf{x}} + \alpha\mathbf{d}) - f(\overline{\mathbf{x}}) = -\alpha\mathbf{b}^T\mathbf{d},$$

which is nonnegative for any $\alpha \geq 0$ if and only if $\mathbf{b}^T\mathbf{d} \leq 0$.
(ii) Let us now assume that (2.54) is satisfied and observe that if $\mathbf{c} = \mathbf{o}$, then Ω_I is a cone, so that a solution is known to exist even in infinite dimension (see Zeidler [269, pp. 553–556]).

If $\mathbf{c}$ is arbitrary, then by Proposition 2.13 there are $\mathbf{x}_1, \ldots, \mathbf{x}_k \in \Omega_I$ such that

$$\Omega_I = \mathcal{C} + \mathrm{conv}\{\mathbf{x}_1, \ldots, \mathbf{x}_k\}, \quad \mathcal{C} = \{\mathbf{x} : \mathsf{B}\mathbf{x} \leq \mathbf{o}\}.$$

Observing that $\mathbf{d} \in \mathcal{C}$ if and only if $2\mathbf{d} \in \mathcal{C}$, we get that $\mathbf{x} \in \Omega_I$ if and only if

$$\mathbf{x} = 2\mathbf{d} + \mathbf{y}, \quad \mathbf{d} \in \mathcal{C}, \quad \mathbf{y} \in \mathrm{Conv}\{\mathbf{x}_1, \ldots, \mathbf{x}_k\}.$$

Thus

$$f(\mathbf{x}) = f(2\mathbf{d} + \mathbf{y}) = \mathbf{d}^T \mathsf{A}\mathbf{d} - 2\mathbf{b}^T\mathbf{d} + \mathbf{d}^T\mathsf{A}\mathbf{d} + 2\mathbf{d}^T\mathsf{A}\mathbf{y} + \frac{1}{2}\mathbf{y}^T\mathsf{A}\mathbf{y} - \mathbf{b}^T\mathbf{y}$$
$$\geq 2f(\mathbf{d}) + (\mathbf{d}^T\mathsf{A}\mathbf{d} + 2\mathbf{d}^T\mathsf{A}\mathbf{y}) - \mathbf{b}^T\mathbf{y}. \tag{2.57}$$

We have already seen that f is bounded from below on $\mathcal{C}$. Moreover, using the Euclidean norm, we get

$$-\mathbf{b}^T\mathbf{y} \geq -\|\mathbf{b}\| \max\{\|\mathbf{x}_1\|, \dots, \|\mathbf{x}_k\|\}$$

and by (2.9)

$$\mathbf{d}^T\mathsf{A}\mathbf{d} + 2\mathbf{d}^T\mathsf{A}\mathbf{y} \geq -(\mathsf{A}\mathbf{y})^T\mathsf{A}^\dagger\mathsf{A}\mathbf{y} = -\mathbf{y}^T\mathsf{A}\mathbf{y} \geq -\|\mathsf{A}\| \max\{\|\mathbf{x}_1\|^2, \dots, \|\mathbf{x}_k\|^2\},$$

where $\mathsf{A}^\dagger$ denotes the Moore–Penrose generalized inverse to A. Thus f is bounded from below on Ω_I and we can use the Frank–Wolfe theorem (see Proposition 2.5(iii)) to finish the proof of (ii).
(iii) The first inclusion of (2.55) holds by Proposition 2.4(ii) for any convex problem. The second one follows by the KKT condition (2.52).
(iv) If A is SPD, then f is strictly convex by Proposition 2.2, so that by Proposition 2.5 there is a unique minimizer of f subject to $\mathbf{x} \in \Omega_I$. □

Remark 2.1. The proof of (ii) follows Dostál [74].

2.5.5 Optimality Conditions for Convex QP Problems

When the cost function is convex, then the necessary conditions of Sect. 2.5.3 are also sufficient.

Proposition 2.17. *Let f be a convex quadratic function defined by (2.3) with a positive semidefinite Hessian matrix A. Then the following statements hold:*
(i) A vector $\overline{\mathbf{x}} \in \Omega_I$ is a solution of (2.49) if and only if

$$(\mathsf{A}\overline{\mathbf{x}} - \mathbf{b})^T\mathbf{d} \geq 0 \tag{2.58}$$

for any feasible direction $\mathbf{d}$ of Ω_I at $\overline{\mathbf{x}}$.
(ii) A vector $\overline{\mathbf{x}} \in \Omega_I$ solves (2.49) if and only if there is $\overline{\boldsymbol{\lambda}} \in \mathbb{R}^m$ such that

$$\overline{\boldsymbol{\lambda}} > \mathbf{o}, \quad \mathsf{A}\overline{\mathbf{x}} - \mathbf{b} + \mathsf{B}^T\overline{\boldsymbol{\lambda}} = \mathbf{o}, \quad \text{and} \quad \overline{\boldsymbol{\lambda}}^T(\mathsf{B}\overline{\mathbf{x}} - \mathbf{c}) = 0. \tag{2.59}$$

Proof. Since f is convex, it follows by Proposition 2.4(i) that each local minimizer is a global minimizer. Moreover, by Proposition 2.15, each minimizer satisfies (2.58) and (2.59). Thus it is enough to prove that the convexity of f, the feasibility condition, and (2.58) or (2.59) are sufficient for $\overline{\mathbf{x}}$ to be a solution of the inequality constrained minimization problem (2.49).
(i) Let us assume that $\overline{\mathbf{x}} \in \Omega_I$ satisfies (2.58) and $\mathbf{x} \in \Omega_I$. Since Ω_I is convex,

it follows that $\mathbf{d} = \mathbf{x} - \overline{\mathbf{x}}$ is a feasible direction of Ω_I at $\overline{\mathbf{x}}$, so that, using Taylor's expansion and the assumptions, we have

$$f(\mathbf{x}) - f(\overline{\mathbf{x}}) = (\mathsf{A}\overline{\mathbf{x}} - \mathbf{b})^T\mathbf{d} + \frac{1}{2}\mathbf{d}^T\mathsf{A}\mathbf{d} \geq 0.$$

(ii) Let us assume that $\overline{\mathbf{x}} \in \Omega_I$ satisfies (2.59) and $\mathsf{B}\mathbf{x} - \mathbf{c} \leq \mathbf{o}$. Then

$$\begin{aligned} L_0(\overline{\mathbf{x}}, \overline{\boldsymbol{\lambda}}) &= f(\overline{\mathbf{x}}) + \overline{\boldsymbol{\lambda}}^T(\mathsf{B}\overline{\mathbf{x}} - \mathbf{c}) = f(\overline{\mathbf{x}}), \\ L_0(\mathbf{x}, \overline{\boldsymbol{\lambda}}) &= f(\mathbf{x}) + \overline{\boldsymbol{\lambda}}^T(\mathsf{B}\mathbf{x} - \mathbf{c}) \leq f(\mathbf{x}), \end{aligned}$$

and

$$\begin{aligned} f(\mathbf{x}) - f(\overline{\mathbf{x}}) &\geq L_0(\mathbf{x}, \overline{\boldsymbol{\lambda}}) - L_0(\overline{\mathbf{x}}, \overline{\boldsymbol{\lambda}}) \\ &= \nabla_{\mathbf{x}} L_0(\overline{\mathbf{x}}, \overline{\boldsymbol{\lambda}})^T(\mathbf{x} - \overline{\mathbf{x}}) + \frac{1}{2}(\mathbf{x} - \overline{\mathbf{x}})^T\mathsf{A}(\mathbf{x} - \overline{\mathbf{x}}) \\ &= (\mathsf{A}\overline{\mathbf{x}} - \mathbf{b} + \mathsf{B}^T\overline{\boldsymbol{\lambda}})^T(\mathbf{x} - \overline{\mathbf{x}}) + \frac{1}{2}(\mathbf{x} - \overline{\mathbf{x}})^T\mathsf{A}(\mathbf{x} - \overline{\mathbf{x}}) \\ &= \frac{1}{2}(\mathbf{x} - \overline{\mathbf{x}})^T\mathsf{A}(\mathbf{x} - \overline{\mathbf{x}}) \geq 0. \end{aligned}$$

□

2.5.6 Optimality Conditions for Bound Constrained Problems

A special case of problem (2.49) is the bound constrained problem

$$\min_{\mathbf{x} \in \Omega_B} f(\mathbf{x}), \quad \Omega_B = \{\mathbf{x} \in \mathbb{R}^n : \mathbf{x} \geq \boldsymbol{\ell}\}, \tag{2.60}$$

where f is a quadratic function defined by (2.3) and $\boldsymbol{\ell} \in \mathbb{R}^n$. The optimality conditions for convex bound constrained problems can be written in the form which is more convenient in some applications.

Proposition 2.18. *Let f be a convex quadratic function defined by (2.3) with a positive semidefinite Hessian A. Then $\overline{\mathbf{x}} \in \Omega_B$ solves (2.60) if and only if*

$$\mathsf{A}\overline{\mathbf{x}} - \mathbf{b} \geq \mathbf{o} \quad \textit{and} \quad (\mathsf{A}\overline{\mathbf{x}} - \mathbf{b})^T(\overline{\mathbf{x}} - \boldsymbol{\ell}) = 0. \tag{2.61}$$

Proof. First observe that denoting $\mathsf{B} = -\mathsf{I}_n$, $\mathbf{c} = -\boldsymbol{\ell}$, and

$$\Omega_I = \{\mathbf{x} \in \mathbb{R}^n : \mathsf{B}\mathbf{x} \leq \mathbf{c}\},$$

the bound constrained problem (2.60) becomes the standard inequality constrained problem (2.49) with $\Omega_I = \Omega_B$. Using Proposition 2.17, it follows that $\overline{\mathbf{x}} \in \Omega_B$ is the solution of (2.60) if and only if there is $\lambda \in \mathbb{R}^n$ such that

$$\lambda \geq \mathbf{o}, \quad \mathsf{A}\overline{\mathbf{x}} - \mathbf{b} - \mathsf{I}\lambda = \mathbf{o}, \quad \text{and} \quad \lambda^T(\overline{\mathbf{x}} - \ell) = 0. \tag{2.62}$$

We complete the proof by observing that (2.61) can be obtained from (2.62) and vice versa by substituting $\boldsymbol{\lambda} = \mathsf{A}\overline{\mathbf{x}} - \mathbf{b}$.

In the proof, we have shown that $\boldsymbol{\lambda} = \nabla f(\overline{\mathbf{x}})$ is a vector of Lagrange multipliers for the constraints $-\mathbf{x} \leq -\ell$, or, equivalently, for $\mathbf{x} \geq \ell$. Notice that the conditions (2.61) require that none of the vectors $\mathbf{s}_i$ is a feasible decrease direction of Ω_B at $\overline{\mathbf{x}}$, where $\mathbf{s}_i$ denotes a vector of the standard basis of $\mathbb{R}^n$ formed by the columns of I_n, $i \in \mathcal{A}(\overline{\mathbf{x}})$.

2.5.7 Min-Max, Dual, and Saddle Point Problems

As in Sect. 2.4.4, we shall assume that A and B are positive definite and full rank matrices, respectively, postponing the analysis of more general problems to Sects. 2.6.4 and 2.7. The assumptions imply that the related KKT system

$$\boldsymbol{\lambda} \geq \mathbf{o}, \tag{2.63}$$
$$\nabla_{\mathbf{x}} L_0(\mathbf{x}, \boldsymbol{\lambda}) = \mathsf{A}\mathbf{x} - \mathbf{b} + \mathsf{B}^T\boldsymbol{\lambda} = \mathbf{o}, \tag{2.64}$$
$$\nabla_{\boldsymbol{\lambda}} L_0(\mathbf{x}, \boldsymbol{\lambda}) = \mathsf{B}\mathbf{x} - \mathbf{c} \leq \mathbf{o}, \tag{2.65}$$
$$\boldsymbol{\lambda}^T(\mathsf{B}\mathbf{x} - \mathbf{c}) = 0 \tag{2.66}$$

has a unique solution $(\widehat{\mathbf{x}}, \widehat{\boldsymbol{\lambda}})$. We shall now associate the KKT system (2.63)–(2.66) with some other extremal problems.

First observe that we can evaluate $\mathbf{x}$ from (2.64) to get the minimizer

$$\mathbf{x} = \mathbf{x}(\boldsymbol{\lambda}) = \mathsf{A}^{-1}\mathbf{b} - \mathsf{A}^{-1}\mathsf{B}^T\boldsymbol{\lambda}$$

of $L_0(\mathbf{x}, \boldsymbol{\lambda})$ with respect to $\mathbf{x}$. We can express explicitly the *dual function*

$$\begin{aligned} \Theta(\boldsymbol{\lambda}) &= \inf_{\mathbf{x} \in \mathbb{R}^n} L_0(\mathbf{x}, \boldsymbol{\lambda}) = \min_{\mathbf{x} \in \mathbb{R}^n} L_0(\mathbf{x}, \boldsymbol{\lambda}) = L_0(\mathbf{x}(\boldsymbol{\lambda}), \boldsymbol{\lambda}) \\ &= -\frac{1}{2}\boldsymbol{\lambda}^T\mathsf{B}\mathsf{A}^{-1}\mathsf{B}^T\boldsymbol{\lambda} + (\mathsf{B}\mathsf{A}^{-1}\mathbf{b} - \mathbf{c})^T\boldsymbol{\lambda} - \frac{1}{2}\mathbf{b}^T\mathsf{A}^{-1}\mathbf{b} \end{aligned} \tag{2.67}$$

and its gradient

$$\nabla\Theta(\boldsymbol{\lambda}) = -\mathsf{B}\mathsf{A}^{-1}\mathsf{B}^T\boldsymbol{\lambda} + (\mathsf{B}\mathsf{A}^{-1}\mathbf{b} - \mathbf{c}). \tag{2.68}$$

Moreover,

$$\mathsf{B}\mathbf{x}(\boldsymbol{\lambda}) - \mathbf{c} = -\mathsf{B}\mathsf{A}^{-1}\mathsf{B}^T\boldsymbol{\lambda} + (\mathsf{B}\mathsf{A}^{-1}\mathbf{b} - \mathbf{c}) = \nabla\Theta(\boldsymbol{\lambda}),$$

so that (2.65) and (2.66) are equivalent to

$$\nabla\Theta(\boldsymbol{\lambda}) \leq \mathbf{o} \quad \text{and} \quad \boldsymbol{\lambda}^T\nabla\Theta(\boldsymbol{\lambda}) = 0,$$

which can be rewritten as

$$-\nabla\Theta(\boldsymbol{\lambda}) \geq \mathbf{o} \quad \text{and} \quad -\boldsymbol{\lambda}^T\nabla\Theta(\boldsymbol{\lambda}) = 0. \tag{2.69}$$

As the Hessian $\mathsf{BA}^{-1}\mathsf{B}^T$ of $-\Theta$ is SPD, we get that (2.69) is equivalent to the condition (2.61) for the minimizer of $-\Theta$ subject to $\boldsymbol{\lambda} \geq \mathbf{o}$, or, equivalent for the maximizer of Θ subject to $\boldsymbol{\lambda} \geq \mathbf{o}$. Thus the KKT couple $(\widehat{\mathbf{x}}, \widehat{\boldsymbol{\lambda}})$ for the inequality constrained problem (2.49) solves the *min-max problem*

$$L_0(\widehat{\mathbf{x}}, \widehat{\boldsymbol{\lambda}}) = \max_{\boldsymbol{\lambda}\geq\mathbf{o}} \min_{\mathbf{x}\in\mathbb{R}^n} L_0(\mathbf{x}, \boldsymbol{\lambda}), \tag{2.70}$$

$\widehat{\boldsymbol{\lambda}}$ solves the *dual problem*

$$\Theta(\widehat{\boldsymbol{\lambda}}) = \max_{\boldsymbol{\lambda}\geq\mathbf{o}} \Theta(\boldsymbol{\lambda}), \tag{2.71}$$

and

$$f(\widehat{\mathbf{x}}) = L_0(\widehat{\mathbf{x}}, \widehat{\boldsymbol{\lambda}}) = \Theta(\widehat{\boldsymbol{\lambda}}). \tag{2.72}$$

As in Sect. 2.4.4, there is yet another equivalent problem which is related to the penalty method. Since

$$\sup_{\boldsymbol{\lambda}\geq\mathbf{o}} L_0(\mathbf{x}, \boldsymbol{\lambda}) = \infty \quad \text{for} \quad \mathbf{x} \notin \Omega_I \quad \text{and} \quad \sup_{\boldsymbol{\lambda}\geq\mathbf{o}} L_0(\mathbf{x}, \boldsymbol{\lambda}) = f(\mathbf{x}) \quad \text{for} \quad \mathbf{x} \in \Omega_I,$$

it follows that the solution $\widehat{\mathbf{x}}$ of the KKT system (2.63)–(2.66) satisfies

$$f(\widehat{\mathbf{x}}) = \min_{\mathbf{x}\in\mathbb{R}^n} \sup_{\boldsymbol{\lambda}\geq\mathbf{o}} L_0(\mathbf{x}, \boldsymbol{\lambda}).$$

Comparing the result with (2.70) and (2.72), we get the *duality relation*

$$\max_{\boldsymbol{\lambda}\geq\mathbf{o}} \min_{\mathbf{x}\in\mathbb{R}^n} L_0(\mathbf{x}, \boldsymbol{\lambda}) = \min_{\mathbf{x}\in\mathbb{R}^n} \sup_{\boldsymbol{\lambda}\geq\mathbf{o}} L_0(\mathbf{x}, \boldsymbol{\lambda}). \tag{2.73}$$

Using the feasibility of $\widehat{\mathbf{x}}$, equations (2.72), and definition (2.67) of the dual function Θ, we also get that $(\widehat{\mathbf{x}}, \widehat{\boldsymbol{\lambda}})$ solves the *saddle point problem* to find $(\widehat{\mathbf{x}}, \widehat{\boldsymbol{\lambda}})$ so that for any $\mathbf{x} \in \mathbb{R}^n$ and $\boldsymbol{\lambda} \geq \mathbf{o}$

$$L_0(\widehat{\mathbf{x}}, \boldsymbol{\lambda}) \leq L_0(\widehat{\mathbf{x}}, \widehat{\boldsymbol{\lambda}}) \leq L_0(\mathbf{x}, \widehat{\boldsymbol{\lambda}}). \tag{2.74}$$

We have obtained again two *bound constrained* problems equivalent to the original equality constrained problem (2.49). The mixed formulation enhances explicitly the Lagrange multipliers and is only bound constrained at the cost of two sets of variables, while the dual formulation enjoys the bound constraints and typically small dimension at the cost of dealing with more complex matrices.

The dual problem may be also better conditioned. For example, if

$$\mathsf{B}_I = [\mathsf{I}, \mathsf{O}], \quad \mathsf{B} \in \mathbb{R}^{m\times n}, \quad \mathcal{I} = \{1, \ldots, m\},$$

then the dual problem reads

$$\max_{\boldsymbol{\lambda}\geq \mathbf{o}} \Theta(\boldsymbol{\lambda}), \quad \Theta(\boldsymbol{\lambda}) = -\frac{1}{2}\boldsymbol{\lambda}^T[\mathsf{A}^{-1}]_{\mathcal{I}\mathcal{I}}\boldsymbol{\lambda} + (\mathsf{B}\mathsf{A}^{-1}\mathbf{b} - \mathbf{c}) - \frac{1}{2}\mathbf{b}^T\mathsf{A}^{-1}\mathbf{b},$$

and the conditioning of the Schur complement

$$\mathsf{S}^{-1} = [\mathsf{A}^{-1}]_{\mathcal{I}\mathcal{I}} = \mathsf{B}\mathsf{A}^{-1}\mathsf{B}^T$$

is typically much better than that of A.

2.6 Equality and Inequality Constrained QP Problems

We shall consider convex QP problems with linear constraints in the form

$$\min_{\mathbf{x}\in\Omega_{IE}} f(\mathbf{x}), \quad \Omega_{IE} = \{\mathbf{x}\in\mathbb{R}^n : \mathsf{B}_{\mathcal{I}*}\mathbf{x} \leq \mathbf{c}_{\mathcal{I}}, \ \mathsf{B}_{\mathcal{E}*}\mathbf{x} = \mathbf{c}_{\mathcal{E}}\}. \tag{2.75}$$

We restrict our attention to convex problem with SPS or SPD Hessian to simplify the exposition and concentrate on the problems relevant in the rest of the book.

The problem (2.75) makes sense even for $\mathcal{I} = \emptyset$ or $\mathcal{E} = \emptyset$; we consider the conditions with the empty set as always satisfied. For example, $\mathcal{E} = \emptyset$ gives

$$\Omega_{IE} = \{\mathbf{x}\in\mathbb{R}^n : \ \mathbf{b}_i^T\mathbf{x} \leq c_i, \ i\in\mathcal{I}\},$$

and the kernel of an "empty" matrix is defined by

$$\mathrm{Ker}\mathsf{B}_{\mathcal{E}*} = \{\mathbf{x}\in\mathbb{R}^n : \ \mathbf{b}_i^T\mathbf{x} = c_i, \ i\in\mathcal{E}\} = \mathbb{R}^n.$$

2.6.1 Existence and Uniqueness

The main results concerning the existence and uniqueness of a solution for the inequality constrained QP problem (2.75) read as follows.

Proposition 2.19. *Let the quadratic function be defined by an SPS matrix* A *and* $\mathbf{b}\in\mathbb{R}^n$. *Let* $\Omega_{IE}\neq\emptyset$ *be defined by linear equality and inequality*

constraints. Let $\mathscr{C}$ denote the cone of recession directions of the nonempty feasible set Ω. Then the following statements hold:
(i) Problem (2.75) has a solution if and only if

$$\mathbf{d}^T\mathbf{b} \le 0 \quad \textit{for} \quad \mathbf{d} \in \mathscr{C} \cap \mathrm{Ker}\mathsf{A}. \tag{2.76}$$

(ii) $\overline{\mathbf{x}} \in \Omega_{IE}$ is a solution of problem (2.75) if and only if

$$\mathbf{d}^T \nabla f(\overline{\mathbf{x}}) = (\mathsf{A}\overline{\mathbf{x}} - \mathbf{b})^T \mathbf{d} \ge 0 \tag{2.77}$$

for any feasible direction $\mathbf{d}$.
(iii) If f is convex and $\overline{\mathbf{x}}, \overline{\mathbf{y}}$ are two minimizers of f subject to $\mathbf{x} \in \Omega$, then

$$\overline{\mathbf{x}} - \overline{\mathbf{y}} \in \mathrm{Ker}\mathsf{A} \cap \mathrm{Span}\{\mathbf{b}\}^{\perp}.$$

(iv) If A is positive definite, then the inequality constrained minimization problem (2.75) has a unique solution.

Proof. (i) First notice that we can assume that Ω is polyhedral cone, so that we can use arguments of Proposition (2.16) to show that f is bounded from bellow on Ω. The solution exists by Proposition 2.5(iii). See also Dostál [74].
(ii)–(iv) Combine Propositions 2.4 and 2.5. □

2.6.2 KKT Conditions

Let us consider the QP problem (2.75) with linear equality and inequality constraints. Combining Propositions 2.8, 2.15, and 2.17 with the above discussion, we can formulate the KKT conditions for equality and inequality constrained problem (2.75).

Theorem 2.1. *Let the problem (2.75) be defined by an SPS matrix $\mathsf{A} \in \mathbb{R}^{n\times n}$ $\mathbf{b} \in \mathbb{R}^n$. Let Ω_{IE} be nonempty. Then the following statements hold:*
(i) $\overline{\mathbf{x}} \in \Omega_{IE}$ is a solution of (2.75) if and only if there is a vector $\overline{\boldsymbol{\lambda}} \in \mathbb{R}^m$ such that

$$\overline{\boldsymbol{\lambda}}_{\mathcal{I}} \ge \mathbf{o}, \quad \mathsf{A}\overline{\mathbf{x}} - \mathbf{b} + \mathsf{B}^T\overline{\boldsymbol{\lambda}} = \mathbf{o}, \quad \text{and} \quad \overline{\boldsymbol{\lambda}}_{\mathcal{I}}^T[\mathsf{B}\overline{\mathbf{x}} - \mathbf{c}]_{\mathcal{I}} = 0. \tag{2.78}$$

(ii) If A is SPD, then $\overline{\mathbf{x}} \in \Omega_{IE}$ is a unique solution of (2.75).
(iii) If $\nabla\mathbf{h}(\mathbf{x})$ is a full rank matrix, then the Lagrange multipliers are unique.

Remark 2.2. We formulated Theorem 2.1 for problems with symmetric positive definite Hessian A to simplify its statement for the cases which appear very frequently in what follows. It is a simple exercise to reformulate it for more general symmetric Hessian matrix in terms of the local solutions and local convexity of the cost function.

2.6.3 Bound and Equality Constrained Problems

Here we consider the partially bound and equality constrained problem

$$\min_{\mathbf{x}\in\Omega_{BE}} f(\mathbf{x}), \quad \Omega_{BE} = \{\mathbf{x}\in\mathbb{R}^n : \mathbf{x}_{\mathcal{I}} \geq \boldsymbol{\ell}_{\mathcal{I}},\ \mathsf{B}_E\mathbf{x} = \mathbf{c}_{\mathcal{E}}\}, \tag{2.79}$$

where f is a convex quadratic function with a symmetric positive semidefinite Hessian $\mathsf{A}\in\mathbb{R}^{n\times n}$, $\mathsf{B}_E\in\mathbb{R}^{q\times n}$ is a matrix with possibly dependent rows, $\mathbf{c}_{\mathcal{E}}\in\mathbb{R}^q$, and $\boldsymbol{\ell}_{\mathcal{I}}\in\mathbb{R}^p$ is a vector of bounds on the first p components of $\mathbf{x}$ which form variables $\mathbf{x}_{\mathcal{I}}$. To unify the references to the corresponding vectors, we denote by $\mathcal{I}=\{1,\ldots,p\}$ and $\mathcal{R}=\{p+1,\ldots,n\}$ the sets of indices of the bound constrained and remaining entries of $\mathbf{x}$, respectively. With

$$\mathsf{B}_{\mathcal{I}} = [-\mathsf{I}_p, \mathsf{O}_{pr}],\ r=n-p,\ \mathsf{B} = \begin{bmatrix}\mathsf{B}_I\\ \mathsf{B}_E\end{bmatrix}, \quad \mathbf{c} = \begin{bmatrix}-\boldsymbol{\ell}_{\mathcal{I}}\\ \mathbf{c}_{\mathcal{E}}\end{bmatrix},$$

and $\mathcal{E}=\{p+1,\ldots,m\}$, the bound and equality constrained problem (2.79) becomes the standard equality and inequality constrained problem (2.75).

We can use the specific form of bound constraints to simplify the KKT conditions, in particular to avoid explicit use of the Lagrange multipliers for the inequality constraints in the KKT conditions. Introducing the Lagrange multipliers

$$\boldsymbol{\lambda} = \begin{bmatrix}\boldsymbol{\lambda}_{\mathcal{I}}\\ \boldsymbol{\lambda}_{\mathcal{E}}\end{bmatrix}, \quad \boldsymbol{\lambda}\in\mathbb{R}^m, \quad \boldsymbol{\lambda}_{\mathcal{E}}\in\mathbb{R}^{m-p},$$

and denoting

$$\mathbf{g} = \mathbf{g}(\mathbf{x},\boldsymbol{\lambda}_{\mathcal{E}}) = \nabla_{\mathbf{x}} L_0(\mathbf{x},\boldsymbol{\lambda}_{\mathcal{E}}) = \mathsf{A}\mathbf{x} - \mathbf{b} + \mathsf{B}_E^T\boldsymbol{\lambda}_{\mathcal{E}},$$

we get

$$\nabla_{\mathbf{x}} L_0(\mathbf{x},\boldsymbol{\lambda}) = \mathsf{A}\mathbf{x} - \mathbf{b} + \mathsf{B}_E^T\boldsymbol{\lambda}_{\mathcal{E}} + \mathsf{B}_I^T\boldsymbol{\lambda}_{\mathcal{I}} = \mathbf{g} + \mathsf{B}_I^T\boldsymbol{\lambda}_{\mathcal{I}} = \mathbf{g} - \begin{bmatrix}\mathsf{I}\\ \mathsf{O}\end{bmatrix}\boldsymbol{\lambda}_{\mathcal{I}},$$

so that by Theorem 2.1 the KKT conditions for problem (2.79) read

$$\mathbf{g}_{\mathcal{I}} \geq \mathbf{o}, \quad \mathbf{g}_{\mathcal{R}} = \mathbf{o}, \quad \text{and} \quad \mathbf{g}_{\mathcal{I}}^T(\mathbf{x}_{\mathcal{I}} - \boldsymbol{\ell}_I) = 0. \tag{2.80}$$

Thus the KKT conditions for bound and equality constrained problems can be conveniently expressed by means of the Lagrangian function for equality constrained problems. The Lagrange multipliers for the inequality constraints may be recovered by $\boldsymbol{\lambda}_{\mathcal{I}} = \mathbf{g}_{\mathcal{I}}$. If $\overline{\mathbf{x}}\in\Omega_{BE}$ and $\mathbf{g} = \nabla_{\mathbf{x}} L_0(\mathbf{x},\overline{\boldsymbol{\lambda}}_{\mathcal{E}})$ satisfy (2.80), then $(\overline{\mathbf{x}},\overline{\boldsymbol{\lambda}}_{\mathcal{E}})$ is called the *KKT pair for bound and equality constraints.*

Application of the duality typically results in problem

$$\max_{\mathbf{x}\in\Omega_{BE}} f(\mathbf{x}) \tag{2.81}$$

with $-f$ convex. Since

$$\max_{\mathbf{x}\in\Omega_{BE}} f(\mathbf{x}) = -\min_{\mathbf{x}\in\Omega_{BE}} -f(\mathbf{x}),$$

we get easily that the KKT conditions for (2.81) read

$$\mathbf{g}_{\mathcal{I}} \leq \mathbf{o}, \quad \mathbf{g}_{\mathcal{R}} = \mathbf{o}, \quad \text{and} \quad \mathbf{g}_{\mathcal{I}}^T(\mathbf{x}_{\mathcal{I}} - \boldsymbol{\ell}_I) = 0. \tag{2.82}$$

The analysis presented above covers also the *partially bound constrained problem* to find

$$\min_{\mathbf{x}\in\Omega_B} f(\mathbf{x}), \quad \Omega_B = \{\mathbf{x} \in \mathbb{R}^n : \ \mathbf{x}_{\mathcal{I}} \geq \boldsymbol{\ell}_{\mathcal{I}}\}. \tag{2.83}$$

Skipping the terms concerning the equality constraints and denoting

$$\mathbf{g} = \nabla f(\mathbf{x}) = \mathsf{A}\mathbf{x} - \mathbf{b},$$

we get the *KKT conditions for the bound constrained problem* (2.83) in the form

$$\mathbf{g}_{\mathcal{I}} \geq \mathbf{o}, \quad \mathbf{g}_{\mathcal{R}} = \mathbf{o}, \quad \text{and} \quad \mathbf{g}_{\mathcal{I}}^T(\mathbf{x}_{\mathcal{I}} - \boldsymbol{\ell}_{\mathcal{I}}) = 0. \tag{2.84}$$

As above, we get that the KKT conditions for partially bound constrained problem

$$\max_{\mathbf{x}\in\Omega_B} f(\mathbf{x}), \quad \Omega_B = \{\mathbf{x} \in \mathbb{R}^n : \ \mathbf{x}_{\mathcal{I}} \geq \boldsymbol{\ell}_{\mathcal{I}}\} \tag{2.85}$$

with f convex read

$$\mathbf{g}_{\mathcal{I}} \leq \mathbf{o}, \quad \mathbf{g}_{\mathcal{R}} = \mathbf{o}, \quad \text{and} \quad \mathbf{g}_{\mathcal{I}}^T(\mathbf{x}_{\mathcal{I}} - \boldsymbol{\ell}_{\mathcal{I}}) = 0. \tag{2.86}$$

2.6.4 Duality for SPD Hessian and Dependent Constraints

Combining the arguments of the previous subsections, in particular Sects. 2.4.4 and 2.5.7, it is an easy exercise to get the dual and mixed formulations of equality and inequality constrained problems under the conditions that A and B are positive definite and full rank matrices, respectively. Since there are important applications where such assumptions are too restrictive, it is helpful to extend the duality theory without them. We first relax the assumptions on the equality constraints, postponing the discussion of more general cases to the next section.

Duality Relations

Let us consider problem (2.75) with A SPD and B with possibly dependent rows. Let us recall that by Proposition 2.2 our assumptions imply that f is strictly convex; therefore we can apply Proposition 2.5 to get that there is a unique solution $\widehat{\mathbf{x}}$ of (2.75).

Let $(\widehat{\mathbf{x}}, \overline{\boldsymbol{\lambda}})$ be a solution of the bound and equality problem (2.75). Then

$$\overline{\boldsymbol{\lambda}}^T(\mathsf{B}\widehat{\mathbf{x}} - \mathbf{c}) = \overline{\boldsymbol{\lambda}}_{\mathcal{E}}^T[\mathsf{B}\widehat{\mathbf{x}} - \mathbf{c}]_{\mathcal{E}} + \overline{\boldsymbol{\lambda}}_{\mathcal{I}}^T[\mathsf{B}\widehat{\mathbf{x}} - \mathbf{c}]_{\mathcal{I}} = 0,$$

so that

$$f(\widehat{\mathbf{x}}) = L_0(\widehat{\mathbf{x}}, \overline{\boldsymbol{\lambda}}). \tag{2.87}$$

Next observe that if $\overline{\boldsymbol{\lambda}}$ is a vector of the Lagrange multipliers of the solution, then $\widehat{\mathbf{x}}$ is fully determined by the second condition of (2.78) which reads

$$\nabla_{\mathbf{x}} L_0(\mathbf{x}, \overline{\boldsymbol{\lambda}}) = \mathbf{o}.$$

Since L_0 is strictly convex, the latter is the gradient condition for the unconstrained minimizer of L_0 with respect to $\mathbf{x}$; therefore

$$L_0(\widehat{\mathbf{x}}, \overline{\boldsymbol{\lambda}}) = \min_{\mathbf{x}\in\mathbb{R}^n} L_0(\mathbf{x}, \overline{\boldsymbol{\lambda}}). \tag{2.88}$$

Recalling the definition of the dual function

$$\Theta(\boldsymbol{\lambda}) = \min_{\mathbf{x}\in\mathbb{R}^n} L_0(\mathbf{x}, \boldsymbol{\lambda})$$

and using (2.87) and (2.88), we get

$$f(\widehat{\mathbf{x}}) = L_0(\widehat{\mathbf{x}}, \overline{\boldsymbol{\lambda}}) = \Theta(\overline{\boldsymbol{\lambda}}). \tag{2.89}$$

Dual and Saddle Point Problems

Let us first present the analysis which enhances our earlier results and admits the constraint matrices with possibly dependent rows.

Proposition 2.20. *Let the quadratic function f and the feasible set $\Omega_{IE} \neq \emptyset$ be defined by the matrices A, B, the vectors $\mathbf{b}, \mathbf{c}$, and the index sets $\mathcal{I}, \mathcal{E}$ from the definition of problem (2.75). We assume that A is SPD, but we admit that B is not necessarily a full row rank matrix, $\mathcal{I} = \emptyset$, or $\mathcal{E} = \emptyset$. Let*

$$\Theta(\boldsymbol{\lambda}) = -\frac{1}{2}\boldsymbol{\lambda}^T\mathsf{B}\mathsf{A}^{-1}\mathsf{B}^T\boldsymbol{\lambda} + \boldsymbol{\lambda}^T(\mathsf{B}\mathsf{A}^{-1}\mathbf{b} - \mathbf{c}) - \frac{1}{2}\mathbf{b}^T\mathsf{A}^{-1}\mathbf{b} \tag{2.90}$$

denote the dual function.

Then the following statements are equivalent:
(i) $(\widehat{\mathbf{x}}, \overline{\boldsymbol{\lambda}})$ *is a KKT pair for problem (2.75).*
(ii) $\widehat{\mathbf{x}}$ *is a unique solution of the primal problem (2.75) and* $\overline{\boldsymbol{\lambda}}$ *is a solution of the dual problem*

$$\max_{\boldsymbol{\lambda} \in \Omega_B} \Theta(\boldsymbol{\lambda}), \qquad \Omega_B = \{\boldsymbol{\lambda} \in \mathbb{R}^m : \ \boldsymbol{\lambda}_{\mathcal{I}} \geq \mathbf{o}\}. \tag{2.91}$$

(iii) $(\widehat{\mathbf{x}}, \overline{\boldsymbol{\lambda}}) \in \mathbb{R}^n \times \mathbb{R}^m$ *with* $\overline{\boldsymbol{\lambda}}_{\mathcal{I}} \geq \mathbf{o}$ *is a saddle point of* L_0 *in the sense that for any* $\mathbf{x} \in \mathbb{R}^n$ *and* $\boldsymbol{\lambda} \in \mathbb{R}^m$ *such that* $\boldsymbol{\lambda}_{\mathcal{I}} \geq \mathbf{o}$*,*

$$L_0(\widehat{\mathbf{x}}, \boldsymbol{\lambda}) \leq L_0(\widehat{\mathbf{x}}, \overline{\boldsymbol{\lambda}}) \leq L_0(\mathbf{x}, \overline{\boldsymbol{\lambda}}). \tag{2.92}$$

Proof. (i) $\Rightarrow$ (ii). Let $(\widehat{\mathbf{x}}, \overline{\boldsymbol{\lambda}})$ be a KKT pair for (2.75), so that $\widehat{\mathbf{x}}$ is a unique solution of (2.75), $f(\widehat{\mathbf{x}}) = \Theta(\overline{\boldsymbol{\lambda}})$ by (2.89), and $(\widehat{\mathbf{x}}, \overline{\boldsymbol{\lambda}})$ is by Proposition 2.19 a solution of

$$\boldsymbol{\lambda}_{\mathcal{I}} \geq \mathbf{o}, \tag{2.93}$$
$$\nabla_{\mathbf{x}} L_0(\mathbf{x}, \boldsymbol{\lambda}) = \mathsf{A}\mathbf{x} - \mathbf{b} + \mathsf{B}^T\boldsymbol{\lambda} = \mathbf{o}, \tag{2.94}$$
$$[\nabla_{\boldsymbol{\lambda}} L_0(\mathbf{x}, \boldsymbol{\lambda})]_{\mathcal{I}} = [\mathsf{B}\mathbf{x} - \mathbf{c}]_{\mathcal{I}} \leq \mathbf{o}, \tag{2.95}$$
$$[\nabla_{\boldsymbol{\lambda}} L_0(\mathbf{x}, \boldsymbol{\lambda})]_{\mathcal{E}} = [\mathsf{B}\mathbf{x} - \mathbf{c}]_{\mathcal{E}} = \mathbf{o}, \tag{2.96}$$
$$\boldsymbol{\lambda}_{\mathcal{I}}^T [\mathsf{B}\mathbf{x} - \mathbf{c}]_{\mathcal{I}} = 0. \tag{2.97}$$

In particular, since A is positive definite, it follows that we can use (2.94) to get

$$\widehat{\mathbf{x}} = \mathsf{A}^{-1}(\mathbf{b} - \mathsf{B}^T\overline{\boldsymbol{\lambda}}).$$

After substituting into (2.95)–(2.97), we get

$$[-\mathsf{B}\mathsf{A}^{-1}\mathsf{B}^T\overline{\boldsymbol{\lambda}} + (\mathsf{B}\mathsf{A}^{-1}\mathbf{b} - \mathbf{c})]_{\mathcal{I}} \leq \mathbf{o}, \tag{2.98}$$
$$[-\mathsf{B}\mathsf{A}^{-1}\mathsf{B}^T\overline{\boldsymbol{\lambda}} + (\mathsf{B}\mathsf{A}^{-1}\mathbf{b} - \mathbf{c})]_{\mathcal{E}} = \mathbf{o}, \tag{2.99}$$
$$\overline{\boldsymbol{\lambda}}_{\mathcal{I}}^T [-\mathsf{B}\mathsf{A}^{-1}\mathsf{B}^T\overline{\boldsymbol{\lambda}} + (\mathsf{B}\mathsf{A}^{-1}\mathbf{b} - \mathbf{c})]_{\mathcal{I}} = 0. \tag{2.100}$$

Denoting $\mathbf{g} = \nabla\Theta(\overline{\boldsymbol{\lambda}})$, we can rewrite the relations (2.98)–(2.100) as

$$\mathbf{g}_{\mathcal{I}} \leq \mathbf{o}, \quad \mathbf{g}_{\mathcal{E}} = \mathbf{o}, \quad \text{and} \quad \overline{\boldsymbol{\lambda}}_{\mathcal{I}}^T \mathbf{g}_{\mathcal{I}} = 0. \tag{2.101}$$

Comparing (2.101) with the KKT conditions (2.86) for the partially bound constrained problem (2.85), we conclude that (2.101) are the KKT conditions for (2.91). Since $\overline{\boldsymbol{\lambda}}_{\mathcal{I}} \geq \mathbf{o}$ by (2.93), we have thus proved that $\overline{\boldsymbol{\lambda}}$ is a feasible vector for problem (2.91) which satisfies the related KKT conditions. Recalling that A is positive definite, so that $\mathsf{B}\mathsf{A}^{-1}\mathsf{B}^T$ is positive semidefinite, we conclude that $\overline{\boldsymbol{\lambda}}$ solves (2.91).

(ii) $\Rightarrow$ (iii). Let $\widehat{\mathbf{x}}$ be a unique solution of the primal problem (2.75) and let $\overline{\boldsymbol{\lambda}}$ be a solution of the dual problem (2.91), so that $f(\widehat{\mathbf{x}}) = \Theta(\overline{\boldsymbol{\lambda}})$ by (2.89). Then $\widehat{\mathbf{x}}$ is feasible, i.e.,

$$[\mathsf{B}\widehat{\mathbf{x}} - \mathbf{c}]_{\mathcal{E}} = \mathbf{o}, \quad [\mathsf{B}\widehat{\mathbf{x}} - \mathbf{c}]_{\mathcal{I}} \leq \mathbf{o},$$

and

$$\sup_{\boldsymbol{\lambda}_{\mathcal{I}} \geq \mathbf{o}} L_0(\widehat{\mathbf{x}}, \boldsymbol{\lambda}) = f(\widehat{\mathbf{x}}) + \sup_{\boldsymbol{\lambda}_{\mathcal{I}} \geq \mathbf{o}} \overline{\boldsymbol{\lambda}}^T (\mathsf{B}\widehat{\mathbf{x}} - \mathbf{c}) = f(\widehat{\mathbf{x}}).$$

Thus

$$\Theta(\overline{\boldsymbol{\lambda}}) = \min_{\mathbf{x} \in \mathbb{R}^n} L_0(\mathbf{x}, \overline{\boldsymbol{\lambda}}) \leq L_0(\widehat{\mathbf{x}}, \overline{\boldsymbol{\lambda}}) \leq \sup_{\boldsymbol{\lambda}_{\mathcal{I}} \geq \mathbf{o}} L_0(\widehat{\mathbf{x}}, \boldsymbol{\lambda}) = f(\widehat{\mathbf{x}}).$$

Since $f(\widehat{\mathbf{x}}) = \Theta(\overline{\boldsymbol{\lambda}})$, it follows that

$$\sup_{\boldsymbol{\lambda}_{\mathcal{I}} \geq \mathbf{o}} L_0(\widehat{\mathbf{x}}, \boldsymbol{\lambda}) = L_0(\widehat{\mathbf{x}}, \overline{\boldsymbol{\lambda}}) = \min_{\mathbf{x} \in \mathbb{R}^n} L_0(\mathbf{x}, \overline{\boldsymbol{\lambda}})$$

and $(\widehat{\mathbf{x}}, \overline{\boldsymbol{\lambda}})$ is the solution of the saddle point problem (2.92).
(iii) $\Rightarrow$ (i). Let us now assume that there is $(\widehat{\mathbf{x}}, \overline{\boldsymbol{\lambda}})$, $\overline{\boldsymbol{\lambda}}_{\mathcal{I}} \geq 0$, such that (2.92) holds for any $\mathbf{x} \in \mathbb{R}^n$ and $\boldsymbol{\lambda} \in \mathbb{R}^m$, $\boldsymbol{\lambda}_{\mathcal{I}} \geq \mathbf{o}$. To show that $(\widehat{\mathbf{x}}, \overline{\boldsymbol{\lambda}})$ is a KKT pair for (2.75), notice that for any $\mathbf{x} \in \mathbb{R}^n$

$$\sup_{\boldsymbol{\lambda}_{\mathcal{I}} \geq \mathbf{o}} L_0(\mathbf{x}, \boldsymbol{\lambda}) = \infty \text{ for } \mathbf{x} \notin \Omega_{IE}$$

and

$$\sup_{\boldsymbol{\lambda}_{\mathcal{I}} \geq \mathbf{o}} L_0(\mathbf{x}, \boldsymbol{\lambda}) = f(\mathbf{x}) \text{ for } \mathbf{x} \in \Omega_{IE}.$$

Since by the assumptions

$$L_0(\widehat{\mathbf{x}}, \boldsymbol{\lambda}) \leq L_0(\widehat{\mathbf{x}}, \overline{\boldsymbol{\lambda}})$$

for any $\boldsymbol{\lambda} \in \mathbb{R}^m$, $\boldsymbol{\lambda}_{\mathcal{I}} \geq \mathbf{o}$, it follows that $\widehat{\mathbf{x}} \in \Omega_{IE}$ and

$$f(\widehat{\mathbf{x}}) = L_0(\widehat{\mathbf{x}}, \overline{\boldsymbol{\lambda}}).$$

The feasibility of $\widehat{\mathbf{x}}$ and the latter equation imply the complementarity condition

$$\overline{\boldsymbol{\lambda}}_{\mathcal{I}}^T [\mathsf{B}\widehat{\mathbf{x}} - \mathbf{c}]_{\mathcal{I}} = 0.$$

Taking into account the right saddle point inequality in (2.92), we get that

$$L_0(\widehat{\mathbf{x}}, \overline{\boldsymbol{\lambda}}) = \min_{\mathbf{x} \in \mathbb{R}^n} L_0(\mathbf{x}, \overline{\boldsymbol{\lambda}}).$$

Thus $\widehat{\mathbf{x}}$ is the unconstrained minimizer of $L_0(\mathbf{x}, \overline{\boldsymbol{\lambda}})$ with respect to $\mathbf{x}$. Since L_0 is a convex function of $\mathbf{x}$, we conclude that $\nabla_{\mathbf{x}} L_0(\widehat{\mathbf{x}}, \overline{\boldsymbol{\lambda}}) = \mathbf{o}$ and $(\widehat{\mathbf{x}}, \overline{\boldsymbol{\lambda}})$ is a KKT pair for problem (2.75). □

2.7 Duality for SPS Hessian and Dependent Constraints

Now let us examine what happens when the matrices A and B in the definition of the problem (2.75) are SPS and rank deficient, respectively. A special case of (2.75) is a linear programming problem with $\mathsf{A} = \mathsf{O}$.

2.7.1 Constrained Dual Problem

First observe that if A is only SPS and $\mathbf{b} \neq \mathbf{o}$, then the cost function f need not be bounded from below. Thus $-\infty$ can be in the range of the dual function Θ. We resolve this problem by keeping Θ quadratic at the cost of introducing equality constraints. An alternative development of duality for special semicoercive QP problems can be found in the papers by Dorn [56, 57].

Proposition 2.21. *Let matrices* A, B, *vectors* $\mathbf{b}, \mathbf{c}$, *and index sets* $\mathcal{I}, \mathcal{E}$ *be those from the definition of problem (2.75) with* A *positive semidefinite and* $\Omega_{IE} \neq \emptyset$. *Let* $\mathsf{R} \in \mathbb{R}^{n \times d}$ *be a full rank matrix such that*

$$\mathrm{Im}\mathsf{R} = \mathrm{Ker}\mathsf{A},$$

let A^+ *denote a symmetric positive semidefinite generalized inverse of* A, *and let*

$$\Theta(\boldsymbol{\lambda}) = -\frac{1}{2}\boldsymbol{\lambda}^T \mathsf{B}\mathsf{A}^+\mathsf{B}^T\boldsymbol{\lambda} + \boldsymbol{\lambda}^T(\mathsf{B}\mathsf{A}^+\mathbf{b} - \mathbf{c}) - \frac{1}{2}\mathbf{b}^T\mathsf{A}^+\mathbf{b}. \tag{2.102}$$

Then the following statements hold:
(i) If $(\overline{\mathbf{x}}, \overline{\boldsymbol{\lambda}})$ *is a KKT pair for (2.75), then* $\overline{\boldsymbol{\lambda}}$ *is a solution of*

$$\max_{\boldsymbol{\lambda} \in \Omega_{BE}} \Theta(\boldsymbol{\lambda}), \quad \Omega_{BE} = \{\boldsymbol{\lambda} \in \mathbb{R}^m : \ \boldsymbol{\lambda}_{\mathcal{I}} \geq \mathbf{o}, \ \mathsf{R}^T\mathsf{B}^T\boldsymbol{\lambda} = \mathsf{R}^T\mathbf{b}\}. \tag{2.103}$$

Moreover, there is $\overline{\boldsymbol{\alpha}} \in \mathbb{R}^d$ *such that* $(\overline{\boldsymbol{\lambda}}, \overline{\boldsymbol{\alpha}})$ *is a KKT pair for problem (2.103) and*

$$\overline{\mathbf{x}} = \mathsf{A}^+(\mathbf{b} - \mathsf{B}^T\overline{\boldsymbol{\lambda}}) + \mathsf{R}\overline{\boldsymbol{\alpha}}. \tag{2.104}$$

(ii) If $(\overline{\boldsymbol{\lambda}}, \overline{\boldsymbol{\alpha}})$ *is a KKT pair for problem (2.103), then* $\overline{\mathbf{x}}$ *defined by (2.104) is a solution of the equality and inequality constrained problem (2.75).*
(iii) If $(\overline{\mathbf{x}}, \overline{\boldsymbol{\lambda}})$ *is a KKT pair for problem (2.75), then*

$$f(\overline{\mathbf{x}}) = \Theta(\overline{\boldsymbol{\lambda}}). \tag{2.105}$$

Proof. (i) Assume that $(\overline{\mathbf{x}}, \overline{\boldsymbol{\lambda}})$ is a KKT pair for (2.75), so that $(\overline{\mathbf{x}}, \overline{\boldsymbol{\lambda}})$ is by Theorem 2.1 a solution of

$$\boldsymbol{\lambda}_{\mathcal{I}} \geq \mathbf{o}, \tag{2.106}$$

$$\nabla_{\mathbf{x}} L_0(\mathbf{x}, \boldsymbol{\lambda}) = \mathsf{A}\mathbf{x} - \mathbf{b} + \mathsf{B}^T\boldsymbol{\lambda} = \mathbf{o}, \tag{2.107}$$

$$[\nabla_{\boldsymbol{\lambda}} L_0(\mathbf{x}, \boldsymbol{\lambda})]_{\mathcal{I}} = [\mathsf{B}\mathbf{x} - \mathbf{c}]_{\mathcal{I}} \leq \mathbf{o}, \tag{2.108}$$

$$[\nabla_{\boldsymbol{\lambda}} L_0(\mathbf{x}, \boldsymbol{\lambda})]_{\mathcal{E}} = [\mathsf{B}\mathbf{x} - \mathbf{c}]_{\mathcal{E}} = \mathbf{o}, \tag{2.109}$$

$$\boldsymbol{\lambda}_{\mathcal{I}}^T[\mathsf{B}\mathbf{x} - \mathbf{c}]_{\mathcal{I}} = 0. \tag{2.110}$$

Notice that given a vector $\boldsymbol{\lambda} \in \mathbb{R}^m$, we can express the condition

$$\mathbf{b} - \mathsf{B}^T\boldsymbol{\lambda} \in \mathrm{Im}\mathsf{A},$$

which guarantees solvability of (2.107) with respect to $\mathbf{x}$, conveniently as

$$\mathsf{R}^T(\mathsf{B}^T\boldsymbol{\lambda} - \mathbf{b}) = \mathbf{o}. \tag{2.111}$$

If the latter condition is satisfied, then we can use any symmetric left generalized inverse A^+ to find all the solutions of (2.107) with respect to $\mathbf{x}$ in the form

$$\mathbf{x}(\boldsymbol{\lambda}, \boldsymbol{\alpha}) = \mathsf{A}^+(\mathbf{b} - \mathsf{B}^T\boldsymbol{\lambda}) + \mathsf{R}\boldsymbol{\alpha}, \ \boldsymbol{\alpha} \in \mathbb{R}^d,$$

where d is the dimension of Ker$\mathsf{A}$. After substituting for $\mathbf{x}$ into (2.108)–(2.110), we get

$$[-\mathsf{B}\mathsf{A}^+\mathsf{B}^T\boldsymbol{\lambda} + (\mathsf{B}\mathsf{A}^+\mathbf{b} - \mathbf{c}) + \mathsf{B}\mathsf{R}\boldsymbol{\alpha}]_{\mathcal{I}} \leq \mathbf{o}, \tag{2.112}$$

$$[-\mathsf{B}\mathsf{A}^+\mathsf{B}^T\boldsymbol{\lambda} + (\mathsf{B}\mathsf{A}^+\mathbf{b} - \mathbf{c}) + \mathsf{B}\mathsf{R}\boldsymbol{\alpha}]_{\mathcal{E}} = \mathbf{o}, \tag{2.113}$$

$$\boldsymbol{\lambda}_{\mathcal{I}}^T[-\mathsf{B}\mathsf{A}^+\mathsf{B}^T\boldsymbol{\lambda} + (\mathsf{B}\mathsf{A}^+\mathbf{b} - \mathbf{c}) + \mathsf{B}\mathsf{R}\boldsymbol{\alpha}]_{\mathcal{I}} = 0. \tag{2.114}$$

The formulae in (2.112)–(2.114) look like something that we have already seen. Indeed, introducing the vector of Lagrange multipliers $\boldsymbol{\alpha}$ for (2.111) and denoting

$$\begin{aligned} \Lambda(\boldsymbol{\lambda}, \boldsymbol{\alpha}) &= \Theta(\boldsymbol{\lambda}) + \boldsymbol{\alpha}^T(\mathsf{R}^T\mathsf{B}^T\boldsymbol{\lambda} - \mathsf{R}^T\mathbf{b}) \\ &= -\frac{1}{2}\boldsymbol{\lambda}^T\mathsf{B}\mathsf{A}^+\mathsf{B}^T\boldsymbol{\lambda} + \boldsymbol{\lambda}^T(\mathsf{B}\mathsf{A}^+\mathbf{b} - \mathbf{c}) - \frac{1}{2}\mathbf{b}^T\mathsf{A}^+\mathbf{b} \\ &\quad + \boldsymbol{\alpha}^T(\mathsf{R}^T\mathsf{B}^T\boldsymbol{\lambda} - \mathsf{R}^T\mathbf{b}), \end{aligned}$$

$$\mathbf{g} = \nabla_{\boldsymbol{\lambda}}\Lambda(\boldsymbol{\lambda}, \boldsymbol{\alpha}) = -\mathsf{B}\mathsf{A}^+\mathsf{B}^T\boldsymbol{\lambda} + (\mathsf{B}\mathsf{A}^+\mathbf{b} - \mathbf{c}) + \mathsf{B}\mathsf{R}\boldsymbol{\alpha},$$

we can rewrite the relations (2.112)–(2.114) as

$$\mathbf{g}_{\mathcal{I}} \leq \mathbf{o}, \quad \mathbf{g}_{\mathcal{E}} = \mathbf{o}, \quad \text{and} \quad \boldsymbol{\lambda}_{\mathcal{I}}^T\mathbf{g}_{\mathcal{I}} = 0. \tag{2.115}$$

Comparing (2.115) with the KKT conditions (2.82) for the bound and equality constrained problem (2.81), we conclude that (2.115) are the KKT conditions for

$$\max \Theta(\boldsymbol{\lambda}) \quad \text{subject to} \quad \mathsf{R}^T\mathsf{B}^T\boldsymbol{\lambda} - \mathsf{R}^T\mathbf{b} = \mathbf{o} \quad \text{and} \quad \boldsymbol{\lambda}_{\mathcal{I}} \geq \mathbf{o}. \tag{2.116}$$

We have thus proved that if $(\overline{\mathbf{x}}, \overline{\boldsymbol{\lambda}})$ solves (2.106)–(2.110), then $\overline{\boldsymbol{\lambda}}$ is a feasible vector for problem (2.116) which satisfies the related KKT conditions. Recalling that A^+ is by the assumption symmetric positive semidefinite, so that $\mathsf{BA}^+\mathsf{B}^T$ is also positive semidefinite, we conclude that $\overline{\boldsymbol{\lambda}}$ solves (2.103). Moreover, we have shown that any solution $\overline{\mathbf{x}}$ can be obtained in the form (2.104) with a KKT pair $(\overline{\boldsymbol{\lambda}}, \overline{\boldsymbol{\alpha}})$, where $\overline{\boldsymbol{\alpha}}$ is a vector of the Lagrange multipliers for (2.103).

(ii) Let $(\overline{\boldsymbol{\lambda}}, \overline{\boldsymbol{\alpha}})$ be a KKT pair for problem (2.103), so that $(\overline{\boldsymbol{\lambda}}, \overline{\boldsymbol{\alpha}})$ satisfies (2.111)–(2.114) and $\overline{\boldsymbol{\lambda}}_{\mathcal{I}} \geq \mathbf{o}$. If we denote

$$\overline{\mathbf{x}} = \mathsf{A}^+(\mathbf{b} - \mathsf{B}^T\overline{\boldsymbol{\lambda}}) + \mathsf{R}\overline{\boldsymbol{\alpha}},$$

we can use (2.112)–(2.114) to verify directly that $\overline{\mathbf{x}}$ is feasible and that $(\overline{\mathbf{x}}, \overline{\boldsymbol{\lambda}})$ satisfies the complementarity conditions, respectively. Finally, using (2.111), we get that there is $\mathbf{y} \in \mathbb{R}^n$ such that

$$\mathbf{b} - \mathsf{B}^T\overline{\boldsymbol{\lambda}} = \mathsf{A}\mathbf{y}.$$

Thus

$$\begin{aligned} \mathsf{A}\overline{\mathbf{x}} - \mathbf{b} + \mathsf{B}^T\overline{\boldsymbol{\lambda}} &= \mathsf{A}(\mathsf{A}^+(\mathbf{b} - \mathsf{B}^T\overline{\boldsymbol{\lambda}}) + \mathsf{R}\overline{\boldsymbol{\alpha}}) - \mathbf{b} + \mathsf{B}^T\overline{\boldsymbol{\lambda}} \\ &= \mathsf{AA}^+\mathsf{A}\mathbf{y} - \mathbf{b} + \mathsf{B}^T\overline{\boldsymbol{\lambda}} = \mathbf{b} - \mathsf{B}^T\overline{\boldsymbol{\lambda}} - \mathbf{b} + \mathsf{B}^T\overline{\boldsymbol{\lambda}} = \mathbf{o}, \end{aligned}$$

which proves that $(\overline{\mathbf{x}}, \overline{\boldsymbol{\lambda}})$ is a KKT pair for (2.75).

(iii) Let $(\overline{\mathbf{x}}, \overline{\boldsymbol{\lambda}})$ be a KKT pair for (2.75). Using the feasibility condition (2.108) and the complementarity condition (2.110), we get

$$\overline{\boldsymbol{\lambda}}^T(\mathsf{B}\overline{\mathbf{x}} - \mathbf{c}) = \overline{\boldsymbol{\lambda}}_{\mathcal{E}}^T[\mathsf{B}\overline{\mathbf{x}} - \mathbf{c}]_{\mathcal{E}} + \overline{\boldsymbol{\lambda}}_{\mathcal{I}}^T[\mathsf{B}\overline{\mathbf{x}} - \mathbf{c}]_{\mathcal{I}} = 0.$$

Hence

$$f(\overline{\mathbf{x}}) = f(\overline{\mathbf{x}}) + \overline{\boldsymbol{\lambda}}^T(\mathsf{B}\overline{\mathbf{x}} - \mathbf{c}) = L_0(\overline{\mathbf{x}}, \overline{\boldsymbol{\lambda}}).$$

Next recall that if $(\overline{\mathbf{x}}, \overline{\boldsymbol{\lambda}})$ is a KKT pair, then

$$\nabla_{\mathbf{x}} L_0(\overline{\mathbf{x}}, \overline{\boldsymbol{\lambda}}) = \mathbf{o}.$$

Since L_0 is convex, the latter is the gradient condition for the unconstrained minimizer of L_0 with respect to $\mathbf{x}$; therefore

$$L_0(\overline{\mathbf{x}}, \overline{\boldsymbol{\lambda}}) = \min_{\mathbf{x} \in \mathbb{R}^n} L_0(\mathbf{x}, \overline{\boldsymbol{\lambda}}) = \Theta(\overline{\boldsymbol{\lambda}}).$$

Thus

$$f(\overline{\mathbf{x}}) = L_0(\overline{\mathbf{x}}, \overline{\boldsymbol{\lambda}}) = \Theta(\overline{\boldsymbol{\lambda}}).$$

□

The result which we have just proved is useful for reformulation of the convex quadratic problems with general inequality constraints to the problems with bound and equality constraints. See Chaps. 13 and 14 for examples.

Since the constant term is not essential in our applications and we formulate our algorithms for minimization problems, we shall consider the function

$$\theta(\boldsymbol{\lambda}) = -\Theta(\boldsymbol{\lambda}) - \frac{1}{2}\mathbf{b}^T\mathsf{A}^+\mathbf{b} = \frac{1}{2}\boldsymbol{\lambda}^T\mathsf{B}\mathsf{A}^+\mathsf{B}^T\boldsymbol{\lambda} - \boldsymbol{\lambda}^T(\mathsf{B}\mathsf{A}^+\mathbf{b} - \mathbf{c}), \qquad (2.117)$$

so that

$$\arg\min_{\lambda\in\Omega_{BE}} \theta(\boldsymbol{\lambda}) = \arg\max_{\lambda\in\Omega_{BE}} \Theta(\boldsymbol{\lambda}).$$

2.7.2 Uniqueness of a KKT Pair

We shall complete our exposition of duality by formulating the results concerning the uniqueness of the solution for the *constrained dual problem*

$$\min_{\boldsymbol{\lambda}\in\Omega_{BE}} \theta(\boldsymbol{\lambda}), \quad \Omega_{BE} = \{\boldsymbol{\lambda} \in \mathbb{R}^m : \boldsymbol{\lambda}_{\mathcal{I}} \geq \mathbf{o}, \mathsf{R}^T\mathsf{B}^T\boldsymbol{\lambda} = \mathsf{R}^T\mathbf{b}\}, \qquad (2.118)$$

where θ is defined by (2.117).

Proposition 2.22. *Let the matrices* A, B, *the vectors* $\mathbf{b}, \mathbf{c}$, *and the index sets* $\mathcal{I}, \mathcal{E}$ *be those from the definition of problem (2.75) with* A *positive semidefinite,* $\Omega_{IE} \neq \emptyset$, *and* $\Omega_{BE} \neq \emptyset$. *Let* $\mathsf{R} \in \mathbb{R}^{n\times d}$ *be a full rank matrix such that*

$$\mathrm{Im}\mathsf{R} = \mathrm{Ker}\mathsf{A}.$$

Then the following statements hold:
(i) If B^T *and* BR *are full column rank matrices, then there is a unique solution* $\widehat{\boldsymbol{\lambda}}$ *of problem (2.118).*
(ii) If $\widehat{\boldsymbol{\lambda}}$ *is a unique solution of the constrained dual problem (2.118),*

$$\mathcal{A} = \{i : [\boldsymbol{\lambda}]_i > 0\} \cup \mathcal{E},$$

and $\mathsf{B}_{\mathcal{A}*}\mathsf{R}$ *is a full column rank matrix, then there is a unique triple* $(\widehat{\mathbf{x}}, \widehat{\boldsymbol{\lambda}}, \widehat{\boldsymbol{\alpha}})$ *such that* $(\widehat{\mathbf{x}}, \widehat{\boldsymbol{\lambda}})$ *solves the primal problem (2.75) and* $(\widehat{\boldsymbol{\lambda}}, \widehat{\boldsymbol{\alpha}})$ *solves the constrained dual problem (2.118). If* $\widehat{\boldsymbol{\lambda}}$ *is known, then*

$$\widehat{\boldsymbol{\alpha}} = (\mathsf{R}^T\mathsf{B}_{\mathcal{A}*}^T\mathsf{B}_{\mathcal{A}*}\mathsf{R})^{-1}\mathsf{R}^T\mathsf{B}_{\mathcal{A}*}^T\left(\mathsf{B}_{\mathcal{A}*}\mathsf{A}^+\mathsf{B}^T\widehat{\boldsymbol{\lambda}} - (\mathsf{B}_{\mathcal{A}*}\mathsf{A}^+\mathbf{b} - \mathbf{c}_{\mathcal{A}})\right) \qquad (2.119)$$

and

$$\widehat{\mathbf{x}} = \mathsf{A}^{+}(\mathbf{b} - \mathsf{B}^T\widehat{\boldsymbol{\lambda}}) + \mathsf{R}\widehat{\boldsymbol{\alpha}}. \tag{2.120}$$

(iii) If B^T *and* $\mathsf{B}_{\mathcal{E}*}\mathsf{R}$ *are full column rank matrices, then there is a unique triple* $(\widehat{\mathbf{x}}, \widehat{\boldsymbol{\lambda}}, \widehat{\boldsymbol{\alpha}})$ *such that* $(\widehat{\mathbf{x}}, \widehat{\boldsymbol{\lambda}})$ *solves the primal problem (2.75) and* $(\widehat{\boldsymbol{\lambda}}, \widehat{\boldsymbol{\alpha}})$ *solves the constrained dual problem (2.118).*

Proof. (i) Let B^T and BR be full column rank matrices. To show that there is a unique solution of (2.118), we examine the Hessian $\mathsf{BA}^{+}\mathsf{B}^T$ of θ. Let $\mathsf{R}^T\mathsf{B}^T\boldsymbol{\lambda} = \mathbf{o}$ and $\mathsf{BA}^{+}\mathsf{B}^T\boldsymbol{\lambda} = \mathbf{o}$. Using the definition of R, it follows that $\mathsf{B}^T\boldsymbol{\lambda} \in \mathrm{Im}\mathsf{A}$. Hence there is $\boldsymbol{\mu} \in \mathbb{R}^n$ such that

$$\mathsf{B}^T\boldsymbol{\lambda} = \mathsf{A}\boldsymbol{\mu}$$

and

$$\boldsymbol{\mu}^T\mathsf{A}\boldsymbol{\mu} = \boldsymbol{\mu}^T\mathsf{AA}^{+}\mathsf{A}\boldsymbol{\mu} = \boldsymbol{\lambda}^T\mathsf{BA}^{+}\mathsf{B}^T\boldsymbol{\lambda} = 0.$$

Thus $\boldsymbol{\mu} \in \mathrm{Ker}\mathsf{A}$ and

$$\mathsf{B}^T\boldsymbol{\lambda} = \mathsf{A}\boldsymbol{\mu} = \mathbf{o}.$$

Since we assume that B^T has independent columns, we conclude that $\boldsymbol{\lambda} = \mathbf{o}$. We have thus proved that the restriction of $\mathsf{BA}^{+}\mathsf{B}^T$ to $\mathrm{Ker}(\mathsf{R}^T\mathsf{B}^T)$ is positive definite, so that $\theta|\mathrm{Ker}\mathsf{R}^T\mathsf{B}^T$ is by Proposition 2.5 strictly convex, and by Corollary 2.3 it is strictly convex on

$$\mathcal{U} = \{\boldsymbol{\lambda} \in \mathbb{R}^m : \ \mathsf{R}^T\mathsf{B}^T\boldsymbol{\lambda} = \mathsf{R}^T\mathbf{b}\}.$$

Since $\Omega_{BE} \neq \emptyset$ and $\Omega_{BE} \subseteq \mathcal{U}$, we have that θ is strictly convex on Ω_{BE}, and it follows by Proposition 2.4 that there is a unique solution $\widehat{\boldsymbol{\lambda}}$ of (2.118).
(ii) Let $\widehat{\boldsymbol{\lambda}}$ be a unique solution of problem (2.118). Since the solution satisfies the related KKT conditions, it follows that there is $\widehat{\boldsymbol{\alpha}}$ such that

$$\mathsf{B}_{\mathcal{A}*}\mathsf{A}^{+}\mathsf{B}^T\widehat{\boldsymbol{\lambda}} - (\mathsf{B}_{\mathcal{A}*}\mathsf{A}^{+}\mathbf{b} - \mathbf{c}_{\mathcal{A}}) - \mathsf{B}_{\mathcal{A}*}\mathsf{R}\widehat{\boldsymbol{\alpha}} = \mathbf{o}.$$

After multiplying on the left by $\mathsf{R}^T\mathsf{B}_{\mathcal{A}*}^T$ and simple manipulations, we get (2.119). The inverse exists and the solution $\widehat{\boldsymbol{\alpha}}$ is unique due to the uniqueness of $\widehat{\boldsymbol{\lambda}}$ and the assumption on the full column rank of $\mathsf{B}_{\mathcal{A}*}\mathsf{R}$.
(iii) If B^T and $\mathsf{B}_{\mathcal{E}*}\mathsf{R}$ are full column rank matrices, then BR is also a full column rank matrix. Hence there is a unique solution $\widehat{\boldsymbol{\lambda}}$ of problem (2.118) by (i). Since $\mathcal{E} \subseteq \mathcal{A}$ and $\mathsf{B}_{\mathcal{E}*}\mathsf{R}$ has independent columns, it follows that $\mathsf{B}_{\mathcal{A}*}\mathsf{R}$ has also independent columns. Thus we can use (ii) to finish the proof. □

The reconstruction formula (2.119) can be modified in order to work whenever the dual problem has a solution $\overline{\boldsymbol{\lambda}}$. The resulting formula obtained by analysis of the related KKT conditions then reads

$$\overline{\boldsymbol{\alpha}} = (\mathsf{R}^T\mathsf{B}_{\mathcal{A}*}^T\mathsf{B}_{\mathcal{A}*}\mathsf{R})^{+}\mathsf{R}^T\mathsf{B}_{\mathcal{A}*}^T\left(\mathsf{B}_{\mathcal{A}*}\mathsf{A}^{+}\mathsf{B}^T\overline{\boldsymbol{\lambda}} - (\mathsf{B}_{\mathcal{A}*}\mathsf{A}^{+}\mathbf{b} - \mathbf{c}_{\mathcal{A}})\right). \tag{2.121}$$

It is rather evident that if we have the Lagrange multiplier $\boldsymbol{\alpha}$ from the solution of the dual problem (2.103), then there is no reason to use formulae (2.119) or (2.121). Notice that their evaluation can be expensive as it requires assembling and solving a possibly large system of linear equations. However, we can have a solution to the modified dual problem, but not the corresponding Lagrange multiplier $\boldsymbol{\alpha}$. For example, the effective solver can require a modification of the dual cost function (2.117) such as preconditioning or regularization that does not change the solution $\widehat{\boldsymbol{\lambda}}$ but changes the multiplier $\boldsymbol{\alpha}$ of the dual problem. In this case, the formulae (2.119) or (2.121) can be helpful.

If we get a multiplier $\boldsymbol{\alpha}$ corresponding to some modified problem, it is possible that it can be used to evaluate effectively the multiplier $\widehat{\boldsymbol{\alpha}}$ of the original problem (2.118). Such procedure is problem dependent. The effective evaluation of the primal solution using the multiplier of a modified dual problem resulting from the application of the FETI methodology is described in Horák et al. [170].

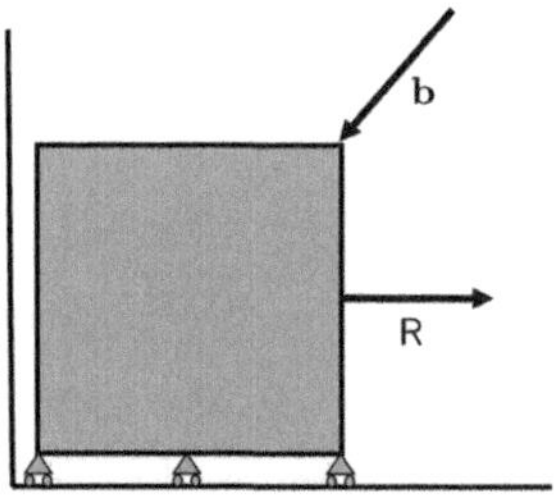

Fig. 2.11. Unique displacement

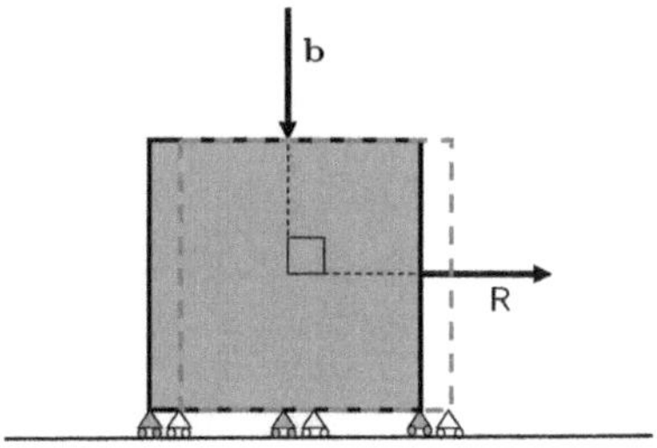

Fig. 2.12. Nonunique displacement

The theory concerning the existence and uniqueness of a solution can be illustrated on a problem to find the displacement $\mathbf{x}$ of an elastic body depicted in Figs. 2.11 and 2.12 under traction $\mathbf{b}$. Assuming that the bodies are discretized by the finite element method, we have a convex QP problem to find a minimum of the discrete quadratic energy function f subject to equality (prescribed displacements) and inequality (nonpenetration) constraints that enforce some limits on the movement of the body. We can distinguish the following cases:

- If the body is fixed to the support, then it does not admit any rigid body motion, f is strictly convex, and the displacements are unique.
- If the body is fixed to the support only in normal direction as in Figs. 2.11 and 2.12, then the existence of a solution depends on the direction of the traction $\mathbf{b}$:

- If the body is not allowed to penetrate the vertical obstacle and the traction $\boldsymbol{b}$ corresponds to Fig. 2.11, then the vector of displacements exists and is unique.
- If the horizontal component of $\boldsymbol{b}$ is zero as in Fig. 2.12, then the solution still exists, but it is not unique.
- If the horizontal component of $\boldsymbol{b}$ in Fig. 2.11 points from the obstacle, then the solution does not exist.
- If the horizontal component of $\boldsymbol{b}$ in Fig. 2.12 is not zero, then the solution does not exist.

If the solution exists and the active constraints are represented by

$$\mathsf{B}_{\mathscr{A}*}\mathbf{x} = \mathbf{o}_{\mathscr{A}}$$

with independent rows, then the corresponding Laplace multipliers, i.e., the reaction forces, are unique. The condition

$$\mathsf{R}^T\mathbf{b} = \mathsf{R}^T\mathsf{B}^T\widehat{\boldsymbol{\lambda}} \tag{2.122}$$

requires that the resulting forces are balanced in the directions of the feasible rigid body motions and $\widehat{\boldsymbol{\lambda}}_{\mathscr{I}} \geq \mathbf{o}$ guarantees that the body is not glued to the obstacle. The condition (2.122) enables the reconstruction of displacements.

If the reaction forces $\mathsf{B}^T\widehat{\boldsymbol{\lambda}}$ determine the components of $\widehat{\boldsymbol{\lambda}}$, then $\widehat{\boldsymbol{\lambda}}$ is uniquely determined by the conditions of equilibrium. Notice that $\mathsf{B}^T\widehat{\boldsymbol{\lambda}}$ can be interpreted as reaction nodal forces and, unlike the Lagrange multipliers, is always uniquely determined by the conditions of equilibrium.

2.8 Separable Quadratic Inequality Constraints

Let us now consider problems with *separable quadratic inequality constraints*

$$\min_{\mathbf{x}\in\Omega_S} f(\mathbf{x}), \quad \Omega_S = \{\mathbf{x} \in \mathbb{R}^n : \mathbf{h}(\mathbf{x}) \leq \mathbf{o}\}. \tag{2.123}$$

Mapping $\mathbf{h} : \mathbb{R}^n \to \mathbb{R}^s$ is defined by its components $h_i : \mathbb{R}^{n_i} \to \mathbb{R}$,

$$h_i(\mathbf{x}) = \mathbf{x}_i^T\mathsf{H}_i\mathbf{x}_i - d_i, \quad i = 1, \ldots, s, \quad {}_i > 0, \tag{2.124}$$

H_i is an SPD matrix, and $\mathbf{x}_i$ are disjoint components of $\mathbf{x}$. The components $\mathbf{x}_i$ are defined by the decomposition of $\mathcal{N} = \{1, \ldots, n\}$ into disjoint subsets $\mathcal{I}_i$, $i = 1, \ldots, s$, so $\mathbf{x}_i = \mathbf{x}_{\mathcal{I}_i}$. We can use $h_i(\mathbf{x}_i) = -1$ to admit unconstrained variables. We also assume that we can evaluate effectively projection on Ω_S.

We use different algorithms for bound constraints and separable quadratic constraints. The bound constraints enjoy a special property—if a bound constraint is active, then the corresponding component of the solution is fully determined. This observation can be used to improve the performance of the algorithms for solving problems with bound constraints. If a more general separable constraint is active in the solution, then the components of the solution are typically not specified.

2.8.1 Feasible Cone and Optimality Conditions

First notice that the functions h_i defined by (2.124) are convex by Proposition 2.2, so that the feasible set

$$\Omega_S = \{\mathbf{x} : h_i(\mathbf{x}_i) \leq 0, \ \mathbf{x} \in \mathbb{R}^n, \ i = 1, \dots, s\}$$

is also convex.

To see that Ω_S is nonempty, let λ_i and $\mathbf{e_i}$ denote any eigenvalue and normalized eigenvector of H_i, so that $\mathsf{H}_i \mathbf{e}_i = \lambda_i \mathbf{e}_i$, and define

$$\mathbf{x}_i = \sqrt{\frac{d_i}{\lambda_i}} \mathbf{e}_i \quad \text{to get} \quad \mathbf{x}_i \mathsf{H}_i \mathbf{x}_i - d_i = 0.$$

The minimizers exist under the assumptions of Proposition 2.5.

We shall characterize the minimizers of f on Ω_S by the *linear cone*

$$\mathscr{L}_{\Omega_S}(\overline{\mathbf{x}}) = \{\mathbf{d} \in \mathbb{R}^n : \mathbf{d}^T \nabla h_i(\overline{\mathbf{x}}) \leq 0, \ i \in \mathscr{A}(\overline{\mathbf{x}})\},$$

which includes feasible decrease directions at $\overline{\mathbf{x}}$. As depicted in Fig. 2.13, if the constraint h is active at the solution $\overline{\mathbf{x}}$, then the gradient of the cost function $\nabla f(\overline{\mathbf{x}})$ points in the opposite direction than the gradient of the constraint $\nabla h(\overline{\mathbf{x}})$, i.e., there is $\lambda \geq 0$ such that

$$\nabla f(\overline{\mathbf{x}}) = -\lambda \nabla h(\overline{\mathbf{x}}).$$

Introducing the *cone of decrease directions at* $\overline{\mathbf{x}}$

$$\mathscr{D}_{\Omega_S}(\overline{\mathbf{x}}) = \{\mathbf{d} \in \mathbb{R}^n : \mathbf{d}^T \nabla f(\overline{\mathbf{x}}) < 0, \ i \in \mathcal{A}(\overline{\mathbf{x}})\},$$

we get a that the minimizer $\overline{\mathbf{x}}$ is fully characterized by the condition

$$\mathscr{D}_{\Omega_S} \cap \mathscr{L}_{\Omega_S} = \emptyset.$$

We conclude that the solution of the problem with separable constraints satisfies

$$\nabla f(\overline{\mathbf{x}}) + \nabla \mathbf{h}(\overline{\mathbf{x}}) \boldsymbol{\lambda} = \mathbf{o}, \qquad \boldsymbol{\lambda} \geq \mathbf{o}, \qquad \boldsymbol{\lambda}^T \mathbf{h}(\overline{\mathbf{x}}) = 0. \tag{2.125}$$

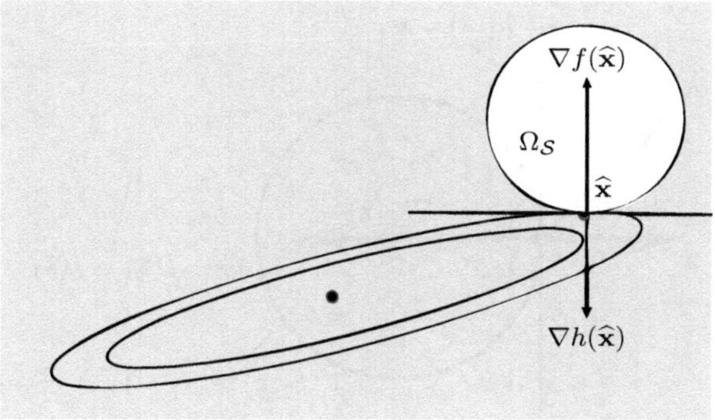

Fig. 2.13. KKT conditions for convex constraints

2.8.2 Separable Inequality and Equality QCQP Problems

In what follows, we shall consider quadratic separable constraints in the form

$$h_i(\mathbf{x}) = (\mathbf{x}_i - \mathbf{y}_i)^T \mathsf{H}_i(\mathbf{x}_i - \mathbf{y}_i) - d_i, \ i = 1, \dots s, \ \mathsf{H}_i \ \text{SPD}, \ d_i > 0, \quad (2.126)$$

where $\mathbf{x}_i$ and $\mathbf{y}_i$ are block components of $\mathbf{x} \in \mathbb{R}^n$ and $\mathbf{y} \in \mathbb{R}^n$. We shall consider elliptic constraints with $\mathsf{H}_i \in \mathbb{R}^2$ and spherical constraints with $\mathsf{H} = \mathsf{H}_1 = a\mathsf{I}$, $a > 0$. The optimization problems with quadratic cost function and separable inequality constraints (2.126) are special cases of Quadratically Constrained Quadratic Programs (QCQP).

Admitting separable constraints increases the scope of the application of our algorithms. In particular, it opens the way to reducing the problems with the non-differentiable term

$$\phi(\mathbf{x}) = f(\mathbf{x}) + j(\mathbf{x}), \quad j(\mathbf{x}) = \sum_{i=1}^{m} \|\mathsf{T}_i \mathbf{x}_i\| = \sum_{i=1}^{m} \max_{\|\boldsymbol{\mu}\| \le 1} \boldsymbol{\mu}^T \mathsf{T}_i \mathbf{x}_i$$

to QCQP problems. The nonsmooth problems arise, e.g., in solving contact problems with friction in mechanics (see, e.g., Chap. 14).

Here, we sum up the conditions satisfied by the solutions of optimization problems with the variables constrained by equality and inequality constraints. More formally, we look for

$$\min_{\mathbf{x} \in \Omega_{IE}} f(\mathbf{x}), \quad \Omega_{IE} = \{\mathbf{x} \in \mathbb{R}^n : [\mathbf{h}(\mathbf{x})]_{\mathcal{I}} \le \mathbf{o}_{\mathcal{I}}, \ [\mathbf{h}(\mathbf{x})]_{\mathcal{E}} = \mathbf{o}_{\mathcal{E}}\}, \quad (2.127)$$

where f is a quadratic function with the SPS Hessian $\mathsf{A} \in \mathbb{R}^{n \times n}$ and the linear term defined by $\mathbf{b} \in \mathbb{R}^n$, $\mathcal{I}$ and $\mathcal{E}$ are disjoint sets of indices which decompose $\{1, \dots, m\}$, the equality constraints are linear, and the inequality constraints are linear or separable convex quadratic.

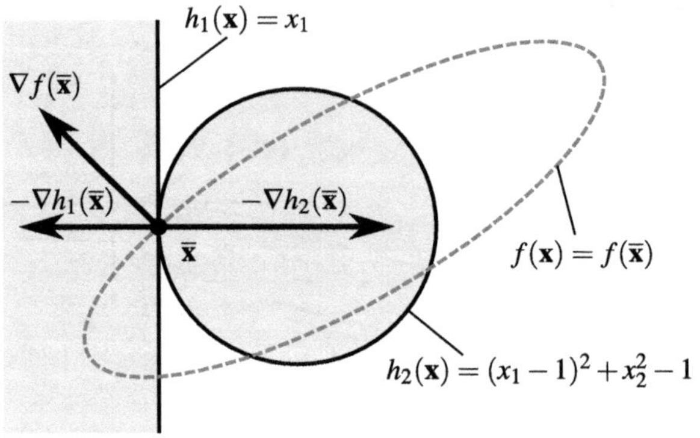

Fig. 2.14. Example of $\Omega_I = \{\overline{\mathbf{x}}\}$, $\mathscr{L}_{\Omega_I}(\overline{\mathbf{x}}) \neq \mathscr{T}_{\Omega_I}(\overline{\mathbf{x}})$

2.8.3 Constraint Qualification

If the inequality constraints h_i that define (2.127) are nonlinear, it can happen that their first-order representation using the gradients ∇h_i is not adequate. The reason is illustrated in Fig. 2.14, where the *linear cone* $\mathscr{L}_{\Omega_I}(\overline{\mathbf{x}})$ of Ω_I at $\overline{\mathbf{x}}$ on the boundary of Ω_I comprises the whole line, while the *tangent cone* $\mathscr{T}_{\Omega_I}(\overline{\mathbf{x}})$ of Ω_I at $\overline{\mathbf{x}}$ on the boundary of Ω_I defined by

$$\mathscr{T}_{\Omega_I}(\overline{\mathbf{x}}) = \{\mathbf{d} \in \mathbb{R}^n : \ \mathbf{d} = \lim_{i\to\infty} \mathbf{d}^i, \ \overline{\mathbf{x}} + \alpha_i \mathbf{d}^i \in \Omega_I, \ \lim_{i\to\infty} \alpha_i = 0, \ \alpha_i > 0\}$$

comprises only one point. To avoid such pathological situations, we shall assume that $\mathscr{L}_{\Omega_I}(\overline{\mathbf{x}}) = \mathscr{T}_{\Omega_I}(\overline{\mathbf{x}})$. This assumption is coined the *Abadie constraint qualification (ACQ)* [153]. The linear constraints satisfy ACQ – we also speak about *linear constraint qualification (LCQ).*

Remark 2.3. We can sometimes check the constraint qualification using additional information about the problem. Recognizing a solution that does not satisfy a nonlinear constraint qualification can be difficult. If the feasible set is empty, it seems reasonable to require that the algorithm finds the 'most feasible solution' that minimizes the feasibility error. See, e.g., Dostál and Beremlijski [76].

2.8.4 Existence and Uniqueness

In our discussion of the existence and uniqueness results for the equality and inequality constrained QP problem (2.127), we restrict our attention to the following proposition, which is a generalization of Proposition 2.19.

Proposition 2.23. *Let the quadratic function f be defined by an SPS matrix* A *and* $\mathbf{b} \in \mathbb{R}^n$. *Let $\Omega_{IE} \neq \emptyset$ be defined by linear equality constraints and linear and separable convex inequality constraints that satisfy Abadie constraints qualification. Let $\mathscr{C}$ denote the cone of recession directions of $\Omega_{\mathcal{IE}}$. Then the following statements hold:*
(i) Problem (2.127) has a solution if and only if

$$\mathbf{d}^T\mathbf{b} \leq 0 \quad \textit{for} \quad \mathbf{d} \in \mathscr{C} \cap \mathrm{Ker}\mathsf{A}. \tag{2.128}$$

(ii) $\overline{\mathbf{x}} \in \Omega_{IE}$ is a solution of problem (2.127) if and only if

$$\mathbf{d}^T\nabla f(\overline{\mathbf{x}}) = (\mathsf{A}\overline{\mathbf{x}} - \mathbf{b})^T\mathbf{d} \geq 0 \tag{2.129}$$

for any feasible direction $\mathbf{d}$.
(iii) If f is convex and $\overline{\mathbf{x}}, \overline{\mathbf{y}}$ are two minimizers of f subject to $\mathbf{x} \in \Omega$, then

$$\overline{\mathbf{x}} - \overline{\mathbf{y}} \in \mathrm{Ker}\mathsf{A} \cap \mathrm{Span}\{\mathbf{b}\}^{\perp}.$$

(iv) If A *is positive definite, then the inequality-constrained minimization problem (2.127) has a unique solution.*

Proof. (i) First, notice that we can assume that Ω is a polyhedral cone so that we can use arguments of Proposition 2.16 to show that f is bounded from bellow on Ω. The solution exists by Proposition 2.5(iii). See also Dostál [74]. (ii)–(iv) Combine Proposition 2.4 and 2.5. □

2.8.5 KKT Conditions for QCQP

If the solution satisfies the ACQ constraint qualification, it satisfies the KKT conditions on the solution of problem (2.127). Let us recall that the equality constraints are defined by

$$h_i(\mathbf{x}) = \mathbf{b}_i^T\mathbf{x} - d_i, \quad i \in \mathcal{E}, \tag{2.130}$$

and the inequality constraints are defined by

$$h_i(\mathbf{x}) = \mathbf{b}_i^T\mathbf{x} - d_i \quad \text{or} \quad h_i(\mathbf{x}) = \mathbf{x}_i^T\mathsf{H}_i\mathbf{x}_i - d_i, \quad i \in \mathcal{I}, \tag{2.131}$$

so that the feasible set is defined by

$$\Omega_{IE} = \{\mathbf{x} \in \mathbb{R}^n : \mathsf{B}_{\mathcal{E}*}\mathbf{x} = \mathbf{c}_{\mathcal{E}}, \quad \mathbf{h}_{\mathcal{I}*}(\mathbf{x}) \leq \mathbf{c}_{\mathcal{I}}\}. \tag{2.132}$$

If there are no quadratic constraints, we define the feasible set Ω_{IE} by

$$\Omega_{IE} = \{\mathbf{x} \in \mathbb{R}^n : \mathsf{B}_{\mathcal{E}*}\mathbf{x} = \mathbf{c}_E, \quad \mathsf{B}_{\mathcal{I}*}\mathbf{x} \leq \mathbf{c}_{\mathcal{I}}\}. \tag{2.133}$$

Combining Propositions 2.8 and 2.17 and Sect. 2.8, we can formulate the KKT conditions for equality and inequality constrained problem (2.127).

Theorem 2.2. *Let the problem (2.127) be defined by an SPS matrix* $\mathsf{A} \in \mathbb{R}^{n \times n}$, *linear equalities* (2.130), *and inequality constrains* (2.131). *Let* $\Omega_{IE} \neq \emptyset$ *satisfy ACQ. Then the following statements hold:*
(i) $\overline{\mathbf{x}} \in \Omega_{IE}$ *solves (2.127) if and only if there is a vector* $\overline{\boldsymbol{\lambda}} \in \mathbb{R}^m$ *such that*

$$\overline{\boldsymbol{\lambda}}_{\mathcal{I}} \geq \mathbf{o}, \quad \nabla f(\overline{\mathbf{x}}) + \nabla \mathbf{h}^T \overline{\boldsymbol{\lambda}} = \mathbf{o}, \quad \text{and} \quad \overline{\boldsymbol{\lambda}}_{\mathcal{I}}^T \mathbf{h}_{\mathcal{I}} = 0. \tag{2.134}$$

(ii) $\overline{\mathbf{x}} \in \Omega_{IE}$ *is a solution of (2.127) with linear constraints* (2.133) *if and only if there is a vector* $\overline{\boldsymbol{\lambda}} \in \mathbb{R}^m$ *such that*

$$\overline{\boldsymbol{\lambda}}_{\mathcal{I}} \geq \mathbf{o}, \quad \mathsf{A}\overline{\mathbf{x}} - \mathbf{b} + \mathsf{B}^T \overline{\boldsymbol{\lambda}} = \mathbf{o}, \quad \text{and} \quad \overline{\boldsymbol{\lambda}}_{\mathcal{I}}^T [\mathsf{B}\overline{\mathbf{x}} - \mathbf{c}]_{\mathcal{I}} = 0. \tag{2.135}$$

(iii) If A *is SPD, then* $\overline{\mathbf{x}} \in \Omega_{IE}$ *is a unique solution of (2.127).*
(iv) If $\nabla \mathbf{h}(\mathbf{x})$ *is a full rank matrix, then the Lagrange multipliers are unique.*

Part II
Basic Algorithms

3
Gradient Methods

We begin our exposition of the main ingredients of optimal algorithms by introducing *gradient methods* (also known as *steepest descent methods*) for solving unconstrained minimization problem

$$\min_{\mathbf{x}\in\mathbb{R}^n} f(\mathbf{x}), \quad f(\mathbf{x}) = \frac{1}{2}\mathbf{x}^T\mathsf{A}\mathbf{x} - \mathbf{x}^T\mathbf{b}, \tag{3.1}$$

where $\mathbf{b} \in \mathbb{R}^n$ and A is an $n \times n$ SPD matrix. Given $\mathbf{x}^k$, the basic gradient method for solving (3.1) generates $\mathbf{x}^{k+1}$ by

$$\mathbf{x}^{k+1} = \mathbf{x}^k - \alpha_k\mathbf{g}^k, \quad \mathbf{g}^k = \nabla f(\mathbf{x}^k), \quad \alpha_k > 0. \tag{3.2}$$

The definition (3.2) assumes that there is a rule for α_k.

The gradient method uses the most straightforward local linear cost function approximation. As a result, the decrease of the cost function in a gradient step is typically modest. Still, simultaneously, the gradient step is robust in sense that large perturbations of the decrease direction $-\mathbf{g}$ are typically the decrease directions of f at $\mathbf{x}$. Here, we estimate the reduction of Euclidean and energy error norms, provide the formula for the optimal step length, and compare the effects of gradient and Newton's steps.

The *Barzialai-Borwain (BB) gradient method* uses the step length based on a local approximation of the inverse matrix by a scalar multiple of the identity matrix, using the gradient information from the previous step. Though the BB steps admit the cost function increase, they are often more effective than the monotonic gradient steps with optimized step length.

The last two sections describe the effect of *momentum* viewed as a tool for enhancing the information from the preceding steps into the algorithms. We overview the basic algorithms and describe the Polyak *heavy ball method*, including analysis.

Z. Dostál, *Optimal Quadratic Programming and QCQP Algorithms with Case Studies*, Springer Optimization and Its Applications 23,
https://doi.org/10.1007/978-3-031-95167-1_3

3.1 Reducing Euclidean Error

Given $\mathbf{x}^k$, the *fixed steplength gradient method* (also called the *steepest descent method*) for solving (3.1) generates $\mathbf{x}^{k+1}$ by

$$\mathbf{x}^{k+1} = \mathbf{x}^k - \alpha \mathbf{g}^k, \quad \mathbf{g}^k = \mathsf{A}\mathbf{x}^k - \mathbf{b}, \quad \alpha > 0. \tag{3.3}$$

The alternative name "steepest descent" is derived from observation that the linear model of f at $\mathbf{x}$ achieves its minimum on the set of all unit vectors

$$\mathcal{U} = \{\mathbf{d} \in \mathbb{R}^n, \|\mathbf{d}\| = 1\}$$

at

$$\widehat{\mathbf{d}} = -\|\nabla f(\mathbf{x})\|^{-1} \nabla f(\mathbf{x}).$$

Indeed, for any $\mathbf{d} \in \mathcal{U}$

$$\nabla f(\mathbf{x})^T \mathbf{d} \geq -\|\nabla f(\mathbf{x})\| \|\mathbf{d}\| = -\|\nabla f(\mathbf{x})\| = \nabla f(\mathbf{x})^T \widehat{\mathbf{d}}.$$

Let us start our analysis by estimating the effect of one gradient step.

Proposition 3.1. *Let $\widehat{\mathbf{x}}$ and f denote the solution and cost function of the problem* (3.1), *respectively. Let $\mathbf{x} \in \mathbb{R}^n$, $\alpha > 0$, and $\mathbf{g} = \nabla f(\mathbf{x})$. Let $\lambda_{\min}$ and $\lambda_{\max}$ denote the extreme eigenvalues of A. Then the following statements hold:*

(i) $\|(\mathbf{x} - \alpha \mathbf{g}) - \widehat{\mathbf{x}}\| \leq \gamma \|(\mathbf{x} - \widehat{\mathbf{x}}\|, \quad \gamma = \gamma(\alpha) = \max\{|1 - \alpha\lambda_{\min}|, |1 - \alpha\lambda_{\max}|\}.$

(ii) If $\alpha < 2/\|\mathsf{A}\|$, then $\gamma < 1$.

(iii) If $\alpha_{opt} = 2/(\lambda_{\min} + \lambda_{\max}),$ *then*

$$\gamma(\alpha_{opt}) = 1 - \frac{2}{\lambda_{\min} + \lambda_{\max}} = \frac{\kappa(\mathsf{A}) - 1}{\kappa(\mathsf{A}) + 1} \leq \gamma(\alpha).$$

Proof. (i) Notice that

$$\begin{aligned}
\|(\mathbf{x} - \alpha \mathbf{g}) - \widehat{\mathbf{x}}\| &= \|\mathbf{x} - \alpha(\mathsf{A}\mathbf{x} - \mathbf{b}) - \widehat{\mathbf{x}}\| \\
&= \|\mathbf{x} - \widehat{\mathbf{x}} - \alpha(\mathsf{A}\mathbf{x} - \mathbf{b}) + \alpha(\mathsf{A}\widehat{\mathbf{x}} - \mathbf{b})\| \\
&= \|(\mathsf{I} - \alpha\mathsf{A})(\mathbf{x} - \widehat{\mathbf{x}})\| \\
&\leq \|\mathsf{I} - \alpha\mathsf{A}\| \|\mathbf{x} - \widehat{\mathbf{x}}\|
\end{aligned}$$

and

$$\|\mathsf{I} - \alpha\mathsf{A}\| = \max_{i=1,\ldots,n} |1 - \alpha\lambda_i| = \max\{|1 - \alpha\lambda_{\min}|, |1 - \alpha\lambda_{\max}|\}.$$

(ii) Substitute into $\gamma(\alpha)$.
(iii): Notice that $\gamma(\alpha)$ achieves minimum when $1 - \lambda_{\min}\alpha = \lambda_{\max}\alpha - 1$. □

The convergence of the gradient method (3.2) with a fixed step length is an easy corollary of Proposition 3.1.

Corollary 3.1. *Let* $0 < a \le b < 2\|\mathsf{A}\|^{-1}$. *Then the gradient method* (3.2) *with* $\alpha \in [a, b]$ *converges to the solution* $\widehat{\mathbf{x}}$ *of* (3.1) *and*

$$\|\mathbf{x}^k - \widehat{\mathbf{x}}\| \le \gamma^k \|\mathbf{x}^0 - \widehat{\mathbf{x}}\|, \quad \gamma = \max\{|1 - a\lambda_{\min}|, |1 - b\lambda_{\max}|\}.$$

The gradient method's convergence is slow as illustrated in Fig. 3.1. Despite its slow convergence for ill-conditioned problems, the steepest descent method is useful as it is easy to implement and uses a robust decrease direction. Figure 3.2 illustrates that even if $\partial\mathbf{g}$ is a large perturbation of the gradient $\mathbf{g}$, the vector $-\widetilde{\mathbf{g}} = -\mathbf{g} + \partial\mathbf{g}$ is still a decreasing direction. In particular, we can see that the nearest non-decreasing direction generated by the perturbations of the Newton direction $\mathbf{d}_N = -\mathsf{A}^{-1}\mathbf{g}$ is much closer to $\mathbf{d}_N$ than those associated with the perturbations of a negative gradient to $-\mathbf{g}$. We shall use this nice property in Chap. 6.

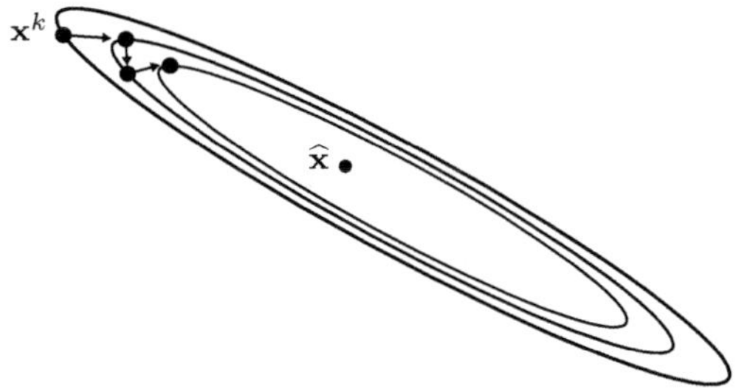

Fig. 3.1. Slow convergence of the steepest descent method

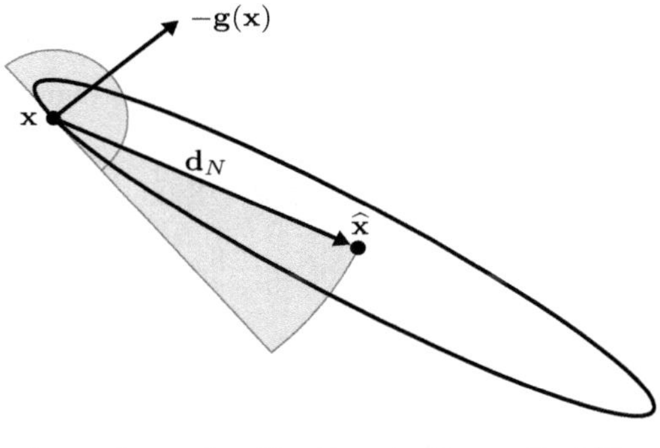

Fig. 3.2. The nearest non-decreasing directions to the gradient/Newton directions

3.2 Reducing the Cost Function

A natural option for α_k maximizes the local decrease, i.e.,

$$\begin{aligned}\alpha_k &= \arg\min_{\alpha\in\mathbb{R}} f(\mathbf{x}^k - \alpha\mathbf{g}^k)\\ &= \arg\min_{\alpha\in\mathbb{R}} \frac{1}{2}\alpha^2(\mathbf{g}^k)^T\mathsf{A}\mathbf{g}^k - \alpha(\mathbf{g}^k)^T(\mathsf{A}\mathbf{x}^\mathbf{k} - \mathbf{b}) \ + \ \text{constant terms.}\end{aligned}$$

Making the derivative

$$f'(\mathbf{x}^k - \alpha\mathbf{g}^k) = \alpha(\mathbf{g}^k)^T\mathsf{A}\mathbf{g} - \|\mathbf{g}\|^2$$

with respect to α equal to zero, we get

$$\alpha_k = \|\mathbf{g}\|^2/(\mathbf{g}^k)^T\mathsf{A}\mathbf{g}. \tag{3.4}$$

The gradient method with the minimizing step length reads:

Algorithm 3.1. Gradient method with optimal step length

Given a symmetric positive definite matrix $\mathsf{A}\in\mathbb{R}^{n\times n}$ *and* $\mathbf{b}\in\mathbb{R}^n$.

Step 0. {Initialization.}
Choose $\mathbf{x}^0\in\mathbb{R}^n$, *set* $\mathbf{g}^0 = \mathsf{A}\mathbf{x}^0 - \mathbf{b}$, $k=0$

Step 1. {Steepest descent loop. }
while $\|\mathbf{g}^k\|$ *is not small*
$\alpha_k = \|\mathbf{g}^k\|^2/(\mathbf{g}^k)^T\mathsf{A}\mathbf{g}^k$
$\mathbf{x}^{k+1} = \mathbf{x}^k - \alpha_k\mathbf{g}^k$
$\mathbf{g}^{k+1} = \mathbf{g}^k - \alpha_k\mathsf{A}\mathbf{g}^k$
$k = k+1$
end while

Step 2. {Return a (possibly approximate) solution.}
$\widetilde{\mathbf{x}} = \mathbf{x}^k$

Let us assess the effect of the gradient step in the energy norm using

$$f(\mathbf{x}) - f(\widehat{\mathbf{x}}) = \frac{1}{2}\|\mathbf{x} - \widehat{\mathbf{x}}\|_\mathsf{A}^2 = \frac{1}{2}\|\mathbf{g} - \widehat{\mathbf{g}}\|_{\mathsf{A}^{-1}}^2 = \frac{1}{2}\mathbf{g}^T\mathsf{A}^{-1}\mathbf{g}. \tag{3.5}$$

Our main tool will be a matrix form of the classical Kantorovich inequality [178], which appeared in Rennie [237]:

Lemma 3.1. *(Kantorovich's Inequality) Let* A *denote an SPD matrix,*

$$0 < m \le \lambda_{\min}(\mathsf{A}) \le \lambda_{\max}(\mathsf{A}) \le M, \quad \|\mathbf{x}\| = 1.$$

Then

$$(\mathsf{A}\mathbf{x},\mathbf{x})(\mathsf{A}^{-1}\mathbf{x},\mathbf{x}) \le \frac{(m+M)^2}{4mM}. \tag{3.6}$$

Proof. Since $h(x) = (M - x)(x - m)x^{-1} \geq 0$ for any $x \in [m, M]$, it follows by (1.32) that

$$h(\mathsf{A}) = (M\mathsf{I} - \mathsf{A})(\mathsf{A} - m\mathsf{I})\mathsf{A}^{-1} = (m + M)\mathsf{I} - \mathsf{A} - mM\mathsf{A}^{-1}$$

is SPS, i.e.,

$$(m + M)\|\mathbf{x}\|^2 \geq \mathbf{x}^T\mathsf{A}\mathbf{x} + mM\mathbf{x}^T\mathsf{A}^{-1}\mathbf{x}.$$

To finish the proof, take to square both sides, assume $\|\mathbf{x}\| = 1$, and apply to the right-hand side $a^2 + b^2 \geq 2ab$. □

Proposition 3.2. *Let* $\mathbf{x}^k$*,* α_k*, and* $\mathbf{g}^k$ *be generated by Algorithm 3.1 for problem* (3.1)*. Then Algorithm 3.1 converges to the solution* $\widehat{\mathbf{x}}$ *and for any* $k \geq 0$

$$\|\mathbf{x}^{k+1} - \widehat{\mathbf{x}}\|_{\mathsf{A}} \leq \gamma^k \|(\mathbf{x}^0 - \widehat{\mathbf{x}}\|_{\mathsf{A}}, \quad \gamma = \frac{\kappa(\mathsf{A}) - 1}{\kappa(\mathsf{A}) + 1}, \tag{3.7}$$

$$f(\mathbf{x}^{k+1}) - f(\widehat{\mathbf{x}}) \leq \gamma \left(f(\mathbf{x}^k) - f(\widehat{\mathbf{x}})\right). \tag{3.8}$$

Proof. First, recall that

$$\mathsf{A}(\mathbf{x}^k - \widehat{\mathbf{x}}) = \mathsf{A}\mathbf{x}^k - \mathbf{b} = \mathbf{g}^k \quad \text{and} \quad \|\mathbf{x}^k - \widehat{\mathbf{x}}\|_{\mathsf{A}}^2 = \|\mathbf{g}^k\|_{\mathsf{A}^{-1}}^2.$$

Thus

$$\begin{aligned}
\|\mathbf{x}^{k+1} - \widehat{\mathbf{x}}\|_{\mathsf{A}}^2 &= \|(\mathbf{x}^k - \alpha_k\mathbf{g}^k) - \widehat{\mathbf{x}}\|_{\mathsf{A}}^2 = \|(\mathsf{I} - \alpha_k\mathsf{A})(\mathbf{x}^k - \widehat{\mathbf{x}})\|_{\mathsf{A}}^2 \\
&= (\mathbf{x}^k - \widehat{\mathbf{x}})^T(\mathsf{I} - \alpha_k\mathsf{A})\mathsf{A}(\mathsf{I} - \alpha_k\mathsf{A})(\mathbf{x}^k - \widehat{\mathbf{x}}) \\
&= (\mathbf{x}^k - \widehat{\mathbf{x}})^T\mathsf{A}(\mathsf{I} - \alpha_k\mathsf{A})\mathsf{A}^{-1}(\mathsf{I} - \alpha_k\mathsf{A})\mathsf{A}(\mathbf{x}^k - \widehat{\mathbf{x}}) \\
&= (\mathbf{g}^k)^T\left(\mathsf{A}^{-1} - 2\alpha_k\mathsf{I} + \alpha_k^2\mathsf{A}\right)\mathbf{g}^k.
\end{aligned}$$

After substituting for α_k from (3.4), we get

$$\begin{aligned}
\|\mathbf{x}^{k+1} - \widehat{\mathbf{x}}\|_{\mathsf{A}}^2 &= (\mathsf{A}^{-1}\mathbf{g}^k, \mathbf{g}^k) - \|\mathbf{g}^k\|^4/(\mathsf{A}\mathbf{g}^k, \mathbf{g}^k) \\
&= \left(1 - \frac{\|\mathbf{g}^k\|^4}{(\mathsf{A}\mathbf{g}^k, \mathbf{g}^k)(\mathsf{A}^{-1}\mathbf{g}^k, \mathbf{g}^k)}\right)(\mathsf{A}^{-1}\mathbf{g}^k, \mathbf{g}^k).
\end{aligned}$$

Using the Kantorovich inequality

$$\frac{(\mathsf{A}\mathbf{g}^k, \mathbf{g}^k)(\mathsf{A}^{-1}\mathbf{g}^k, \mathbf{g}^k)}{\|\mathbf{g}^k\|^4} \leq \frac{(\lambda_{\min} + \lambda_{\max})^2}{4\lambda_{\min}\lambda_{\max}}$$

and simple manipulations, we get

$$\|\mathbf{x}^{k+1} - \widehat{\mathbf{x}}\|_{\mathsf{A}}^2 \leq \left(1 - \frac{4\lambda_{\min}\lambda_{\max}}{(\lambda_{\min} + \lambda_{\max})^2}\right)\|\mathbf{x}^k - \widehat{\mathbf{x}}\|_{\mathsf{A}}^2 = \gamma^2\|\mathbf{x}^k - \widehat{\mathbf{x}}\|_{\mathsf{A}}^2.$$

The rest follows by (3.5). □

3.3 Barzilai–Borwein Gradient Method

The estimates in the Sects. 3.1 and 3.2 indicate that the gradient methods are slow even with optimal step length. Since the steepest descent method is known since the times of Cauchy [32], it was surprising when Barzilai and Borwein [11] came in 1988 with new step length rules promising faster convergence.

Barzilai and Borwain were inspired by the Newton method for solving a scalar nonlinear equation $f(x) = 0$. Recall that the Newton method with the secant approximation of $f'(x)$ reads

$$x^{k+1} = x^k - \frac{1}{f'(x^k)} f(x^k) \approx x^k - \frac{x^k - x^{k-1}}{f(x^k) - f(x^{k-1})} f(x^k).$$

Applying the same method to solving $g(x) = f'(x) = 0$, we get

$$x^{k+1} = x^k - \frac{1}{g'(x^k)} g(x^k) \approx x^k - \frac{x^k - x^{k-1}}{g(x^k) - g(x^{k-1})} g(x^k) = x^k - \frac{1}{\sigma_k} g(x^k),$$

so the secant method can be viewed as the gradient descent method with

$$\sigma_k = \frac{g(x^k) - g(x^{k-1})}{x^k - x^{k-1}} \approx f''(x^k). \tag{3.9}$$

If $\mathbf{g}$ maps $\mathbb{R}^n$ to $\mathbb{R}^n$, we can replace (3.9) by the *secant equation*

$$\sigma_k(\mathbf{x}^k - \mathbf{x}^{k-1}) = \mathbf{g}^k - \mathbf{g}^{k-1} = \mathsf{A}(\mathbf{x}^k - \mathbf{x}^{k-1}) \tag{3.10}$$

and solve it in the sense of least squares. Denoting

$$\mathbf{s}^k = \mathbf{x}^k - \mathbf{x}^{k-1}, \quad \mathsf{A}\mathbf{s}^k = \mathbf{g}^k - \mathbf{g}^{k-1},$$

we get the strictly convex minimization problem

$$\begin{aligned}\sigma_k &= \arg\min_\sigma \|\sigma\mathbf{s}^k - \mathbf{g}^k\|^2 \\ &= \arg\min_\sigma \left((\mathbf{s}^k, \mathbf{s}^k)\sigma^2 - 2(\mathsf{A}\mathbf{s}^k, \mathbf{s}^k)\sigma + (\mathsf{A}\mathbf{s}^k, \mathsf{A}\mathbf{s}^k)\right)\end{aligned}$$

with the unique solution

$$\sigma_k = (\mathsf{A}\mathbf{s}^k, \mathbf{s}^k)/(\mathbf{s}^k, \mathbf{s}^k). \tag{3.11}$$

Comparing (3.9) with (3.10) suggests that σ_k defines a (very rough) approximation of the cost function's Hessian by a multiple of identity, i.e.,

$$\mathsf{A} \approx \sigma_k \mathsf{I}.$$

Thus, we can define the Barzilai-Borwein step

$$\mathbf{x}^{k+1} = \mathbf{x}^k - \alpha_k^{\mathrm{BB1}} \mathbf{g}^k, \quad \alpha_k^{\mathrm{BB1}} = \sigma_k^{-1} = \frac{(\mathbf{s}^k, \mathbf{s}^k)}{(\mathsf{A}\mathbf{s}^k, \mathbf{s}^k)}, \tag{3.12}$$

as an approximate Newton step

$$-\alpha_k^{\mathrm{BB1}} \mathbf{g}^k \approx -\mathsf{A}^{-1} \mathbf{g}^k.$$

Alternatively, we can formulate the least square problem (3.10) as

$$\mathbf{x}^k - \mathbf{x}^{k-1} = \alpha_k (\mathbf{g}^k - \mathbf{g}^{k-1}). \tag{3.13}$$

We get the strictly convex minimization problem

$$\begin{aligned} \alpha_k &= \arg\min_{\alpha} \|\mathbf{s}^k - \alpha \mathbf{g}^k\|^2 \\ &= \arg\min_{\alpha} \left((\mathsf{A}\mathbf{s}^k, \mathsf{A}\mathbf{s}^k)\alpha^2 - 2(\mathsf{A}\mathbf{s}^k, \mathbf{s}^k)\alpha + \mathbf{s}^k, \mathbf{s}^k) \right) \end{aligned}$$

with the unique solution

$$\alpha_k^{\mathrm{BB2}} = \frac{(\mathsf{A}\mathbf{s}^k, \mathbf{s}^k)}{(\mathsf{A}\mathbf{s}^k, \mathsf{A}\mathbf{s}^k)}. \tag{3.14}$$

To examine the relations between α_k^{BB2} and optimal coefficients α_k defined in Step 1 of Algorithm 3.1, let us recall that by (3.12)

$$\mathbf{s}^k = \mathbf{x}^k - \mathbf{x}^{k-1} = \alpha_{k-1}^{\mathrm{BB1}} \mathbf{g}^{k-1}.$$

After substituting into (3.12), we get

$$\alpha_k^{\mathrm{BB1}} = \frac{(\mathbf{g}^{k-1}, \mathbf{g}^{k-1})}{(\mathsf{A}\mathbf{g}^{k-1}, \mathbf{g}^{k-1})} = \alpha_{k-1}.$$

Similarly, we can interpret the Barzilai–Borwein coefficients α_k^{BB2} in terms of optimal gradient coefficients α_k defined in Step 1 of Algorithm 3.1 for

$$f_{\mathsf{A}}(\mathbf{x}) = \frac{1}{2}\mathbf{x}^T \mathsf{A}^2 \mathbf{x} - (\mathsf{A}\mathbf{b})^T \mathbf{x}.$$

The functions f and f_{A} have the same minimizers as their gradients satisfy

$$\nabla f_{\mathsf{A}} = \mathsf{A} \nabla f,$$

but the conditioning of the Hessian A^2 of f_{A} is much worse that of the Hessian A of f. We shall use the coefficients α_k^{BB1} in what follows.

The convergence of the Barzilai–Borwein method was proved by Raydan [236]. Later, Dai and Liao [45] proved its R-linear convergence.

3.4 Exploiting Momentum

Figure 3.1 reveals severe drawbacks of the basic gradient method, particularly slow convergence and oscillatory behavior. The picture suggests that the performance of the gradient method could be improved by a kind of a 'ravine step' combining the gradient direction with the *momentum*

$$\mathbf{s}^k = \mathbf{x}^k - \mathbf{x}^{k-1}.$$

The first gradient method with momentum is the *heavy ball method* by Polyak [228]. He proposed the iterations

$$\mathbf{x}^{k+1} = \mathbf{x}^k - \alpha \mathbf{g}^k + \beta \mathbf{s}^k, \quad \mathbf{g}^k = \nabla f(\mathbf{x}^k), \quad \mathbf{x}^1 = \mathbf{x}^0 - \alpha \mathbf{g}^0, \tag{3.15}$$

and proved that with α, β defined by the extreme eigenvalues of A as

$$\alpha = \frac{4}{\left(\sqrt{\lambda_{\max}} + \sqrt{\lambda_{\min}}\right)^2} \quad \text{and} \quad \beta = \left(\frac{\sqrt{\lambda_{\max}} - \sqrt{\lambda_{\min}}}{\sqrt{\lambda_{\max}} + \sqrt{\lambda_{\min}}}\right)^2, \tag{3.16}$$

the algorithm converges to the solution $\widehat{\mathbf{x}}$ with the convergence rate given by

$$\left\|\mathbf{x}^k - \widehat{\mathbf{x}}\right\| \le \left(\frac{\sqrt{\kappa(\mathsf{A})} - 1}{\sqrt{\kappa(\mathsf{A})} + 1} + \varepsilon_k\right)^k \left\|\mathbf{x}^0 - \widehat{\mathbf{x}}\right\|, \quad \lim_{k \to \infty} \varepsilon_k = 0. \tag{3.17}$$

The result shows that the heavy ball method with fixed parameters outperforms the gradient method in the late stage of computations, even if the latter uses optimal step lengths varying in each step.

A popular modification of the heavy ball algorithm is due to Nesterov [218], who suggested computing the gradient after applying the momentum. He suggested iterations

$$\mathbf{x}^{k+1} = \mathbf{x}^k - \alpha_k \nabla f(\mathbf{x}^k - \beta_k \mathbf{s}^k) - \beta_k \mathbf{s}^k \tag{3.18}$$

with the rules for α_k and β_k that guarantee an asymptotically faster convergence than the gradient method with the optimal step length. Recently, Oviedo et al. [223] presented interesting numerical experiments with the algorithm inspired by Nesterov. The algorithm reads

$$\mathbf{x}^{k+1} = \mathbf{x}^k - \alpha_k \nabla f(\mathbf{x}^k) - \gamma_k \mathbf{s}^k - \beta_k \mathsf{A}\mathbf{s}^k, \tag{3.19}$$

$$(\alpha_k, \beta_k, \gamma_k) = \arg \min_{(\alpha,\beta,\gamma) \in \mathbb{R}^3} \left\|\nabla f\left(\mathbf{x}^k - \alpha \mathbf{g}^k - \beta \mathbf{s}^k - \gamma \mathsf{A}\mathbf{s}^k\right)\right\|. \tag{3.20}$$

We shall comment on the above modifications in the next chapter.

3.5 Analysis of Heavy Ball Method

Let us prove the asymptotic estimate of the rate convergence (3.17) of the heavy ball method (3.15). Our goal is to eliminate ε_k in (3.17). Using the notation

$$\mathbf{g}^k = \mathsf{A}\mathbf{x}^k - \mathbf{b} = \mathsf{A}(\mathbf{x}^k - \widehat{\mathbf{x}}) \quad \text{and} \quad \mathbf{b} = \mathsf{A}\widehat{\mathbf{x}},$$

we get the error vector in the form

$$\begin{aligned}\mathbf{x}^{k+1} - \widehat{\mathbf{x}} &= \mathbf{x}^k - \widehat{\mathbf{x}} - \alpha\mathsf{A}(\mathbf{x}^k - \widehat{\mathbf{x}}) + \beta(\mathbf{x}^k - \mathbf{x}^{k-1})\\ &= (\mathsf{I} - \alpha\mathsf{A})(\mathbf{x}^k - \widehat{\mathbf{x}}) + \beta(\mathbf{x}^k - \widehat{\mathbf{x}}) - \beta(\mathbf{x}^{k-1} - \widehat{\mathbf{x}})\\ &= \big((1+\beta)\mathsf{I} - \alpha\mathsf{A}\big)(\mathbf{x}^k - \widehat{\mathbf{x}}) - \beta(\mathbf{x}^{k-1} - \widehat{\mathbf{x}}).\end{aligned}$$

To get the standard matrix form of algorithm (3.15), we concatenate two generations of iterations into one column to get

$$\begin{bmatrix}\mathbf{x}^{k+1} - \widehat{\mathbf{x}}\\ \mathbf{x}^k - \widehat{\mathbf{x}}\end{bmatrix} = \begin{bmatrix}(1+\beta)\mathsf{I} - \alpha\mathsf{A} & -\beta\mathsf{I}\\ \mathsf{I} & \mathsf{O}\end{bmatrix}\begin{bmatrix}\mathbf{x}^k - \widehat{\mathbf{x}}\\ \mathbf{x}^{k-1} - \widehat{\mathbf{x}}\end{bmatrix} = \mathsf{T}\begin{bmatrix}\mathbf{x}^k - \widehat{\mathbf{x}}\\ \mathbf{x}^{k-1} - \widehat{\mathbf{x}}\end{bmatrix}.$$

Since A is SPD, we can use the spectral decomposition (1.28) of A to reduce the analysis to the case with A diagonal, i.e.,

$$\mathsf{A} = \operatorname{diag}(\lambda_1, \dots, \lambda_n), \quad \lambda_{\max} = \lambda_1 \geq \dots \geq \lambda_n = \lambda_{\min} > 0.$$

Thus $\mathsf{T} \in \mathbb{R}^{2n\times 2n}$ comprises three diagonal and one zero $n \times n$ blocks.

We shall further simplify the structure of T without changing its norm by assuming that n is even and applying the permutation matrices to exchange, for each even i, $i = 2, 4, \dots, n$, the ith and $(i+n-1)$th rows and columns to transfer T into the block-diagonal form. For $n = 2$ and

$$\operatorname{diag}(d_1, d_2) = (1+\beta)\mathsf{I} - \alpha\mathsf{A},$$

the transformation exchanges the 2nd and 3rd rows and columns, as illustrated by

$$\begin{bmatrix} d_1 & 0 & -\beta & 0\\ 0 & d_2 & 0 & -\beta\\ 1 & 0 & 0 & 0\\ 0 & 1 & 0 & 0\end{bmatrix} \;\underset{\rightarrow}{\text{rows}}\; \begin{bmatrix} d_1 & 0 & -\beta & 0\\ 1 & 0 & 0 & 0\\ 0 & d_2 & 0 & -\beta\\ 0 & 1 & 0 & 0\end{bmatrix} \;\underset{\rightarrow}{\text{columns}}\; \left[\begin{array}{cc|cc} d_1 & -\beta & 0 & 0\\ 1 & 0 & 0 & 0\\ \hline 0 & 0 & d_2 & -\beta\\ 0 & 0 & 1 & 0\end{array}\right],$$

Thus

$$\mathsf{T} = \operatorname{diag}(\mathsf{T}_1, \dots, \mathsf{T}_n), \quad \mathsf{T}_i = \begin{bmatrix}1+\beta-\alpha\lambda_i & -\beta\\ 1 & 0\end{bmatrix}, \quad \lambda_i = \lambda_i(\mathsf{A}).$$

The block structure of T simplifies the analysis of the spectral radius $\varrho(\mathsf{T})$ as

$$\begin{aligned}\varrho(\mathsf{T}) &= \max\{|\theta| : \theta \in \sigma(\mathsf{T})\} && (3.21)\\ &= \max\{|\theta_i| : \theta_i \in \sigma(\mathsf{T}_i), \quad k = 1, \dots, n\}. && (3.22)\end{aligned}$$

The eigenvalues of T_i solve the characteristic equation

$$\det(\mathsf{T}_i - \theta\,\mathsf{I}) = \det\begin{bmatrix} 1+\beta-\alpha\lambda_i-\theta & -\beta \\ 1 & -\theta \end{bmatrix} = 0.$$

Its explicit form and solutions read

$$\begin{aligned}&\theta^2 - (1+\beta-\alpha\lambda_i)\theta + \beta = 0, && (3.23)\\ &\theta_{i\pm} = \frac{1}{2}\left(1+\beta-\alpha\lambda_i \pm \sqrt{(1+\beta-\alpha\lambda_i)^2 - 4\beta}\right). && (3.24)\end{aligned}$$

If $\theta_i = \theta_{i+}$ is complex, then $\overline{\theta}_i = \theta_{i-}$ and we get a nice formula

$$|\theta_i|^2 = \theta_i\overline{\theta}_i = \frac{1}{4}\left((1+\beta-\alpha\lambda_i)^2 + |(1+\beta-\alpha\lambda_i)^2 - 4\beta|\right) = \beta. \qquad (3.25)$$

The eigenvectors $\mathbf{u}_i$ and $\mathbf{v}_i$ corresponding to θ_i and $\overline{\theta}_i$ read

$$\mathbf{u}_i = \begin{bmatrix}\theta_i \\ 1\end{bmatrix} \quad \text{and} \quad \mathbf{v}_i = \begin{bmatrix}\overline{\theta}_i \\ 1\end{bmatrix}, \qquad (3.26)$$

respectively. Let us use this observation to analyze T_i.

Spectral Decomposition Based Estimate

Let us first assume that the eigenvalues θ_i of T_i satisfy $\theta_i \neq \overline{\theta}_i$.

Lemma 3.2. *Let* $0 < \beta < 1$, α, $\delta_{\min} > 0$,

$$\mathsf{T}_i = \begin{bmatrix} 1+\beta-\alpha\lambda_i & -\beta \\ 1 & 0 \end{bmatrix}, \quad \lambda_i = \lambda_i(\mathsf{A}), \quad i = 1\dots n,$$

let α, β *and* $\delta_{\min}$ *satisfy*

$$(1+\beta-\alpha\lambda_i)^2 - 4\beta \le -\delta_{\min}^2, \qquad (3.27)$$

and denote

$$\delta_i = \sqrt{4\beta - (1+\beta-\alpha\lambda_i)^2} \ge \delta_{\min}.$$

Then the following statements hold.
(i) T_i *has distinct complex conjugate eigenvalues* $\overline{\theta}_i$ *and* θ_i.
(ii) T_i *satisfies*

$$\varrho(\mathsf{T}_i) = |\theta_i| = \sqrt{\beta}. \tag{3.28}$$

(iii) There is an invertible matrix U_i *of the eigenvectors of* T_i *such that*

$$\mathsf{T}_i = \mathsf{U}_i \mathsf{D}_i \mathsf{U}_i^{-1}, \quad \mathsf{D}_i = \mathrm{diag}(\theta_i, \overline{\theta}_i). \tag{3.29}$$

(iv) The norms of the factors of T_i *are bounded as*

$$\|\mathsf{U}_i^{-1}\| \le \delta_i^{-1} \le \delta_{\min}^{-1}, \quad \|\mathsf{U}_i\| \le \|\mathsf{U}_i\|_\infty = 2, \quad \|\mathsf{D}_i\| = \sqrt{\beta}, \tag{3.30}$$

$$\|\mathsf{T}_i^k\| \le 2\delta_{\min}^{-1}(\sqrt{\beta})^k. \tag{3.31}$$

(v) If (3.27) *holds, then* (3.28)–(3.31) *hold for all* $i = 1, \ldots n$.

Proof. (i), (ii) Examining the formula (3.24), we can check that θ_i has nonzero imaginary part if and only if α and β satisfies (3.27). It follows that if α and β satisfy (3.27), then the roots have nonzero imaginary parts, $\theta_i \neq \overline{\theta}_i$, and

$$\varrho(\mathsf{T}_i) = |\theta_i| = \sqrt{\beta}.$$

(iii) Let u_i, v_i denote the eigenvectors corresponding to θ_i and $\overline{\theta}_i$. Then

$$\mathsf{U}_i = [\mathbf{u}_i, \mathbf{v}_i] = \begin{bmatrix} \theta_i & \overline{\theta}_i \\ 1 & 1 \end{bmatrix}$$

is invertible and

$$\mathsf{T}_i \mathsf{U}_i = \mathsf{U}_i \mathsf{D}_i.$$

After multiplying the above equation by U_i^{-1} on the right, we get (3.29).
(iv) Notice that

$$\mathsf{U}_i^{-1} = \begin{bmatrix} \theta_i & \overline{\theta}_i \\ 1 & 1 \end{bmatrix}^{-1} = (\theta_i - \overline{\theta}_i)^{-1} \begin{bmatrix} 1 & -\overline{\theta}_i \\ -1 & \theta_i \end{bmatrix} = (2\delta_i)^{-1} \begin{bmatrix} 1 & -\overline{\theta}_i \\ -1 & \theta_i \end{bmatrix}.$$

so that

$$\|\mathsf{U}_i^{-1}\| \le \|\mathsf{U}_i^{-1}\|_\infty \le \delta_i^{-1}.$$

Since

$$\|\mathsf{U}_i\| \le 2, \quad \|\mathsf{D}_i\| = \sqrt{\beta}, \quad \text{and} \quad \delta_i \ge \delta_{\min},$$

it follows that

$$\|\mathsf{T}_i^k\| \le \|\mathsf{U}_i\| \|\mathsf{D}_i\|^k \|\mathsf{U}_i^{-1}\|_\infty \le 2\delta_i^{-1}\sqrt{\beta^k} \le 2\delta_{\min}^{-1}\sqrt{\beta^k}. \tag{3.32}$$

The statement follows by (iii) and the properties of the norm. The rest is easy. □.

Proposition 3.3. *Let* $\{\mathbf{x}^k\}$ *be generated by the heavy ball method* (3.15) *for the solution of* (3.1) *with the parameters* α *and* β *that satisfy* (3.27). *Then* $\{\mathbf{x}^k\}$ *converges to the solution* $\widehat{\mathbf{x}}$ *of* (3.1) *R-linearly and*

$$\left\|\begin{bmatrix}\mathbf{x}^k-\widehat{\mathbf{x}}\\ \mathbf{x}^{k-1}-\widehat{\mathbf{x}}\end{bmatrix}\right\| \le 2\delta_{\min}^{-1}(\sqrt{\beta})^k\left\|\begin{bmatrix}\mathbf{x}^1-\widehat{\mathbf{x}}\\ \mathbf{x}^0-\widehat{\mathbf{x}}\end{bmatrix}\right\|. \tag{3.33}$$

Proof. See (3.32).

Schur Decomposition-Based Estimate

Proposition 3.3 does not support the application of β defined by (3.16). To admit such β, assume that α and β satisfy only

$$(1+\beta-\alpha\lambda_i)^2-4\beta\le 0,\quad i=1,\dots n, \tag{3.34}$$

and recall that

$$1+\beta+\alpha\lambda_i=\theta_i+\overline{\theta}_i=2\mathrm{Re}\theta_i \quad \text{and} \quad \mathsf{T}_i=\begin{bmatrix}2\mathrm{Re}\theta_i & -|\theta_i|^2\\ 1 & 0\end{bmatrix}.$$

To estimate $\|\mathsf{T}_i^k\|$, we apply the Schur decomposition to get

$$\mathsf{T}_i=\mathsf{U}_i\mathsf{S}_i\mathsf{U}_i^*,\quad \mathsf{S}_i=\begin{bmatrix}\theta_i & s_{12}^i\\ 0 & \overline{\theta}_i\end{bmatrix},\quad \mathsf{U}_i=[\mathbf{u}_i,\mathbf{u}_i^{\perp}],\quad \mathsf{U}_i^*\mathsf{U}_i=\mathsf{I}. \tag{3.35}$$

It is easy to check that $\mathbf{u}_i$ is a normalized eigenvector corresponding to θ_i and $\mathbf{u}_i^{\perp}$ is a normalized vector orthogonal to $\mathbf{u}_i$. Using (3.26), we can get explicit expressions for both vectors in the form

$$\mathbf{u}_i=\frac{1}{\sqrt{1+\beta}}\begin{bmatrix}\theta_i\\ 1\end{bmatrix},\quad \mathbf{u}_i^{\perp}=\frac{1}{\sqrt{1+\beta}}\begin{bmatrix}1\\ -\overline{\theta}_i\end{bmatrix}.$$

Since

$$\mathsf{S}_i=\mathsf{U}_i^*\mathsf{T}_i\mathsf{U}_i,$$

we get

$$s_{12}^i=\mathbf{u}_i^*\mathsf{T}_i\mathbf{u}_i^{\perp}=\frac{1+2\overline{\theta}_iRe\theta+\beta^2}{1+\beta},\quad |s_{12}^i|\le\frac{(1+\beta)^2}{1+\beta}=1+\beta,$$

and the Schur decomposition

$$\mathsf{T}_i = \mathsf{U}_i \begin{bmatrix} \theta_i & s_{12}^i \\ 0 & \overline{\theta}_{i,} \end{bmatrix} \mathsf{U}_i^*. \tag{3.36}$$

Denoting

$$\mathsf{S} = \begin{bmatrix} \sigma & 1+\sigma^2 \\ 0 & \sigma \end{bmatrix}, \quad \sigma = \sqrt{\beta},$$

we get

$$\|\mathsf{T}_i^k\| = \|\mathsf{S}_i^k\| \le \|\mathsf{S}^k\|_\infty, \quad i = 1, \dots, n; \quad k = 1, 2, \dots,$$

$$\mathsf{S}^k = \begin{bmatrix} \sigma^k & k\sigma^{k-1}(1+\sigma^2) \\ 0 & \sigma^k \end{bmatrix}, \tag{3.37}$$

$$\|\mathsf{T}_i^k\| \le \|\mathsf{S}^k\|_\infty = \sigma^k + k\sigma^{k-1}(1+\sigma^2) \le (2k+1)\,\sigma^{k-1},$$

and

$$\lim_{k\to\infty} \sqrt[k]{\|\mathsf{T}_i^k\|} = \sigma = \sqrt{\beta}.$$

We can sum up the above observations into the following proposition.

Proposition 3.4. *Let α and β satisfy* (3.34)*. The following statements hold.*
(i) There is an orthogonal matrix U_i such that

$$\mathsf{T}_i = \mathsf{U}_i \mathsf{S}_i \mathsf{U}_i^*, \quad \mathsf{S}_i = \begin{bmatrix} \theta_i & (1+\overline{\theta}_i)^2 \\ 0 & \overline{\theta}, \end{bmatrix}. \tag{3.38}$$

(ii) The norms of the powers of T_i are bounded and

$$\|\mathsf{T}_i^k\| \le (2k+1)\sqrt{\beta^{k-1}}. \tag{3.39}$$

(iii) The iterates $\{\mathbf{x}^k\}$ converge to the solution $\widehat{\mathbf{x}}$ of (3.1) *and*

$$\left\| \begin{bmatrix} \mathbf{x}^k - \widehat{\mathbf{x}} \\ \mathbf{x}^{k-1} - \widehat{\mathbf{x}} \end{bmatrix} \right\| \le (2k+1)\sqrt{\beta^{k-1}} \left\| \begin{bmatrix} \mathbf{x}^1 - \widehat{\mathbf{x}} \\ \mathbf{x}^0 - \widehat{\mathbf{x}} \end{bmatrix} \right\|. \tag{3.40}$$

(iv) If α and β are defined by (3.16)*, then*

$$\left\| \begin{bmatrix} \mathbf{x}^k - \widehat{\mathbf{x}} \\ \mathbf{x}^{k-1} - \widehat{\mathbf{x}} \end{bmatrix} \right\| \le (2k+1)\left(\frac{\sqrt{\kappa}(\mathsf{A}) - 1}{\sqrt{\kappa}(\mathsf{A}) + 1} \right)^{k-1} \left\| \begin{bmatrix} \mathbf{x}^1 - \widehat{\mathbf{x}} \\ \mathbf{x}^0 - \widehat{\mathbf{x}} \end{bmatrix} \right\|. \tag{3.41}$$

□

Remark 3.1. If the parameters satisfy (3.27), then the R-linear estimate (3.33) gives for $2k+1 \ge \delta_{\min}^{-1}$ a better bound than the general estimate (3.40). The classical analysis uses the bound on the spectrum and the spectral radius formula to get (3.17). However, unspecified ε_k in (3.17) spoils any attempt to get relevant quantitative information on the process in the finite stage of computations. See, e.g., Pták [235], or Dostál [58], [59].

3.6 Comments

We presented the gradient methods for minimizing a quadratic function with three step-length options. The first two, the classical steepest descent method with fixed and maximum decrease step lengths, are well-understood standard parts of the applications of gradient methods. We recall them here as we use them effectively for solving various quadratic programming problems.

We also include two modifications of the basic gradient method. The Barzilai-Borwein method uses the step length from the previous step and is still the subject of active research. The results include new step length rules, see, e.g., Yuan [268], convergence results, e.g., Dai and Liao [45], and interesting applications. The R-linear convergence rate estimates include potentially large constants depending on the conditioning of the Hessian matrix and the problem's dimension, as observed by Li and Sun [203]. They also showed the limits of Barzilai and Borwein's method by proving that for any SPD matrix A, there is $\mathbf{x}^0$ such that

$$\|\mathbf{g}^{k+1}\| \geq \frac{\kappa(\mathsf{A}) - 1}{\kappa(\mathsf{A}) + 1}\|\mathbf{g}^k\|, \quad k = 1, 2, \dots .$$

The above rather pessimistic theoretical results do not explain an often observed good performance of the spectral method, especially in the early stage of computations. The simple observations of Fletcher [132] indicate that the method first tends to reduce the components of residua associated with the small eigenvalues, i.e., the smooth components of the error vector. In particular, if $\lambda_i < 2\lambda_{\min}$, then the corresponding coordinate $g_i^{k+1} = \mathbf{e}_i^T \mathbf{g}^{k+1}$, where $\mathbf{e}_i$ denotes a unit eigenvector of A corresponding to λ_i, decreases as

$$|g_i^{k+1}| \leq |1 - \alpha_k \lambda_i||g_i^k| < |g_i^k|, \quad 0 < \lambda_{\max}^{-1} \leq \alpha_k \leq \lambda_{\min}^{-1},$$

with α_k often near $\lambda_{\min}^{-1}$. This conjecture complies with numerical experiments indicating fast convergence in the initial stage of computations (see, e.g., Asmundis et al. [3]). The arguments presented above indicate that we can use spectral algorithms to generate useful initial approximations and exploit their robust decrease directions. An application in the classification problem is in Chap. 16.

In the last part of this chapter, we introduced the concept of momentum and described Polyak's Heavy Ball Method [228], the very first method with momentum using fixed parameters. Similarly to the spectral methods, the methods with momentum are the subject of exciting research motivated by problems in AI. For the discussion of gradient methods with momentum in AI, see the book by Watt, Borhani, and Katsaggelos [262, Appendix A]. The proof of convergence is based on [115].

4
Conjugate Gradients as Direct Method

Let us introduce the *conjugate gradient (CG) method* for solving unconstrained optimization problems

$$\min_{\mathbf{x}\in\mathbb{R}^n} f(\mathbf{x}), \quad f(\mathbf{x}) = \frac{1}{2}\mathbf{x}^T\mathsf{A}\mathbf{x} - \mathbf{x}^T\mathbf{b}, \tag{4.1}$$

where $\mathbf{b}$ is a given column n-vector, and A is an $n \times n$ an SPD matrix. Problem (4.1) is equivalent to the solution of a system of linear equations $\mathsf{A}\mathbf{x} = \mathbf{b}$, see Sect. 2.2.2.

We are interested in solving (4.1) as it appears in auxiliary problems of algorithms for optimization problems with constraints. Here, we consider the problems with n modestly large so that it is possible to solve (4.1) in a finite number of CG steps, thus providing an alternative to the Cholesky factorization introduced in Sect. 1.5. We postpone the application of CG to large problems to Chap. 8.

The main asset of the CG method is its capability to provide cheap improving approximations to the solution $\widehat{\mathbf{x}}$ until it gets close or finds it. Notice that the Cholesky method does not provide any approximation to the solution of (4.1) before it returns the final solution.

The basic idea of the CG method is to generate the vectors of the conjugate basis of $\mathbb{R}^n$ that reduces a general quadratic form to a diagonal form. This trick reduces (4.1) to the sequence of its scalar variants. The conjugate gradient method is our primary tool for treating the auxiliary unconstrained problems generated by the algorithms for solving QP problems.

The CG method uses momentum to preserve the information from all proceeding steps. The cost of solving $n \times n$ system by direct CG method is proportional to n-times number of non-zero entries of A.

Z. Dostál, *Optimal Quadratic Programming and QCQP Algorithms with Case Studies*, Springer Optimization and Its Applications 23,
https://doi.org/10.1007/978-3-031-95167-1_4

4.1 Conjugate Directions and Minimization

The conjugate gradient method, an ingenious and powerful engine of our algorithms, is based on simple observations. In this section, we examine the first one, namely, that it is possible to reduce the solution of (4.1) to the solution of a sequence of one-dimensional problems.

Let $\mathsf{A} \in \mathbb{R}^{n\times n}$ be an SPD matrix, and let us assume that there are nonzero n-vectors $\mathbf{p}^1, \dots, \mathbf{p}^n$ such that

$$(\mathbf{p}^i, \mathbf{p}^j)_{\mathsf{A}} = (\mathbf{p}^i)^T \mathsf{A}\mathbf{p}^j = 0 \quad \text{for} \quad i \neq j.$$

We call such vectors A-*conjugate* or briefly *conjugate*. Specializing the arguments of Sect. 1.8, we get that $\mathbf{p}^1, \dots, \mathbf{p}^n$ are independent. Thus $\mathbf{p}^1, \dots, \mathbf{p}^n$ form the basis of $\mathbb{R}^n$ and any $\mathbf{x} \in \mathbb{R}^n$ can be written in the form

$$\mathbf{x} = \xi_1\mathbf{p}^1 + \dots + \xi_n\mathbf{p}^n.$$

Substituting into f and using the conjugacy results in

$$\begin{aligned} f(\mathbf{x}) &= \left(\frac{1}{2}\xi_1^2(\mathbf{p}^1)^T\mathsf{A}\mathbf{p}^1 - \xi_1\mathbf{b}^T\mathbf{p}^1\right) + \dots + \left(\frac{1}{2}\xi_n^2(\mathbf{p}^n)^T\mathsf{A}\mathbf{p}^n - \xi_n\mathbf{b}^T\mathbf{p}^n\right) \\ &= f(\xi_1\mathbf{p}^1) + \dots + f(\xi_n\mathbf{p}^n). \end{aligned}$$

Thus we have decomposed the original problem (4.1) into n one-dimensional problems

$$f(\widehat{\mathbf{x}}) = \min_{\mathbf{x}\in\mathbb{R}^n} f(\mathbf{x}) = \min_{\xi_1\in\mathbb{R}} f(\xi_1\mathbf{p}^1) + \dots + \min_{\xi_n\in\mathbb{R}} f(\xi_n\mathbf{p}^n).$$

Since

$$\left.\frac{\mathrm{d}f(\xi\mathbf{p}^i)}{\mathrm{d}\xi}\right|_{\xi_i} = \xi_i(\mathbf{p}^i)^T\mathsf{A}\mathbf{p}^i - \mathbf{b}^T\mathbf{p}^i = 0,$$

the solution $\widehat{\mathbf{x}}$ of (4.1) is given by

$$\widehat{\mathbf{x}} = \xi_1\mathbf{p}^1 + \dots + \xi_n\mathbf{p}^n, \quad \xi_i = \mathbf{b}^T\mathbf{p}^i/(\mathbf{p}^i)^T\mathsf{A}\mathbf{p}^i,\ i = 1, \dots, n. \tag{4.2}$$

If the dimension of the problem (4.1) is large, the task to evaluate $\widehat{\mathbf{x}}$ may be too ambitious. In this case, it may be useful to modify the procedure so that it can find an approximation $\widetilde{\mathbf{x}}$ to the solution $\widehat{\mathbf{x}}$ for (4.1) using some initial guess $\mathbf{x}^0$ and a few vectors $\mathbf{p}^1, \dots, \mathbf{p}^k$, $k \ll n$. A natural choice for the approximation $\widetilde{\mathbf{x}}$ is the minimizer $\mathbf{x}^k$ of f in $\mathcal{S}^k = \mathbf{x}^0 + \mathrm{Span}\{\mathbf{p}^1, \dots, \mathbf{p}^k\}$. To find it, notice that any $\mathbf{x} \in \mathcal{S}^k$ can be written in the form

$$\mathbf{x} = \mathbf{x}^0 + \xi_1\mathbf{p}^1 + \dots + \xi_k\mathbf{p}^k,$$

so, after substituting into f and using that $\mathbf{p}^1, \dots, \mathbf{p}^k$ are conjugate, we get

$$f(\mathbf{x}) = f(\mathbf{x}^0) + \left(\frac{1}{2}\xi_1^2(\mathbf{p}^1)^T \mathsf{A}\mathbf{p}^1 + \xi_1(\mathsf{A}\mathbf{x}^0 - \mathbf{b})^T \mathbf{p}^1\right) + \dots$$
$$+ \left(\frac{1}{2}\xi_k^2(\mathbf{p}^k)^T \mathsf{A}\mathbf{p}^k + \xi_k(\mathsf{A}\mathbf{x}^0 - \mathbf{b})^T \mathbf{p}^k\right).$$

Denoting $\mathbf{g}^0 = \mathbf{g}(\mathbf{x}^0) = \nabla f(\mathbf{x}^0) = \mathsf{A}\mathbf{x}^0 - \mathbf{b}$ and

$$f_0(\mathbf{x}) = \frac{1}{2}\mathbf{x}^T \mathsf{A}\mathbf{x} + \mathbf{x}^T \mathbf{g}^0,$$

we have

$$f(\mathbf{x}) = f(\mathbf{x}^0) + f_0(\xi_1 \mathbf{p}^1) + \dots + f_0(\xi_k \mathbf{p}^k)$$

and

$$f(\mathbf{x}^k) = \min_{\mathbf{x} \in \mathcal{S}^k} f(\mathbf{x}) = f(\mathbf{x}^0) + \min_{\xi_1 \in \mathbb{R}} f_0(\xi_1 \mathbf{p}^1) + \dots + \min_{\xi_k \in \mathbb{R}} f_0(\xi_k \mathbf{p}^k). \tag{4.3}$$

We have thus again reduced our problem to solving a sequence of simple one-dimensional problems. The approximation $\mathbf{x}^k$ is given by

$$\mathbf{x}^k = \mathbf{x}^0 + \xi_1 \mathbf{p}^1 + \dots + \xi_k \mathbf{p}^k, \quad \xi_i = -(\mathbf{g}^0)^T \mathbf{p}^i / (\mathbf{p}^i)^T \mathsf{A}\mathbf{p}^i, \ i = 1, \dots, k, \tag{4.4}$$

as

$$\left.\frac{\mathrm{d}f(\xi \mathbf{p}^i)}{\mathrm{d}\xi}\right|_{\xi_i} = \xi_i (\mathbf{p}^i)^T \mathsf{A}\mathbf{p}^i + (\mathbf{g}^0)^T \mathbf{p}^i = 0.$$

Since by (4.3) for $k \geq 1$

$$f(\mathbf{x}^k) = \min_{\mathbf{x} \in \mathcal{S}^k} f(\mathbf{x}) = f(\mathbf{x}^{k-1}) + \min_{\xi \in \mathbb{R}} f_0(\xi \mathbf{p}^k), \tag{4.5}$$

we can generate the approximations $\mathbf{x}^k$ iteratively. The *conjugate direction method* starts from an arbitrary initial guess $\mathbf{x}^0$. If $\mathbf{x}^{k-1}$ is given, then $\mathbf{x}^k$ is generated by the formula

$$\mathbf{x}^k = \mathbf{x}^{k-1} - \alpha_k \mathbf{p}^k, \quad \alpha_k = (\mathbf{g}^0)^T \mathbf{p}^i / (\mathbf{p}^i)^T \mathsf{A}\mathbf{p}^i. \tag{4.6}$$

Thus $f(\mathbf{x}^{k-1} + \xi \mathbf{p}^k)$ achieves its minimum at $\xi = -\alpha_k$ and the procedure guarantees that the *successive iterates* $\mathbf{x}^k$ *minimize* f *over a progressively expanding manifold* $\mathcal{S}^k$ *that eventually includes the global minimum of* f.

The coefficients α_k can be evaluated by alternative formulae. For example, using Corollary 2.9 and the definition of $\mathcal{S}^k$, we get

$$(\mathbf{g}^k)^T \mathbf{p}^i = 0, \quad i = 1, \dots, k. \tag{4.7}$$

Since for $i \geq 1$

$$\begin{aligned}\mathbf{g}^i &= \mathsf{A}\mathbf{x}^i - \mathbf{b} = \mathsf{A}(\mathbf{x}^{i-1} - \alpha_i \mathbf{p}^i) - \mathbf{b} = (\mathsf{A}\mathbf{x}^{i-1} - \mathbf{b}) - \alpha_i \mathsf{A}\mathbf{p}^i \\ &= \mathbf{g}^{i-1} - \alpha_i \mathsf{A}\mathbf{p}^i,\end{aligned}$$

we get for $k \geq 1$ and $i = 1, \ldots, k-1$, by using the conjugacy, that

$$(\mathbf{g}^i)^T \mathbf{p}^k = (\mathbf{g}^{i-1})^T \mathbf{p}^k - \alpha_i (\mathbf{p}^i)^T \mathsf{A}\mathbf{p}^k = (\mathbf{g}^{i-1})^T \mathbf{p}^k.$$

Thus

$$(\mathbf{g}^0)^T \mathbf{p}^k = (\mathbf{g}^1)^T \mathbf{p}^k = \cdots = (\mathbf{g}^{k-1})^T \mathbf{p}^k$$

and

$$\alpha_k = \frac{(\mathbf{g}^0)^T \mathbf{p}^k}{(\mathbf{p}^k)^T \mathsf{A}\mathbf{p}^k} = \cdots = \frac{(\mathbf{g}^{k-1})^T \mathbf{p}^k}{(\mathbf{p}^k)^T \mathsf{A}\mathbf{p}^k}. \tag{4.8}$$

Combining the latter formula with the Taylor expansion, we get

$$f(\mathbf{x}^k) = f(\mathbf{x}^{k-1}) - \frac{1}{2} \frac{((\mathbf{g}^{k-1})^T \mathbf{p}^k)^2}{(\mathbf{p}^k)^T \mathsf{A}\mathbf{p}^k}. \tag{4.9}$$

We have yet to discuss how to get the vectors $\mathbf{p}^1, \ldots, \mathbf{p}^n$. Are we able to generate them efficiently? A positive answer in the next section is key to the success of the conjugate gradient method.

4.2 Generating Conjugate Directions and Krylov Spaces

Let us recall how to generate conjugate directions with the *Gramm–Schmidt procedure*. Let us first suppose that $\mathbf{p}^1, \ldots, \mathbf{p}^k$ are nonzero conjugate directions, $1 \leq k < n$, and let us examine how to use $\mathbf{h}^k \notin \operatorname{Span}\{\mathbf{p}^1, \ldots, \mathbf{p}^k\}$ to generate a new member $\mathbf{p}^{k+1}$ in the form

$$\mathbf{p}^{k+1} = \mathbf{h}^k + \beta_{k1} \mathbf{p}^1 + \cdots + \beta_{kk} \mathbf{p}^k. \tag{4.10}$$

Since $\mathbf{p}^{k+1}$ should be conjugate to $\mathbf{p}^1, \ldots, \mathbf{p}^k$, we get

$$\begin{aligned}0 = (\mathbf{p}^i)^T \mathsf{A}\mathbf{p}^{k+1} &= (\mathbf{p}^i)^T \mathsf{A}\mathbf{h}^k + \beta_{k1} (\mathbf{p}^i)^T \mathsf{A}\mathbf{p}^1 + \cdots + \beta_{kk} (\mathbf{p}^i)^T \mathsf{A}\mathbf{p}^k \\ &= (\mathbf{p}^i)^T \mathsf{A}\mathbf{h}^k + \beta_{ki} (\mathbf{p}^i)^T \mathsf{A}\mathbf{p}^i, \quad i = 1, \ldots, k.\end{aligned}$$

Thus

$$\beta_{ki} = -\frac{(\mathbf{p}^i)^T \mathsf{A}\mathbf{h}^k}{(\mathbf{p}^i)^T \mathsf{A}\mathbf{p}^i}, \quad i = 1, \ldots, k. \tag{4.11}$$

Obviously

$$\operatorname{Span}\{\mathbf{p}^1, \ldots, \mathbf{p}^{k+1}\} = \operatorname{Span}\{\mathbf{p}^1, \ldots, \mathbf{p}^k, \mathbf{h}^k\}.$$

Therefore, given any independent vectors $\mathbf{h}^0, \dots, \mathbf{h}^{k-1}$, we can start from $\mathbf{p}^1 = \mathbf{h}^0$ and use (4.10) and (4.11) to construct a set of mutually A-conjugate directions $\mathbf{p}^1, \dots, \mathbf{p}^k$ such that

$$\mathrm{Span}\{\mathbf{h}^0, \dots, \mathbf{h}^{i-1}\} = \mathrm{Span}\{\mathbf{p}^1, \dots, \mathbf{p}^i\},\ i = 1, \dots, k.$$

For $\mathbf{h}^0, \dots, \mathbf{h}^{k-1}$ arbitrary, the construction is increasingly expensive as it requires both the storage for the vectors $\mathbf{p}^1, \dots, \mathbf{p}^k$ and heavy calculations including evaluation of $k(k+1)/2$ scalar products. However, it turns out that we can adapt the procedure so that it generates the conjugate basis of the *Krylov spaces* very efficiently, with

$$\mathcal{K}^k = \mathcal{K}^k(\mathsf{A}, \mathbf{g}^0) = \mathrm{Span}\{\mathbf{g}^0, \mathsf{A}\mathbf{g}^0, \dots, \mathsf{A}^{k-1}\mathbf{g}^0\},\ k = 1, \dots, n,$$

and $\mathbf{g}^0 = \mathsf{A}\mathbf{x}^0 - \mathbf{b}$ defined by a suitable initial vector $\mathbf{x}^0$ and $\mathcal{K}^0 = \{\mathbf{o}\}$. The powerful method is again based on a few simple observations.

First assume that $\mathbf{p}^1, \dots, \mathbf{p}^i$ form a conjugate basis of $\mathcal{K}^i, i = 1, \dots, k$, and observe that if $\mathbf{x}^k$ denotes the minimizer of f on $\mathbf{x}^0 + \mathcal{K}^k$, then by Corollary 2.9 the gradient $\mathbf{g}^k = \nabla f(\mathbf{x}^k)$ is orthogonal to the Krylov space $\mathcal{K}^k$, that is,

$$(\mathbf{g}^k)^T \mathbf{x} = 0 \quad \text{for any} \quad \mathbf{x} \in \mathcal{K}^k.$$

In particular, if $\mathbf{g}^k \neq \mathbf{o}$, then

$$\mathbf{g}^k \notin \mathcal{K}^k.$$

Since $\mathbf{g}^k \in \mathcal{K}^{k+1}$, we can use (4.10) with $\mathbf{h}^k = \mathbf{g}^k$ to expand any conjugate basis of $\mathcal{K}^k$ to the conjugate basis of $\mathcal{K}^{k+1}$. Obviously

$$\mathcal{K}^k(\mathsf{A}, \mathbf{g}^0) = \mathrm{Span}\{\mathbf{g}^0, \dots, \mathbf{g}^{k-1}\}.$$

Next observe that for any $\mathbf{x} \in \mathcal{K}^{k-1}$ and $k \geq 1$

$$\mathsf{A}\mathbf{x} \in \mathcal{K}^k,$$

or briefly $\mathsf{A}\mathcal{K}^{k-1} \subseteq \mathcal{K}^k$. Since $\mathbf{p}^i \in \mathcal{K}^i \subseteq \mathcal{K}^{k-1}$, $i = 1, \dots, k-1$, we have

$$(\mathsf{A}\mathbf{p}^i)^T \mathbf{g}^k = (\mathbf{p}^i)^T \mathsf{A}\mathbf{g}^k = 0,\ i = 1, \dots, k-1.$$

It follows that

$$\beta_{ki} = -\frac{(\mathbf{p}^i)^T \mathsf{A}\mathbf{g}^k}{(\mathbf{p}^i)^T \mathsf{A}\mathbf{p}^i} = 0,\ i = 1, \dots, k-1.$$

Summing up, if we have a set of such conjugate vectors $\mathbf{p}^1, \dots, \mathbf{p}^k$ that

$$\mathrm{Span}\{\mathbf{p}^1, \dots, \mathbf{p}^i\} = \mathcal{K}^i,\ i = 1, \dots k,$$

then the formula (4.10) applied to $\mathbf{p}^1, \dots, \mathbf{p}^k$ and $\mathbf{h}^k = \mathbf{g}^k$ simplifies to

$$\mathbf{p}^{k+1} = \mathbf{g}^k + \beta_k \mathbf{p}^k \tag{4.12}$$

with

$$\beta_k = \beta_{kk} = -\frac{(\mathbf{p}^k)^T \mathsf{A} \mathbf{g}^k}{(\mathbf{p}^k)^T \mathsf{A} \mathbf{p}^k}. \tag{4.13}$$

Finally, observe that the orthogonality of $\mathbf{g}^k$ to $\mathrm{Span}\{\mathbf{p}^1, \dots, \mathbf{p}^k\}$ and (4.12) imply that

$$\|\mathbf{p}^{k+1}\| \geq \|\mathbf{g}^k\|. \tag{4.14}$$

In particular, if $\mathbf{g}^{k-1} \neq \mathbf{o}$, then $\mathbf{p}^k \neq \mathbf{o}$, so the formula (4.13) is well defined provided $\mathbf{g}^{k-1} \neq \mathbf{o}$.

4.3 Conjugate Gradient Method

In the previous two sections, we found that we can use the conjugate directions to reduce problem (4.1) to a sequence of one-dimensional problems and that the conjugate directions can be generated very efficiently. The famous *conjugate gradient (CG) method* combines these two observations.

The algorithm starts from an initial guess $\mathbf{x}^0$, $\mathbf{g}^0 = \mathsf{A}\mathbf{x}^0 - \mathbf{b}$, and $\mathbf{p}^1 = \mathbf{g}^0$. If $\mathbf{x}^{k-1}$ and $\mathbf{g}^{k-1}$ are given, $k \geq 1$, it first checks if $\mathbf{x}^{k-1}$ is the solution. If not, then the algorithm generates

$$\begin{aligned}
\mathbf{x}^k &= \mathbf{x}^{k-1} - \alpha_k \mathbf{p}^k \quad \text{with} \quad \alpha_k = (\mathbf{g}^{k-1})^T \mathbf{p}^k / (\mathbf{p}^k)^T \mathsf{A} \mathbf{p}^k \\
\mathbf{g}^k &= \mathsf{A}\mathbf{x}^k - \mathbf{b} = \mathsf{A}(\mathbf{x}^{k-1} - \alpha_k \mathbf{p}^k) - \mathbf{b} = \mathbf{g}^{k-1} - \alpha_k \mathsf{A} \mathbf{p}^k.
\end{aligned}$$

Finally the new conjugate direction $\mathbf{p}^{k+1}$ is generated by (4.12) and (4.13).

The decision if $\mathbf{x}^{k-1}$ is an acceptable solution is typically based on the value of $\|\mathbf{g}^{k-1}\|$, so we must evaluate the norm of the gradient at each step. We can also use the norm to replace the scalar products involving the gradient in the definition of α_k and β_k. To find the formulae, let us replace k in (4.12) by $k-1$ and multiply the result by $(\mathbf{g}^{k-1})^T$. Using the orthogonality, we get

$$(\mathbf{g}^{k-1})^T \mathbf{p}^k = \|\mathbf{g}^{k-1}\|^2 + \beta_{k-1} (\mathbf{g}^{k-1})^T \mathbf{p}^{k-1} = \|\mathbf{g}^{k-1}\|^2, \tag{4.15}$$

so by (4.8)

$$\alpha_k = \|\mathbf{g}^{k-1}\|^2 / (\mathbf{p}^k)^T \mathsf{A} \mathbf{p}^k. \tag{4.16}$$

Using that $\alpha_k > 0$ for $\mathbf{g}^{k-1} \neq \mathbf{o}$ and that by (4.15)

$$\mathsf{A}\mathbf{p}^k = \frac{1}{\alpha_k} (\mathbf{g}^{k-1} - \mathbf{g}^k) \quad \text{and}$$

we can find an alternative formula for β_k as

$$\beta_k = -\frac{(\mathbf{p}^k)^T \mathsf{A} \mathbf{g}^k}{(\mathbf{p}^k)^T \mathsf{A} \mathbf{p}^k} = \frac{\|\mathbf{g}^k\|^2}{\alpha_k (\mathbf{p}^k)^T \mathsf{A} \mathbf{p}^k} = \frac{\|\mathbf{g}^k\|^2}{\|\mathbf{g}^{k-1}\|^2}. \tag{4.17}$$

We present the complete CG method as Algorithm 4.1.

Algorithm 4.1. Conjugate gradient method (CG)

Given a symmetric positive definite matrix $\mathsf{A} \in \mathbb{R}^{n\times n}$ *and* $\mathbf{b} \in \mathbb{R}^n$.

Step 0. {*Initialization.*}
Choose $\mathbf{x}^0 \in \mathbb{R}^n$, *set* $\mathbf{g}^0 = \mathsf{A}\mathbf{x}^0 - \mathbf{b}$, $\mathbf{p}^1 = \mathbf{g}^0$, $k = 1$
Step 1. {*Conjugate gradient loop.*}
while $\|\mathbf{g}^{k-1}\| > 0$
$\alpha_k = \|\mathbf{g}^{k-1}\|^2/(\mathbf{p}^k)^T\mathsf{A}\mathbf{p}^k$
$\mathbf{x}^k = \mathbf{x}^{k-1} - \alpha_k\mathbf{p}^k$
$\mathbf{g}^k = \mathbf{g}^{k-1} - \alpha_k\mathsf{A}\mathbf{p}^k$
$\beta_k = \|\mathbf{g}^k\|^2/\|\mathbf{g}^{k-1}\|^2 = -(\mathsf{A}\mathbf{p}^k)^T\mathbf{g}^k/((\mathbf{p}^k)^T\mathsf{A}\mathbf{p}^k)$
$\mathbf{p}^{k+1} = \mathbf{g}^k + \beta_k\mathbf{p}^k$
$k = k + 1$
***end** while*
Step 2. {*Return the solution.*}
$\widehat{\mathbf{x}} = \mathbf{x}^k$

We can implement each step of the CG method with just one matrix–vector multiplication. This multiplication by the Hessian matrix A typically dominates the cost of the step. Only one generation of vectors $\mathbf{x}^k, \mathbf{p}^k$, and $\mathbf{g}^k$ is typically stored, so the memory requirements are modest.

Let us recall that the algorithm finds at each step the minimizer $\mathbf{x}^k$ of f on $\mathbf{x}^0 + \mathcal{K}^k = \mathbf{x}^0 + \mathcal{K}^k(\mathsf{A}, \mathbf{g}^0)$ and expands the conjugate basis of $\mathcal{K}^k$ to that of $\mathcal{K}^{k+1}$ provided $\mathbf{g}^k \neq \mathbf{o}$. Since the dimension of $\mathcal{K}^k$ is less than or equal to k, it follows that for some $k \leq n$

$$\mathcal{K}^k = \mathcal{K}^{k+1}.$$

Since $\mathbf{g}^k \in \mathcal{K}^{k+1}$ and $\mathbf{g}^k$ is orthogonal to $\mathcal{K}^k$, Algorithm 4.1 finds the solution $\widehat{\mathbf{x}}$ of (4.1) in at most n steps. We can sum up the most important properties of Algorithm 4.1 into the following theorem.

Theorem 4.1. *Let* $\{\mathbf{x}^k\}$ *be generated by Algorithm 4.1 to find the solution* $\widehat{\mathbf{x}}$ *of (4.1) starting from* $\mathbf{x}^0 \in \mathbb{R}^n$. *Then the algorithm is well defined and there is* $k \leq n$ *such that* $\mathbf{x}^k = \widehat{\mathbf{x}}$. *Moreover, the following statements hold for* $i = 1, \ldots, k$:

(i) $f(\mathbf{x}^i) = \min\{f(\mathbf{x}) : \mathbf{x} \in \mathbf{x}^0 + \mathcal{K}^i(\mathsf{A}, \mathbf{g}^0)\}$.
(ii) $\|\mathbf{p}^{i+1}\| \geq \|\mathbf{g}^i\|$.
(iii) $(\mathbf{g}^i)^T\mathbf{g}^j = 0$ *for* $i \neq j$.
(iv) $(\mathbf{p}^i)^T\mathsf{A}\mathbf{p}^j = 0$ *for* $i \neq j$.
(v) $\mathcal{K}^i(\mathsf{A}, \mathbf{g}^0) = \mathrm{Span}\{\mathbf{g}^0, \ldots, \mathbf{g}^{i-1}\} = \mathrm{Span}\{\mathbf{p}^1, \ldots, \mathbf{p}^i\}$.

It is usually sufficient to find $\mathbf{x}^k$ such that $\|\mathbf{g}^k\|$ is small. For example, given a small $\varepsilon > 0$, we can consider $\mathbf{g}^k$ small if

$$\|\mathbf{g}^k\| \leq \varepsilon\|\mathbf{b}\|.$$

Then $\widetilde{\mathbf{x}} = \mathbf{x}^k$ is an approximate solution which satisfies

$$\|\mathsf{A}(\widetilde{\mathbf{x}} - \widehat{\mathbf{x}})\| \leq \varepsilon\|\mathbf{b}\|, \quad \|\widetilde{\mathbf{x}} - \widehat{\mathbf{x}}\| \leq \varepsilon\lambda_{\min}(\mathsf{A})^{-1},$$

where $\lambda_{\min}(\mathsf{A})$ denotes the least eigenvalue of A. It is easy to check that the approximate solution $\widetilde{\mathbf{x}}$ solves the perturbed problem

$$\min_{\mathbf{x}\in\mathbb{R}^n} \widetilde{f}(\mathbf{x}) = \frac{1}{2}\mathbf{x}^T\mathsf{A}\mathbf{x} - \widetilde{\mathbf{b}}^T\mathbf{x}, \quad \widetilde{\mathbf{b}} = \mathbf{b} + \mathbf{g}^k.$$

What is "small" depends on the problem solved. To keep our exposition general, we often do not specify the meaning of "small" in what follows. Of course $\mathbf{g}^k = \mathbf{o}$ is always considered small.

Given an approximation $\mathbf{x}^0$ of the solution $\widehat{\mathbf{x}}$, we can use k conjugate gradient iterations to find an improved approximation $\mathbf{x}^k$. Repeating the procedure with $\mathbf{x}^0 = \mathbf{x}^k$, we get the *restarted conjugate gradient method.*

4.4 Comments

The conjugate gradient method was preceded by the conjugate direction method proposed by Fox, Huskey, and Wilkinson [135]. They did not elaborate on the generation of the conjugate basis. Hestenes and Stiefel showed in their seminal paper [164] that it is possible to effectively generate the conjugate basis of Krylov space by applying the Gram-Schmidt orthogonalization procedure of Sect. 1.9 to its natural basis $\mathbf{b}, \mathsf{A}\mathbf{b}, \ldots, \mathsf{A}^{k-1}\mathbf{b}$. They also proved all the results presented in this chapter. An alternative implementation was proposed by Lanczos [199].

The CG method is optimal because in exact arithmetic and starting from zero, it generates the kth iterates that minimize the cost function f on the k-dimensional Krylov subspace. Since all the gradient methods discussed in Chap. 3 generate the iterates in the same Krylov space, they cannot find smaller cost value in a given number of steps than the CG method; however, they can generate the same iterations. In particular, the algorithm (3.18) with the parameters α_k, β_k chosen to minimize f generates in exact arithmetic the same iterates as CG (see, e.g., Bertsekas [16, Example 1.6.5]). Moreover, the alternative gradient methods can be more resistant to the effect of rounding errors and other perturbations.

The conjugate gradient method is the most important iterative method for solving huge problems. We shall develop this comment in Chap. 8.

5
Gradient Projection

The Euclidean projection on the feasible set is an essential ingredient of the practical algorithms for solving QP and QCQP problems. We restrict our attention to the feasible sets defined by separable convex constraints that admit practical projection evaluation. Recall that we can often reduce more general inequality constraints to separable constraints by the duality.

To combine the gradient projection with CG or other iterative methods for solving linear problems, we need nontrivial bounds on the decrease of f along the projected gradient path in terms of bounds on the spectrum of its Hessian matrix A. While such results are standard for the gradient path (see Chap. 3), their generalization to the projected gradient path is more sophisticated.

Here, we present the results concerning the decrease of f along the projected-gradient path to the extent necessary for developing optimal algorithms for solving QP and QCQP problems with separable constraints. The primary tool of the analysis is the dominating function associated with the cost function. We pay special attention to over-relaxation.

5.1 Separable Convex Constraints and Projections

Our goal is to get insight into the effect of the projected-gradient step on solving the problem to find

$$\widehat{\mathbf{x}} = \arg\min_{\mathbf{x}\in\Omega} f(\mathbf{x}), \tag{5.1}$$

where $f = \frac{1}{2}\mathbf{x}^T\mathsf{A}\mathbf{x} - \mathbf{x}^T\mathbf{b}$, $\mathsf{A} \in \mathbb{R}^{n\times n}$ denotes an SPD matrix, $\mathbf{b}, \mathbf{x} \in \mathbb{R}^n$,

Z. Dostál, *Optimal Quadratic Programming and QCQP Algorithms with Case Studies*, Springer Optimization and Its Applications 23,
https://doi.org/10.1007/978-3-031-95167-1_5

$$\mathbf{x} = [\mathbf{x}_1^T, \dots, \mathbf{x}_s^T]^T, \quad \mathbf{x}_i = [x_1^i, \dots, x_{n_i}^i], \quad n_1 + \dots + n_s = n,$$

and

$$\Omega = \Omega_1 \times \dots \times \Omega_s$$

denotes a closed convex set defined by separable convex constraints

$$h_i : \mathbb{R}^{n_i} \to \mathbb{R}, \quad \Omega_i = \{\mathbf{x}_i \in \mathbb{R}^{n_i} : h_i(\{\mathbf{x}_i) \leq 0\}, \quad i = 1, \dots, s.$$

We are especially interested in the problems defined by the *bound constraints*

$$h_i(x_i) = \ell_i - x_i, \quad \ell_i, x_i \in \mathbb{R},$$

arising in the dual formulation of problems with more general inequality constraints, and the *separable elliptic constraints*

$$h_i(\mathbf{x}_i) = (\mathbf{x}_i - \mathbf{y}_i)^T \mathsf{H}_i (\mathbf{x}_i - \mathbf{y}_i) - d_i, \quad \mathbf{x}_i, \mathbf{y}_i \in \mathbb{R}^2, \quad d_i > 0, \tag{5.2}$$

where H_i is an SPD matrix. Such constraints appear in the dual formulation of some non-smooth optimization problems (see, e.g., the dissipative term in the variational formulation of contact problems with Tresca friction in Sect. 14.3 or [114, Chap. 12]). We also consider the *spherical constraints*

$$h_i(\mathbf{x}_i) = \|\mathbf{x}_i - \mathbf{y}_i\|^2 - d_i, \quad \mathbf{x}_i, \mathbf{y}_i \in \mathbb{R}^m, \quad d_i > 0.$$

In what follows, we denote by P_Ω the *Euclidean projection* to Ω, so that

$$P_\Omega(\mathbf{x}) = \arg\min_{\mathbf{y} \in \Omega} \|\mathbf{x} - \mathbf{y}\|.$$

Since the constraints that define Ω are separable, we can define P_Ω block-wise by

$$P_{\Omega_i}(\mathbf{x}_i) = \arg\min_{\mathbf{y} \in \Omega_i} \|\mathbf{x}_i - \mathbf{y}\|, \quad P_\Omega(\mathbf{x}) = \left[P_{\Omega_1}(\mathbf{x}_1)^T, \dots, P_{\Omega_s}(\mathbf{x}_s)^T\right]^T. \tag{5.3}$$

The action of P_Ω is especially easy to calculate for the spherical or bound constraints. As illustrated by Fig. 5.1, the components of the projection $P_{\Omega_B}(\mathbf{x})$ of $\mathbf{x}$ onto

$$\Omega_B = \{\mathbf{x} \in \mathbb{R}^n : x_i \geq \ell_i, \ i = 1, \dots, n\}$$

are given by

$$[P_{\Omega_B}(\mathbf{x})]_i = \max\{\ell_i, x_i\}, \ i = 1, \dots, n.$$

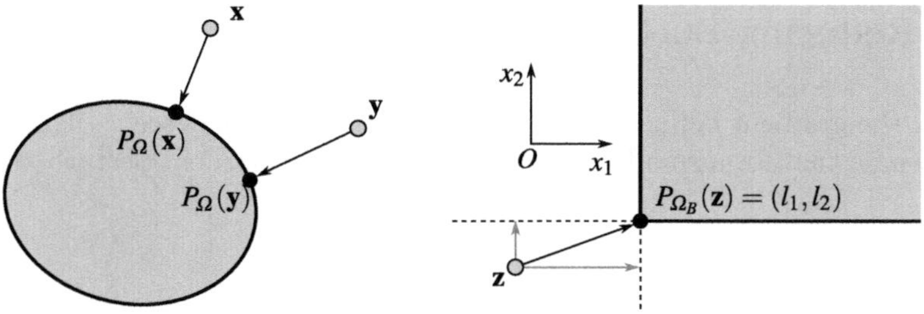

Fig. 5.1. Euclidean projections onto convex sets

The gradient projection is an important ingredient of minimization algorithms. A typical step of the gradient projection method is in Fig. 5.2.

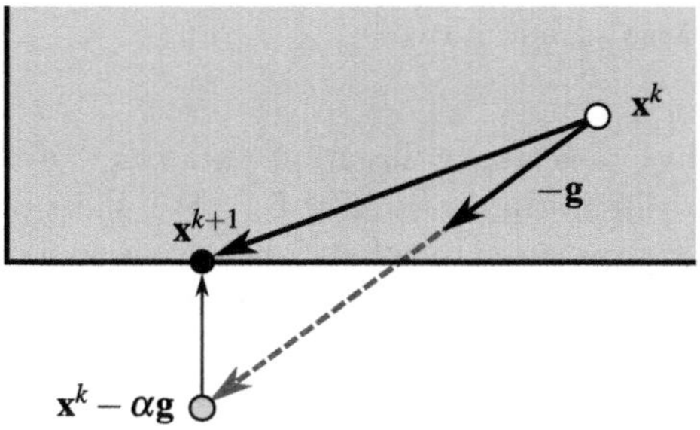

Fig. 5.2. Gradient projection step

We can alternatively describe the gradient projection on bound constrained set using the increment defined for $\mathbf{x} \in \Omega_B$ component-wise by the $(\alpha-)$*reduced gradient* $\widetilde{\mathbf{g}}_\alpha(\mathbf{x})$ as

$$[\widetilde{\mathbf{g}}_\alpha(\mathbf{x})]_i = \min\left\{g_i, \frac{x_i - \ell_i}{\alpha}\right\}. \tag{5.4}$$

Thus

$$P_{\Omega_B}(\mathbf{x} - \alpha\mathbf{g}(\mathbf{x})) = \mathbf{x} - \alpha\widetilde{\mathbf{g}}_\alpha(\mathbf{x}) \quad \text{and} \quad \|\widetilde{\mathbf{g}}_\alpha(\mathbf{x})\| \leq \|\mathbf{g}(\mathbf{x})\|. \tag{5.5}$$

A nonzero reduced gradient defines a feasible decrease direction applicable in stopping tests. More discussion on the stopping criterion is in Sect. 10.1.

5.2 Reducing Euclidean Error

Since the gradient defines the direction of the steepest descent, it is natural to assume that for a smallstep length, the negative gradient perturbed by the

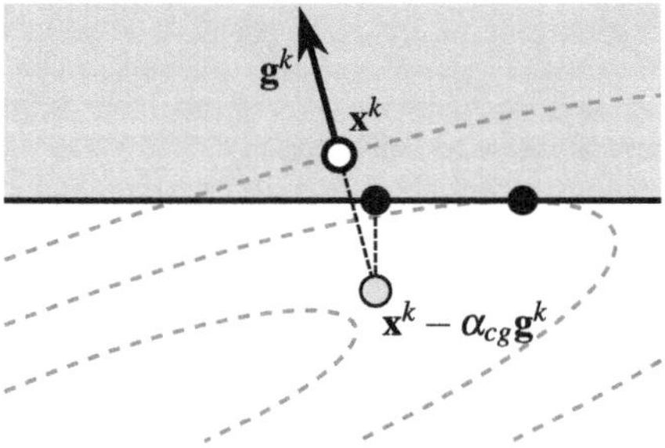

Fig. 5.3. Fixed steplength gradient step

projection P_Ω defines a decrease direction and a new iterate, which is closer to $\widehat{\mathbf{x}}$ in the Euclidean norm, as in Fig. 5.3. We shall prove a quantitative refinement of this conjecture. In what follows, we restrict our attention to the analysis of the fixed step-length gradient iteration

$$\mathbf{x}^{k+1} = P_\Omega(\mathbf{x}^k - \alpha\mathbf{g}^k), \tag{5.6}$$

where $\mathbf{g}^k = \nabla f(\mathbf{x}^k)$.

Which values of α guarantee that the iterates defined by the fixed gradient projection step (5.6) approach the solution $\widehat{\mathbf{x}}$ of problem (5.1) in the Euclidean norm? Somewhat surprisingly, we find that the projected gradient step reduces the distance of $\mathbf{x}$ from the solution $\widehat{\mathbf{x}}$ with the same bound as the gradient step for the unconstrained problem (3.1).

Let us start with the following lemma.

Lemma 5.1. *Let Ω denote a closed convex set, let $\widehat{\mathbf{x}} \in \Omega$ solves* (5.1), *and denote $\widehat{\mathbf{g}} = \nabla f(\widehat{\mathbf{x}})$. Then for any $\alpha > 0$*

$$P_\Omega(\widehat{\mathbf{x}} - \alpha\widehat{\mathbf{g}}) = \widehat{\mathbf{x}}. \tag{5.7}$$

Proof. The statement is trivial for $\widehat{\mathbf{g}} = \mathbf{o}$. Assume that

$$\mathbf{y} = \widehat{\mathbf{x}} + \mathbf{d} \in \Omega \quad \text{and} \quad \widehat{\mathbf{g}} \neq \mathbf{o}.$$

Since $\widehat{\mathbf{x}}$ solves (5.1), it follows that

$$\widehat{\mathbf{g}}^T\mathbf{d} \geq 0$$

and for any $\alpha > 0$

$$\begin{aligned}\|\widehat{\mathbf{x}}-\alpha\widehat{\mathbf{g}}-\mathbf{y}\|^2 &= \|\widehat{\mathbf{x}}-\alpha\widehat{\mathbf{g}}-(\widehat{\mathbf{x}}+\mathbf{d})\|^2 = \|\alpha\widehat{\mathbf{g}}+\mathbf{d}\|^2\\ &= \|\alpha\widehat{\mathbf{g}}\|^2+2\alpha\widehat{\mathbf{g}}^T\mathbf{d}+\|\mathbf{d}\|^2 \geq \|\alpha\widehat{\mathbf{g}}\|^2+\|\mathbf{d}\|^2.\end{aligned}$$

Thus the minimum is achieved at $\mathbf{y}=\widehat{\mathbf{x}}$ and

$$P_\Omega(\widehat{\mathbf{x}}-\alpha\widehat{\mathbf{g}}) = \arg\min_{\mathbf{y}\in\Omega}\|\widehat{\mathbf{x}}-\alpha\widehat{\mathbf{g}}-\mathbf{y}\| = \widehat{\mathbf{x}}. \tag{5.8}$$

□.

The following proposition extends the estimates of Proposition 3.1 to the projected gradient step.

Proposition 5.1. *Let Ω denote a closed convex set, $\mathbf{x}\in\Omega$, and $\mathbf{g}=\nabla f(\mathbf{x})$. Then for any $\alpha>0$*

$$\|P_\Omega(\mathbf{x}-\alpha\mathbf{g})-\widehat{\mathbf{x}}\| \leq \eta_E\|\mathbf{x}-\widehat{\mathbf{x}}\|, \tag{5.9}$$

where η_E is defined by the extreme eigenvalues of A as

$$\eta_E = \max\{|1-\alpha\lambda_{\min}|, |1-\alpha\lambda_{\max}|\}. \tag{5.10}$$

Proof. First recall that by Lemma 5.1

$$P_\Omega(\widehat{\mathbf{x}}-\alpha\widehat{\mathbf{g}}) = \widehat{\mathbf{x}}.$$

Using that the projection P_Ω is nonexpansive by Corollary 2.7, the definition of gradient, the relations between the norm of a symmetric matrix and its spectrum (1.30), and the observation that if λ_i are the eigenvalues of A, then $1-\alpha\lambda_i$ are the eigenvalues of $\mathsf{I}-\alpha\mathsf{A}$ (see also (1.32)), we get

$$\begin{aligned}\|P_\Omega(\mathbf{x}-\alpha\mathbf{g})-\widehat{\mathbf{x}}\| &= \|P_\Omega(\mathbf{x}-\alpha\mathbf{g})-P_\Omega(\widehat{\mathbf{x}}-\alpha\widehat{\mathbf{g}})\|\\ &\overset{C.2.7}{\leq} \|(\mathbf{x}-\alpha\mathbf{g})-(\widehat{\mathbf{x}}-\alpha\widehat{\mathbf{g}})\|\\ &= \|(\mathbf{x}-\widehat{\mathbf{x}})-\alpha(\mathbf{g}-\widehat{\mathbf{g}})\| = \|(\mathbf{x}-\widehat{\mathbf{x}})-\alpha\mathsf{A}(\mathbf{x}-\widehat{\mathbf{x}})\|\\ &= \|(\mathsf{I}-\alpha\mathsf{A})(\mathbf{x}-\widehat{\mathbf{x}})\|\\ &\leq \max\{|1-\alpha\lambda_{\min}|, |1-\alpha\lambda_{\max}|\}\|\mathbf{x}-\widehat{\mathbf{x}}\|.\end{aligned}$$

□

We call η_E the *coefficient of Euclidean contraction.* If $\alpha\in(0,2\|\mathsf{A}\|^{-1})$, then $\eta_E<1$. Using the same arguments as in the proof of Proposition 3.1, we get that η_E is minimized by

$$\alpha_E^{opt} = \frac{2}{\lambda_{\min}+\lambda_{\max}}, \qquad \eta_E^{opt} = \frac{\lambda_{\max}-\lambda_{\min}}{\lambda_{\max}+\lambda_{\min}} = \frac{\kappa(\mathsf{A})-1}{\kappa(\mathsf{A})+1}. \tag{5.11}$$

Notice that the estimate (5.9) does not guarantee any bound on the decrease of the cost function. We shall study this topic in Sect. 5.8.

5.3 Quadratic Function with Identity Hessian

Which values of α guarantee the decrease of f in each iterate defined by the fixed gradient projection step (5.6)? How much? To answer these questions, we shall carry out some analysis for a special function

$$F(\mathbf{x}) = \frac{1}{2}\mathbf{x}^T\mathbf{x} - \mathbf{c}^T\mathbf{x}, \quad \mathbf{x} \in \mathbb{R}^n, \tag{5.12}$$

which is defined by a fixed $\mathbf{c} \in \mathbb{R}^n$, $\mathbf{c} = [c_i]$. We can also define F componentwise as

$$F(\mathbf{x}) = \sum_{i=1}^{n} F_i(x_i), \quad F_i(x_i) = \frac{1}{2}x_i^2 - c_i x_i, \quad \mathbf{x} = [x_i]. \tag{5.13}$$

The Hessian and the gradient of F are expressed by

$$\nabla^2 F(\mathbf{x}) = \mathsf{I} \quad \text{and} \quad \mathbf{g} = \nabla F(\mathbf{x}) = \mathbf{x} - \mathbf{c}, \quad \mathbf{g} = [g_i], \tag{5.14}$$

respectively. Thus $\mathbf{c} = \mathbf{x} - \mathbf{g}$ and for any $\mathbf{x} \in \mathbb{R}^n$

$$\|\mathbf{x} - \mathbf{c}\|^2 = \|\mathbf{x}\|^2 - 2\mathbf{c}^T\mathbf{x} + \|\mathbf{c}\|^2 \overset{(5.12)}{=} 2F(\mathbf{x}) + \|\mathbf{c}\|^2. \tag{5.15}$$

Let us denote

$$\widehat{\mathbf{c}} = P_\Omega(\mathbf{c}) = P_\Omega(\mathbf{x} - \mathbf{g})$$

and notice that by Corollary 2.6 for any $\mathbf{x} \in \Omega$

$$\begin{aligned}\|\mathbf{x} - \mathbf{c}\|^2 &= \|(\mathbf{x} - \widehat{\mathbf{c}}) + (\widehat{\mathbf{c}} - \mathbf{c})\|^2 = \|\mathbf{x} - \widehat{\mathbf{c}}\|^2 + 2(\mathbf{x} - \widehat{\mathbf{c}})^T(\widehat{\mathbf{c}} - \mathbf{c}) + \|\widehat{\mathbf{c}} - \mathbf{c}\|^2 \\ &\overset{C.2.6}{\geq} \|\mathbf{x} - \widehat{\mathbf{c}}\|^2 + \|\widehat{\mathbf{c}} - \mathbf{c}\|^2.\end{aligned}$$

Using (5.15), we get that for any $\mathbf{x} \in \Omega$

$$2F(\mathbf{x}) = \|\mathbf{x} - \mathbf{c}\|^2 - \|\mathbf{c}\|^2 \overset{C.2.7}{\geq} \|\widehat{\mathbf{c}} - \mathbf{c}\|^2 - \|\mathbf{c}\|^2 = 2F\left(\widehat{\mathbf{c}}\right). \tag{5.16}$$

We have thus proved that if $\mathbf{y} \in \Omega$, then, as also illustrated in Fig. 5.4,

$$F(\widehat{\mathbf{c}}) = F\big(P_\Omega(\mathbf{x} - \mathbf{g})\big) \leq F(\mathbf{y}). \tag{5.17}$$

We are especially interested in the analysis of F along the projected-gradient path

$$p(\mathbf{x}, \alpha) = P_\Omega\big(\mathbf{x} - \alpha\nabla F(\mathbf{x})\big),$$

where $\alpha \geq 0$ and $\mathbf{x} \in \Omega$ is fixed. A geometric illustration of the projected-gradient path for the feasible set of a bound-constrained problem is in Fig. 5.5.

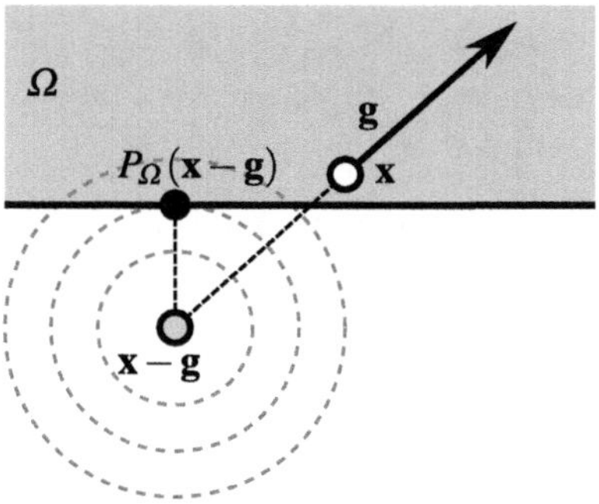

Fig. 5.4. Minimizer of F in Ω

5.4 Decrease Along the Projected-Gradient Path

Let us prove a simple property of an infinite half-closed interval that opens the way to the proof of over-relaxation of the gradient projection step for bound constraints.

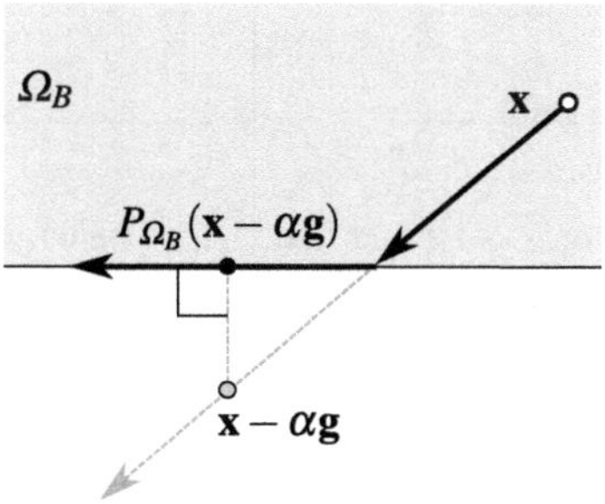

Fig. 5.5. Projected-gradient path

Lemma 5.2. *Let $\ell, c \in \mathbb{R}$ and $\Omega_B = [\ell, \infty)$. Let F and g be defined by*

$$F(x) = \frac{1}{2}x^2 - cx \quad \text{and} \quad g = x - c.$$

Then for any $\delta \in [0, 1]$

$$F\big(P_{\Omega_B}(x - (2-\delta)g)\big) \le F\big(P_{\Omega_B}(x - \delta g)\big). \tag{5.18}$$

Proof. First assume that $x \ge \ell$ is fixed and denote

$$g = F'(x) = x - c, \quad \widetilde{g}(0) = 0, \quad \widetilde{g}(\alpha) = \min\{(x-\ell)/\alpha, g\}, \quad \alpha > 0.$$

For convenience, let us define for $\alpha \in (0, 2]$

$$F\big(P_{\Omega_B}(x-\alpha g)\big) = F(x) + \Phi(\alpha), \quad \Phi(\alpha) = -\alpha \widetilde{g}(\alpha) g + \frac{\alpha^2}{2}\left(\widetilde{g}(\alpha)\right)^2.$$

Alternatively, we can define Φ by

$$\Phi(\alpha) = \begin{cases} \Phi_F(\alpha) & \text{for} \quad \alpha \in (0, \overline{\alpha}], \\ \Phi_A(\alpha) & \text{for} \quad \alpha \in [\overline{\alpha}, 2], \end{cases}$$

where $\overline{\alpha} = 2$ if $g \le 0$ and $\overline{\alpha} = \min\{(x-\ell)/g,\, 2\}$ if $g > 0$,

$$\Phi_F(\alpha) = \left(-\alpha + \frac{\alpha^2}{2}\right) g^2, \quad \text{and} \quad \Phi_A(\alpha) = -g(x-\ell) + \frac{1}{2}(x-\ell)^2.$$

We are interested in $\alpha \in (0, 2]$. See also Fig. 5.6.

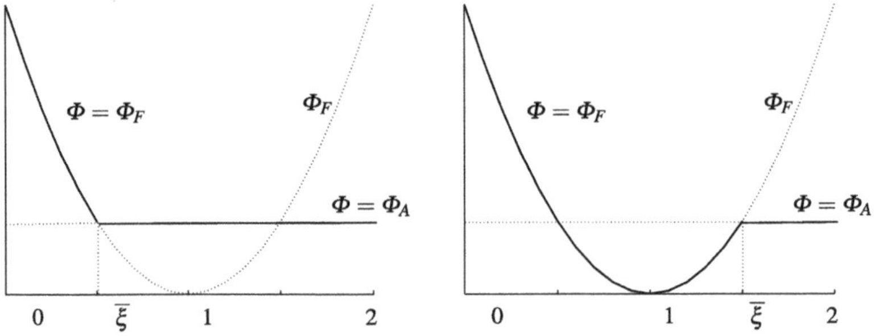

Fig. 5.6. Graphs of $\Phi(\alpha)$ for $\overline{\xi} < 1$ (*left*) and $\overline{\xi} > 1$ (*right*) when $g > 0$

It follows that for any α

$$\Phi_F(2-\alpha) = \left(-(2-\alpha) + \frac{(2-\alpha)^2}{2}\right) g^2 = \Phi_F(\alpha), \tag{5.19}$$

and if $g \le 0$, then

$$\Phi(\alpha) = \Phi_F(\alpha) = \Phi_F(2-\alpha) = \Phi(2-\alpha).$$

Let us now assume that $g > 0$ and denote $\overline{\xi} = (x-\ell)/g$. Simple analysis shows that if $\overline{\xi} \in [0,1]$, then Φ is nonincreasing on $[0, 2]$ and (5.18) is satisfied for $\alpha \in [0,1]$. To finish the proof of (5.18), notice that if $1 < \overline{\xi}$, then

$$\Phi(\alpha) = \Phi_F(\alpha), \quad \alpha \in [0,1], \quad \Phi(\alpha) \le \Phi_F(\alpha), \quad \alpha \in [1,2],$$

so that we can use (5.19) to get that for $\alpha \in [0,1]$

$$\Phi(2-\alpha) \le \Phi_F(2-\alpha) = \Phi_F(\alpha) = \Phi(\alpha).$$

□

We can extend (5.18) to the product of infinite half-closed intervals.

Proposition 5.2. *Let us define $\Omega_B \subset \mathbb{R}^n$ as a direct product of infinite half-closed intervals. Then for any $0 < \delta < 1$*

$$F\big(P_{\Omega_B}(\mathbf{x} - (2-\delta)\mathbf{g}(\mathbf{x}))\big) \le F\big(P_{\Omega_B}(\mathbf{x} - \delta\mathbf{g}(\mathbf{x}))\big). \tag{5.20}$$

Proof. If $n = 1$, then the statement reduces to Lemma 5.2. To prove the statement for $n > 1$, notice that F is the sum of the functions that satisfy the required relations. □

5.5 Dominating Function

Let us assume that Ω is convex,

$$0 < \delta\|\mathsf{A}\| \le 1,$$

and let $\mathbf{x} \in \Omega$ be fixed so that we can define the *dominating function*

$$F_\delta(\mathbf{y}) = \delta f(\mathbf{y}) + \frac{1}{2}(\mathbf{y}-\mathbf{x})^T(\mathsf{I} - \delta\mathsf{A})(\mathbf{y}-\mathbf{x}), \quad \mathbf{y} \in \mathbb{R}^n. \tag{5.21}$$

It is defined so that

$$F_\delta(\mathbf{x}) = \delta f(\mathbf{x}), \quad \nabla F_\delta(\mathbf{x}) = \delta\nabla f(\mathbf{x}) = \delta\mathbf{g}(\mathbf{x}), \quad \text{and} \quad \nabla^2 F_\delta(\mathbf{y}) = \mathsf{I}. \tag{5.22}$$

Moreover, for any $\mathbf{y} \in \mathbb{R}^n$

$$\delta f(\mathbf{y}) \le F_\delta(\mathbf{y}). \tag{5.23}$$

It follows that

$$\delta f\left(P_\Omega(\mathbf{x} - \delta\mathbf{g}(\mathbf{x}))\right) - \delta f(\widehat{\mathbf{x}}) \overset{(5.23)}{\le} F_\delta\left(P_\Omega(\mathbf{x} - \delta\mathbf{g}(\mathbf{x}))\right) - \delta f(\widehat{\mathbf{x}}) \tag{5.24}$$

and

$$\nabla F_\delta(\mathbf{y}) \overset{(5.21)}{=} \delta\nabla f(\mathbf{y}) + (\mathsf{I} - \delta\mathsf{A})(\mathbf{y}-\mathbf{x}) = \mathbf{y} - (\mathbf{x} - \delta\mathbf{g}(\mathbf{x})). \tag{5.25}$$

Using (5.17) and (5.22), we get that for any $\mathbf{z} \in \Omega$

$$F_\delta\left(P_\Omega(\mathbf{x} - \delta\mathbf{g}(\mathbf{x}))\right) \le F_\delta(\mathbf{z}). \tag{5.26}$$

Now, we are ready to prove the following lemma.

Lemma 5.3. *Let* A *denote the Hessian of the cost function f defined by* (5.1), $\mathbf{x} \in \Omega_B$, $\overline{\alpha} \in (0, 2\|\mathsf{A}\|^{-1})$, *and* $\widehat{\alpha} = \min\{\overline{\alpha}, 2\|\mathsf{A}\|^{-1} - \overline{\alpha}\} \le \|\mathsf{A}\|^{-1}$.

Then

$$f(P_{\Omega_B}(\mathbf{x} - \overline{\alpha}\boldsymbol{\varphi}(\mathbf{x}) \leq f(\mathbf{x}) - \frac{\widehat{\alpha}}{2}\widetilde{\boldsymbol{\varphi}}_{\widehat{\alpha}}(\mathbf{x})^T\boldsymbol{\varphi}(\mathbf{x}). \tag{5.27}$$

Proof. Using the definition of the dominating function and its properties, Proposition 5.2, and the above inequalities, we get

$$\begin{aligned}
\widehat{\alpha}f(P_{\Omega_B}(\mathbf{x} - \overline{\alpha}\boldsymbol{\varphi}(\mathbf{x}) &\overset{(5.23)}{\leq} F_{\widehat{\alpha}}(P_{\Omega_B}(\mathbf{x} - \overline{\alpha}\boldsymbol{\varphi}(\mathbf{x})))\\
&\overset{P.\ 5.2}{\leq} F_{\widehat{\alpha}}(P_{\Omega_B}(\mathbf{x} - \widehat{\alpha}\boldsymbol{\varphi}(\mathbf{x}))) \overset{(5.5)}{=} F_{\widehat{\alpha}}\left(\mathbf{x} - \widehat{\alpha}\widetilde{\boldsymbol{\varphi}}_{\widehat{\alpha}}(\mathbf{x})\right)\\
&= F_{\widehat{\alpha}}(\mathbf{x}) - \widehat{\alpha}\widetilde{\boldsymbol{\varphi}}_{\widehat{\alpha}}(\mathbf{x})^T\widehat{\alpha}\boldsymbol{\varphi}(\mathbf{x}) + \frac{\widehat{\alpha}^2}{2}\|\widetilde{\boldsymbol{\varphi}}_{\widehat{\alpha}}(\mathbf{x})\|^2\\
&\leq F_{\widehat{\alpha}}(\mathbf{x}) - \frac{\widehat{\alpha}^2}{2}\widetilde{\boldsymbol{\varphi}}_{\widehat{\alpha}}(\mathbf{x})^T\boldsymbol{\varphi}(\mathbf{x})\\
&= \widehat{\alpha}f(\mathbf{x}) - \frac{\widehat{\alpha}^2}{2}\widetilde{\boldsymbol{\varphi}}_{\widehat{\alpha}}(\mathbf{x})^T\boldsymbol{\varphi}(\mathbf{x}).
\end{aligned}$$

To get (5.27), divide the above chain by $\widehat{\alpha}$. □

5.6 The Schöberl Lemma

The following lemma and its nice proof for bound constraints are due to Schöberl [244, 245]. We shall formulate it for a more general convex set Ω so that we can use it also for separable constraints.

Lemma 5.4. *Let Ω be a closed convex set, let $\lambda_{\min}$ denote the smallest eigenvalue of* A,

$$\mathbf{g} = \nabla f(\mathbf{x}), \quad \mathbf{x} \in \Omega, \quad \delta \in (0, \|\mathsf{A}\|^{-1}],$$

and let

$$\widehat{\mathbf{x}} = \arg\min_{\mathbf{x}\in\Omega} f(\mathbf{x})$$

denote a unique solution of (5.1). Then

$$F_\delta\left(P_\Omega(\mathbf{x} - \delta\mathbf{g})\right) - \delta f(\widehat{\mathbf{x}}) \leq \delta(1 - \delta\lambda_{\min})\left(f(\mathbf{x}) - f(\widehat{\mathbf{x}})\right). \tag{5.28}$$

Proof. Let us denote

$$[\widehat{\mathbf{x}}, \mathbf{x}] = \{\mathbf{x} + t\mathbf{d} : t \in [0, 1]\} = \mathrm{Conv}\{\widehat{\mathbf{x}}, \mathbf{x}\} \subseteq \Omega \quad \text{and} \quad \mathbf{d} = \widehat{\mathbf{x}} - \mathbf{x}.$$

Using

$$0 < \lambda_{\min}\delta \leq \|\mathsf{A}\|\delta \leq 1, \ \lambda_{\min}\|\mathbf{d}\|^2 \leq \mathbf{d}^T\mathsf{A}\mathbf{d},$$

substitution $t = \delta\lambda_{\min}$, and simple manipulations, we get

$$
\begin{aligned}
F_\delta\left(P_\Omega(\mathbf{x}-\delta\mathbf{g})\right) - \delta f(\widehat{\mathbf{x}}) &\overset{(5.26)}{=} \min\{F_\delta(\mathbf{y}) - \delta f(\widehat{\mathbf{x}}) :\ \mathbf{y}\in\Omega\} \\
&\le \min\{F_\delta(\mathbf{y}) - \delta f(\widehat{\mathbf{x}}) :\ \mathbf{y}\in[\widehat{\mathbf{x}},\mathbf{x}]\} \\
&= \min\{F_\delta(\mathbf{x}+t\mathbf{d}) - \delta f(\mathbf{x}+\mathbf{d}) :\ t\in[0,1]\} \\
&\overset{(5.22)}{=} \min\{\delta t\mathbf{d}^T\mathbf{g} + \frac{t^2}{2}\|\mathbf{d}\|^2 - \delta\mathbf{d}^T\mathbf{g} - \frac{\delta}{2}\mathbf{d}^T\mathsf{A}\mathbf{d} :\ t\in[0,1]\} \\
&\overset{subst.}{\le} \delta^2\lambda_{\min}\mathbf{d}^T\mathbf{g} + \frac{1}{2}\delta^2\lambda_{\min}^2\|\mathbf{d}\|^2 - \delta\mathbf{d}^T\mathbf{g} - \frac{\delta}{2}\mathbf{d}^T\mathsf{A}\mathbf{d} \\
&\le \delta^2\lambda_{\min}\mathbf{d}^T\mathbf{g} + \frac{1}{2}\delta^2\lambda_{\min}\mathbf{d}^T\mathsf{A}\mathbf{d} - \delta\mathbf{d}^T\mathbf{g} - \frac{\delta}{2}\mathbf{d}^T\mathsf{A}\mathbf{d} \\
&= \delta(\delta\lambda_{\min} - 1)(\mathbf{d}^T\mathbf{g} + \frac{1}{2}\mathbf{d}^T\mathsf{A}\mathbf{d}) \\
&= \delta(\delta\lambda_{\min} - 1)\left(f(\mathbf{x}+\mathbf{d}) - f(\mathbf{x})\right) \\
&= \delta(1 - \delta\lambda_{\min})\left(f(\mathbf{x}) - f(\widehat{\mathbf{x}})\right).
\end{aligned}
$$

□

5.7 Subsymmetric Sets

The proof of the over-relaxation estimate for bound constrained sets used the inequality (5.20) for infinite half-closed intervals. The inequality is satisfied by some other separable sets; we call such sets *sub-symmetric*.

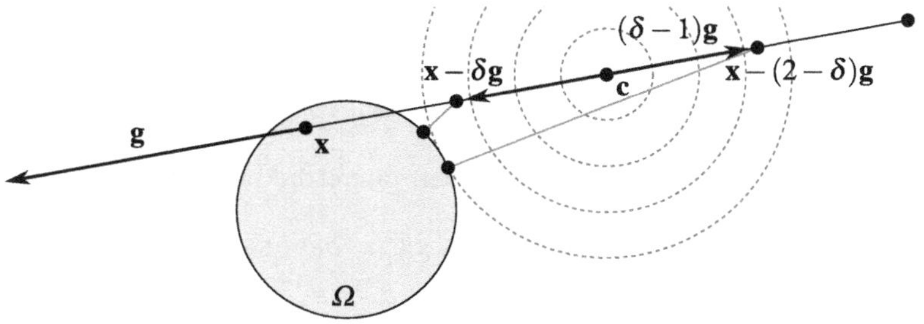

Fig. 5.7. The condition which defines a subsymmetric set

Definition 1. A closed convex set $\Omega\subseteq\mathbb{R}^n$ is *subsymmetric* if for F defined by (5.12), $\mathbf{x}\in\Omega$, $\delta\in[0,1]$, $\mathbf{c}\in\mathbb{R}^n$, and $\mathbf{g}=\mathbf{x}-\mathbf{c}$

$$F\big(P_\Omega(\mathbf{x}-(2-\delta)\mathbf{g})\big) \le F\big(P_\Omega(\mathbf{x}-\delta\mathbf{g})\big). \tag{5.29}$$

The bound-constrained set Ω_B is subsymmetric by Proposition 5.2. Fig. 5.7 indicates that the sphere is subsymmetric. Notice that it is easy to evaluate effectively the projection on the sphere and product of spheres. The following lemma by Bouchala and Vodstrčil guarantees that both ellipses and spheres are subsymmetric sets [26].

Lemma 5.5. *Let $\Omega \subset \mathbb{R}^2$ be defined by an elliptic constraint*

$$\Omega = \{\mathbf{x} \in \mathbb{R}^2 : h(\mathbf{x}) \leq 0\}, \quad h(\mathbf{x}) = (\mathbf{x}-\mathbf{y})^T \mathsf{H}(\mathbf{x}-\mathbf{y}) - d, \quad \mathbf{y} \in \mathbb{R}^2, \quad d > 0,$$

where H is SPD. Then Ω is subsymmetric.

Proof. We can assume that the ellipse Ω is centered at the origin, i.e.,

$$\Omega = \left\{\mathbf{x} = \begin{bmatrix} x_1 \\ x_2 \end{bmatrix} \in \mathbb{R}^2 : \frac{x_1^2}{a^2} + \frac{x_2^2}{b^2} \leq 1\right\},$$

where $a, b \in \mathbb{R}$ and $a, b > 0$. We prove the following three cases:

1. $\mathbf{x} + \alpha\mathbf{g} \in \Omega$, $\mathbf{x} - \alpha\mathbf{g} \in \Omega$,
2. $\mathbf{x} + \alpha\mathbf{g} \in \Omega$, $\mathbf{x} - \alpha\mathbf{g} \notin \Omega$,
3. $\mathbf{x} + \alpha\mathbf{g} \notin \Omega$, $\mathbf{x} - \alpha\mathbf{g} \notin \Omega$.

Ad 1. In this case,

$$\|P_\Omega(\mathbf{x} + \alpha\mathbf{g}) - \mathbf{x}\| = \|\alpha\mathbf{g}\| = \|P_\Omega(\mathbf{x} - \alpha\mathbf{g}) - \mathbf{x}\|.$$

Ad 2. Since $[b\cos t, a\sin t]$ is the outer normal vector of the ellipse

$$\partial\Omega = \{[a\cos t, b\sin t] \in \mathbb{R}^2 : \ t \in \mathbb{R}\}$$

at $[a\cos t, b\sin t]$, we can write $\mathbf{x} + \alpha\mathbf{g}$ and $\mathbf{x} - \alpha\mathbf{g}$ in the form

$$\begin{aligned} \mathbf{x} + \alpha\mathbf{g} &= s[a\cos t_1, b\sin t_1]^T, \\ \mathbf{x} - \alpha\mathbf{g} &= [a\cos t_2, b\sin t_2]^T + k[b\cos t_2, a\sin t_2]^T, \end{aligned} \tag{5.30}$$

where $0 \leq s \leq 1$, $k > 0$, $t_1, t_2 \in \mathbb{R}$. It can be verified directly that

$$\|P_\Omega(\mathbf{x} + \alpha g) - \mathbf{x}\|^2 - \|P_\Omega(\mathbf{x} - \alpha g) - \mathbf{x}\|^2 = abk\big(1 - s\cos(t_2 - t_1)\big) \geq 0.$$

Ad 3. Now we can write

$$\begin{aligned} \mathbf{x} + \alpha\mathbf{g} &= [a\cos t_1, b\sin t_1]^T + s[b\cos t_1, a\sin t_1]^T, \\ \mathbf{x} - \alpha\mathbf{g} &= [a\cos t_2, b\sin t_2]^T + k[b\cos t_2, a\sin t_2]^T, \end{aligned} \tag{5.31}$$

where $s, k > 0$, and we can check that

$$\|P_\Omega(\mathbf{x} + \alpha\mathbf{g}) - \mathbf{x}\|^2 - \|P_\Omega(\mathbf{x} - \alpha\mathbf{g}) - \mathbf{x}\|^2 = ab(k - s)\big(1 - \cos(t_1 - t_2)\big).$$

It remains to prove that $k \geq s$. Due to the symmetry of the ellipse, we can assume $0 < t_2 \leq t_1 < \pi$. Now we consider such $t_0 \in (t_1, \pi]$ that the line given by two points

$$\mathbf{x} + \alpha\mathbf{g} = [a\cos t_1, b\sin t_1]^T + s[b\cos t_1, a\sin t_1]^T \quad \text{and} \quad [a\cos t_0, b\sin t_0]^T$$

is a tangent to the ellipse. It is easy to calculate that

$$s = \frac{ab\big(1 - \cos(t_0 - t_1)\big)}{b^2\cos t_0\cos t_1 + a^2\sin t_0\sin t_1}, \qquad k \geq \frac{ab\big(1 - \cos(t_0 - t_2)\big)}{b^2\cos t_0\cos t_2 + a^2\sin t_0\sin t_2},$$

and that the function

$$f(t) := \frac{ab\big(1 - \cos(t_0 - t)\big)}{b^2\cos t_0\cos t + a^2\sin t_0\sin t}$$

has a nonpositive derivative (and therefore f is nonincreasing) on the interval $[t_2, t_1]$. This implies $k \geq s$. □

If Ω is a closed sphere in $\mathbb{R}^m$ and $\mathbf{x} \in \Omega$, then the projected gradient path's vectors $\mathbf{y} = P_\Omega(\mathbf{x} - \delta\mathbf{g})$ are on the plane defined by $\mathbf{x}$, $\mathbf{c}$, and the center of the sphere, see Bouchala et al. [27]. It follows that the m-dimensional sphere is also subsymmetric. We are ready to prove the main result of this section.

Proposition 5.3. *Let $\Omega \subset \mathbb{R}^n$ denote a direct product of sub-symmetric sets. Then Ω is subsymmetric, i.e.,*

$$F\big(P_\Omega(\mathbf{x} - (2 - \delta)\mathbf{g})\big) \leq F\big(P_\Omega(\mathbf{x} - \delta\mathbf{g})\big). \tag{5.32}$$

Proof. Let

$$\Omega = \Omega_1 \times \cdots \times \Omega_s, \tag{5.33}$$

where Ω_i are subsymmetric sets. If $s = 1$, then the statement is obvious.

To prove the statement for $s > 1$, first observe that for any $\mathbf{y} \in \mathbb{R}^N$

$$[P_\Omega(\mathbf{y})]_i = P_{\Omega_i}(\mathbf{y}_i), \; i = 1, \ldots, s.$$

Using the separable representation of F, we can rewrite F in the form

$$F(\mathbf{x}) = \sum_{i=1}^{s} F_i(\mathbf{x}_i), \; i = 1, \ldots, s,$$

which complies with (5.33). To complete the proof, use this representation and Lemmas 5.2 and 5.5 to get

$$\begin{aligned}F\big(P_\Omega(\mathbf{x}-(2-\delta)\mathbf{g})\big) &= \sum_{i=1}^{s} F_i\big(P_{\Omega_i}(\mathbf{x}_i-(2-\delta)\mathbf{g}_i)\big)\\ &\le \sum_{i=1}^{s} F_i\big(P_{\Omega_i}(\mathbf{x}_i-\delta\mathbf{g}_i)\big) = F\big(P_\Omega(\mathbf{x}-\delta\mathbf{g})\big).\end{aligned}$$

□

5.8 Reducing Energy Error

Now we are ready to give an estimate of the decrease of the cost function

$$f(\mathbf{x}) = \frac{1}{2}\mathbf{x}^T\mathsf{A}\mathbf{x} - \mathbf{b}^T\mathbf{x}$$

along the projected-gradient path. The idea of the proof is to replace f with a dominating function F.

Theorem 5.1. *Let Ω denote the product of subsymmetric convex sets. Let $\widehat{\mathbf{x}}$ denote the unique solution of (5.1), $\mathbf{g} = \nabla f(\mathbf{x})$, $\mathbf{x} \in \Omega$, and let $\lambda_{\min}$ denote the smallest eigenvalue of* A.

If $\overline{\alpha} \in (0, 2\|\mathsf{A}\|^{-1}]$, then

$$f\left(P_\Omega(\mathbf{x}-\overline{\alpha}\mathbf{g})\right) - f(\widehat{\mathbf{x}}) \le \eta_f\left(f(\mathbf{x}) - f(\widehat{\mathbf{x}})\right), \tag{5.34}$$

where

$$\eta_f = \eta_f(\alpha) = 1 - \widehat{\alpha}\lambda_{\min} \tag{5.35}$$

denote the cost function reduction coefficient and $\widehat{\alpha} = \min\{\alpha, 2\|\mathsf{A}\|^{-1} - \alpha\}$.

Proof. Let us first assume that $0 < \widehat{\alpha}\|\mathsf{A}\| \le 1$ and let $\mathbf{x} \in \Omega$ be arbitrary but fixed, so that we can use Lemma 5.4 to get

$$\begin{aligned}\widehat{\alpha}\big(f\left(P_\Omega(\mathbf{x}-\widehat{\alpha}\mathbf{g})\right) - \widehat{\alpha}f(\widehat{\mathbf{x}})\big) &\le F_{\widehat{\alpha}}\left(P_\Omega(\mathbf{x}-\widehat{\alpha}\mathbf{g})\right) - \widehat{\alpha}f(\widehat{\mathbf{x}})\\ &\le \widehat{\alpha}(1-\widehat{\alpha}\lambda_{\min})\left(f(\mathbf{x}) - f(\widehat{\mathbf{x}})\right).\end{aligned} \tag{5.36}$$

In combination with (5.24), this proves (5.34) for $0 < \widehat{\alpha} \le \|\mathsf{A}\|^{-1}$.

To prove the statement for $\overline{\alpha} \in (\|\mathsf{A}\|^{-1}, 2\|\mathsf{A}\|^{-1}]$ and subsymmetric Ω, let us define

$$h(\mathbf{y}) = \|\mathsf{A}\|^{-1}f(\mathbf{y}), \quad \delta = \widehat{\alpha}\|\mathsf{A}\| \in (0,1], \quad \overline{\alpha} = (2-\delta)\|\mathsf{A}\|^{-1}.$$

It follows that

$$\delta\nabla h(\mathbf{x}) = \delta\|\mathsf{A}\|^{-1}\nabla f(\mathbf{x}) = \delta\|\mathsf{A}\|^{-1}\mathbf{g} = \widehat{\alpha}\mathbf{g}.$$

Thus we can rescale the gradient and apply the definition of subsymmetry (5.29) to get

$$\begin{aligned} F_{\widehat{\alpha}}(P_\Omega(\mathbf{x}-\overline{\alpha}\mathbf{g})) &= F_{\widehat{\alpha}}(P_\Omega(\mathbf{x}-(2-\delta)\|\mathsf{A}\|^{-1}\|\mathsf{A}\|\nabla h(\mathbf{x})) \\ &= F_{\widehat{\alpha}}(P_\Omega(\mathbf{x}-(2-\delta)\nabla h(\mathbf{x})) \\ &\le F_{\widehat{\alpha}}(P_\Omega(\mathbf{x}-\delta\nabla h(\mathbf{x})) \\ &= F_{\widehat{\alpha}}(P_\Omega(\mathbf{x}-\widehat{\alpha}\mathbf{g})). \end{aligned}$$

To finish the proof, we use again Lemma 5.4 to get

$$\begin{aligned} \widehat{\alpha}(f(P_\Omega(\mathbf{x}-\overline{\alpha}\mathbf{g})-\widehat{\alpha}f(\widehat{\mathbf{x}})) &\le F_{\widehat{\alpha}}(P_\Omega(\mathbf{x}-\widehat{\alpha}\mathbf{g}))-\widehat{\alpha}f(\widehat{\mathbf{x}}) \\ &\le \widehat{\alpha}(1-\widehat{\alpha}\lambda_{\min})\,(f(\mathbf{x})-f(\widehat{\mathbf{x}}))\,. \end{aligned} \tag{5.37}$$

□

The estimate (10.24) gives the best value

$$\eta_f^{opt} = 1-\kappa(\mathsf{A})^{-1}$$

for $\alpha = \|\mathsf{A}\|^{-1}$ with $\kappa(\mathsf{A}) = \|\mathsf{A}\|\|\mathsf{A}^{-1}\|$.

5.9 Comments

While the decrease of the cost function with the optimal gradients step is a classical result of numerical optimization that has been known since the times of Kantorovich (see Sect. 3.2 or Bertsekas [16, Prop. 3.1] and references therein), it seems that the first results concerning the decrease of the cost function along the projected-gradient path are due to Schöberl [22, 85, 244], who found the bound on the rate of convergence of the cost function in the energy norm for the gradient projection method with the fixed steplength $\alpha \in (0, \|\mathsf{A}\|^{-1}]$ in terms of bounds on the spectrum of the Hessian matrix A. Schöberl considered bound constraints. Later Kučera [197] proved the same bound for separable convex constraints and used the result in the algorithm with the rate of convergence.

A successful application of the projections with a longer steplength to the solution of contact problems by Farhat and Avery in their FETI–DP code [102] suggested the extension of the estimate for the bound constraints to $\alpha \in [0, 2\|\mathsf{A}\|^{-1}]$ by Dostál [73] and Dostál et al. [103]. The latter proof used a simple property of interval that motivated the concept of a subsymmetric set. Bouchala and Vodstrčil in [26] and Bouchala et al. [28] extended the estimates to the spheres and ellipses by proving that they are subsymmetric. Let us mention that there are convex sets that are not subsymmetric [26], but no example of a convex set for which the estimates do not hold is known. The estimates of the decrease of quadratic cost function along the projected-gradient path are essential for the development of practical monotonic algorithms that combine Euclidean projections with CG or other methods for solving linear problems—see Chaps. 10–12. Let us recall that Luo and Tseng [208] proved the linear rate of convergence of the cost function for the gradient projection method earlier, even for more general problems. Still, they did not make any attempt to specify the constants.

6 From Penalty to Exact Augmented Lagrangians

We shall consider two basic methods for solving the problem

$$\min_{\mathbf{x}\in\Omega_E} f(\mathbf{x}), \quad \Omega_E = \{\mathbf{x} \in \mathbb{R}^n : \mathsf{B}\mathbf{x} = \mathbf{c}\}, \tag{6.1}$$

where $f(\mathbf{x}) = \frac{1}{2}\mathbf{x}^T\mathsf{A}\mathbf{x} - \mathbf{x}^T\mathbf{b}, \mathbf{b}$ is a given column n-vector, A is an $n \times n$ SPD matrix, $\mathsf{B} \in \mathbb{R}^{m\times n}$, and $\mathbf{c} \in \mathrm{Im}\mathsf{B}$. We assume that $\mathsf{B} \neq \mathsf{O}$ is not a full column rank matrix so that $\mathrm{Ker}\mathsf{B} \neq \{\mathbf{o}\}$, but we allow dependent rows of B. Using a simple observation of Sect. 9.4.3, we can extend the results to solving problems with A positive definite on $\mathrm{Ker}\mathsf{B}$. We shall consider two ways to reduce the problem (6.1) to a sequence of unconstrained problems.

The first method uses a scalar multiple of a *penalization function* h, i.e., a function with zero values on the feasible region and large values outside the feasible region to approximate the original equality-constrained problem by the unconstrained minimization problem with the modified (penalized) cost function $f(\mathbf{x}) + \frac{\varrho}{2}h(\mathbf{x})$. The resulting *penalty method* presented in Sect. 6.2 is probably the most straightforward way to reduce the equality-constrained problems to the unconstrained ones.

The penalty parameter must be sufficiently large to enforce the equality constraints with an acceptable error, but the large penalty can be a source of many problems. A starting point for developing practical algorithms for solving (6.1) is the *exact augmented Lagrangian method* that works with small penalty parameters. The augmented Lagrangian method repeatedly applies the penalty method to the Lagrangian function and replaces the increase of penalty parameters by accumulating the corrections in the Lagrange multipliers. In Chap. 9, we shall show that the augmented Lagrangian method with an iterative solution of auxiliary unconstrained problems can solve some classes of problems with linear complexity.

Z. Dostál, *Optimal Quadratic Programming and QCQP Algorithms with Case Studies*, Springer Optimization and Its Applications 23,
https://doi.org/10.1007/978-3-031-95167-1_6

6.1 Review of Alternative Methods

Before studying augmented Lagrangian methods, let us briefly review alternative methods for solving the equality-constrained problem (6.1).

Using Proposition 2.8, it follows that (6.1) is equivalent to the solution of the *saddle point system of linear equations*

$$\begin{bmatrix} \mathsf{A} & \mathsf{B}^T \\ \mathsf{B} & \mathsf{O} \end{bmatrix} \begin{bmatrix} \mathbf{x} \\ \boldsymbol{\lambda} \end{bmatrix} = \begin{bmatrix} \mathbf{b} \\ \mathbf{c} \end{bmatrix}. \tag{6.2}$$

If B is a full row rank matrix, we can solve (6.2) by the *Gauss elimination* with a suitable pivoting strategy or by a *symmetric factorization* which takes into account that (6.2) is indefinite. Alternatively, we can use *MINRES*, a Krylov space method which generates the iterates minimizing the Euclidean norm of the residual in the Krylov space. The performance of MINRES depends on the distribution of the spectrum of the KKT system (6.2), similar to the performance of the CG method. A comprehensive review of the methods for solving saddle point systems with many references is in Benzi, Golub, and Liesen [14]; see also Elman, Sylvester, and Wathen [120].

We can also reduce (6.2) to an SPD case. If B is a full row rank matrix, and if we can evaluate the action of A^{-1} effectively, we can multiply the first block row in (6.2) by $\mathsf{B}\mathsf{A}^{-1}$, subtract the second row from the result, and change the signs to obtain the symmetric positive definite *Schur complement system*

$$\mathsf{B}\mathsf{A}^{-1}\mathsf{B}^T\boldsymbol{\lambda} = \mathsf{B}\mathsf{A}^{-1}\mathbf{b} - \mathbf{c}, \tag{6.3}$$

which can be solved by the methods described in Chap. 8. The method is also known as the *range-space method.* Let us point out that if we solve (6.3) by the CG method, then it is not necessary to evaluate $\mathsf{B}\mathsf{A}^{-1}\mathsf{B}^T$ explicitly. Using the CG method and the left generalized inverse of Sect. 1.4, the method can be extended to A positive semidefinite and B with dependent rows. We shall see that the range-space method is closely related to the Uzawa-type methods that we shall study in Chap. 9.

Alternatively, we can use the *null-space* method, provided we have a basis Z of $\mathrm{Ker}\mathsf{B}$ and a feasible $\mathbf{x}_0$,

$$\mathsf{B}\mathbf{x}_0 = \mathbf{c}.$$

Observing that $\Omega_E = \{\mathbf{x}_0 + \mathsf{Z}\mathbf{y} : \mathbf{y} \in \mathbb{R}^d\}$, we can substitute into (6.1) to get

$$\min_{\mathbf{x}\in\Omega_E} f(\mathbf{x}) = \min_{\mathbf{y}\in\mathbb{R}^d} f(\mathbf{x}_0 + \mathsf{Z}\mathbf{y}) = \frac{1}{2}\mathbf{y}^T\mathsf{Z}^T\mathsf{A}\mathsf{Z}\mathbf{y} - (\mathbf{b} - \mathsf{A}\mathbf{x}_0)^T\mathsf{Z}\mathbf{y} + \frac{1}{2}\mathbf{x}_0^T\mathsf{A}\mathbf{x}_0,$$

so that we can evaluate $\mathbf{y}$ by solving the gradient equation

$$\mathsf{Z}^T\mathsf{A}\mathsf{Z}\mathbf{y} = \mathsf{Z}^T(\mathbf{b} - \mathsf{A}\mathbf{x}_0).$$

If the resulting system is solved by the CG method, then the method can be directly applied to the problems with SPS A and B with dependent rows. Results concerning the application of domain decomposition methods can be found in the monograph by Toselli and Widlund [256].

A class of algorithms which is important for our exposition is based on the mixed formulation

$$L_0(\widehat{\mathbf{x}}, \widehat{\boldsymbol{\lambda}}) = \max_{\boldsymbol{\lambda} \in \mathbb{R}^m} \min_{\mathbf{x} \in \mathbb{R}^n} L_0(\mathbf{x}, \boldsymbol{\lambda})$$

for the problems with full row rank B. As an example let us recall the *Arrow–Hurwitz algorithm* 6.1, which exploits the first-order approximation of L_0 given by

$$L_0(\mathbf{x} + \alpha \mathbf{d}, \boldsymbol{\lambda} + r\boldsymbol{\delta}) \approx L_0(\mathbf{x}, \boldsymbol{\lambda}) + \alpha \nabla_{\mathbf{x}} L_0(\mathbf{x}, \boldsymbol{\lambda})\mathbf{d} + r \nabla_{\boldsymbol{\lambda}} L_0(\mathbf{x}, \boldsymbol{\lambda})\boldsymbol{\delta}$$

to improve the approximations of the solution $\widehat{\mathbf{x}}$ by taking small steps in the direction opposite to the gradient

$$\nabla_{\mathbf{x}} L_0(\mathbf{x}, \boldsymbol{\lambda}) = \mathsf{A}\mathbf{x} - \mathbf{b} + \mathsf{B}^T \boldsymbol{\lambda},$$

and to improve the approximations of the Lagrange multipliers $\widehat{\boldsymbol{\lambda}}$ by taking small steps in the direction

$$\nabla_{\boldsymbol{\lambda}} L_0(\mathbf{x}, \boldsymbol{\lambda}) = \mathsf{B}\mathbf{x} - \mathbf{c}.$$

Algorithm 6.1. Arrow-Hurwitz algorithm

Given a symmetric positive definite matrix $\mathsf{A} \in \mathbb{R}^{n \times n}$*, a matrix* $\mathsf{B} \in \mathbb{R}^{m \times n}$ *with the nonempty kernel,* $\mathbf{b} \in \mathbb{R}^n$*, and* $\mathbf{c} \in \mathrm{Im}\mathsf{B}$.

Step 0. *{Initialization.}*
Choose $\boldsymbol{\lambda}^0 \in \mathbb{R}^m$, $\mathbf{x}^{-1} \in \mathbb{R}^n$, $\alpha > 0$, $r > 0$

for $k = 0, 1, 2, \ldots$
Step 1. *{Reducing the value of* L_0 *in* $\mathbf{x}$ *direction.}*
$\mathbf{x}^k = \mathbf{x}^{k-1} - \alpha \nabla_{\mathbf{x}} L_0(\mathbf{x}^{k-1}, \boldsymbol{\lambda}^k) = \mathbf{x}^{k-1} - \alpha(\mathsf{A}\mathbf{x} - \mathbf{b} + \mathsf{B}^T \boldsymbol{\lambda})$
Step 2. *{Increasing the value of* L_0 *in* $\boldsymbol{\lambda}$ *direction.}*
$\boldsymbol{\lambda}^{k+1} = \boldsymbol{\lambda}^k + r \nabla_{\boldsymbol{\lambda}} L_0(\mathbf{x}^k, \boldsymbol{\lambda}^k) = \boldsymbol{\lambda}^k + r(\mathsf{B}\mathbf{x}^k - \mathbf{c})$
***end** for*

The Arrow–Hurwitz algorithm is known to converge for sufficiently small steplengths α and r. Even though its convergence is known to be slow, the algorithm has found its applications due to the low cost of the iterations and minimal memory requirements. It can be considered an extreme case of the inexact Uzawa-type algorithms treated in Chap. 9.

6.2 Penalty Method

Probably the simplest way to reduce the equality-constrained problem (6.1) to the unconstrained one is to add a suitable term that penalizes violations of the constraints. In this section we consider the *quadratic penalty method* which approximates the solution $\widehat{\mathbf{x}}$ of (6.1) by the solution $\widehat{\mathbf{x}}_\varrho$ of

$$\min_{\mathbf{x}\in\mathbb{R}^n} f_\varrho(\mathbf{x}), \quad f_\varrho(\mathbf{x}) = f(\mathbf{x}) + \frac{\varrho}{2}\|\mathsf{B}\mathbf{x} - \mathbf{c}\|^2, \tag{6.4}$$

where $\varrho \geq 0$ is the *penalty parameter* and $\|\mathsf{B}\mathbf{x} - \mathbf{c}\|^2$ is the *penalty function.*

Intuitively, if ϱ is large, then the solution $\widehat{\mathbf{x}}_\varrho$ of (6.4) can hardly be far from the solution of (6.1). Indeed, if ϱ were infinite, then the minimizer of f_ϱ would solve the equality-constrained problem (6.1). The effect of the penalty term is illustrated in Fig. 6.1. Notice that $\widehat{\mathbf{x}}_\varrho$ is typically near the feasible set but does not belong to it. That is why our penalty method is also called the *exterior penalty method.*

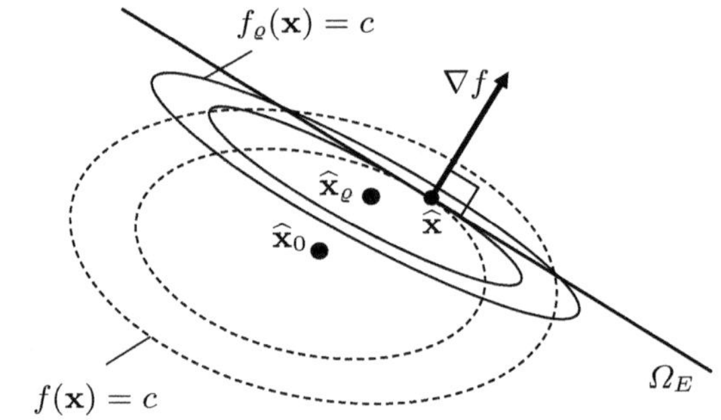

Fig. 6.1. The effect of the quadratic penalty

In the following sections, we shall often use the more general *augmented Lagrangian* penalty function $L : \mathbb{R}^{n+m+1} \rightarrow \mathbb{R}$ which is defined by

$$L(\mathbf{x}, \boldsymbol{\lambda}, \varrho) = f(\mathbf{x}) + (\mathsf{B}\mathbf{x} - \mathbf{c})^T\boldsymbol{\lambda} + \frac{\varrho}{2}\|\mathsf{B}\mathbf{x} - \mathbf{c}\|^2 = L_0(\mathbf{x}, \boldsymbol{\lambda}) + \frac{\varrho}{2}\|\mathsf{B}\mathbf{x} - \mathbf{c}\|^2, \tag{6.5}$$

where $\boldsymbol{\lambda} \in \mathbb{R}^m$ is arbitrary and $L_0(\mathbf{x}, \boldsymbol{\lambda}) = L(\mathbf{x}, \boldsymbol{\lambda}, 0)$ is the Lagrangian function (2.19). Notice that $f_\varrho(\mathbf{x}) = L(\mathbf{x}, \mathbf{o}, \varrho)$. Since $\varrho\|\mathsf{B}\mathbf{x} - \mathbf{c}\|^2$ and $(\mathsf{B}\mathbf{x} - \mathbf{c})^T\boldsymbol{\lambda}$ vanish when $\mathsf{B}\mathbf{x} = \mathbf{c}$, it follows that

$$f(\mathbf{x}) = f_\varrho(\mathbf{x}) = L(\mathbf{x}, \boldsymbol{\lambda}, \varrho)$$

for any $\mathbf{x} \in \Omega_E, \boldsymbol{\lambda} \in \mathbb{R}^m$, and $\varrho \geq 0$.

6.2.1 Minimizing Augmented Lagrangian

Let us start with the modified problem

$$\min_{\mathbf{x}\in\Omega_E} L(\mathbf{x}, \boldsymbol{\lambda}, \varrho). \tag{6.6}$$

Since the gradient of the augmented Lagrangian is given by

$$\nabla_{\mathbf{x}} L(\mathbf{x}, \boldsymbol{\lambda}, \varrho) = \mathsf{A}\mathbf{x} - \mathbf{b} + \mathsf{B}^T(\boldsymbol{\lambda} + \varrho(\mathsf{B}\mathbf{x} - \mathbf{c})), \tag{6.7}$$

it follows that the KKT system for (6.6) reads

$$\begin{bmatrix} \mathsf{A}_\varrho & \mathsf{B}^T \\ \mathsf{B} & \mathsf{O} \end{bmatrix} \begin{bmatrix} \mathbf{x} \\ \boldsymbol{\lambda} \end{bmatrix} = \begin{bmatrix} \mathbf{b} + \varrho\mathsf{B}^T\mathbf{c} \\ \mathbf{c} \end{bmatrix}, \tag{6.8}$$

where $\mathsf{A}_\varrho = \mathsf{A} + \varrho\mathsf{B}^T\mathsf{B}$. Eliminating $\mathbf{x}$, we get that any multiplier $\overline{\lambda}$ satisfies

$$\mathsf{B}\mathsf{A}_\varrho^{-1}\mathsf{B}^T\overline{\boldsymbol{\lambda}} = \mathsf{B}\mathsf{A}_\varrho^{-1}(\mathbf{b} + \varrho\mathsf{B}^T\mathbf{c}) - \mathbf{c}. \tag{6.9}$$

Moreover, if we substitute $\mathsf{B}\mathbf{x} = \mathbf{c}$ into (6.7), we get that (6.8) is equivalent to the KKT system (6.2) in sense that the $\mathbf{x}$-components of the saddle points of L_0 are exactly those of L. This result is not surprising as

$$L(\mathbf{x}, \boldsymbol{\lambda}, \varrho) = f_\varrho(\mathbf{x}) + (\mathsf{B}\mathbf{x} - \mathbf{c})^T\boldsymbol{\lambda}$$

is the Lagrangian for the penalized equality-constrained problem

$$\min_{\mathbf{x}\in\Omega_E} f_\varrho(\mathbf{x}).$$

To see how the penalty method enforces the feasibility, let us assume that $\boldsymbol{\lambda} \in \mathbb{R}^m$ is fixed, and let us denote by $\widehat{\mathbf{x}}_0$ and $\widehat{\mathbf{x}}_\varrho$ the minimizers of $L_0(\mathbf{x}, \boldsymbol{\lambda})$ and $L(\mathbf{x}, \boldsymbol{\lambda}, \varrho)$, respectively. Then the solution $\widehat{\mathbf{x}}$ satisfies

$$L_0(\widehat{\mathbf{x}}_\varrho, \boldsymbol{\lambda}) + \frac{\varrho}{2}\|\mathsf{B}\widehat{\mathbf{x}}_\varrho - \mathbf{c}\|^2 = L(\widehat{\mathbf{x}}_\varrho, \boldsymbol{\lambda}, \varrho) \le L(\widehat{\mathbf{x}}, \boldsymbol{\lambda}, \varrho) = f(\widehat{\mathbf{x}}),$$

so that, using $L_0(\widehat{\mathbf{x}}_0, \boldsymbol{\lambda}) \le L_0(\widehat{\mathbf{x}}_\varrho, \boldsymbol{\lambda})$, we get

$$\|\mathsf{B}\widehat{\mathbf{x}}_\varrho - \mathbf{c}\|^2 \le \frac{2}{\varrho}(f(\widehat{\mathbf{x}}) - L_0(\widehat{\mathbf{x}}_\varrho, \boldsymbol{\lambda})) \le \frac{2}{\varrho}(f(\widehat{\mathbf{x}}) - L_0(\widehat{\mathbf{x}}_0, \boldsymbol{\lambda})).$$

It follows that the feasibility error $\|\mathsf{B}\widehat{\mathbf{x}}_\varrho - \mathbf{c}\|$ can be made arbitrarily small. We shall give stronger or easier computable bounds later in this section.

To see how $\widehat{\mathbf{x}}_\varrho$ satisfies the first block equation of the KKT conditions (6.2), let us recall that the gradient of the augmented Lagrangian is given by (6.7) and denote

$$\widetilde{\boldsymbol{\lambda}} = \boldsymbol{\lambda} + \varrho(\mathsf{B}\widehat{\mathbf{x}}_\varrho - \mathbf{c}). \tag{6.10}$$

Then

$$\mathsf{A}\widehat{\mathbf{x}}_\varrho - \mathbf{b} + \mathsf{B}^T\widetilde{\boldsymbol{\lambda}} = \nabla_{\mathbf{x}} L_0(\widehat{\mathbf{x}}_\varrho, \widetilde{\boldsymbol{\lambda}}) = \nabla_{\mathbf{x}} L(\widehat{\mathbf{x}}_\varrho, \boldsymbol{\lambda}, \varrho) = \mathbf{o},$$

so that $(\widehat{\mathbf{x}}_\varrho, \widetilde{\boldsymbol{\lambda}})$ satisfies the first block equation of the KKT conditions exactly. Moreover, if $\boldsymbol{\lambda}$ is an approximation of a vector of Lagrange multipliers of the solution of (6.1), then our observations indicate that $\widetilde{\boldsymbol{\lambda}}$ is a better approximation. Using Proposition 2.12, we conclude that $(\widehat{\mathbf{x}}_\varrho, \widetilde{\boldsymbol{\lambda}})$ can approximate the KKT pair of (6.1) with arbitrarily small error.

6.2.2 An Optimal Feasibility Error Estimate for Homogeneous Constraints

Let us first examine the feasibility error of an approximate solution $\mathbf{x}$ of the problem

$$\min_{\mathbf{x}\in\mathbb{R}^n} f_\varrho(\mathbf{x}), \quad f_\varrho(\mathbf{x}) = f(\mathbf{x}) + \frac{\varrho}{2}\|\mathsf{B}\mathbf{x}\|^2, \tag{6.11}$$

where f and B are from the definition of problem (6.1) and $\varrho > 0$. We assume that $\mathbf{x}$ satisfies

$$\|\nabla f_\varrho(\mathbf{x})\| \le \varepsilon\|\mathbf{b}\|, \tag{6.12}$$

where $\varepsilon > 0$ is a small number.

Notice that our $\mathbf{x}$ can be considered as an approximation to the solution $\widehat{\mathbf{x}}$ of the equality constrained problem (6.1) in case that the equality constraints are homogeneous, i.e., $\mathbf{c} = \mathbf{o}$. To check that $\mathbf{x}$ satisfies approximately the first part of the KKT conditions (6.2), observe that

$$\nabla f_\varrho(\mathbf{x}) = (\mathsf{A} + \varrho\mathsf{B}^T\mathsf{B})\mathbf{x} - \mathbf{b}.$$

After denoting $\boldsymbol{\lambda} = \varrho\mathsf{B}\mathbf{x}$ and $\mathbf{g} = \nabla f_\varrho(\mathbf{x})$, we get

$$\mathsf{A}\mathbf{x} - \mathbf{b} + \mathsf{B}^T\boldsymbol{\lambda} = \mathbf{g}, \tag{6.13}$$

which can be considered as an approximation of the first block equation of the KKT conditions (6.2).

We discuss the feasibility error in the following theorem.

Theorem 6.1. *Let* A, B, *and* $\mathbf{b}$ *be those of the definition of problem (6.1) with* B *not necessarily a full rank matrix, let* $\lambda_{\min} = \lambda_{\min}(\mathsf{A}) > 0$ *denote the smallest eigenvalue of* A, *and let* $\varepsilon \ge 0$ *and* $\varrho > 0$.

If $\mathbf{x}$ *is an approximate solution of (6.11) such that*

$$\|\nabla f_\varrho(\mathbf{x})\| \le \varepsilon\|\mathbf{b}\|,$$

then

$$\|\mathsf{B}\mathbf{x}\| \le \frac{1+\varepsilon}{\sqrt{\lambda_{\min}\varrho}}\|\mathbf{b}\|. \tag{6.14}$$

Proof. Let us denote

$$\mathsf{A}_\varrho = \mathsf{A} + \varrho \mathsf{B}^T\mathsf{B}$$

and notice that for any $\mathbf{x}, \mathbf{d} \in \mathbb{R}^n$

$$\begin{aligned}
f_\varrho(\mathbf{x}+\mathbf{d}) &= f_\varrho(\mathbf{x}) + \mathbf{g}^T\mathbf{d} + \frac{1}{2}\mathbf{d}^T\mathsf{A}_\varrho\mathbf{d} \\
&\geq f_\varrho(\mathbf{x}) - \|\mathbf{g}\|\|\mathbf{d}\| + \frac{1}{2}\lambda_{\min}\|\mathbf{d}\|^2 \\
&\geq \min_{\xi\in\mathbb{R}} \left(f_\varrho(\mathbf{x}) - \|\mathbf{g}\|\xi + \frac{1}{2}\lambda_{\min}\xi^2 \right) \\
&= f_\varrho(\mathbf{x}) - \frac{1}{2\lambda_{\min}}\|\mathbf{g}\|^2,
\end{aligned}$$

where $\mathbf{g} = \nabla f_\varrho(\mathbf{x})$. Recalling that by (2.10)

$$\min_{\mathbf{d}\in\mathbb{R}^n} f_\varrho(\mathbf{x}+\mathbf{d}) = \min_{\mathbf{y}\in\mathbb{R}^n} f_\varrho(\mathbf{y}) = -\frac{1}{2}\mathbf{b}^T\mathsf{A}_\varrho^{-1}\mathbf{b},$$

we get

$$0 \geq -\frac{1}{2}\mathbf{b}^T\mathsf{A}_\varrho^{-1}\mathbf{b} \geq f(\mathbf{x}) + \frac{\varrho}{2}\|\mathsf{B}\mathbf{x}\|^2 - \frac{1}{2\lambda_{\min}}\|\mathbf{g}\|^2.$$

Let us now assume that $\mathbf{x}$ satisfies $\|\mathbf{g}\| \leq \varepsilon\|\mathbf{b}\|$. After substituting into the last inequality and using (2.10), (1.30), and the properties of the Euclidean norm, we get

$$\begin{aligned}
0 &\geq f(\mathbf{x}) + \frac{\varrho}{2}\|\mathsf{B}\mathbf{x}\|^2 - \frac{1}{2\lambda_{\min}}\|\mathbf{g}\|^2 \\
&\geq \min_{\mathbf{y}\in\mathbb{R}^n} f(\mathbf{y}) + \frac{\varrho}{2}\|\mathsf{B}\mathbf{x}\|^2 - \frac{\varepsilon^2}{2\lambda_{\min}}\|\mathbf{b}\|^2 \\
&= -\frac{1}{2}\mathbf{b}^T\mathsf{A}^{-1}\mathbf{b} + \frac{\varrho}{2}\|\mathsf{B}\mathbf{x}\|^2 - \frac{\varepsilon^2}{2\lambda_{\min}}\|\mathbf{b}\|^2 \\
&\geq -\frac{1}{2\lambda_{\min}}\|\mathbf{b}\|^2 + \frac{\varrho}{2}\|\mathsf{B}\mathbf{x}\|^2 - \frac{\varepsilon^2}{2\lambda_{\min}}\|\mathbf{b}\|^2 \\
&\geq \frac{\varrho}{2}\|\mathsf{B}\mathbf{x}\|^2 - \frac{1+\varepsilon^2}{2\lambda_{\min}}\|\mathbf{b}\|^2.
\end{aligned}$$

Equation (6.14) can be obtained by simple manipulations with application of $1+\varepsilon^2 \leq (1+\varepsilon)^2$. □

An interesting feature of Theorem 6.1 is that *the estimate is independent of the constraint matrix* B. In particular, the estimate (6.14) is valid even if B has dependent rows. The assumption that the constraints are homogeneous was used to get that the unconstrained minimum of f_ϱ is not positive.

6.2.3 Approximation Error and Convergence

Using the feasibility estimate (6.14) of the previous subsection and an error bound on the violation of the first block equation of the KKT conditions (2.45), we can bound the approximation error of the penalty method for homogeneous constraints.

Theorem 6.3. *Let problem (6.1) be defined by* $\mathsf{A}, \mathsf{B}, \mathbf{b}$, *and* $\mathbf{c} = \mathbf{o}$, *with* $\mathsf{B} \neq \mathsf{O}$ *not necessarily a full rank matrix, let* $(\widehat{\mathbf{x}}, \boldsymbol{\lambda}_{\mathrm{LS}})$ *denote the least square KKT pair for (6.1) with* $\mathbf{c} = \mathbf{o}$, *let* $\lambda_{\min}$ *denote the least eigenvalue of* A, *let* $\overline{\sigma}_{\min}$ *denote the least nonzero singular value of* B, *let* $\varepsilon > 0$, $\varrho > 0$, *and let*

$$\boldsymbol{\lambda} = \varrho \mathsf{B}\mathbf{x}. \tag{6.15}$$

If $\mathbf{x}$ *is such that*

$$\|\nabla f_{\varrho}(\mathbf{x})\| \leq \varepsilon \|\mathbf{b}\|,$$

then

$$\|\mathbf{x} - \widehat{\mathbf{x}}\| \leq \varepsilon \frac{\kappa(\mathsf{A}) + 1}{\lambda_{\min}} \|\mathbf{b}\| + \frac{1 + \varepsilon}{\sqrt{\varrho}} \frac{\kappa(\mathsf{A})}{\overline{\sigma}_{\min} \sqrt{\lambda_{\min}}} \|\mathbf{b}\| \tag{6.16}$$

and

$$\|\boldsymbol{\lambda} - \boldsymbol{\lambda}_{\mathrm{LS}}\| \leq \frac{1}{\overline{\sigma}_{\min}} (\varepsilon \kappa(\mathsf{A}) \|\mathbf{b}\| + \frac{1 + \varepsilon}{\sqrt{\varrho}} \frac{\|\mathsf{A}\|}{\overline{\sigma}_{\min} \sqrt{\lambda_{\min}}} \|\mathbf{b}\|). \tag{6.17}$$

Proof. Let us denote $\mathbf{g} = \nabla f_{\varrho}(\mathbf{x})$ and $\mathbf{e} = \mathsf{B}\mathbf{x}$, so that

$$\mathsf{A}\mathbf{x} + \mathsf{B}^T \boldsymbol{\lambda} = \mathbf{b} + \mathbf{g} \quad \text{and} \quad \mathsf{B}\mathbf{x} = \mathbf{e}, \tag{6.18}$$

and notice that by the assumptions $\boldsymbol{\lambda} \in \mathrm{Im}\mathsf{B}$. Assuming that

$$\|\mathbf{g}\| = \|\nabla f_{\varrho}(\mathbf{x})\| \leq \varepsilon \|\mathbf{b}\|,$$

it follows by Theorem 6.1 that

$$\|\mathsf{B}\mathbf{x}\| \leq \frac{1 + \varepsilon}{\sqrt{\lambda_{\min} \varrho}} \|\mathbf{b}\|.$$

Substituting into the estimates (2.47) and (2.48) of Proposition 2.12, we get (6.16) and (6.17). □

Notice that the error bounds (6.16) and (6.17) depend on the representation of Ω_E, namely, on the constraint matrix B.

The performance of the penalty method can also be described in terms of convergence. Let $\varepsilon_k > 0$ denote a sequence converging to zero, let $\varrho_k > 0$ denote an increasing unbounded sequence, let $\mathbf{g}^k = \nabla f_{\varrho_k}(\mathbf{x}^k)$, and let $\mathbf{x}^k$ satisfy

$$\|\mathbf{g}^k\| = \|\nabla f_{\varrho_k}(\mathbf{x}^k)\| \leq \varepsilon_k \|\mathbf{b}\|.$$

Let us denote

$$\boldsymbol{\lambda}^k = \varrho_k \mathsf{B}\mathbf{x}^k.$$

Then by (6.16) there is a constant C_1 dependent on A and C_2 dependent on A, B such that

$$\|\mathbf{x}^k - \widehat{\mathbf{x}}\| \leq \varepsilon_k C_1 \|\mathbf{b}\| + \frac{1+\varepsilon_k}{\sqrt{\varrho_k}} C_2 \|\mathbf{b}\|,$$

and by (6.17) there are constants C_3 and C_4 dependent on A, B such that

$$\|\boldsymbol{\lambda}^k - \boldsymbol{\lambda}_{\mathrm{LS}}\| \leq \varepsilon_k C_3 \|\mathbf{b}\| + \frac{1+\varepsilon_k}{\sqrt{\varrho_k}} C_4 \|\mathbf{b}\|. \qquad (6.19)$$

It follows that $\boldsymbol{\lambda}^k$ converges to $\boldsymbol{\lambda}_{\mathrm{LS}}$ and $\mathbf{x}^k$ converges to $\widehat{\mathbf{x}}$.

6.2.4 Improved Feasibility Error Estimate

We shall now give a feasibility error estimate for the penalty approximation of (6.1), which is valid for nonhomogeneous constraints with $\mathbf{c} \neq \mathbf{o}$. Our new bound on the error is proportional to ϱ^{-1}, but *dependent on* B and $\mathbf{c}$.

Theorem 6.4. *Let* $\mathsf{A}, \mathsf{B}, \mathbf{b}$, *and* $\mathbf{c}$ *be those of the definition of problem (6.1) with* $\mathsf{B} \neq \mathsf{O}$ *not necessarily a full rank matrix, let* $\overline{\beta}_{\min} > 0$ *denote the smallest nonzero eigenvalue of* $\mathsf{B}^T\mathsf{A}^{-1}\mathsf{B}$, *let* ε *denote a given positive number, and let* $\varrho > 0$.

If $\mathbf{x}$ *is such that*

$$\|\nabla f_\varrho(\mathbf{x})\| \leq \varepsilon \|\mathbf{b}\|,$$

then the feasibility error satisfies

$$\|\mathsf{B}\mathbf{x} - \mathbf{c}\| \leq (1 + \overline{\beta}_{\min}\varrho)^{-1}((1+\varepsilon)\|\mathsf{B}\mathsf{A}^{-1}\|\|\mathbf{b}\| + \|\mathbf{c}\|). \qquad (6.20)$$

Proof. Let us recall that for any vector $\mathbf{x}$

$$\nabla f_\varrho(\mathbf{x}) = (\mathsf{A} + \varrho \mathsf{B}^T\mathsf{B})\mathbf{x} - \mathbf{b} - \varrho \mathsf{B}^T\mathbf{c},$$

so that, after denoting $\mathbf{g} = \nabla f_\varrho(\mathbf{x})$ and $\mathsf{A}_\varrho = \mathsf{A} + \varrho \mathsf{B}^T\mathsf{B}$,

$$\mathbf{x} = \mathsf{A}_\varrho^{-1}(\mathbf{g} + \mathbf{b} + \varrho \mathsf{B}^T\mathbf{c}).$$

It follows that

$$\mathsf{B}\mathbf{x} = \mathsf{B}\mathsf{A}_\varrho^{-1}(\mathbf{g} + \mathbf{b}) + \varrho \mathsf{B}\mathsf{A}_\varrho^{-1}\mathsf{B}^T\mathbf{c}.$$

Using equation (1.49) of Lemma 1.6 and simple manipulations, we get

$$\begin{aligned}\mathsf{B}\mathbf{x} - \mathbf{c} &= \mathsf{BA}_\varrho^{-1}(\mathbf{g} + \mathbf{b}) + \varrho(\mathsf{I} + \varrho\mathsf{BA}^{-1}\mathsf{B}^T)^{-1}\mathsf{BA}^{-1}\mathsf{B}^T\mathbf{c} - \mathbf{c} \\ &= \mathsf{BA}_\varrho^{-1}(\mathbf{g} + \mathbf{b}) + (\mathsf{I} + \varrho\mathsf{BA}^{-1}\mathsf{B}^T)^{-1}((\mathsf{I} + \varrho\mathsf{BA}^{-1}\mathsf{B}^T) - \mathsf{I})\mathbf{c} - \mathbf{c} \\ &= \mathsf{BA}_\varrho^{-1}(\mathbf{g} + \mathbf{b}) - (\mathsf{I} + \varrho\mathsf{BA}^{-1}\mathsf{B}^T)^{-1}\mathbf{c}.\end{aligned}$$

To finish the proof, use the assumptions that $\mathbf{c} \in \mathrm{Im}\mathsf{B}$ and $\|\mathbf{g}\| \leq \varepsilon\|\mathbf{b}\|$, Lemma 1.8, and the properties of norms. □

Numerical experiments illustrating (6.20) are in Dostál and Horák [78, 79].

6.3 Exact Augmented Lagrangian Method

Because of its simplicity and intuitive appeal, the penalty method is often used in computations. However, a good solution approximation may require a very large penalty parameter. The remedy can be based on the observation that having a minimizer $\mathbf{x}^k$ of the augmented Lagrangian (6.6), we can modify its linear term in such a way that the unconstrained minimum of $L(\mathbf{x}, \boldsymbol{\lambda}^{k+1}, 0)$ *without the penalization term* is achieved again at $\mathbf{x}^k$ as

$$\begin{aligned}\nabla L(\mathbf{x}^k, \boldsymbol{\lambda}^k, \varrho) &= \mathsf{A}_\varrho\mathbf{x}^k - \mathbf{b} + \mathsf{B}^T\boldsymbol{\lambda}^k = \mathsf{A}\mathbf{x}^k - \mathbf{b} + \mathsf{B}^T\mathsf{B}\mathbf{x}^k + \mathsf{B}^T\boldsymbol{\lambda}^k \\ &= \mathsf{A}\mathbf{x}^k - \mathbf{b} + \mathsf{B}^T(\boldsymbol{\lambda}^k + \varrho\mathsf{B}\mathbf{x}^k) = \nabla L(\mathbf{x}^k, \boldsymbol{\lambda}^k + \varrho\mathsf{B}\mathbf{x}^k, 0) = \mathbf{o}.\end{aligned}$$

Denoting $\boldsymbol{\lambda}^{k+1} = \boldsymbol{\lambda}^k + \varrho\mathsf{B}\mathbf{x}^k$, we get $\mathbf{x}^{k+1}$ by minimizing $L(\mathbf{x}, \boldsymbol{\lambda}^{k+1}, \varrho)$ It is easy to check that

$$L(\mathbf{x}^k, \boldsymbol{\lambda}^k, \varrho) \leq L(\mathbf{x}^{k+1}, \boldsymbol{\lambda}^{k+1}, \varrho) \leq L(\widehat{\mathbf{x}}, \widehat{\boldsymbol{\lambda}}, \varrho). \tag{6.21}$$

The procedure is illustrated in Fig. 6.2.

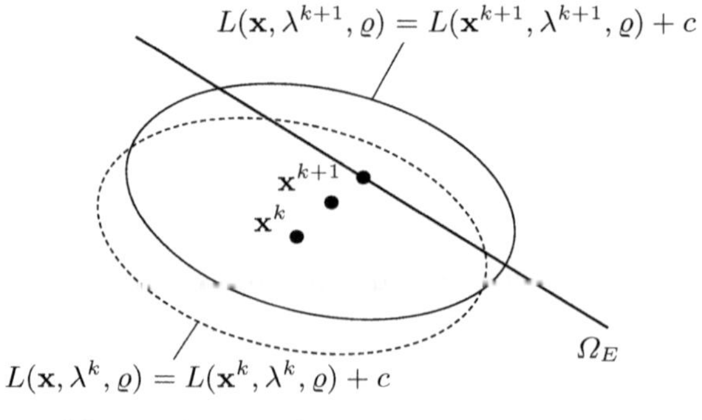

Fig. 6.2. Augmented Lagrangian iteration

The result is the well-known classical *augmented Lagrangian algorithm*, also named the *method of multipliers*, which was proposed by Hestenes [163] and Powell [233].

In this section, we present as the augmented Lagrangian algorithm a little more general algorithm; its special cases are the classical Uzawa algorithm [2] and the original algorithm by Hestenes and Powell. We review and slightly extend the well-known arguments presented, e.g., in the classical monographs by Bertsekas [15] and Glowinski and Le Tallec [144].

6.3.1 Algorithm

The augmented Lagrangian algorithm is based similarly on the Arrow–Hurwitz algorithm 6.1, on the mixed formulation (2.37) of the equality-constrained problem (6.1). However, the augmented Lagrangian algorithm differs from the Arrow–Hurwitz algorithm applied to the penalized problem (6.4) in Step 1, where the former algorithm assigns $\mathbf{x}^k$ the minimizer of $L(\mathbf{x}, \lambda^k, \varrho_k)$ with respect to $\mathbf{x} \in \mathbb{R}^n$. Both algorithms use the same update rule for the Lagrange multipliers in Step 2. Here, we present a variant of the augmented Lagrangian algorithm whose special cases are the original *Uzawa algorithm* [2], which corresponds to $\varrho_k = 0, k = 0, 1, \ldots$, and the original *method of multipliers*, which corresponds to $r_k = \varrho_k$. Our augmented Lagrangian algorithm reads as follows.

Algorithm 6.2. Exact augmented Lagrangian algorithm

Given a symmetric positive definite matrix $\mathsf{A} \in \mathbb{R}^{n\times n}$, $\mathsf{B} \in \mathbb{R}^{m\times n}$, $\mathbf{b} \times \mathbb{R}^n$, *and* $\mathbf{c} \in \mathrm{Im}\mathsf{B}$.

Step 0. *{Initialization.}*
  Choose $\boldsymbol{\lambda}^0 \in \mathbb{R}^m$, $r > 0$, $r_k \geq r$, $\varrho_k \geq 0$
  for $k = 0, 1, 2, \ldots$
Step 1. *{Minimization with respect to* $\mathbf{x}$.*}*
  $\mathbf{x}^k = \arg\min\{L(\mathbf{x}, \boldsymbol{\lambda}^k, \varrho_k) : \mathbf{x} \in \mathbb{R}^n\}$
Step 2. *{Updating the Lagrange multipliers.}*
  $\boldsymbol{\lambda}^{k+1} = \boldsymbol{\lambda}^k + r_k(\mathsf{B}\mathbf{x}^k - \mathbf{c})$
  end *for*

Since $\mathbf{x}^k$ is the unconstrained minimizer of the Lagrangian L with respect to its first variable, it follows that

$$\nabla_{\mathbf{x}} L(\mathbf{x}^k, \boldsymbol{\lambda}^k, \varrho_k) = (\mathsf{A} + \varrho_k \mathsf{B}^T\mathsf{B})\mathbf{x}^k - \mathbf{b} - \varrho_k \mathsf{B}^T\mathbf{c} + \mathsf{B}^T\boldsymbol{\lambda}^k = \mathbf{o},$$

so that Step 1 of Algorithm 6.2 can be implemented by solving the system

$$(\mathsf{A} + \varrho_k \mathsf{B}^T\mathsf{B})\mathbf{x}^k = \mathbf{b} + \varrho_k \mathsf{B}^T\mathbf{c} - \mathsf{B}^T\boldsymbol{\lambda}^k. \tag{6.22}$$

To understand the algorithm better, we shall examine its alternative formulation, which we obtain after eliminating $\mathbf{x}^k$ or $\boldsymbol{\lambda}^k$ from Algorithm 6.2. Thus denoting for any $\varrho \in \mathbb{R}$

$$\mathsf{A}_\varrho = \mathsf{A} + \varrho \mathsf{B}^T \mathsf{B},$$

we can use (6.22) to get

$$\mathbf{x}^k = \mathsf{A}_{\varrho_k}^{-1}(\mathbf{b} + \varrho_k \mathsf{B}^T \mathbf{c} - \mathsf{B}^T \boldsymbol{\lambda}^k).$$

After substituting for $\mathbf{x}^k$ into Step 2 of Algorithm 6.2 and simple manipulations, we can rewrite our augmented Lagrangian algorithm as

$$\text{Choose } \boldsymbol{\lambda}^0 \in \mathbb{R}^{\mathrm{m}}, \tag{6.23}$$
$$\boldsymbol{\lambda}^{k+1} = \boldsymbol{\lambda}^k - r_k(\mathsf{B}\mathsf{A}_{\varrho_k}^{-1}\mathsf{B}^T\boldsymbol{\lambda}^k - \mathsf{B}\mathsf{A}_{\varrho_k}^{-1}(\mathbf{b} + \varrho_k \mathsf{B}^T\mathbf{c}) + \mathbf{c}). \tag{6.24}$$

To understand the formula (6.24), notice that

$$f_\varrho(\mathbf{x}) = \frac{1}{2}\mathbf{x}^T \mathsf{A}_\varrho \mathbf{x} - (\mathbf{b} + \varrho \mathsf{B}^T \mathbf{c})^T \mathbf{x} + \frac{\varrho}{2}\|\mathbf{c}\|^2.$$

Using the formula (2.35) for the dual function Θ for problem (6.1), we can check that the explicit expression for the dual function Θ_ϱ for the minimum of $f_\varrho(\mathbf{x})$ subject to $\mathbf{x} \in \Omega_E$ reads

$$\begin{aligned}\Theta_\varrho(\boldsymbol{\lambda}) = & -\frac{1}{2}\boldsymbol{\lambda}^T \mathsf{B}\mathsf{A}_\varrho^{-1}\mathsf{B}^T\boldsymbol{\lambda} + (\mathsf{B}\mathsf{A}_\varrho^{-1}(\mathbf{b} + \varrho\mathsf{B}^T\mathbf{c}) - \mathbf{c})\boldsymbol{\lambda} \\ & - \frac{1}{2}(\mathbf{b} + \varrho\mathsf{B}^T\mathbf{c})^T\mathsf{A}^{-1}(\mathbf{b} + \varrho\mathsf{B}^T) + \frac{\varrho}{2}\|\mathbf{c}\|^2.\end{aligned}$$

It follows that

$$\nabla\Theta_{\varrho_k}(\boldsymbol{\lambda}^k) = -\mathsf{B}\mathsf{A}_{\varrho_k}^{-1}\mathsf{B}^T\boldsymbol{\lambda}^k + \mathsf{B}\mathsf{A}_{\varrho_k}^{-1}(\mathbf{b} + \varrho_k\mathsf{B}^T\mathbf{c}) - \mathbf{c}. \tag{6.25}$$

Comparing the latter formula with (6.24), we conclude that

$$\boldsymbol{\lambda}^{k+1} = \boldsymbol{\lambda}^k + r_k \nabla\Theta_{\varrho_k}(\boldsymbol{\lambda}^k).$$

Thus the augmented Lagrangian algorithm may be interpreted as the gradient method for maximization of the dual function Θ_ϱ for the penalized problem (6.1) with the steplength r_k.

Alternatively, we can eliminate $\boldsymbol{\lambda}^k$ from Algorithm 6.2 to get

$$\mathbf{x}^0 = \mathsf{A}_{\varrho_0}^{-1}(\mathbf{b} + \varrho_k\mathsf{B}^T\mathbf{c} - \mathsf{B}^T\boldsymbol{\lambda}^0), \tag{6.26}$$
$$\mathbf{x}^{k+1} = \mathbf{x}^k - r_k\mathsf{A}_{\varrho_k}^{-1}\mathsf{B}^T(\mathsf{B}\mathbf{x}^k - \mathbf{c}). \tag{6.27}$$

6.3.2 Convergence of Lagrange Multipliers

Let us first recall that, by Proposition 2.10(iii) and the discussion at the end of Sect. 2.4.2, any Lagrange multiplier $\overline{\boldsymbol{\lambda}}$ of the equality constrained problem (6.1) can be expressed as

$$\overline{\boldsymbol{\lambda}} = \boldsymbol{\lambda}_{\mathrm{LS}} + \boldsymbol{\delta}, \qquad \boldsymbol{\lambda}_{\mathrm{LS}} \in \mathrm{Im}\mathsf{B}, \qquad \boldsymbol{\delta} \in \mathrm{Ker}\mathsf{B}^T,$$

where $\boldsymbol{\lambda}_{\mathrm{LS}}$ is the Lagrange multiplier with the minimal Euclidean norm. If we denote by P and $\mathsf{Q} = \mathsf{I} - \mathsf{P}$ the orthogonal projectors on $\mathrm{Im}\mathsf{B}$ and $\mathrm{Ker}\mathsf{B}^T$, respectively, then the components of $\overline{\boldsymbol{\lambda}}$ are given by

$$\boldsymbol{\lambda}_{\mathrm{LS}} = \mathsf{P}\overline{\boldsymbol{\lambda}}, \qquad \boldsymbol{\nu} = \mathsf{Q}\overline{\boldsymbol{\lambda}}.$$

To simplify the notations, we assume that $\varrho_k = \varrho$ and $r_k = r$.

To study the convergence of $\boldsymbol{\lambda}^k$ generated by Algorithm 6.2, let $\boldsymbol{\lambda}^0 \in \mathbb{R}^m$, let us denote

$$\overline{\boldsymbol{\lambda}} = \boldsymbol{\lambda}_{\mathrm{LS}} + \mathsf{Q}\boldsymbol{\lambda}^0,$$

and observe that

$$\begin{aligned}
\boldsymbol{\lambda}^0 - \overline{\boldsymbol{\lambda}} &= \mathsf{P}\boldsymbol{\lambda}^0 + \mathsf{Q}\boldsymbol{\lambda}^0 - \boldsymbol{\lambda}_{\mathrm{LS}} - \mathsf{Q}\boldsymbol{\lambda}^0 = \mathsf{P}(\boldsymbol{\lambda}^0 - \boldsymbol{\lambda}_{\mathrm{LS}}) \in \mathrm{Im}\mathsf{B},\\
\boldsymbol{\lambda}^{k+1} - \overline{\boldsymbol{\lambda}} &= (\boldsymbol{\lambda}^k - \overline{\boldsymbol{\lambda}}) - r(\mathsf{B}\mathsf{A}_\varrho^{-1}\mathsf{B}^T\boldsymbol{\lambda}^k - \mathsf{B}\mathsf{A}_\varrho^{-1}(\mathbf{b} + \varrho\mathsf{B}^T\mathbf{c}) + \mathbf{c})\\
&= (\boldsymbol{\lambda}^k - \overline{\boldsymbol{\lambda}}) - r\mathsf{B}\mathsf{A}_\varrho^{-1}\mathsf{B}^T(\boldsymbol{\lambda}^k - \overline{\boldsymbol{\lambda}}),
\end{aligned}$$

where we used $\mathsf{P}\boldsymbol{\lambda}_{\mathrm{LS}} = \boldsymbol{\lambda}_{\mathrm{LS}}$ and (6.9). It follows that

$$\boldsymbol{\lambda}^{k+1} - \overline{\boldsymbol{\lambda}} = (\mathsf{I} - r\mathsf{B}\mathsf{A}_\varrho^{-1}\mathsf{B}^T)(\boldsymbol{\lambda}^k - \overline{\boldsymbol{\lambda}}) \tag{6.28}$$

and

$$\boldsymbol{\lambda}^{k+1} - \overline{\boldsymbol{\lambda}} \in \mathrm{Im}\mathsf{B}, \quad k = 0, 1, \ldots.$$

Therefore the analysis of convergence of $\boldsymbol{\lambda}^k$ reduces to the analysis of the spectrum of the restriction of the iteration matrix $\mathsf{I} - r\mathsf{B}\mathsf{A}_\varrho^{-1}\mathsf{B}^T$ to its invariant subspace $\mathrm{Im}\mathsf{B}$.

Using (1.32) and Lemma 1.7, we get that the eigenvalues μ_i of the iteration matrix are related to the eigenvalues $\overline{\beta}_i$ of $\mathsf{B}\mathsf{A}^{-1}\mathsf{B}^T|\mathrm{Im}\mathsf{B}$ by

$$\mu_i = 1 - \frac{r\overline{\beta}_i}{1 + \varrho\overline{\beta}_i} = \frac{1 + (\varrho - r)\overline{\beta}_i}{1 + \varrho\overline{\beta}_i},$$

so that

$$\|(\mathsf{I} - r\mathsf{B}\mathsf{A}_\varrho^{-1}\mathsf{B}^T)|\mathrm{Im}\mathsf{B}\| = \max_{\substack{i\in\{1,\ldots,m\}\\ \beta_i > 0}} \frac{|1 + (\varrho - r)\beta_i|}{1 + \varrho\beta_i}.$$

Denoting

$$R(\varrho, r) = \max_{\substack{i \in \{1,\dots,m\} \\ \beta_i > 0}} \frac{|1 + (\varrho - r)\beta_i|}{1 + \varrho\beta_i} \tag{6.29}$$

and using that the norm is submultiplicative, we get for $R(\varrho, r) < 1$ the linear rate of convergence

$$\|\boldsymbol{\lambda}^{k+1} - \overline{\boldsymbol{\lambda}}\| \leq R(\varrho, r)\|\boldsymbol{\lambda}^k - \overline{\boldsymbol{\lambda}}\|. \tag{6.30}$$

We have thus reduced the study of convergence of Algorithm 6.2 to the analysis of $R(\varrho, r)$. We shall formulate the result on the convergence of the Lagrange multipliers in the following theorem.

Theorem 6.5. *Let $\boldsymbol{\lambda}^k$, $k = 0, 1, \dots$, denote the sequence of vectors generated by Algorithm 6.2 for problem (6.1) with a given ϱ_k and r_k starting from a given vector $\boldsymbol{\lambda}^0 \in \mathbb{R}^m$. Let $\boldsymbol{\lambda}_{\mathrm{LS}}$ denote the least square Lagrange multiplier, let P denote the orthogonal projector on* ImB, *let $\beta_{\max}$ denote the largest eigenvalue of $\mathsf{B}\mathsf{A}^{-1}\mathsf{B}^T$, and denote*

$$\overline{\boldsymbol{\lambda}} = \boldsymbol{\lambda}_{\mathrm{LS}} + (\mathbf{I} - \mathsf{P})\boldsymbol{\lambda}^0.$$

If there are $\varepsilon > 0$ and $M > 0$ such that

$$\varepsilon \leq r_k \leq \frac{2}{\beta_{\max}} + 2\varrho_k - \varepsilon \leq M, \tag{6.31}$$

then $\boldsymbol{\lambda}^k$ converge to $\overline{\boldsymbol{\lambda}}$ and the rate of convergence is at least linear, i.e., there is $R < 1$ such that

$$\|\boldsymbol{\lambda}^{k+1} - \overline{\boldsymbol{\lambda}}\| \leq R\|\boldsymbol{\lambda}^k - \overline{\boldsymbol{\lambda}}\|.$$

Proof. Elementary, but a bit laborious analysis of $R(\varrho_k, r_k)$, where R is defined by (6.29), reveals that if ϱ_k, r_k satisfy (6.31), then

$$\sup_{k=0,1,\dots} R(\varrho_k, r_k) = R < 1.$$

We finish the proof by substituting this result into (6.30). □

It is possible to use different arguments to prove convergence of Algorithm 6.2 under more relaxed conditions. For example, Glowinski and Le Tallec [144] give the condition

$$0 < \varepsilon \leq r_k \leq 2\varrho_k.$$

6.3.3 Effect of the Steplength

Let us now examine possible options of the steplength $r = r(\varrho)$ as a function of ϱ, including their effect on $R(\varrho, r)$. We shall denote by $\overline{\beta}_{\min}$ the smallest nonzero eigenvalues of $\mathsf{BA}^{-1}\mathsf{B}^T$, i.e., the smallest eigenvalue of $\mathsf{BA}^{-1}\mathsf{B}^T|\mathrm{Im}\mathsf{B}$. Our examples are from Glowinski and Le Tallec [144].

Optimal Choice of r with $\varrho = 0$

In this case, which corresponds to the original Uzawa algorithm,

$$R(\varrho, r) = R(0, r) = \max_{\substack{i\in\{1,\dots,m\}\\ \beta_i>0}} |1 - r\beta_i|, \tag{6.32}$$

so that the best choice of r is given by

$$\begin{aligned} R(0, r_{\mathrm{opt}}) &= \min_r \max_{\substack{i\in\{1,\dots,m\}\\ \beta_i>0}} |1 - r\beta_i| = \min_r \max_{\substack{i\in\{1,\dots,m\}\\ \beta_i>0}} \{1 - r\beta_i, r\beta_i - 1\} \\ &= \min_r \max\{1 - r\overline{\beta}_{\mathrm{min}}, r\beta_{\mathrm{max}} - 1\}. \end{aligned}$$

A simple analysis reveals that r_{opt} satisfies

$$1 - r\overline{\beta}_{\mathrm{min}} = r\beta_{\mathrm{max}} - 1.$$

Solving the last equation with respect to r, we get

$$r_{\mathrm{opt}} = \frac{2}{\overline{\beta}_{\mathrm{min}} + \beta_{\mathrm{max}}}$$

and

$$R(0, r_{\mathrm{opt}}) = 1 - r_{\mathrm{opt}}\overline{\beta}_{\mathrm{min}} = \frac{\beta_{\mathrm{max}} - \overline{\beta}_{\mathrm{min}}}{\beta_{\mathrm{max}} + \overline{\beta}_{\mathrm{min}}} = \frac{\beta_{\mathrm{max}}/\overline{\beta}_{\mathrm{min}} - 1}{\beta_{\mathrm{max}}/\overline{\beta}_{\mathrm{min}} + 1}.$$

This is the optimal rate of convergence of the gradient method of Sect. 3.1 applied to the dual function. Inspection of (6.32) reveals that $R(0, r) < 1$ requires that $1 - r\beta_{\mathrm{max}} > -1$, i.e.,

$$r < 2/\beta_{\mathrm{max}},$$

so that r_{opt} is typically near the bound which guarantees the convergence.

Choice $r = \varrho$

In this case, which is natural from the point of view of our analysis of the penalty method,

$$R(\varrho, \varrho) = \max_{\substack{i\in\{1,\dots,m\}\\ \beta_i>0}} \frac{1}{1 + \varrho\beta_i} = \frac{1}{1 + \varrho\beta_{\mathrm{min}}}. \tag{6.33}$$

An interesting feature of this choice is that

$$\lim_{\varrho\to\infty} R(\varrho,\varrho) = 0,$$

so that by increasing ϱ, it is possible to achieve arbitrary preconditioning effect.

Choice $r = (1+\delta)\varrho$

Let us now consider the choice $r = (1+\delta)\varrho$ with $\delta > -1$. In this case

$$R(\varrho,(1+\delta)\varrho) = \max_{\substack{i\in\{1,\dots,m\}\\ \beta_i>0}} \frac{|1-\delta\varrho\beta_i|}{1+\varrho\beta_i},$$

so that $r > 0$ and $R(\varrho,(1+\delta)\varrho) < 1$ if and only if $-1 < \delta < 1 + 2/(\varrho\beta_{\max})$. Moreover

$$\lim_{\varrho\to\infty} R(\varrho,(1+\delta)\varrho) = |\delta|.$$

The preconditioning effect, which can be achieved by increasing the penalty parameter, is limited when $\delta \neq 0$.

Optimal Steplength for a Given ϱ

If ϱ is given, then the optimal steplength $r_{\text{opt}}(\varrho)$ is given by

$$r_{\text{opt}} = \arg\min_{r\ge 0}\left(\max_{\substack{i\in\{1,\dots,m\}\\ \beta_i>0}} \frac{|1+(\varrho-r)\beta_i|}{1+\varrho\beta_i}\right).$$

To find it, let us denote

$$\varphi_i(r) = \frac{1+(\varrho-r)\beta_i}{1+\varrho\beta_i} = \frac{(1+\varrho\beta_i)-r\beta_i}{1+\varrho\beta_i},$$

and observe that if $\beta_i > 0$, then $\varphi_i(r)$ is decreasing. Since $\varphi_i(0) = 1$, it follows that for $r \ge 0$

$$\max_{\substack{i\in\{1,\dots,m\}\\ \beta_i>0}} \varphi_i(r) = \max_{\substack{i\in\{1,\dots,m\}\\ \beta_i>0}} \frac{1+(\varrho-r)\beta_i}{1+\varrho\beta_i} = \frac{1+(\varrho-r)\overline{\beta}_{\min}}{1+\varrho\overline{\beta}_{\min}}.$$

Similarly $-\varphi_i(0) = -1$, and if $\beta_i > 0$, then $-\varphi_i(r)$ is increasing. Therefore

$$\max_{\substack{i\in\{1,\dots,m\}\\ \beta_i>0}} -\varphi_i(r) = \max_{\substack{i\in\{1,\dots,m\}\\ \beta_i>0}} -\frac{1+(\varrho-r)\beta_i}{1+\varrho\beta_i} = -\frac{1+(\varrho-r)\overline{\beta}_{\max}}{1+\varrho\overline{\beta}_{\max}}$$

for nonnegative r. Since the both maxima are nonnegative on the positive interval $[\varrho+1/\beta_{\max}, \varrho+1/\beta_{\min}]$, it follows that the optimal choice $r_{\text{opt}}(\varrho)$ is a nonnegative solution of

$$\frac{1+(\varrho-r)\overline{\beta}_{\min}}{1+\varrho\overline{\beta}_{\min}} = -\frac{1+(\varrho-r)\beta_{\max}}{1+\varrho\beta_{\max}}.$$

Carrying out the computations, we get that

$$r_{\text{opt}}(\varrho) = \varrho + \frac{2+\varrho(\overline{\beta}_{\min}+\beta_{\max})}{2\varrho\overline{\beta}_{\min}\beta_{\max}+\overline{\beta}_{\min}+\beta_{\max}}.$$

We conclude that the optimal steplength $r_{\text{opt}}(\varrho)$ based on the estimate (6.30) is *longer* than the penalization parameter ϱ, but $r_{\text{opt}}(\varrho)$ approaches ϱ for large values of ϱ as

$$\lim_{\varrho\to\infty} r_{\text{opt}}(\varrho)/\varrho = 1.$$

This observation agrees with the above discussion and our analysis of the penalty method in Sect. 6.2.1, which suggests that $r=\varrho$ is a suitable step length for large ϱ.

Given $\mathbf{x}^k$ which minimizes $L(\mathbf{x},\boldsymbol{\lambda}^k,\varrho)$ with respect to $\mathbf{x}\in\mathbb{R}^n$, then $\mathbf{x}^k$ satisfies

$$\mathsf{A}\mathbf{x}^k + \mathsf{B}^T(\boldsymbol{\lambda}^k + \varrho(\mathsf{B}\mathbf{x}^k - \mathbf{c})) - \mathbf{b} = \mathbf{o}.$$

Thus the choice $r=\varrho$ results in

$$\nabla L_0(\mathbf{x}^k, \boldsymbol{\lambda}^{k+1}) = \mathbf{o},$$

so that it is optimal also in the sense that

$$\varrho = \arg\min_{r\ge 0} \|\mathsf{A}\mathbf{x}^k + \mathsf{B}^T(\boldsymbol{\lambda}^k + r(\mathsf{B}\mathbf{x}^k - \mathbf{c})) - \mathbf{b}\|.$$

Maximizing the Gradient Ascent

In Sect. 6.3.1, we have shown that Algorithm 6.2 may be interpreted as a gradient algorithm applied to the maximization of the dual function Θ_ϱ. Thus, it seems natural to define r_k by maximizing the quadratic function

$$\phi(r) = \Theta_{\varrho_k}(\boldsymbol{\lambda}^k + r\mathbf{g}^k),$$

where $\mathbf{g}^k = \nabla\Theta_{\varrho_k}(\boldsymbol{\lambda}^k)$, with respect to r. Denoting $\mathsf{A}_{\varrho_k} = \mathsf{A} + \varrho_k\mathsf{B}^T\mathsf{B}$,

$$\mathsf{F}_{\varrho_k} = \mathsf{B}\mathsf{A}_{\varrho_k}^{-1}\mathsf{B}^T, \qquad \text{and} \qquad \mathbf{d} = \mathbf{d}_{\varrho_k} = \mathsf{B}\mathsf{A}_{\varrho_k}^{-1}(\mathbf{b} + \varrho_k\mathsf{B}^T\mathbf{c}) - \mathbf{c},$$

we can write

$$\phi(r) = -\frac{1}{2}(\boldsymbol{\lambda}^k + r\mathbf{g}^k)^T\mathsf{F}_{\varrho_k}(\boldsymbol{\lambda}^k + r\mathbf{g}^k) + (\boldsymbol{\lambda}^k + r\mathbf{g}^k)^T\mathbf{d},$$

so that the maximizer satisfies

$$\frac{d}{dr}\phi(r) = -r(\mathbf{g}^k)^T\mathsf{F}_{\varrho_k}\mathbf{g}^k - (\mathbf{g}^k)^T(\mathsf{F}_{\varrho_k}\boldsymbol{\lambda}^k - \mathbf{d}) = -r(\mathbf{g}^k)^T\mathsf{F}_{\varrho_k}\mathbf{g}^k + (\mathbf{g}^k)^T\mathbf{g}^k = 0.$$

Thus, we can use the steepest ascent formula

$$r_k = \frac{\|\mathbf{g}^k\|^2}{(\mathbf{g}^k)^T\mathsf{F}_{\varrho_k}\mathbf{g}^k}, \tag{6.34}$$

which may be applied to obtain the largest increase of Θ_{ϱ_k} in step k. For large ϱ_k, we get by (1.55) and (1.56) that r_k is close to ϱ_k in agreement with the optimal choice of the steplength based on the estimate (6.30). Notice that the steplength based on (6.34) depends on the current iteration.

6.3.4 Convergence of the Feasibility Error

To estimate the feasibility error $\|\mathsf{B}\mathbf{x}^k - \mathbf{c}\|$, let us multiply equation (6.27) by B and then subtract $\mathbf{c}$ from both sides of the result to get

$$\mathsf{B}\mathbf{x}^{k+1} - \mathbf{c} = \mathsf{B}\mathbf{x}^k - \mathbf{c} - r_k\mathsf{B}\mathsf{A}_{\varrho_k}^{-1}\mathsf{B}^T(\mathsf{B}\mathbf{x}^k - \mathbf{c}),$$

where $\mathsf{A}_{\varrho_k} = \mathsf{A} + \varrho_k\mathsf{B}^T\mathsf{B}$. It follows that

$$\|\mathsf{B}\mathbf{x}^{k+1} - \mathbf{c}\| \le \|(\mathsf{I} - r_k\mathsf{B}\mathsf{A}_{\varrho_k}^{-1}\mathsf{B}^T)|\mathrm{Im}\mathsf{B}\|\|\mathsf{B}\mathbf{x}^k - \mathbf{c}\|, \tag{6.35}$$

so that, under the assumptions of Theorem 6.5, we can use the same arguments to prove the linear convergence of the feasibility error.

Theorem 6.6. *Let $\mathbf{x}^k, k = 0, 1, \ldots,$ be generated by Algorithm 6.2 for problem (6.1) with given $\varrho_k, r_k,$ and $\boldsymbol{\lambda}^0 \in \mathbb{R}^m$. Let $\beta_{\max}$ denote the largest eigenvalue of $\mathsf{B}\mathsf{A}^{-1}\mathsf{B}^T$. If there are $\varepsilon > 0$ and $M > 0$ such that*

$$\varepsilon \le r_k \le \frac{2}{\beta_{\max}} + 2\varrho_k - \varepsilon \le M, \tag{6.36}$$

then the feasibility error $\|\mathsf{B}\mathbf{x}^k - \mathbf{c}\|$ converges to zero, and the rate of convergence is at least linear, i.e., there is $R < 1$ such that

$$\|\mathsf{B}\mathbf{x}^{k+1} - \mathbf{c}\| \le R\|\mathsf{B}\mathbf{x}^k - \mathbf{c}\|. \tag{6.37}$$

6.3.5 Convergence of Primal Variables

Having the proofs of convergence of the Lagrange multipliers and the feasibility error, we may use Proposition 2.12 to prove the convergence of the primal variables.

Theorem 6.7. *Let $\mathbf{x}^k, k = 0, 1, \ldots,$ be generated by Algorithm 6.2 for problem (6.1) with given $\varrho_k, r_k,$ and $\boldsymbol{\lambda}^0 \in \mathbb{R}^m$, let $\widehat{\mathbf{x}}$ denote the unique solution of (6.1), let $\overline{\sigma}_{\min}$ denote the least nonzero singular value of B, and let $\beta_{\max}$ denote the largest eigenvalue of $\mathsf{BA}^{-1}\mathsf{B}^T$.*

If there are $\varepsilon > 0$ and $M > 0$ such that

$$\varepsilon \leq r_k \leq \frac{2}{\beta_{\max}} + 2\varrho_k - \varepsilon \leq M, \tag{6.38}$$

then $\|\mathbf{x}^k - \widehat{\mathbf{x}}\|$ converges to zero and the rate of convergence is at least R-linear, i.e., there is $R < 1$ such that

$$\|\mathbf{x}^k - \widehat{\mathbf{x}}\| \leq R^k \frac{\kappa(\mathsf{A})\|B\|}{\overline{\sigma}_{\min}} \|\mathbf{x}^0 - \widehat{\mathbf{x}}\|. \tag{6.39}$$

Proof. First, recall that by the assumptions and (6.37),

$$\|\mathsf{B}\mathbf{x}^k - \mathbf{c}\| \leq R^k \|\mathsf{B}\mathbf{x}^0 - \mathbf{c}\| = R^k \|\mathsf{B}(\mathbf{x}^0 - \widehat{\mathbf{x}})\| \leq R^k \|\mathsf{B}\| \|\mathbf{x}^0 - \widehat{\mathbf{x}}\|,$$

where $R < 1$. Using (2.47), we get that

$$\|\mathbf{x}^k - \widehat{\mathbf{x}}\| \leq \frac{\kappa(\mathsf{A})}{\overline{\sigma}_{\min}} \|\mathsf{B}\mathbf{x}^k - \mathbf{c}\| \leq R^k \frac{\kappa(\mathsf{A})\|\mathsf{B}\|}{\overline{\sigma}_{\min}} \|\mathbf{x}^0 - \widehat{\mathbf{x}}\|.$$

We have used the fact that $\nabla_{\mathbf{x}} L(\mathbf{x}^k, \boldsymbol{\lambda}^k, \varrho) = \mathbf{o}$. □

6.3.6 Implementation

Since it is possible to approximate the solution of (6.1) with a single step of the penalty method, the above discussion suggests taking $r_k = \varrho_k = \varrho$ as large as possible.

The auxiliary problems in Step 1 can be effectively solved by the Cholesky factorization

$$\mathsf{LL}^T = \mathsf{A}_\varrho,$$

which should be carried out after each update of ϱ, and the multiplication of a vector $\boldsymbol{\lambda}$ by $\mathsf{BA}_\varrho^{-1}\mathsf{B}^T$ should be carried out as

$$\mathsf{BA}_\varrho^{-1}\mathsf{B}^T\boldsymbol{\lambda} = \mathsf{B}(\mathsf{L}^{-T}(\mathsf{L}^{-1}(\mathsf{B}^T\boldsymbol{\lambda}))).$$

Multiplication by the inverse factors should be implemented as the solution for the related triangular systems. Since the sensitivity to round-off errors is greater when ϱ is large, the algorithm should be implemented in double precision.

Application of iterative solvers can hardly be efficient for the implementation of Step 1 of Algorithm 6.2, where an exact solution is required, but it can be very efficient for the implementation of inexact augmented Lagrangian algorithms discussed in Chap. 9.

6.4 Comments and References

The augmented Lagrangian method was proposed independently by Powell [233] and Hestenes [163] for problems with a general cost function subject to general equality constraints. Comprehensive analysis of the augmented Lagrangian method (called the Lagrange multiplier method), including the asymptotically exact minimization of auxiliary problems, was presented in the monograph by Bertsekas [15]. Applications to the solution of boundary value problems are discussed in Glowinski and Fortin [134] and Glowinski and Le Tallec [144]. Helpful discussion on Augmented Lagrangian techniques for solving practical optimization problems can be found in Birgin and Martínez [19].

The augmented Lagrangian algorithm for equality constraints was also used to solve problems with inequality constraints. In their seminal paper on contact problems with friction, Laursen and Simo [202] used the corrections of primal variables and Lagrange multipliers ('return mapping') to obtain the solution of the problem with inequality constraints. We shall introduce an alternative approach in Chap. 12. See also Hunek [173].

7
Active Sets with Finite Termination

We shall now be concerned with the *bound constrained problem* to find

$$\min_{\mathbf{x}\in\Omega_B} f(\mathbf{x}), \quad \Omega_B = \{\mathbf{x} \in \mathbb{R}^n : \ \mathbf{x} \geq \boldsymbol{\ell}\}, \tag{7.1}$$

with $f(\mathbf{x}) = \frac{1}{2}\mathbf{x}^T\mathsf{A}\mathbf{x} - \mathbf{x}^T\mathbf{b}$, $\boldsymbol{\ell}$ and $\mathbf{b}$ given column n-vectors, and A an $n \times n$ symmetric positive definite matrix. To include the possibility that not all the components of $\mathbf{x}$ are constrained, we admit $\ell_i = -\infty$. Here, we present basic *working set* based algorithms that can find the solution in a finite number of iterations. The analysis of these algorithms is typically based on arguments of a combinatorial nature.

The *exact working (active) set method* of Sect. 7.2 reduces the solution of (7.1) to a sequence of unconstrained problems that are defined by the constraints that are assumed to be active at the solution. See also Algorithm 7.1. We prove the finite termination of the algorithm by the combinatorial arguments.

The *Polyak algorithm* is a variant of the working set method that solves the auxiliary linear problems by the conjugate gradient method. The algorithm expands the active set, typically by one index, whenever it generates an unfeasible iterate, and it reduces the active set only after it finds the exact solution of an auxiliary unconstrained problem. We describe the algorithm in Sect. 7.3. See Algorithm 7.2 for the formal description.

The *looking ahead Polyak algorithm* is based on the observation that it is possible to recognize the incorrect active set before solving the auxiliary unconstrained problem. The algorithm accepts inexact solutions of auxiliary unconstrained problems and preserves the finite termination property of the original Polyak Algorithm 7.3. We describe the algorithm in Sect. 7.4.2. A more efficient expansion of the active set by the gradient projection is considered in Chap. 10.

Z. Dostál, *Optimal Quadratic Programming and QCQP Algorithms with Case Studies*, Springer Optimization and Its Applications 23,
https://doi.org/10.1007/978-3-031-95167-1_7

7.1 KKT Conditions and Related Inequalities

Since Ω_B is closed and convex and f is assumed to be strictly convex quadratic, the solution $\widehat{\mathbf{x}}$ of problem (7.1) exists and is necessarily unique by Proposition 2.5(i). Here, we introduce some definitions and notations that enable us to exploit the special form of the KKT conditions in developing our algorithms.

By Proposition 2.18, the KKT conditions read

$$\widehat{\mathbf{x}} \in \Omega_B, \quad \mathsf{A}\widehat{\mathbf{x}} - \mathbf{b} \geq \mathbf{o} \quad \text{and} \quad (\mathsf{A}\widehat{\mathbf{x}} - \mathbf{b})^T(\widehat{\mathbf{x}} - \boldsymbol{\ell}) = 0,$$

or componentwise

$$\widehat{x}_i = \ell_i \Rightarrow \widehat{g}_i \geq 0 \quad \text{and} \quad \widehat{x}_i > \ell_i \Rightarrow \widehat{g}_i = 0, \quad i = 1, \dots, n, \tag{7.2}$$

where $\widehat{g}_i = [\mathsf{A}\widehat{\mathbf{x}} - \mathbf{b}]_i$. Notice that $\widehat{g}_i$ are the components of the vector of Lagrange multipliers for the bound constraints.

The KKT conditions (7.2) determine three important subsets of the set $\mathcal{N} = \{1, 2, \dots, n\}$ of all indices. The set of all indices for which $x_i = \ell_i$ is called an *active set* of $\mathbf{x}$. We denote it by $\mathcal{A}(\mathbf{x})$, so

$$\mathcal{A}(\mathbf{x}) = \{i \in \mathcal{N} : \ x_i = \ell_i\}.$$

Its complement

$$\mathcal{F}(\mathbf{x}) = \{i \in \mathcal{N} : \ x_i \neq \ell_i\}$$

and subsets

$$\mathcal{B}(\mathbf{x}) = \{i \in \mathcal{N} : \ x_i = \ell_i \text{ and } g_i > 0\}, \quad \mathcal{B}_0(\mathbf{x}) = \{i \in \mathcal{N} : \ x_i = \ell_i \text{ and } g_i \geq 0\}$$

are called a *free set*, a *binding set*, and a *weakly binding set*, respectively. Thus we can rewrite the KKT conditions in the form

$$\mathbf{g}_{\mathcal{A}}(\widehat{\mathbf{x}}) \geq \mathbf{o}_{\mathcal{A}} \quad \text{and} \quad \mathbf{g}_{\mathcal{F}}(\widehat{\mathbf{x}}) = \mathbf{o}_{\mathcal{F}}.$$

Notice that

$$\widehat{\mathbf{y}} = \arg \min_{\widehat{\mathcal{B}} \subseteq \mathcal{B}(\mathbf{y})} f(\mathbf{y}) = \widehat{\mathbf{x}}, \quad \widehat{\mathcal{B}} = \mathcal{B}(\widehat{\mathbf{x}}). \tag{7.3}$$

Using the subsets of $\mathcal{N}$, we can decompose the part of the gradient $\mathbf{g}(\mathbf{x}) = \mathsf{A}\mathbf{x} - \mathbf{b}$ which violates the KKT conditions into the *free gradient* $\boldsymbol{\varphi}$ and the *chopped gradient* $\boldsymbol{\beta}$ that are defined by

$$\begin{aligned} \varphi_i(\mathbf{x}) &= g_i(\mathbf{x}) \text{ for } i \in \mathcal{F}(\mathbf{x}), \quad \varphi_i(\mathbf{x}) = 0 \text{ for } i \in \mathcal{A}(\mathbf{x}), \\ \beta_i(\mathbf{x}) &= 0 \text{ for } i \in \mathcal{F}(\mathbf{x}), \quad \beta_i(\mathbf{x}) = g_i^-(\mathbf{x}) \text{ for } i \in \mathcal{A}(\mathbf{x}), \end{aligned}$$

where $g_i^- = \min\{g_i, 0\}$. Introducing the *projected gradient*

$$\mathbf{g}^P(\mathbf{x}) = \boldsymbol{\varphi}(\mathbf{x}) + \boldsymbol{\beta}(\mathbf{x}), \tag{7.4}$$

we can write the Karush–Kuhn–Tucker conditions (7.2) conveniently as

$$\mathbf{g}^P(\mathbf{x}) = \mathbf{o}. \tag{7.5}$$

Obviously $\boldsymbol{\beta}(\mathbf{x})$ and $\boldsymbol{\varphi}(\mathbf{x})$ are orthogonal and $-\boldsymbol{\beta}(\mathbf{x})$ and $-\boldsymbol{\varphi}(\mathbf{x})$ are feasible decrease directions of f at $\mathbf{x}$. See also Fig. 7.1.

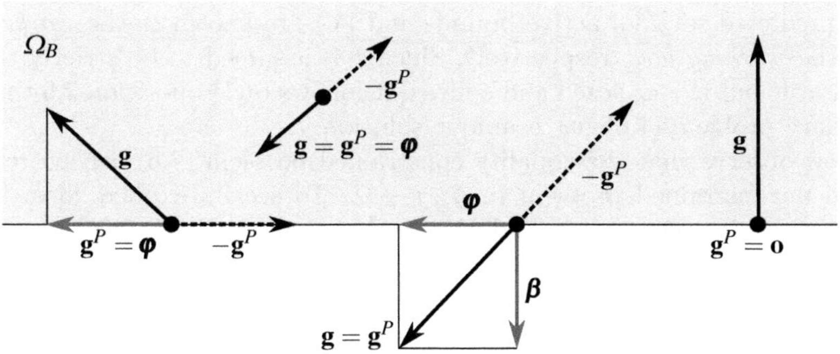

Fig. 7.1. Gradient splitting

The projected gradient models the function f only locally and does not take into account near bounds. Such information is partly captured by the α-reduced projected gradient defined by (5.5).

7.2 The Working Set Method with Exact Solutions

The basic idea of the *working set method*, or, as it is often called less correctly, the *active set method*, is to reduce the solution of an inequality-constrained problem to the solution of a sequence of auxiliary equality constrained problems which are defined by a subset of the set $\mathcal{N} = \{1, \dots, n\}$ of all indices of the constraints. This task would be very simple if we knew in advance which inequality constraints are active in the solution, as we could replace the relevant inequalities with equalities, ignore the other inequalities, and solve the resulting equality-constrained problem. As this is usually not the case, the working set method starts by making a guess which inequality constraints will be active in the solution, and if this guess turns out to be incorrect, it exploits the gradient and Lagrange multiplier information obtained by the trial minimization to define the next prediction.

7.2.1 Auxiliary Problems

If the working set method is applied to (7.1), it exploits the auxiliary equality constrained problems

$$\min_{\mathbf{y}\in\mathcal{W}_\mathcal{I}} f(\mathbf{y}), \tag{7.6}$$

where $\mathcal{I} \subseteq \mathcal{N}$ denotes the set of indices of bounds ℓ_i that are predicted to be active in the solution, and

$$\mathcal{W}_\mathcal{I} = \{\mathbf{y} :\ y_i = \ell_i,\ i \in \mathcal{I}\}.$$

The predicted set $\mathcal{I}$ of active bounds and $\mathcal{W}_\mathcal{I}$ are known as the *working set* and the *working face*, respectively. Since f is assumed to be strictly convex quadratic and $\mathcal{W}_\mathcal{I}$ is closed and convex, it follows by Proposition 2.5 that the auxiliary problem (7.6) has a unique solution $\widehat{\mathbf{y}}$.

Now observe that the equality constrained problem (7.6) can be reduced to an unconstrained problem in $y_j, j \notin \mathcal{I}$. To see its explicit form in the nontrivial cases $\mathcal{W}_\mathcal{I} \neq \{\boldsymbol{\ell}\}$ and $\mathcal{W}_\mathcal{I} \neq \mathbb{R}^n$, assume that $\emptyset \subsetneq \mathcal{J} \subsetneq \mathcal{N}$, and denote $\mathcal{J} = \mathcal{N}\backslash\mathcal{I}$, so that, after possibly rearranging the indices, we get

$$\mathbf{y} = \begin{bmatrix}\mathbf{y}_\mathcal{I}\\ \mathbf{y}_\mathcal{J}\end{bmatrix}, \quad \mathbf{b} = \begin{bmatrix}\mathbf{b}_\mathcal{I}\\ \mathbf{b}_\mathcal{J}\end{bmatrix}, \quad \text{and} \quad \mathsf{A} = \begin{bmatrix}\mathsf{A}_{\mathcal{I}\mathcal{I}} & \mathsf{A}_{\mathcal{I}\mathcal{J}}\\ \mathsf{A}_{\mathcal{J}\mathcal{I}} & \mathsf{A}_{\mathcal{J}\mathcal{J}}\end{bmatrix}. \tag{7.7}$$

Thus for any $\mathbf{y} \in \mathbb{R}^n$

$$f(\mathbf{y}) = \frac{1}{2}\mathbf{y}_\mathcal{J}^T\mathsf{A}_{\mathcal{J}\mathcal{J}}\mathbf{y}_\mathcal{J} + \mathbf{y}_\mathcal{J}^T\mathsf{A}_{\mathcal{J}\mathcal{I}}\mathbf{y}_\mathcal{I} + \frac{1}{2}\mathbf{y}_\mathcal{I}^T\mathsf{A}_{\mathcal{I}\mathcal{I}}\mathbf{y}_\mathcal{I} - \mathbf{y}_\mathcal{J}^T\mathbf{b}_\mathcal{J} - \mathbf{y}_\mathcal{I}^T\mathbf{b}_\mathcal{I}.$$

Since $\mathbf{y} \in \mathcal{W}_\mathcal{I}$ if and only if $\mathbf{y}_\mathcal{I} = \boldsymbol{\ell}_\mathcal{I}$, we have for any $\mathbf{y} \in \mathcal{W}_\mathcal{I}$

$$f(\mathbf{y}) = \frac{1}{2}\mathbf{y}_\mathcal{J}^T\mathsf{A}_{\mathcal{J}\mathcal{J}}\mathbf{y}_\mathcal{J} - \mathbf{y}_\mathcal{J}^T(\mathbf{b}_\mathcal{J} - \mathsf{A}_{\mathcal{J}\mathcal{I}}\boldsymbol{\ell}_\mathcal{I}) + \frac{1}{2}\boldsymbol{\ell}_\mathcal{I}^T\mathsf{A}_{\mathcal{I}\mathcal{I}}\boldsymbol{\ell}_\mathcal{I} - \mathbf{b}_\mathcal{I}^T\boldsymbol{\ell}_\mathcal{I}.$$

Thus the solution $\widehat{\mathbf{y}}$ of (7.6) has the components $\widehat{\mathbf{y}}_\mathcal{I} = \boldsymbol{\ell}_\mathcal{I}$ and

$$\widehat{\mathbf{y}}_\mathcal{J} = \arg\min_{\mathbf{y}_\mathcal{J}\in\mathbb{R}^m} f_\mathcal{J}(\mathbf{y}_\mathcal{J}),\ \ f_\mathcal{J}(\mathbf{y}_\mathcal{J}) = \frac{1}{2}\mathbf{y}_\mathcal{J}^T\mathsf{A}_{\mathcal{J}\mathcal{J}}\mathbf{y}_\mathcal{J} - \mathbf{y}_\mathcal{J}^T(\mathbf{b}_\mathcal{J} - \mathsf{A}_{\mathcal{J}\mathcal{I}}\boldsymbol{\ell}_\mathcal{I}). \tag{7.8}$$

Since

$$\nabla f_\mathcal{J}(\mathbf{y}_\mathcal{J}) = \mathsf{A}_{\mathcal{J}\mathcal{J}}\mathbf{y}_\mathcal{J} - (\mathbf{b}_\mathcal{J} - \mathsf{A}_{\mathcal{J}\mathcal{I}}\boldsymbol{\ell}_\mathcal{I})$$

and $\nabla f_\mathcal{J}(\widehat{\mathbf{y}}_\mathcal{J}) = \mathbf{o}$, we get that $\widehat{\mathbf{y}}_\mathcal{J}$ satisfies

$$\mathsf{A}_{\mathcal{J}\mathcal{J}}\widehat{\mathbf{y}}_\mathcal{J} = \mathbf{b}_\mathcal{J} - \mathsf{A}_{\mathcal{J}\mathcal{I}}\boldsymbol{\ell}_\mathcal{I}. \tag{7.9}$$

We can check easily that (7.9) has a unique solution. Indeed, since $\mathsf{A}_{\mathcal{J}\mathcal{J}}$ is a submatrix of a positive definite matrix A, we get by Cauchy's interlacing inequalities (1.26) that $\mathsf{A}_{\mathcal{J}\mathcal{J}}$ is also positive definite.

7.2.2 Algorithm

The *working set method with exact solutions* of auxiliary problems starts from an arbitrary $\mathbf{x}^0 \in \Omega_B$ and $\mathcal{I}^0 = \mathcal{A}(\mathbf{x}^0)$. Assuming that $\mathbf{x}^k$ is known, we first check if $\mathbf{x}^k$ is the solution of (7.1) by evaluating the KKT conditions

$$\mathbf{g}^P(\mathbf{x}^k) = \boldsymbol{\beta}(\mathbf{x}^k) + \boldsymbol{\varphi}(\mathbf{x}^k) = \mathbf{o}.$$

If $\boldsymbol{\varphi}(\mathbf{x}^k) \neq \mathbf{o}$, then we find the solution $\widehat{\mathbf{y}}$ of the auxiliary problem (7.6) by solving (7.9). If $\widehat{\mathbf{y}} \in \Omega_B$, we define the next iteration by the *feasible step*

$$\mathbf{x}^{k+1} = \widehat{\mathbf{y}}$$

and set $\mathcal{I}^{k+1} = \mathcal{A}(\mathbf{x}^{k+1})$, else we define $\mathbf{x}^{k+1}$ by an *expansion step* so that

$$f(\mathbf{x}^{k+1}) \leq f(\mathbf{x}^k) \quad \text{and} \quad \mathcal{A}(\mathbf{x}^{k+1}) \supsetneq \mathcal{I}^k, \tag{7.10}$$

and then set $\mathcal{I}^{k+1} = \mathcal{A}(\mathbf{x}^{k+1})$.

If $\boldsymbol{\varphi}(\mathbf{x}^k) = \mathbf{o}$, then we define the *release step* $\mathbf{x}^{k+1} = \mathbf{x}^k - \alpha_{opt}\boldsymbol{\beta}(\mathbf{x}^k)$, where $\alpha_{opt} = \arg\min f(\mathbf{x}^k - \alpha\boldsymbol{\beta}(\mathbf{x}^k)),\ \alpha > 0$. A new active set is defined by $\mathcal{I}^{k+1} = \mathcal{B}(\mathbf{x}^{k+1})$. The basic working set algorithm in the form convenient for analysis reads as follows.

Algorithm 7.1. The working set method with exact solutions

Given a symmetric positive definite matrix $\mathsf{A} \in \mathbb{R}^{n\times n}$ *and n-vectors* $\mathbf{b}, \boldsymbol{\ell}$.

Step 0. {*Initialization.*}
  Choose $\mathbf{x}^0 \in \Omega_B$, *set* $\mathcal{I}^0 = \mathcal{A}(\mathbf{x}^0)$, $k = 0$
  while $\|\mathbf{g}^P(\mathbf{x}^k)\| > 0$
Step 1. {*Minimization in face* $\mathcal{W}_{\mathcal{I}^k}$. }
   if $\boldsymbol{\varphi}(\mathbf{x}^k) \neq \mathbf{o}$
    $\widehat{\mathbf{y}} = \arg\min_{\mathbf{y}\in\mathcal{W}_{\mathcal{I}^k}} f(\mathbf{y})$
    if $\widehat{\mathbf{y}} \in \Omega_B$
Step 2. {*Feasible step.*}
     $\mathbf{x}^{k+1} = \widehat{\mathbf{y}}$
    else
Step 3. {*Expansion step.*}
     Set $\mathbf{x}^{k+1}$ *so that* $f(\mathbf{x}^{k+1}) \leq f(\mathbf{x}^k)$ *and* $\mathcal{A}(\mathbf{x}^{k+1}) \supsetneq \mathcal{I}^k$
    end *if*
Step 4. {*Release step.*}
   else
    if $\boldsymbol{\beta}(\mathbf{x}^k) \neq \mathbf{o}$
     $\mathbf{x}^{k+1} = \mathbf{x}^k - \alpha_{opt}\boldsymbol{\beta}(\mathbf{x}^k),\ \mathcal{I}^{k+1} = \mathcal{A}(\mathbf{x}^{k+1}),$
   $k = k + 1$
  end *while*
  $\widehat{\mathbf{x}} = \mathbf{x}^k$

To implement the algorithm, we should specify the expansion step in more detail. For example, if $\mathbf{x}^k \in \Omega_B$ and

$$\mathbf{d} = \begin{bmatrix} \mathbf{o}_{\mathcal{I}^k} \\ \mathbf{d}_{\mathcal{J}^k} \end{bmatrix}, \quad \mathbf{d}_{\mathcal{J}^k} = \mathbf{x}^k_{\mathcal{J}^k} - \widehat{\mathbf{y}}_{\mathcal{J}^k},$$

we can observe that $-\mathbf{d}$ is a feasible decrease direction and that $f(\mathbf{x}^k - \alpha\mathbf{d})$ is a decreasing function of α for $\alpha \in [0,1]$. Thus we can look for $\mathbf{x}^{k+1}$ in the form $\mathbf{x}^{k+1} = \mathbf{x}^k - \alpha\mathbf{d}$, $\alpha \in (0,1]$. A possible choice of α is given by

$$\alpha_f = \arg\min_{\alpha\in(0,1]} \{f(\mathbf{x}^k - \alpha\mathbf{d}) : \mathbf{x}^k - \alpha\mathbf{d} \in \Omega_B\}, \tag{7.11}$$

which can be evaluated by using

$$\alpha_f = \min\{\alpha_m, 1\}, \quad \alpha_m = \min\{(x_i^k - \ell_i)/d_i : d_i > 0,\ i \in \mathcal{N}\}. \tag{7.12}$$

Notice that if $\widehat{\mathbf{y}} \notin \Omega_B$, then the step length α_f necessarily results in the expansion of the working set but typically only by one index (Fig. 7.2).

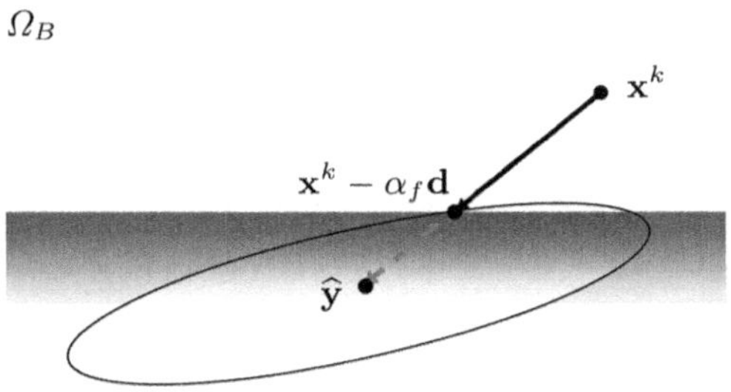

Fig. 7.2. Feasible steplength

We can improve the algorithm's performance by following $-\mathbf{d}$ or the projected-gradient path as long as the cost function decreases. Notice that the function $f(P_{\Omega_B}(\mathbf{y} - \alpha\mathbf{d}))$ is a piece-wise quadratic function in α.

Notice that $-\mathbf{d}$ is the Newton direction that allows the largest decrease of f in the current working set, but Fig. 7.3 shows that α_f may be the best steplength in the direction of $-\mathbf{d}$! We refer interested readers to Moré and Toraldo [217], Nocedal and Wright [220, Sect. 16.4], or to the discussion of the projected-gradient path in Conn et al. [41, Sect. 12.1.3]. Alternatively, we can apply the fixed-steplength reduced gradient projection which is described in Chap. 5.

The algorithm assumes by default that Step 1 is carried out by a direct method such as a matrix factorization, in which economies are possible by updating rather than recomputing the factorizations to account for gradual changes in the working set.

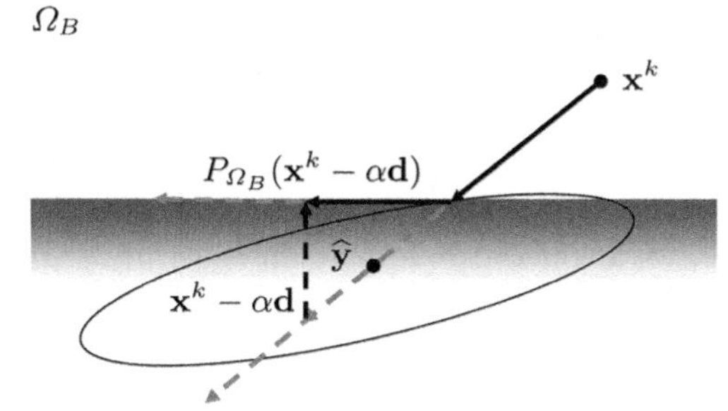

Fig. 7.3. Projected best unconstrained decrease path

7.2.3 Finite Termination

The analysis of the working set method can be based on the following finite termination property.

Theorem 7.2. *Let Algorithm 7.1 be applied to find the solution* $\hat{\mathbf{x}}$ *of (7.1) starting from* $\mathbf{x}^0 \in \Omega_B$. *Then there is* k *such that* $\mathbf{x}^k = \hat{\mathbf{x}}$.

Proof. Since each expansion step adds at least one index to the current working set, and the number of indices in the working set cannot exceed n, it follows that there are at most n consecutive expansion steps. Thus after each consecutive series of expansion steps, the algorithm either finds the solution of (7.1) and we are finished, or generates a minimizer on the current face. However, since $f(\mathbf{x}^k)$ is a nonincreasing sequence such that

$$f(\mathbf{x}^{k+1}) < f(\mathbf{x}^k)$$

whenever $\mathbf{x}^{k+1}$ is generated by a feasible or release step, it follows that no face with the feasible minimizer can reappear. The number of different working sets being finite, we conclude that the working set method finds the solution of (7.1) in a finite number of steps. □

Since the number of different working sets is 2^n and there can be at most n expanding steps for each feasible step, the proof of Theorem 7.2 gives that the number N of iterations of the working set method is bounded by

$$\overline{N} = n2^n. \tag{7.13}$$

This bound is very pessimistic and gives poor theoretical support for practical computations, especially if we consider the high cost of the iterations.

The bound can be essentially improved for special problems. For example, if $\mathbf{x}^0 = \boldsymbol{\ell} = \mathbf{o}$ and the Hessian A of f is an M-matrix, then it is possible to show that Algorithm 5.1 generates only feasible steps and finds the solution in at most $n - p$ iterations, where p is the number of positive entries in $\mathbf{b}$. For more details see Diamond [50].

7.3 Polyak's Algorithm

If we solve the auxiliary problems (7.6) by the conjugate gradient method, it is reasonable not to wait with the feasibility test until their solution is found but to modify the working set whenever unfeasible CG iteration is generated. This observation was enhanced in the *Polyak algorithm* [229], the starting point of our development of, in a sense, optimal algorithms.

7.3.1 Basic Algorithm

The new ingredient of the Polyak algorithm is replacing the minimization in the face by a sequence of *conjugate gradient steps* defined by

$$\mathbf{x}^k = \mathbf{x}^{k-1} - \alpha_{cg}\mathbf{p}^k, \tag{7.14}$$

where $\mathbf{p}^k$ denotes the recurrently constructed conjugate direction introduced in Sect. 4.2, and α_{cg} is the minimizer of $f(\mathbf{x}^{k-1} - \xi\mathbf{p}^k)$ on the feasible set.

The algorithm starts from an arbitrary feasible $\mathbf{x}^0$ by assigning $\mathcal{I}^0 = \mathcal{B}_0(\mathbf{x}^0)$ and initializing the conjugate gradient loop by $\mathbf{p}^1 = \varphi(\boldsymbol{x}^0)$ (for details see Algorithm 7.2 or Sect. 4.2) for the minimization in $\mathcal{W}_{\mathcal{I}^0}$.

Assuming that $\mathbf{x}^k$ is known, we first check if $\mathbf{x}^k$ solves (7.1) by testing $\mathbf{g}^P(\mathbf{x}^k) = \mathbf{o}$ or the auxiliary problem (7.6) by testing $\varphi(\mathbf{x}^k) = \mathbf{o}$. If $\mathbf{g}^P(\mathbf{x}^k) = \mathbf{o}$, we are finished; if $\varphi(\mathbf{x}^k) = \mathbf{o}$, we reduce the working set to $\mathcal{I}^k = \mathcal{B}_0(\mathbf{x}^k)$ and initialize the new conjugate gradient loop. If $\varphi(\mathbf{x}^k) \neq \mathbf{o}$, we either start the new conjugate gradient loop or use the conjugate direction $\mathbf{p}^{k+1}$ to generate a new iterate.

The conjugate gradient loop starts (or restarts) from $\mathbf{p}^{k+1} = \varphi(\mathbf{x}^k)$ whenever $\mathbf{x}^k$ is generated by the expansion step, release step, or $k = 0$. If $\mathbf{p}^k$ is known, then $\mathbf{p}^{k+1}$ is given by the formulae

$$\mathbf{p}^{k+1} = \varphi(\mathbf{x}^k) - \beta\mathbf{p}^k \quad \text{and} \quad \beta = \varphi(\mathbf{x}^k)^T \mathsf{A}\mathbf{p}^k/(\mathbf{p}^k)^T\mathsf{A}\mathbf{p}^k, \tag{7.15}$$

obtained by specialization of those introduced in Sect. 4.2. Let us recall that the conjugate directions

$$\mathbf{p}^{k+1}, \mathbf{p}^{k+2}, \ldots$$

that are generated by the recurrence (7.15) from the restart $\mathbf{x}^k$ are A-orthogonal.

The basic Polyak algorithm for solving the strictly convex bound constrained quadratic programming problems takes the form shown by the following algorithm, where we omitted the indices of the vectors that we do not refer to in what follows.

Our description of Algorithm 7.2 does not explicitly use the working sets; the information about the current working set is enhanced in the iterates

Algorithm 7.2. Polyak's algorithm

Given a symmetric positive definite matrix $\mathsf{A} \in \mathbb{R}^{n\times n}$ *and n-vectors* $\mathbf{b}, \boldsymbol{\ell}$.

Step 0. {*Initialization.*}
  Choose $\mathbf{x}^0 \in \Omega_B$, *set* $\mathbf{g} = \mathsf{A}\mathbf{x}^0 - \mathbf{b}$, $\mathbf{p} = \mathbf{g}^P(\mathbf{x}^0)$, $k = 0$
  while $\|\mathbf{g}^P(\mathbf{x}^k)\| > 0$
  if $\|\boldsymbol{\varphi}(\mathbf{x}^k)\| > 0$
Step 1. {*Trial conjugate gradient step.*}
  $\alpha_{cg} = \mathbf{g}^T\mathbf{p}/\mathbf{p}^T\mathsf{A}\mathbf{p}, \ \mathbf{y} = \mathbf{x}^k - \alpha_{cg}\mathbf{p}$
  $\alpha_f = \max\{\alpha : \ \mathbf{x}^k - \alpha\mathbf{p} \in \Omega_B\} = \min\{(x_i^k - \ell_i)/p_i : p_i > 0\}$
  if $\alpha_{cg} \le \alpha_f$
Step 2. {*Conjugate gradient step.*}
  $\mathbf{x}^{k+1} = \mathbf{y}, \ \mathbf{g} = \mathbf{g} - \alpha_{cg}\mathsf{A}\mathbf{p},$
  $\beta = \boldsymbol{\varphi}(\mathbf{y})^T\mathsf{A}\mathbf{p}/\mathbf{p}^T\mathsf{A}\mathbf{p}, \mathbf{p} = \boldsymbol{\varphi}(\mathbf{y}) - \beta\mathbf{p}$
  else
Step 3. {*Expansion step.*}
  $\mathbf{x}^{k+1} = \mathbf{x}^k - \alpha_f\mathbf{p}, \ \mathbf{g} = \mathbf{g} - \alpha_f\mathsf{A}\mathbf{p}, \ \mathbf{p} = \boldsymbol{\varphi}(\mathbf{x}^{k+1})$
  end *if*
  else
Step 4. {*Leaving the face after finding the minimizer – release step.*}
  $\mathbf{d} = \boldsymbol{\beta}(\mathbf{x}^k), \ \alpha_{cg} = \mathbf{g}^T\mathbf{d}/\mathbf{d}^T\mathsf{A}\mathbf{d},$
  $\mathbf{x}^{k+1} = \mathbf{x}^k - \alpha_{cg}\mathbf{d}, \ \mathbf{g} = \mathbf{g} - \alpha_{cg}\ \mathsf{A}\mathbf{d}, \ \mathbf{p} = \boldsymbol{\varphi}(\mathbf{x}^{k+1})$
  end *if*
  $k = k + 1$
  end *while*
Step 5. {*Return solution.*}
  $\widehat{\mathbf{x}} = \mathbf{x}^k$

$\mathbf{x}^k$ and the conjugate directions $\mathbf{p}^k$. Let us recall that the properties of the unconstrained conjugate gradient method are summarized in Theorem 4.1.

7.3.2 Finite Termination

Theorem 7.3. *Let Polyak's Algorithm 7.2 be applied to find the solution* $\widehat{\mathbf{x}}$ *of (7.1) starting from* $\mathbf{x}^0 \in \Omega_B$. *Then there is k such that* $\mathbf{x}^k = \widehat{\mathbf{x}}$.

Proof. First notice that by Theorem 4.1, there can be at most n consecutive conjugate gradient iterations before the minimizer in a face or unfeasible iterate is found. Moreover, each unfeasible iterate expands the active set so that there cannot be more than n^2 consecutive iterates before generating the minimizer of a working set. Since the cost function decreases, the working sets with the minimizers can be generated at most once. Since a set with n elements has a finite number of different subsets, the Polyak algorithm implemented in exact arithmetic must find the solution. The arguments show that the Polyak algorithm is a special implementation of the active set method. The statement follows by Theorem 7.2. □

The arguments of Sect. 7.2.3 can be used to show that the number of iterations of Polyak's algorithm is bounded by

$$\overline{N} = n^2 2^n. \tag{7.16}$$

Let us emphasize here that this bound is very pessimistic and can be improved, at least for particular problems. For suitable tools for such analysis, see, e.g., Bomze et al. [23] and the references therein.

7.3.3 Characteristics of Polyak's Algorithm

The Polyak algorithm suffers from several drawbacks. The first is related to an unpleasant consequence of applying the CG step with the feasible steplength. Since the working set is typically expanded by one index only, there is little chance that the number of iterations will be small when many indices of the binding set of the solution do not belong to $\mathcal{B}(\mathbf{x}^0)$.

Another drawback concerns the basic approach combining the CG method, which is now understood as an efficient iterative method for approximate solution of linear systems [6, 150, 240], and the finite termination strategy, which is based on combinatorial reasoning that requires exact solution of the auxiliary problems.

Finally, as we have seen above, the combinatorial arguments give extremely poor bounds on the number of iterations that are necessary to find the solution of (7.1). The bound (7.16) does not take into account any information on the spectrum of A and does not indicate that the algorithm could effectively solve problems with some hundreds of variables.

7.4 Inexact Polyak's Algorithm

In this section, we consider the variants of Polyak's algorithm, which accept inexact solutions of auxiliary problems but preserve the finite termination property.

7.4.1 Looking Ahead and Estimate

Let us first show that it is possible to solve the auxiliary problems (7.6) exactly to preserve the finite termination property of the Polyak algorithm. The key observation is that if $\mathbf{x}^{k+1} \in \Omega_B$ satisfies

$$f(\mathbf{x}^{k+1}) < \min\{f(\mathbf{x}) : \ \mathbf{x} \in \mathcal{W}_{\mathcal{I}}\}, \tag{7.17}$$

then the working set $\mathcal{I}$ cannot appear again as long as $\{f(\mathbf{x}^k)\}$ is nonincreasing. We shall use this simple observation to define the precision control test and reduction of the active set.

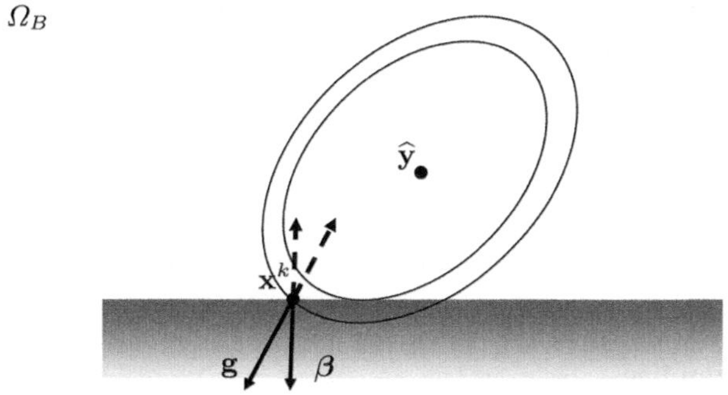

Fig. 7.4. Release directions at $\mathbf{x}^k$

Given $\mathbf{x}^k \in \mathcal{W}_\mathcal{I}$, we can try to find $\mathbf{x}^{k+1}$ which satisfies (7.17) in the form $\mathbf{x}^{k+1} = \mathbf{x}^k - \alpha\mathbf{d}$ with a given $\mathbf{d}$; if we are successful, we call $\mathbf{d}$ the *release direction of* $\mathcal{W}_\mathcal{I}$ *at* $\mathbf{x}^k$. Using the release direction, we can reduce the active set in a way that prevents its appearance in any following iteration.

Let us examine the conditions for $\mathbf{d}$, typically obtained by reducing the components of $\nabla f(\mathbf{x}^k)$, to be a release direction. Such situation is depicted in Fig. 7.4 with $\mathbf{d} = \mathbf{g}(\mathbf{x}^k)$ and $\mathbf{d} = \boldsymbol{\beta}(\mathbf{x}^k)$. We can see that it is not necessary to explore the face until we find the minimum, i.e.,

$$\boldsymbol{\beta}(\mathbf{x}^k) = \mathbf{g}(\mathbf{x}^k).$$

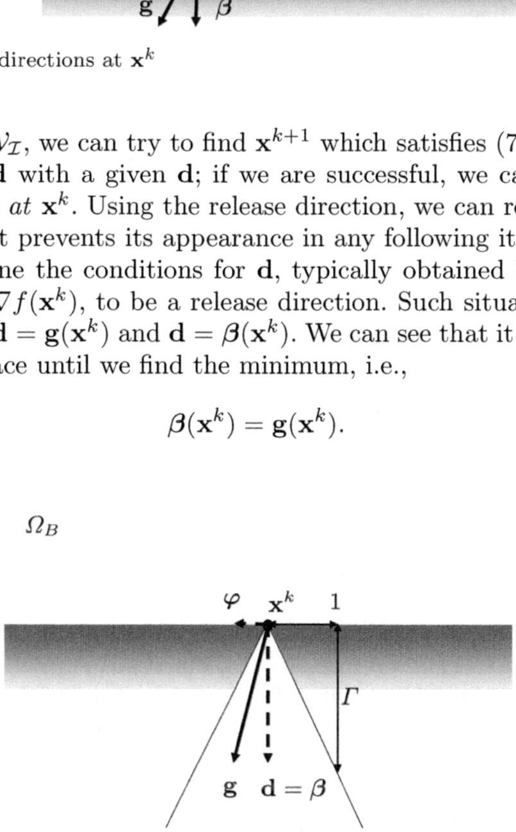

Fig. 7.5. The gradient and $\mathbf{d} = \boldsymbol{\beta}(\mathbf{x})$ that satisfy the release condition (7.18)

The following lemma gives a criterion that can be used to control the precision of the minimization in faces. The cone of feasible decrease directions specified by $\Gamma > 0$ is illustrated in Fig. 7.5.

Lemma 7.4. *Let* $\mathcal{I} = \mathcal{A}(\mathbf{x})$ *and*

$$\Gamma \geq \kappa(\mathsf{A})^{1/2},$$

where $\kappa(\mathsf{A})$ *denotes the spectral condition number of* A. *Denote* $\mathbf{g} = \nabla f(\mathbf{x})$ *and suppose that* $\mathbf{d}$ *satisfies*

$$\mathbf{g}^T\mathbf{d} \geq \|\mathbf{d}\|^2 \quad \text{and} \quad \|\mathbf{d}\| > \Gamma\|\varphi(\mathbf{x})\|. \tag{7.18}$$

Then the vector $\mathbf{y} = \mathbf{x} - \|\mathsf{A}\|^{-1}\mathbf{d}$ *satisfies*

$$f(\mathbf{y}) < \min\{f(\mathbf{x}) : \ \mathbf{x} \in \mathcal{W}_{\mathcal{I}}\}. \tag{7.19}$$

Proof. Let $\mathbf{x}, \Gamma$, and $\mathbf{d}$ satisfy the assumptions of Lemma 7.4 and notice that $\mathbf{g}^T\mathbf{d} \geq \|\mathbf{d}\|^2$ implies

$$f(\mathbf{y}) - f(\mathbf{x}) = \frac{1}{2}\|\mathsf{A}\|^{-2}\mathbf{d}^T\mathsf{A}\mathbf{d} - \|\mathsf{A}\|^{-1}\mathbf{d}^T\mathbf{g} \leq -\frac{1}{2}\|\mathsf{A}\|^{-1}\|\mathbf{d}\|^2. \tag{7.20}$$

Denoting $\mathcal{J} = \mathcal{F}(\mathbf{x})$, we have that $\|\mathbf{g}_{\mathcal{J}}\| = \|\varphi(\mathbf{x})\|$ and by the assumptions

$$\|\mathbf{d}\|^2 > \kappa(\mathsf{A})\|\mathbf{g}_{\mathcal{J}}\|^2. \tag{7.21}$$

Substituting (7.21) into (7.20) then yields

$$f(\mathbf{y}) - f(\mathbf{x}) < -\frac{1}{2}\|\mathsf{A}^{-1}\|\|\mathbf{g}_{\mathcal{J}}\|^2. \tag{7.22}$$

Now denote by $\overline{\mathbf{x}}$ and $\overline{\mathbf{g}}$ the minimizer of $f(\mathbf{x})$ on $\mathcal{W}_{\mathcal{I}}$ and the corresponding gradient vector, respectively. Direct computations yield

$$f(\mathbf{x}) - f(\overline{\mathbf{x}}) = \frac{1}{2}(\mathbf{x} - \overline{\mathbf{x}})^T\mathsf{A}(\mathbf{x} - \overline{\mathbf{x}}) + \overline{\mathbf{g}}^T(\mathbf{x} - \overline{\mathbf{x}}). \tag{7.23}$$

If we now rearrange the indices and take into account that

$$\overline{\mathbf{g}}_{\mathcal{J}} = 0 \quad \text{and} \quad \mathbf{x}_{\mathcal{I}} = \overline{\mathbf{x}}_{\mathcal{I}},$$

we can further simplify the right-hand side of (7.23) to get

$$f(\mathbf{x}) - f(\overline{\mathbf{x}}) = \frac{1}{2}(\mathbf{x}_{\mathcal{J}} - \overline{\mathbf{x}}_{\mathcal{J}})^T\mathsf{A}_{\mathcal{J}\mathcal{J}}(\mathbf{x}_{\mathcal{J}} - \overline{\mathbf{x}}_{\mathcal{J}}). \tag{7.24}$$

To express $\mathbf{x}_{\mathcal{J}} - \overline{\mathbf{x}}_{\mathcal{J}}$ in terms of $\mathbf{g}_{\mathcal{J}}$, we can use the rearrangement (7.7) to get

$$\begin{bmatrix} \mathbf{g}_{\mathcal{I}} - \overline{\mathbf{g}}_{\mathcal{I}} \\ \mathbf{g}_{\mathcal{J}} \end{bmatrix} = \begin{bmatrix} \mathsf{A}_{\mathcal{I}\mathcal{I}} & \mathsf{A}_{\mathcal{I}\mathcal{J}} \\ \mathsf{A}_{\mathcal{J}\mathcal{I}} & \mathsf{A}_{\mathcal{J}\mathcal{J}} \end{bmatrix} \begin{bmatrix} \mathbf{o} \\ \mathbf{x}_{\mathcal{J}} - \overline{\mathbf{x}}_{\mathcal{J}} \end{bmatrix}. \tag{7.25}$$

In particular, since $\mathsf{A}_{\mathcal{J}\mathcal{J}}$ is also positive definite, it follows that

$$\mathbf{x}_{\mathcal{J}} - \overline{\mathbf{x}}_{\mathcal{J}} = \mathsf{A}_{\mathcal{J}\mathcal{J}}^{-1}\mathbf{g}_{\mathcal{J}}$$

and by (7.24)

$$f(\mathbf{x}) - f(\overline{\mathbf{x}}) = \frac{1}{2}\mathbf{g}_{\mathcal{J}}^T\mathsf{A}_{\mathcal{J}\mathcal{J}}^{-1}\mathbf{g}_{\mathcal{J}}. \tag{7.26}$$

Taking into account the interlacing properties of the spectra of principal submatrices of symmetric matrices (1.26), we get

$$\frac{1}{2}\mathbf{g}_{\mathcal{J}}^T\mathsf{A}_{\mathcal{J}\mathcal{J}}^{-1}\mathbf{g}_{\mathcal{J}} \le \frac{1}{2}\|\mathsf{A}_{\mathcal{J}\mathcal{J}}^{-1}\|\|\mathbf{g}_{\mathcal{J}}\|^2 \le \frac{1}{2}\|\mathsf{A}^{-1}\|\|\mathbf{g}_{\mathcal{J}}\|^2, \tag{7.27}$$

so that by (7.22) and (7.27)

$$\begin{aligned} f(\mathbf{y}) - f(\overline{\mathbf{x}}) &= (f(\mathbf{y}) - f(\mathbf{x})) + (f(\mathbf{x}) - f(\overline{\mathbf{x}})) \\ &< -\frac{1}{2}\|\mathsf{A}^{-1}\|\|\mathbf{g}_{\mathcal{J}}\|^2 + \frac{1}{2}\|\mathsf{A}^{-1}\|\|\mathbf{g}_{\mathcal{J}}\|^2 = 0. \end{aligned} \tag{7.28}$$

□

7.4.2 Looking Ahead Polyak's Algorithm

Using Lemma 7.4, we can modify Polyak's algorithm to accept the approximate solution of the auxiliary problems while preserving its finite termination property. We only need to change the precision control of auxiliary problems and specify the release direction. The *looking ahead Polyak algorithm* reads as follows.

To see that Algorithm 7.3 deserves its name, denote $\mathbf{d} = \boldsymbol{\beta}(\mathbf{x}^k)$ and assume that

$$\mathbf{x}^k \in \Omega_B, \quad \|\boldsymbol{\beta}(\mathbf{x}^k)\| > \Gamma\|\boldsymbol{\varphi}(\mathbf{x}^k)\|, \quad \text{and} \quad \Gamma \ge \kappa(\mathsf{A})^{1/2}, \tag{7.29}$$

so that $\mathbf{d}$ and Γ satisfy the assumptions of Lemma 7.4. Observing that α_{cg} minimizes $f(\mathbf{x}^k - \alpha\mathbf{d})$ with respect to α, we get for $\mathbf{x}^{k+1} = \mathbf{x}^k - \alpha_{cg}\mathbf{d}$ that

$$f(\mathbf{x}^{k+1}) \le f(\mathbf{x}^k - \|\mathsf{A}\|^{-1}\mathbf{d}) < \min\{f(\mathbf{x}) : \mathbf{x} \in \mathcal{W}_{\mathcal{A}(\mathbf{x}^k)}\}.$$

Moreover, since $\mathbf{x}^k - \alpha\mathbf{d} \in \Omega_B$ for any $\alpha \ge 0$, we have $\mathbf{x}^{k+1} \in \Omega_B$. Thus, the algorithm can recognize the face without the global solution before having a solution for the auxiliary problem, i.e., it "looks ahead".

The same reasoning as above can be carried out with $\mathbf{d} = \mathbf{g}^-(\mathbf{x}^k)$ or with some other nonzero vector $\mathbf{d}$ which satisfies the assumptions of Lemma 7.4. However, we found no significant evidence that there is a better choice than $\mathbf{d} = \boldsymbol{\beta}(\mathbf{x}^k)$.

Algorithm 7.3. Looking ahead Polyak's algorithm

Given a symmetric positive definite matrix $\mathsf{A} \in \mathbb{R}^{n \times n}$, *n-vectors* $\mathbf{b}, \boldsymbol{\ell}$.

Step 0. {*Initialization.*}
Choose $\mathbf{x}^0 \in \Omega_B$, $\Gamma \geq \kappa(\mathsf{A})^{1/2}$, *set* $\mathbf{g} = \mathsf{A}\mathbf{x}^0 - \mathbf{b}$, $\mathbf{p} = \mathbf{g}^P(\mathbf{x}^0)$, $k = 0$
while $\|\mathbf{g}^P(\mathbf{x}^k)\| > 0$
if $\Gamma\|\boldsymbol{\varphi}(\mathbf{x}^k)\| \geq \|\boldsymbol{\beta}(\mathbf{x}^k)\|$
Step 1. {*Trial conjugate gradient step.*}
$\alpha_{cg} = \mathbf{g}^T\mathbf{p}/\mathbf{p}^T\mathsf{A}\mathbf{p}$, $\mathbf{y} = \mathbf{x}^k - \alpha_{cg}\mathbf{p}$
$\alpha_f = \max\{\alpha : \mathbf{x}^k - \alpha\mathbf{p} \in \Omega_B\} = \min\{(x_i^k - \ell_i)/p_i : p_i > 0\}$
if $\alpha_{cg} \leq \alpha_f$
Step 2. {*Conjugate gradient step.*}
$\mathbf{x}^{k+1} = \mathbf{y}, \mathbf{g} = \mathbf{g} - \alpha_{cg}\mathsf{A}\mathbf{p}$,
$\beta = \boldsymbol{\varphi}(\mathbf{y})^T\mathsf{A}\mathbf{p}/\mathbf{p}^T\mathsf{A}\mathbf{p}$, $\mathbf{p} = \boldsymbol{\varphi}(\mathbf{y}) - \beta\mathbf{p}$
else
Step 3. {*Expansion step.*}
$\mathbf{x}^{k+1} = \mathbf{x}^k - \alpha_f\mathbf{p}, \mathbf{g} = \mathbf{g} - \alpha_f\mathsf{A}\mathbf{p}, \mathbf{p} = \boldsymbol{\varphi}(\mathbf{x}^{k+1})$
end *if*
else
Step 4. {*Leaving the face in the release direction.*}
$\mathbf{d} = \boldsymbol{\beta}(\mathbf{x}^k)$, $\alpha_{cg} = \mathbf{g}^T\mathbf{d}/\mathbf{d}^T\mathsf{A}\mathbf{d}$,
$\mathbf{x}^{k+1} = \mathbf{x}^k - \alpha_{cg}\mathbf{d}$, $\mathbf{g} = \mathbf{g} - \alpha_{cg}\mathsf{A}\mathbf{d}$, $\mathbf{p} = \boldsymbol{\varphi}(\mathbf{x}^{k+1})$
end *else*
$k = k + 1$
end *while*
Step 5. {*Return solution.*}
$\widehat{\mathbf{x}} = \mathbf{x}^k$

7.4.3 Properties of Modified Polyak's Algorithms

Theorem 7.5. *Let the looking ahead Polyak Algorithm 7.3 be applied to find the solution* $\widehat{\mathbf{x}}$ *of (7.1) starting from* $\mathbf{x}^0 \in \Omega_B$. *Then there is k such that* $\mathbf{x}^k = \widehat{\mathbf{x}}$.

Proof. Since by Theorem 4.1 there can be at most n consecutive conjugate gradient iterations before the unconstrained minimizer is found, it follows that there can be at most n consecutive proportional conjugate gradient iterations. Moreover, since each proportioning step releases at least one index from the working set, which has at most n elements, we have that there can be at most n^2 iterations without an expansion step.

Now observe that $\{f(\mathbf{x}^k)\}$ is nonincreasing. Since the chain of CG and expansion iterations can be interrupted only after finding the iteration $\mathbf{x}^k$ which either solves (7.1), i.e., $\boldsymbol{\beta}(\mathbf{x}^k) = \boldsymbol{\varphi}(\mathbf{x}^k) = \mathbf{o}$, or is not proportional, i.e., satisfies $\|\boldsymbol{\beta}(\mathbf{x}^k)\| > \Gamma\|\boldsymbol{\varphi}(\mathbf{x}^k)\|$ with $\Gamma \geq \kappa(\mathsf{A})^{1/2}$, it follows by Lemma 7.4 that the associated active set $\mathcal{A}(\mathbf{x}^k)$ cannot be generated again in the following iterations. Since the number of all subsets of $\mathcal{N} = \{1, \ldots, n\}$ is bounded, and by Lemma 7.4 every iteration in the face with the solution is proportional when $\Gamma \geq \kappa(\mathsf{A})^{1/2}$, we conclude that the algorithm must generate $\mathbf{x}^k = \widehat{\mathbf{x}}$ in a finite number of steps. □

Our experience [62] indicates that our modifications of the Polyak algorithm outperforms the original, but a little analysis shows that they suffer from many drawbacks described in Sect. 7.3.3.

7.5 Comments and References

Since the conjugate gradient method was introduced in the celebrated paper by Hestenes and Stiefel [164] as a method for the solution of systems of linear equations, it seems that Polyak [229] was the first researcher who proposed to use the conjugate gradient method to minimize the quadratic cost function subject to bound constraints. Polyak observed that after the conjugate gradient method starts the minimization of f on a given working set, it finds the minimizer or generates a decrease direction that can be used to expand the active set. Polyak specified what to do in both cases in order to find the solution in a finite number of steps.

Though Polyak assumed that the auxiliary problems are solved exactly, O'Leary [222] observed that this assumption can be replaced by refining the accuracy in the process of solution. The presentation of Looking ahead Polyak's algorithm with an inexact solution of auxiliary problems and the finite termination property is based on Dostál [62], and [64].

Many authors fought with the unpleasant consequence of the Polyak strategy, which yields a lower bound on the number of iterations in terms of the difference between the number of active constraints in the initial approximation and the solution. Dembo and Tulowitzski [48] proposed the conjugate gradient projection algorithm which could add and drop many constraints in an iteration. Later, Yang and Tolle [267] further developed this algorithm with backtracking so that they were able to prove its finite termination property. The efficient iterative algorithm MPRGP based on Polyak is in Chap. 10.

Part III
Optimal Algorithms

8
Conjugate Gradients as Iterative Method

Let us consider the application of *conjugate gradient method* presented in Chap. 4 to solving large unconstrained optimization problems

$$\min_{\mathbf{x}\in\mathrm{Im}\mathsf{A}} f(\mathbf{x}), \quad f(\mathbf{x}) = \frac{1}{2}\mathbf{x}^T\mathsf{A}\mathbf{x} - \mathbf{x}^T\mathbf{b}, \tag{8.1}$$

where $\mathbf{b}$ is a given n-vector, and A is an $n \times n$ an SPD or SPS matrix. We are especially interested in the problems that arise as auxiliary problems in solving constrained QP problems. The dimension n of these problems can reach billions, making it possible to carry out only a relatively small number of iterations. In this case, we cannot rely on the finite termination property of the CG algorithm, and we need a theory that provides information about the quality of the iterations.

The development of practical algorithms for large problems uses the results on the convergence rate in terms of the bounds on the spectrum of A. Here, we first assume that A is SPD. We describe the classical convergence theory based on the limits on the values of residual polynomials on the spectrum of A. In particular, we show that if the spectrum of A is in a positive interval $[a_{\min}, a_{\max}]$, then there is a bound on a number of conjugate gradient iterations in terms of $a_{\max}/a_{\min}$ that are necessary to solve the problem (8.1) to a given relative precision.

Then, we describe two methods to improve the conditioning of A, particularly the standard preconditioning method and preconditioning by the conjugate projector, which reduces the iterations to a subspace. The letter one does not transform variables, so it preserves the structure of bound and other separable inequality constraints. The latter is essential for solving optimization problems with inequality constraints.

Z. Dostál, *Optimal Quadratic Programming and QCQP Algorithms with Case Studies*, Springer Optimization and Its Applications 23,
https://doi.org/10.1007/978-3-031-95167-1_8

Finally, we modify the CG algorithm for solving (8.1) with an SPS matrix A and $\mathbf{b} \notin \mathrm{Im}\mathsf{A}$ and adapt our arguments to prove nearly the same convergence results as above for the least square problems.

8.1 Rate of Convergence and Optimality

Although the conjugate gradient method finds by Theorem 4.1 the exact solution $\widehat{\mathbf{x}}$ of (8.1) in a number of steps which does not exceed the dimension of the problem, it turns out that it can often produce a sufficiently accurate approximation $\widetilde{\mathbf{x}}$ of $\widehat{\mathbf{x}}$ in a much smaller number of steps than required for exact termination. Thus we can also consider the conjugate gradients as an iterative method. This section presents the results that substantiate this claim and help us identify the favorable cases.

8.1.1 Min-max Estimate

Let us denote the *solution error* as

$$\mathbf{e} = \mathbf{e}(\mathbf{x}) = \mathbf{x} - \widehat{\mathbf{x}}$$

and observe that

$$\mathbf{g}(\widehat{\mathbf{x}}) = \mathsf{A}\widehat{\mathbf{x}} - \mathbf{b} = \mathbf{o}.$$

It follows that

$$\mathbf{g}^k = \mathsf{A}\mathbf{x}^k - \mathbf{b} = \mathsf{A}\mathbf{x}^k - \mathsf{A}\widehat{\mathbf{x}} = \mathsf{A}(\mathbf{x}^k - \widehat{\mathbf{x}}) = \mathsf{A}\mathbf{e}^k,$$

so in particular

$$\mathcal{K}^k(\mathsf{A}, \mathbf{g}^0) = \mathrm{Span}\{\mathbf{g}^0, \mathsf{A}\mathbf{g}^0, \dots, \mathsf{A}^{k-1}\mathbf{g}^0\} = \mathrm{Span}\{\mathsf{A}\mathbf{e}^0, \dots, \mathsf{A}^k\mathbf{e}^0\}.$$

We start our analysis of the solution error by using the Taylor expansion (2.4) to obtain the identity

$$\begin{aligned} f(\mathbf{x}) - f(\widehat{\mathbf{x}}) &= f(\widehat{\mathbf{x}} + (\mathbf{x} - \widehat{\mathbf{x}})) - f(\widehat{\mathbf{x}}) \\ &= f(\widehat{\mathbf{x}}) + \mathbf{g}(\widehat{\mathbf{x}})^T(\mathbf{x} - \widehat{\mathbf{x}}) + \frac{1}{2}\|\mathbf{x} - \widehat{\mathbf{x}}\|_{\mathsf{A}}^2 - f(\widehat{\mathbf{x}}) \\ &= \frac{1}{2}\|\mathbf{x} - \widehat{\mathbf{x}}\|_{\mathsf{A}}^2 = \frac{1}{2}\|\mathbf{e}\|_{\mathsf{A}}^2. \end{aligned}$$

Combining the latter identity with Theorem 4.1, we get

$$\|\mathbf{e}^k\|_{\mathsf{A}}^2 = 2(f(\mathbf{x}^k) - f(\widehat{\mathbf{x}})) = \min_{\mathbf{x}\in\mathbf{x}^0+\mathcal{K}^k(\mathsf{A},\mathbf{g}^0)} 2(f(\mathbf{x}) - f(\widehat{\mathbf{x}}))$$
$$= \min_{\mathbf{x}\in\mathbf{x}^0+\mathcal{K}^k(\mathsf{A},\mathbf{g}^0)} \|\mathbf{x} - \widehat{\mathbf{x}}\|_{\mathsf{A}}^2 = \min_{\mathbf{x}\in\mathbf{x}^0+\mathcal{K}^k(\mathsf{A},\mathbf{g}^0)} \|\mathbf{e}(\mathbf{x})\|_{\mathsf{A}}^2.$$

Since any $\mathbf{x} \in \mathbf{x}^0 + \mathcal{K}^k(\mathsf{A}, \mathbf{g}^0)$ may be written in the form

$$\mathbf{x} = \mathbf{x}^0 + \xi_1\mathbf{g}^0 + \xi_2\mathsf{A}\mathbf{g}^0 + \cdots + \xi_k\mathsf{A}^{k-1}\mathbf{g}^0 = \mathbf{x}^0 + \xi_1\mathsf{A}\mathbf{e}^0 + \cdots + \xi_k\mathsf{A}^k\mathbf{e}^0,$$

it follows that

$$\mathbf{x} - \widehat{\mathbf{x}} = \mathbf{e}^0 + \xi_1\mathsf{A}\mathbf{e}^0 + \cdots + \xi_k\mathsf{A}^k\mathbf{e}^0 = p(\mathsf{A})\mathbf{e}^0,$$

where p denotes the polynomial defined for any $x \in \mathbb{R}$ by

$$p(x) = 1 + \xi_1 x + \xi_2 x^2 + \cdots + \xi_k x^k.$$

Thus denoting by $\mathcal{P}^k$ the set of all kth degree polynomials p which satisfy $p(0) = 1$, we have

$$\|\mathbf{e}^k\|_{\mathsf{A}}^2 = \min_{\mathbf{x}\in\mathbf{x}^0+\mathcal{K}^k(\mathsf{A},\mathbf{g}^0)} \|\mathbf{e}(\mathbf{x})\|_{\mathsf{A}}^2 = \min_{p\in\mathcal{P}^k} \|p(\mathsf{A})\mathbf{e}^0\|_{\mathsf{A}}^2. \tag{8.2}$$

We shall now derive a bound on the expression on the right-hand side of (8.2) that depends on the spectrum of A but is independent of the direction of the initial error $\mathbf{e}^0$. Let a spectral decomposition of A be written as $\mathsf{A} = \mathsf{U}\mathsf{D}\mathsf{U}^T$, where U is an orthogonal matrix and $\mathsf{D} = \mathrm{diag}(\lambda_1, \ldots, \lambda_n)$ is a diagonal matrix defined by the eigenvalues of A. Since we assume A to be positive definite, the square root of A is well defined by

$$\mathsf{A}^{\frac{1}{2}} = \mathsf{U}\mathsf{D}^{\frac{1}{2}}\mathsf{U}^T.$$

Using $p(\mathsf{A}) = \mathsf{U}p(\mathsf{D})\mathsf{U}^T$, it is also easy to check that

$$\mathsf{A}^{\frac{1}{2}}p(\mathsf{A}) = p(\mathsf{A})\mathsf{A}^{\frac{1}{2}}.$$

Moreover, for any vector $\mathbf{v} \in \mathbb{R}^n$

$$\|\mathbf{v}\|_{\mathsf{A}}^2 = \mathbf{v}^T\mathsf{A}\mathbf{v} = \mathbf{v}^T\mathsf{A}^{\frac{1}{2}}\mathsf{A}^{\frac{1}{2}}\mathbf{v} = (\mathsf{A}^{\frac{1}{2}}\mathbf{v})^T\mathsf{A}^{\frac{1}{2}}\mathbf{v} = \|\mathsf{A}^{\frac{1}{2}}\mathbf{v}\|^2.$$

Using the latter identities, (8.2), and the properties of norms, we get

$$\|\mathbf{e}^k\|_A^2 = \min_{p\in\mathcal{P}^k} \|p(\mathsf{A})\mathbf{e}^0\|_{\mathsf{A}}^2 = \min_{p\in\mathcal{P}^k} \|\mathsf{A}^{\frac{1}{2}}p(\mathsf{A})\mathbf{e}^0\|^2 = \min_{p\in\mathcal{P}^k} \|p(\mathsf{A})\mathsf{A}^{\frac{1}{2}}\mathbf{e}^0\|^2$$
$$\le \min_{p\in\mathcal{P}^k} \|p(\mathsf{A})\|^2\|\mathsf{A}^{\frac{1}{2}}\mathbf{e}^0\|^2 = \min_{p\in\mathcal{P}^k} \|p(\mathsf{D})\|^2\|\mathbf{e}^0\|_{\mathsf{A}}^2.$$

Since

$$\|p(\mathsf{D})\| = \max_{i \in \{1,\ldots,n\}} |p(\lambda_i)|,$$

we can write

$$\|\mathbf{e}^k\|_{\mathsf{A}} \le \min_{p \in \mathcal{P}^k} \max_{i \in \{1,\ldots,n\}} |p(\lambda_i)| \|\mathbf{e}^0\|_{\mathsf{A}}. \tag{8.3}$$

8.1.2 Estimate in the Condition Number

The estimate (8.3) reduces the analysis of convergence of the CG method to the study of approximation of the zero function on the spectrum of A by a kth degree polynomial with the value one at the origin. This result helps us identify favorable cases when the conjugate gradient method is effective. For example, if the spectrum of A is clustered around a single point ξ, then the minimization by the CG should be very effective because $|(1 - x/\xi)^k|$ is small near ξ. We shall use (8.3) to get a "global" estimate of the rate of convergence of the conjugate gradient method in terms of the condition number of A.

Theorem 8.1. *Let $\{\mathbf{x}^k\}$ be generated by Algorithm 4.1 to find the solution $\widehat{\mathbf{x}}$ of (8.1) starting from $\mathbf{x}^0 \in \mathbb{R}^n$. Then the error*

$$\mathbf{e}^k = \mathbf{x}^k - \widehat{\mathbf{x}}$$

satisfies

$$\|\mathbf{e}^k\|_{\mathsf{A}} \le 2 \left(\frac{\sqrt{\kappa(\mathsf{A})} - 1}{\sqrt{\kappa(\mathsf{A})} + 1} \right)^k \|\mathbf{e}^0\|_{\mathsf{A}}, \tag{8.4}$$

where $\kappa(\mathsf{A})$ denotes the spectral condition number of A.

Proof. First notice that if $\mathcal{P}^k$ is the set of all kth degree polynomials p such that $p(0) = 1$, then for any $t \in \mathcal{P}^k$

$$\min_{p \in \mathcal{P}^k} \max_{\lambda \in [\lambda_{\min}, \lambda_{\max}]} |p(\lambda)| \le \max_{\lambda \in [\lambda_{\min}, \lambda_{\max}]} |t(\lambda)|. \tag{8.5}$$

A natural choice for t is the kth (weighted and shifted) Chebyshev polynomial on the interval $[\lambda_{\min}, \lambda_{\max}]$

$$t_k(\lambda) = T_k\left(\frac{2\lambda - \lambda_{\max} - \lambda_{\min}}{\lambda_{\max} - \lambda_{\min}}\right) / T_k\left(-\frac{\lambda_{\max} + \lambda_{\min}}{\lambda_{\max} - \lambda_{\min}}\right),$$

where $T_k(x)$ denotes the Chebyshev polynomial of the first kind on the interval $[-1, 1]$ given by

$$T_k(x) = \frac{1}{2}\left(x + \sqrt{x^2 - 1}\right)^k + \frac{1}{2}\left(x - \sqrt{x^2 - 1}\right)^k.$$

This t_k is known to minimize the right-hand side of (8.5) (see, e.g., Szegö [252]). Obviously $t_k \in \mathcal{P}^k$, so that we can use its well-known properties to get

$$\max_{\lambda \in [\lambda_{\min}, \lambda_{\max}]} |t_k(\lambda)| = 1/T_k\left(\frac{\lambda_{\max}+\lambda_{\min}}{\lambda_{\max}-\lambda_{\min}}\right).$$

Simple manipulations then show that

$$T_k\left(\frac{\lambda_{\max}+\lambda_{\min}}{\lambda_{\max}-\lambda_{\min}}\right) = \frac{1}{2}\left(\frac{\sqrt{\kappa(\mathsf{A})}+1}{\sqrt{\kappa(\mathsf{A})}-1}\right)^k + \frac{1}{2}\left(\frac{\sqrt{\kappa(\mathsf{A})}-1}{\sqrt{\kappa(\mathsf{A})}+1}\right)^k.$$

Thus for any $\lambda \in [\lambda_{\min}, \lambda_{\max}]$

$$|p_k(\lambda)| \le 2\left(\frac{\sqrt{\kappa(\mathsf{A})}-1}{\sqrt{\kappa(\mathsf{A})}+1}\right)^k.$$

Substituting this bound into (8.3) then gives the required result. □

The estimate (8.4) can be improved for some special distributions of the eigenvalues. For example, if the spectrum of A is in a positive interval $[a_{\min}, a_{\max}]$ except for m isolated eigenvalues $\lambda_1, \ldots, \lambda_m$, then we can use special polynomials $p \in \mathcal{P}^{k+m}$ of the form

$$p(\lambda) = \left(1-\frac{\lambda}{\lambda_1}\right)\cdots\left(1-\frac{\lambda}{\lambda_m}\right)q(\lambda),\ q \in \mathcal{P}^k$$

to get the estimate

$$\|\mathbf{e}^{k+m}\|_{\mathsf{A}} \le 2\left(\frac{\sqrt{\widetilde{\kappa}(\mathsf{A})}-1}{\sqrt{\widetilde{\kappa}(\mathsf{A})}+1}\right)^k \|\mathbf{e}^0\|_{\mathsf{A}}, \tag{8.6}$$

where $\widetilde{\kappa} = a_{\max}/a_{\min}$.

If the spectrum of A is distributed in two positive intervals $[a_{\min}, a_{\max}]$ and $[a_{\min}+d, a_{\max}+d]$, $d > 0$, then

$$\|\mathbf{e}^{k}\|_{\mathsf{A}} \le 2\left(\frac{\sqrt{\overline{\kappa}(\mathsf{A})}-1}{\sqrt{\overline{\kappa}(\mathsf{A})}+1}\right)^k \|\mathbf{e}^0\|_{\mathsf{A}}, \tag{8.7}$$

where $\overline{\kappa} = 4a_{\max}/a_{\min}$ approximates the *effective condition number* of a matrix A with the spectrum σ in two intervals,

$$\sigma \subseteq [a_{\min}, a_{\max}] \cup [a_{\min}+d, a_{\max}+d].$$

An interesting feature of the estimates (8.6) and (8.7) is that the *upper bound is independent of the values of some eigenvalues or d*. The proofs of the above and some other interesting estimates can be found in papers by Axelsson [5] and Axelsson and Lindskøg [8].

Notice that the estimate (8.4) for the first step of the conjugate gradient method may give a worse bound than the estimate (3.8) for the gradient method.The reason is that (8.4) captures the global performance of the conjugate gradient method, in particular, its capability to exploit the information from the previous steps. In contrast, (3.8) is based on analysis of just one step, in agreement with the one-step information used by the gradient method.

8.1.3 Optimality

Theorem 8.1 implies an easy optimality result concerning the number of iterations of the CG algorithm. To formulate it, let $\mathcal{T}$ denote any set of indices and assume that for any $t \in \mathcal{T}$ there is defined the problem

$$\text{minimize } f_t(\mathbf{x}), \quad f_t(\mathbf{x}) = \tfrac{1}{2}\mathbf{x}^T \mathsf{A}_t \mathbf{x} - \mathbf{b}_t^T \mathbf{x},$$

with $\mathsf{A}_t \in \mathbb{R}^{n_t \times n_t}$ SPD, and $\mathbf{b}_t, \mathbf{x} \in \mathbb{R}^{n_t}$. Moreover, assume that the condition number of A_t is uniformly bounded. Then the *CG algorithm starting from* $\mathbf{x}_t^0 = \mathbf{o}$ *finds* $\mathbf{x}_t^k$ *such that*

$$\|\mathsf{A}_t \mathbf{x}_t^k - \mathbf{b}_t\| \le \epsilon \|\mathbf{b}_t\|$$

at $O(1)$ *iterations.*

8.2 Preconditioned Conjugate Gradients

The above analysis of the CG method shows that its rate of convergence depends on the distribution of the eigenvalues of the Hessian A of f. In particular, we argued that CG converges fast if the eigenvalues of A are clustered around one point, i.e., if the condition number $\kappa(\mathsf{A})$ is close to one. Here, we show that we can reduce our minimization problem to this favorable case if we have an SPD matrix M such that $\mathsf{M}^{-1}\mathbf{x}$ can be quickly evaluated for any $\mathbf{x}$ and M approximates A in the sense that $\mathsf{M}^{-1}\mathsf{A}$ is close to the identity.

First assume that M is available in the form

$$\mathsf{M} = \widetilde{\mathsf{L}}\widetilde{\mathsf{L}}^T,$$

so that $\mathsf{M}^{-1}\mathsf{A}$ is similar to $\widetilde{\mathsf{L}}^{-1}\mathsf{A}\widetilde{\mathsf{L}}^{-T}$ and close to the identity. Then

$$f(\mathbf{x}) = \frac{1}{2}(\widetilde{\mathsf{L}}^T\mathbf{x})^T(\widetilde{\mathsf{L}}^{-1}\mathsf{A}\widetilde{\mathsf{L}}^{-T})(\widetilde{\mathsf{L}}^T\mathbf{x}) - (\widetilde{\mathsf{L}}^{-1}\mathbf{b})^T(\widetilde{\mathsf{L}}^T\mathbf{x})$$

and we can replace (8.1) by the *preconditioned problem* to find

$$\min_{\mathbf{y}\in\mathbb{R}^n} \overline{f}(\mathbf{y}), \tag{8.8}$$

where we substituted $\mathbf{y} = \widetilde{\mathsf{L}}^T\mathbf{x}$ and set

$$\overline{f}(\mathbf{y}) = \frac{1}{2}\mathbf{y}^T(\widetilde{\mathsf{L}}^{-1}\mathsf{A}\widetilde{\mathsf{L}}^{-T})\mathbf{y} - (\widetilde{\mathsf{L}}^{-1}\mathbf{b})^T\mathbf{y}.$$

The solution $\widehat{\mathbf{y}}$ of the preconditioned problem (8.8) is related to the solution $\widehat{\mathbf{x}}$ of the original problem by

$$\widehat{\mathbf{x}} = \widetilde{\mathsf{L}}^{-T}\widehat{\mathbf{y}}.$$

If the CG algorithm is applied directly to the preconditioned problem (8.8) with a given $\mathbf{y}^0 \in \mathbb{R}^n$, then the algorithm is initialized by

$$\mathbf{y}^0 = \widetilde{\mathsf{L}}^T\mathbf{x}^0, \quad \overline{\mathbf{g}}^0 = \widetilde{\mathsf{L}}^{-1}\mathsf{A}\widetilde{\mathsf{L}}^{-T}\mathbf{y}^0 - \widetilde{\mathsf{L}}^{-1}\mathbf{b} = \widetilde{\mathsf{L}}^{-1}\mathbf{g}^0, \quad \text{and } \overline{\mathbf{p}}^1 = \overline{\mathbf{g}}^0;$$

the iterates are defined by

$$\begin{aligned}
\overline{\alpha}_k &= \|\overline{\mathbf{g}}^{k-1}\|^2/(\overline{\mathbf{p}}^k)^T\widetilde{\mathsf{L}}^{-1}\mathsf{A}\widetilde{\mathsf{L}}^{-T}\overline{\mathbf{p}}^k,\\
\mathbf{y}^k &= \mathbf{y}^{k-1} - \overline{\alpha}_k\overline{\mathbf{p}}^k,\\
\overline{\mathbf{g}}^k &= \overline{\mathbf{g}}^{k-1} - \overline{\alpha}_k\widetilde{\mathsf{L}}^{-1}\mathsf{A}\widetilde{\mathsf{L}}^T\overline{\mathbf{p}}^k,\\
\overline{\beta}_k &= \|\overline{\mathbf{g}}^k\|^2/\|\overline{\mathbf{g}}^{k-1}\|^2,\\
\overline{\mathbf{p}}^{k+1} &= \overline{\mathbf{g}}^k + \overline{\beta}_k\overline{\mathbf{p}}^k.
\end{aligned}$$

Substituting

$$\mathbf{y}^k = \widetilde{\mathsf{L}}^T\mathbf{x}^k, \quad \overline{\mathbf{g}}^k = \widetilde{\mathsf{L}}^{-1}\mathbf{g}^k, \quad \text{and} \quad \overline{\mathbf{p}}^k = \widetilde{\mathsf{L}}^T\mathbf{p}^k,$$

and denoting

$$\mathbf{z}^k = \widetilde{\mathsf{L}}^{-T}\widetilde{\mathsf{L}}^{-1}\mathbf{g}^k = \mathsf{M}^{-1}\mathbf{g}^k,$$

we obtain the *preconditioned conjugate gradient algorithm* (PCG) in the original variables. Notice that the PCG algorithm does not explicitly exploit the Cholesky factorization of the preconditioner M.

The *pseudoresiduals* $\mathbf{z}^k$ are typically obtained by solving $\mathsf{M}\mathbf{z}^k = \mathbf{g}^k$. If M is a good approximation of A, then $\mathbf{z}^k$ is close to the error vector $\mathbf{e}^k$. The rate of convergence of the PCG algorithm depends on the condition number of the Hessian of the transformed function $\overline{f}$, i.e., on $\kappa(\mathsf{M}^{-1}\mathsf{A}) = \kappa(\widetilde{\mathsf{L}}^{-1}\mathsf{A}\widetilde{\mathsf{L}}^{-T})$. Thus, the efficiency of the preconditioned conjugate gradient method depends critically on the choice of a preconditioner, which should balance the cost of its application with the preconditioning effect. We refer interested readers to specialized books like Saad [240] or Axelsson [6] for more information. Since the choice of the preconditioner is problem-dependent, we limit our attention here to the brief discussion of a few general strategies.

Algorithm 8.1. Preconditioned conjugate gradient method (PCG)

Given a symmetric positive definite matrix $\mathsf{A} \in \mathbb{R}^{n\times n}$*, its symmetric positive definite approximation* $\mathsf{M} \in \mathbb{R}^{n\times n}$*, and* $\mathbf{b} \in \mathbb{R}^n$*.*

Step 0. *{Initialization.}*
Choose $\mathbf{x}^0 \in \mathbb{R}^n$*, set* $\mathbf{g}^0 = \mathsf{A}\mathbf{x}^0 - \mathbf{b}$, $\mathbf{z}^0 = \mathsf{M}^{-1}\mathbf{g}^0$, $\mathbf{p}^1 = \mathbf{z}^0$, $k = 1$
Step 1. *{Conjugate gradient loop.}*
while $\|\mathbf{g}^{k-1}\|$ *is not small*
$\alpha_k = (\mathbf{z}^{k-1})^T\mathbf{g}^{k-1}/(\mathbf{p}^k)^T\mathsf{A}\mathbf{p}^k$
$\mathbf{x}^k = \mathbf{x}^{k-1} - \alpha_k\mathbf{p}^k$
$\mathbf{g}^k = \mathbf{g}^{k-1} - \alpha_k\mathsf{A}\mathbf{p}^k$
$\mathbf{z}^k = \mathsf{M}^{-1}\mathbf{g}^k$
$\beta_k = (\mathbf{z}^k)^T\mathbf{g}^k/(\mathbf{z}^{k-1})^T\mathbf{g}^{k-1}$
$\mathbf{p}^{k+1} = \mathbf{z}^k + \beta_k\mathbf{p}^k$
$k = k + 1$
***end** while*
Step 2. *{Return a (possibly approximate) solution.}*
$\widetilde{\mathbf{x}} = \mathbf{x}^k$

We can define the most simple preconditioners through the decomposition

$$\mathsf{A} = \mathsf{D} + \mathsf{E} + \mathsf{E}^T,$$

where D is the diagonal of A and E is its strict lower part with the entries $[\mathsf{E}]_{ij} = [\mathsf{A}]_{ij}$ for $i > j$ and $[\mathsf{E}]_{ij} = 0$ otherwise.

The *Jacobi preconditioner* $\mathsf{M}_J = \mathsf{D}$ is the easiest one to implement, but its efficiency is very limited. Better approximation of A can be achieved by choosing the *block diagonal Jacobi preconditioner*

$$\mathsf{M}_{BJ} = \begin{bmatrix} \mathsf{A}_{11} & \mathsf{O} & \dots & \mathsf{O} \\ \mathsf{O} & \mathsf{A}_{22} & \dots & \mathsf{O} \\ . & . & \dots & . \\ \mathsf{O} & \mathsf{O} & \dots & \mathsf{A}_{kk} \end{bmatrix}, \tag{8.9}$$

where A_{ii} are diagonal blocks of A (see, e.g., Greenbaum [150, Sect. 10.5]). The *pseudoresiduals* $\mathbf{z}^k$ are typically obtained by solving $\mathsf{A}_{ii}\mathbf{z}_i^k = \mathbf{g}_i^k$.

We can often achieve good results with the *symmetric Gauss-Seidel* preconditioner

$$\mathsf{M}_{SGS} = (\mathsf{D} + \mathsf{E})\mathsf{D}^{-1}(\mathsf{D} + \mathsf{E}^T).$$

Notice that

$$\widetilde{\mathsf{L}} = (\mathsf{D} + \mathsf{E})\mathsf{D}^{-\frac{1}{2}}$$

is a regular lower triangular matrix, so we have the triangular factorization

$$\mathsf{M}_{SGS} = \widetilde{\mathsf{L}}\widetilde{\mathsf{L}}^T$$

for free.

More generally, the factorized preconditioners can be produced by *incomplete Cholesky (IC)*, which neglects some fill-in elements in the factor L. When the elements of L are neglected because they are smaller than a certain threshold, we call the factorization "IC-by-value"; when they are omitted because they do not belong to a certain sparsity pattern, we call it "IC-by position". See for example, Axelsson [6] or Saad [240]. The drawback of this method is that it can fail in the generation of diagonal entries.

8.3 Preconditioning by Conjugate Projector (Deflation)

So far, we have assumed that the preconditioners to a symmetric positive definite matrix A are nonsingular matrices that approximate A. In this section, we describe an alternative strategy that is useful when we can find the minimizer $\mathbf{x}^0$ of f over a subspace $\mathcal{U}$ of $\mathbb{R}^n$. We shall show that, in this case, we can get the preconditioning effect by reducing the conjugate gradient iterations to the conjugate complement of $\mathcal{U}$.

8.3.1 Conjugate Projectors

Our main tools will be the projectors with conjugate range and kernel. We shall use the fundamental relations introduced in Sect. 1.3 and some observations reviewed in this subsection.

Let $\mathsf{A} \in \mathbb{R}^{n\times n}$ be a symmetric positive definite matrix. A projector P is an *A-conjugate projector* or briefly a *conjugate projector* if $\mathrm{Im}\mathsf{P}$ is A-conjugate to $\mathrm{Ker}\mathsf{P}$, or equivalently

$$\mathsf{P}^T\mathsf{A}(\mathsf{I}-\mathsf{P}) = \mathsf{P}^T\mathsf{A} - \mathsf{P}^T\mathsf{A}\mathsf{P} = \mathsf{O}.$$

It follows that $\mathsf{Q} = \mathsf{I} - \mathsf{P}$ is also a conjugate projector,

$$\mathsf{P}^T\mathsf{A} = \mathsf{A}\mathsf{P} = \mathsf{P}^T\mathsf{A}\mathsf{P}, \quad \text{and} \quad \mathsf{Q}^T\mathsf{A} = \mathsf{A}\mathsf{Q} = \mathsf{Q}^T\mathsf{A}\mathsf{Q}. \tag{8.10}$$

Let us denote

$$\mathcal{V} = \mathrm{Im}\mathsf{Q}.$$

If $\mathbf{x} \in \mathsf{A}\mathcal{V}$, then $\mathbf{y} = \mathsf{Q}\mathbf{x}$ satisfies $\mathbf{y} \in \mathcal{V}$ and

$$\mathsf{Q}^T\mathsf{A}\mathsf{Q}\mathbf{x} = \mathsf{A}\mathsf{Q}\mathbf{x} = \mathsf{A}\mathbf{y},$$

so that

$$\mathsf{Q}^T\mathsf{A}\mathsf{Q}(\mathsf{A}\mathcal{V}) \subseteq \mathsf{A}\mathcal{V}. \tag{8.11}$$

Thus $\mathsf{A}\mathcal{V}$ is an invariant subspace of $\mathsf{Q}^T\mathsf{A}\mathsf{Q}$.

The following lemma shows that the mapping which assigns to each $\mathbf{x} \in \mathsf{A}\mathcal{V}$ the vector $\mathsf{Q}\mathbf{x} \in \mathcal{V}$ is expansive as in Fig. 8.1.

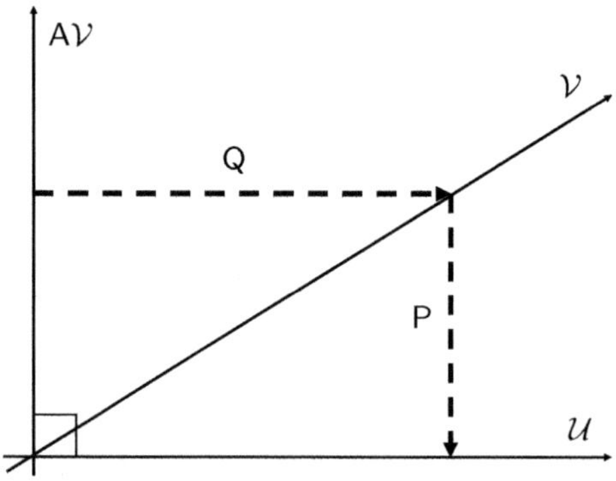

Fig. 8.1. Geometric illustration of Lemma 8.4

Lemma 8.4. *Let* Q *denote a conjugate projector on* $\mathcal{V}$ *and* $\mathbf{x} \in \mathsf{A}\mathcal{V}$*. Then*

$$\|\mathsf{Q}\mathbf{x}\| \geq \|\mathbf{x}\|.$$

Proof. For any $\mathbf{x} \in \mathsf{A}\mathcal{V}$, there is $\mathbf{y} \in \mathbb{R}^n$ such that $\mathbf{x} = \mathsf{AQ}\mathbf{y}$. It follows that

$$\mathsf{Q}^T\mathbf{x} = \mathsf{Q}^T\mathsf{AQ}\mathbf{y} = \mathsf{AQ}\mathbf{y} = \mathbf{x}.$$

Since $\mathbf{x}^T\mathsf{Q}\mathbf{x} = \mathbf{x}^T\mathsf{Q}^T\mathbf{x} = \|\mathbf{x}\|^2$, we have

$$\begin{aligned}
\|\mathsf{Q}\mathbf{x}\|^2 &= \mathbf{x}^T\mathsf{Q}^T\mathsf{Q}\mathbf{x} = \mathbf{x}^T\left((\mathsf{Q}^T - \mathsf{I}) + \mathsf{I}\right)((\mathsf{Q} - \mathsf{I}) + \mathsf{I})\mathbf{x} \\
&= \|(\mathsf{Q} - \mathsf{I})\mathbf{x}\|^2 + \mathbf{x}^T(\mathsf{Q}^T - \mathsf{I})\mathbf{x} + \mathbf{x}^T(\mathsf{Q} - \mathsf{I})\mathbf{x} + \|\mathbf{x}\|^2 \\
&= \|(\mathsf{Q} - \mathsf{I})\mathbf{x}\|^2 + \|\mathbf{x}\|^2 \\
&\geq \|\mathbf{x}\|^2.
\end{aligned}$$

□

8.3.2 Minimization in Subspace

Let us assume that $\mathcal{U}$ is the subspace spanned by the columns of a full column rank matrix $\mathsf{U} \in \mathbb{R}^{n \times n}$ and notice that $\mathsf{U}^T\mathsf{AU}$ is regular. Indeed, if

$$\mathsf{U}^T\mathsf{AU}\mathbf{x} = \mathbf{o},$$

then

$$\|\mathsf{U}\mathbf{x}\|_{\mathsf{A}}^2 = \mathbf{x}^T(\mathsf{U}^T\mathsf{AU}\mathbf{x}) = 0,$$

so $\mathbf{x} = \mathbf{o}$ by the assumptions that U is the full column rank matrix and A is SPD. Thus we can define

$$\mathsf{P} = \mathsf{U}(\mathsf{U}^T\mathsf{A}\mathsf{U})^{-1}\mathsf{U}^T\mathsf{A}.$$

It is easy to check directly that P is a conjugate (A-orthogonal) projector onto $\mathcal{U}$ as

$$\mathsf{P}^2 = \mathsf{U}(\mathsf{U}^T\mathsf{A}\mathsf{U})^{-1}\mathsf{U}^T\mathsf{A}\mathsf{U}(\mathsf{U}^T\mathsf{A}\mathsf{U})^{-1}\mathsf{U}^T\mathsf{A} = \mathsf{P}$$

and

$$\mathsf{P}^T\mathsf{A}(\mathsf{I}-\mathsf{P}) = \mathsf{A}\mathsf{U}(\mathsf{U}^T\mathsf{A}\mathsf{U})^{-1}\mathsf{U}^T\mathsf{A}(\mathsf{I} - \mathsf{U}(\mathsf{U}^T\mathsf{A}\mathsf{U})^{-1}\mathsf{U}^T\mathsf{A}) = \mathsf{O}.$$

Since any vector $\mathbf{x} \in \mathcal{U}$ can be written in the form $\mathbf{x} = \mathsf{U}\mathbf{y}, \mathbf{y} \in \mathbb{R}^m$, and

$$\mathsf{P}\mathbf{x} = \mathsf{U}(\mathsf{U}^T\mathsf{A}\mathsf{U})^{-1}\mathsf{U}^T\mathsf{A}\mathsf{U}\mathbf{y} = \mathsf{U}\mathbf{y} = \mathbf{x},$$

it follows that

$$\mathcal{U} = \mathrm{Im}\mathsf{P}.$$

The conjugate projector P onto $\mathcal{U}$ can be used for the solution of

$$\min_{\mathbf{x}\in\mathcal{U}} f(\mathbf{x}) = \min_{\mathbf{y}\in R^m} f(\mathsf{U}\mathbf{y}) = \min_{\mathbf{y}\in R^m} \frac{1}{2}\mathbf{y}^T\mathsf{U}^T\mathsf{A}\mathsf{U}\mathbf{y} - \mathbf{b}^T\mathsf{U}\mathbf{y}.$$

Using the gradient argument of Proposition 2.1, we get that the minimizer $\mathbf{y}^0$ of the latter problem satisfies

$$\mathsf{U}^T\mathsf{A}\mathsf{U}\mathbf{y}^0 = \mathsf{U}^T\mathbf{b}, \tag{8.12}$$

so that the minimizer $\mathbf{x}^0$ of f over $\mathcal{U}$ satisfies

$$\mathbf{x}^0 = \mathsf{U}\mathbf{y}^0 = \mathsf{U}(\mathsf{U}^T\mathsf{A}\mathsf{U})^{-1}\mathsf{U}^T\mathbf{b} = \mathsf{P}\mathsf{A}^{-1}\mathbf{b}. \tag{8.13}$$

Our assumption concerning the ability to find the minimum of f over $\mathcal{U}$ effectively amounts to the belief that we can solve (8.12). Notice that we can evaluate the product $\mathsf{P}\mathsf{A}^{-1}\mathbf{b}$ without solving any system of linear equations with the matrix A.

8.3.3 Conjugate Gradients in Conjugate Complement

In the next step, we shall use the conjugate projectors P and $\mathsf{Q} = \mathsf{I} - \mathsf{P}$,

$$\mathsf{P}^T\mathsf{A}\mathsf{Q} = \mathsf{O},$$

to decompose our minimization problem (8.1) into the minimization on $\mathcal{U}$ and the minimization on $\mathcal{V} = \mathrm{Im}\mathsf{Q}$. We shall use the conjugate gradient method to solve the latter problem.

Two observations are needed to exploit the special structure of our problem. First, using Lemma 8.4, $\dim \mathcal{V} = \dim \mathsf{A}\mathcal{V}$, and (1.2), we get that the mapping which assigns to each $\mathbf{x} \in \mathsf{A}\mathcal{V}$ a vector $\mathsf{Q}\mathbf{x} \in \mathcal{V}$ is an isomorphism, so that

$$\mathsf{Q}(\mathsf{A}\mathcal{V}) = \mathcal{V}.$$

Second, using (8.10) and (8.13), we get

$$\mathbf{g}^0 = \mathsf{A}\mathbf{x}^0 - \mathbf{b} = \mathsf{APA}^{-1}\mathbf{b} - \mathbf{b} = \mathsf{P}^T\mathbf{b} - \mathbf{b} = -\mathsf{Q}^T\mathbf{b}. \tag{8.14}$$

Using both observations, we get

$$\begin{aligned}\min_{\mathbf{x}\in\mathbb{R}^n} f(\mathbf{x}) &= \min_{\mathbf{y}\in\mathcal{U},\mathbf{z}\in\mathcal{V}} f(\mathbf{y}+\mathbf{z}) = \min_{\mathbf{y}\in\mathcal{U}} f(\mathbf{y}) + \min_{\mathbf{z}\in\mathcal{V}} f(\mathbf{z})\\ &= f(\mathbf{x}^0) + \min_{\mathbf{z}\in\mathcal{V}} f(\mathbf{z}) = f(\mathbf{x}^0) + \min_{\mathbf{x}\in\mathsf{A}\mathcal{V}} \frac{1}{2}\mathbf{x}^T\mathsf{Q}^T\mathsf{AQ}\mathbf{x} - \mathbf{b}^T\mathsf{Q}\mathbf{x}\\ &= f(\mathbf{x}^0) + \min_{\mathbf{x}\in\mathsf{A}\mathcal{V}} \frac{1}{2}\mathbf{x}^T\mathsf{Q}^T\mathsf{AQ}\mathbf{x} + (\mathbf{g}^0)^T\mathbf{x},\end{aligned} \tag{8.15}$$

where $\mathbf{x}^0$ is determined by (8.13).

It remains to solve the minimization problem (8.15). First observe that using Lemma 8.4, we get that $\mathsf{Q}^T\mathsf{AQ}|\mathsf{A}\mathcal{V}$ is positive definite. Since by (8.14) $\mathbf{g}^0 \in \mathrm{Im}\mathsf{Q}^T$,

$$\mathrm{Im}\mathsf{Q}^T = \mathrm{Im}(\mathsf{Q}^T\mathsf{A}) = \mathrm{Im}(\mathsf{AQ}) = \mathsf{A}\mathcal{V}, \tag{8.16}$$

and $\mathsf{A}\mathcal{V}$ is an invariant subspace of $\mathsf{Q}^T\mathsf{AQ}$, we can use the procedure described in Sect. 4.2 to generate $\mathsf{Q}^T\mathsf{AQ}$-conjugate vectors $\mathbf{p}^1, \dots, \mathbf{p}^k$ of

$$\mathcal{K}^k = \mathcal{K}^k(\mathsf{Q}^T\mathsf{AQ}, \mathbf{g}^0) = \mathrm{Span}\{\mathbf{g}^0, \mathsf{Q}^T\mathsf{AQ}\mathbf{g}^0, \dots, (\mathsf{Q}^T\mathsf{AQ})^{k-1}\mathbf{g}^0\}.$$

It simply follows that

$$\mathbf{q}^1 = \mathsf{Q}\mathbf{p}^1, \mathbf{q}^2 = \mathsf{Q}\mathbf{p}^2, \dots$$

are A-conjugate vectors of $\mathcal{V}$. Using (4.14), $\mathbf{p}^i \in \mathsf{A}\mathcal{V}$, and Lemma 8.4, it is easy to see that

$$\|\mathbf{q}^k\| \geq \|\mathbf{p}^k\| \geq \|\mathbf{g}^{k-1}\|,$$

so that we can generate a new conjugate direction $\mathbf{q}^k$ whenever $\mathbf{g}^{k-1} \neq \mathbf{o}$. We can sum up the most important properties of the CG algorithm with the preconditioning by the conjugate projector into the following theorem.

Theorem 8.5. *Let $\mathbf{x}^k$ be generated by Algorithm 8.2 to find the solution $\widehat{\mathbf{x}}$ of (3.1) with a full column rank matrix $\mathsf{U} \in \mathbb{R}^{n\times m}$. Then the algorithm is well defined and there is $k \leq n - m$ such that $\mathbf{x}^k = \widehat{\mathbf{x}}$. Moreover, the following statements hold for $i = 1, \dots, k$:*

(i) $f(\mathbf{x}^i) = \min\{f(\mathbf{x}) : \mathbf{x} \in \mathcal{U} + \mathsf{Q}\mathcal{K}^i(\mathsf{Q}^T\mathsf{A}\mathsf{Q}, \mathbf{g}^0)\}$.
(ii) $\|\mathbf{q}^i\| \geq \|\mathbf{g}^{i-1}\|$.
(iii) $(\mathbf{q}^i)^T\mathsf{A}\mathbf{q}^j = 0$ *for* $i > j$.
(iv) $(\mathbf{q}^i)^T\mathsf{A}\mathbf{x} = 0$ *for* $\mathbf{x} \in \mathcal{U}$.

The complete conjugate gradient algorithm with the preconditioning by the conjugate projector reads as follows:

Algorithm 8.2. Conjugate gradients with projector preconditioning (CGPP)

Given a symmetric positive definite matrix $\mathsf{A} \in \mathbb{R}^{n\times n}$*, a full column rank matrix* $\mathsf{U} \in \mathbb{R}^{n\times m}$*, and* $\mathbf{b} \in \mathbb{R}^n$.

Step 0. *{Initialization.}*
$\mathsf{P} = \mathsf{U}(\mathsf{U}^T\mathsf{A}\mathsf{U})^{-1}\mathsf{U}^T\mathsf{A}, \mathsf{Q} = \mathsf{I} - \mathsf{P}$
$\mathbf{x}^0 = \mathsf{P}\mathsf{A}^{-1}\mathbf{b} = \mathsf{U}(\mathsf{U}^T\mathsf{A}\mathsf{U})^{-1}\mathsf{U}^T\mathbf{b}$
$k = 1, \mathbf{g}^0 = \mathsf{A}\mathbf{x}^0 - \mathbf{b}, \mathbf{q}^1 = \mathsf{Q}\mathbf{g}^0$
Step 1. *{Conjugate gradient loop. }*
while $\|\mathbf{g}^{k-1}\| > 0$
$\alpha_k = (\mathbf{g}^{k-1})^T\mathbf{q}^k/(\mathbf{q}^k)^T\mathsf{A}\mathbf{q}^k$
$\mathbf{x}^k = \mathbf{x}^{k-1} - \alpha_k\mathbf{q}^k$
$\mathbf{g}^k = \mathbf{g}^{k-1} - \alpha_k\mathsf{A}\mathbf{q}^k$
$\beta_k = (\mathbf{g}^k)^T\mathsf{A}\mathbf{q}^k/(\mathbf{q}^k)^T\mathsf{A}\mathbf{q}^k$
$\mathbf{q}^{k+1} = \mathsf{Q}\mathbf{g}^k + \beta_k\mathbf{q}^k$
$k = k + 1$
end while
Step 2. *{Return a (possibly approximate) solution.}*
$\widetilde{\mathbf{x}} = \mathbf{x}^k$

8.3.4 Preconditioning Effect

As we have seen in the previous section, we can consider the iterations of Algorithm 8.2 as the conjugate gradient iterations for the minimization of

$$f_{0,\mathsf{Q}}(\mathbf{x}) = \frac{1}{2}\mathbf{x}^T\mathsf{Q}^T\mathsf{A}\mathsf{Q}\mathbf{x} + (\mathbf{g}^0)^T\mathbf{x}$$

that generate the iterations

$$\mathbf{x}^k \in \mathcal{K}^k(\mathsf{Q}^T\mathsf{A}\mathsf{Q}, \mathbf{g}^0) \subseteq \mathsf{A}\mathcal{V}.$$

Thus only the positive definite restriction $\mathsf{Q}^T\mathsf{A}\mathsf{Q}|\mathsf{A}\mathcal{V}$ of $\mathsf{Q}^T\mathsf{A}\mathsf{Q}$ to $\mathsf{A}\mathcal{V}$ takes part in the process of solution, and the rate of convergence may be estimated by the spectral condition number $\kappa(\mathsf{Q}^T\mathsf{A}\mathsf{Q}|\mathsf{A}\mathcal{V})$ of $\mathsf{Q}^T\mathsf{A}\mathsf{Q}|\mathsf{A}\mathcal{V}$.

It is rather easy to see that

$$\kappa(\mathsf{Q}^T\mathsf{AQ}|\mathsf{A}\mathcal{V}) \leq \kappa(\mathsf{A}).$$

Indeed, denoting by $\lambda_1 \geq \cdots \geq \lambda_n$ the eigenvalues of A, we can observe that if $\mathbf{x} \in \mathsf{A}\mathcal{V}$ and $\|\mathbf{x}\| = 1$, then by Lemma 8.4

$$\mathbf{x}^T\mathsf{Q}^T\mathsf{A}\mathsf{Q}\mathbf{x} \geq (\mathsf{Q}\mathbf{x})^T\mathsf{A}(\mathsf{Q}\mathbf{x})/\|\mathsf{Q}\mathbf{x}\|^2 \geq \lambda_n$$

and

$$\mathbf{x}^T\mathsf{Q}^T\mathsf{A}\mathsf{Q}\mathbf{x} \leq \mathbf{x}^T\mathsf{Q}^T\mathsf{A}\mathsf{Q}\mathbf{x} + \mathbf{x}^T\mathsf{P}^T\mathsf{A}\mathsf{P}\mathbf{x} = \mathbf{x}^T\mathsf{A}\mathbf{x} \leq \lambda_1. \tag{8.17}$$

To see the preconditioning effect of the algorithm in more detail, let us denote by $\mathcal{E}$ the m-dimensional subspace spanned by the eigenvectors corresponding to the m smallest eigenvalues $\lambda_{n-m+1} \geq \cdots \geq \lambda_n$ of A, and let $\mathsf{R}_{\mathsf{A}\mathcal{U}}$ and $\mathsf{R}_{\mathcal{E}}$ denote the orthogonal projectors on $\mathsf{A}\mathcal{U}$ and $\mathcal{E}$, respectively. Let

$$\gamma = \|\mathsf{R}_{\mathsf{A}\mathcal{U}} - \mathsf{R}_{\mathcal{E}}\|$$

denote the *gap* between $\mathsf{A}\mathcal{U}$ and $\mathcal{E}$. We can evaluate it provided we have matrices U and E whose columns form the orthonormal bases of $\mathsf{A}\mathcal{U}$ and $\mathcal{E}$, respectively. Using Sect. 1.12, we get that if σ is the least singular value of $\mathsf{U}^T\mathsf{E}$, then

$$\gamma = \sqrt{1-\sigma^2} \leq 1.$$

Theorem 8.6. *Let $\mathcal{U}, \mathcal{V}, \mathsf{Q}$ be those of Algorithm 8.2, let $\lambda_1 \geq \cdots \geq \lambda_n$ denote the eigenvalues of A, and let $\overline{\lambda}_{\min}$ denote the least nonzero eigenvalue of $\mathsf{Q}^T\mathsf{AQ}$. Then*

$$\lambda_n \leq \sqrt{(1-\gamma^2)\lambda_{n-m}^2 + \gamma^2\lambda_n^2} \leq \overline{\lambda}_{\min} \tag{8.18}$$

and

$$\kappa(\mathsf{Q}^T\mathsf{AQ}|\mathsf{A}\mathcal{V}) \leq \frac{\lambda_1}{\sqrt{(1-\gamma^2)\lambda_{n-m}^2 + \gamma^2\lambda_n^2}}.$$

Proof. Let $\mathbf{x} \in \mathsf{A}\mathcal{V}$, $\|\mathbf{x}\| = 1$, so that $\|\mathsf{Q}\mathbf{x}\| \geq 1$ by Lemma 8.4. Observing that $\mathrm{Im}\mathsf{R}\mathcal{E}$ and $\mathrm{Im}(\mathsf{I} - \mathsf{R}_{\mathcal{E}})$ are orthogonal invariant subspaces of A, we get that

$$\begin{aligned}
\|\mathsf{AQ}\mathbf{x}\|^2 &= \|\mathsf{A}(\mathsf{I}-\mathsf{R}_{\mathcal{E}})\mathsf{Q}\mathbf{x}\|^2 + \|\mathsf{A}\mathsf{R}_{\mathcal{E}}\mathsf{Q}\mathbf{x}\|^2 \\
&\geq \lambda_{n-m}^2\|(\mathsf{I}-\mathsf{R}_{\mathcal{E}})\mathsf{Q}\mathbf{x}\|^2 + \lambda_n^2\|\mathsf{R}_{\mathcal{E}}\mathsf{Q}\mathbf{x}\|^2 \\
&\geq \left(\lambda_{n-m}^2\|(\mathsf{I}-\mathsf{R}_{\mathcal{E}})\mathsf{Q}\mathbf{x}\|^2 + \lambda_n^2\|\mathsf{R}_{\mathcal{E}}\mathsf{Q}\mathbf{x}\|^2\right)/\|\mathsf{Q}\mathbf{x}\|^2 \\
&\geq \lambda_{n-m}^2(1-\xi^2) + \lambda_n^2\xi^2,
\end{aligned} \tag{8.19}$$

where $\xi = \|\mathsf{Q}\mathbf{x}\|^{-1}\|\mathsf{R}_{\mathcal{E}}\mathsf{Q}\mathbf{x}\|$. We have used that

$$\|(\mathsf{I}-\mathsf{R}_{\mathcal{E}})\mathsf{Q}\mathbf{x}\|^2 + \|\mathsf{R}_{\mathcal{E}}\mathsf{Q}\mathbf{x}\|^2 = \|\mathsf{Q}\mathbf{x}\|^2.$$

Since $\mathrm{Im}\mathsf{Q} = \mathcal{V}$, it follows by the definition of $\mathsf{R}_{\mathsf{A}\mathcal{U}}$ that $\mathsf{R}_{\mathsf{A}\mathcal{U}}\mathsf{Q} = \mathsf{O}$ and

$$\xi = \|\mathsf{Q}\mathbf{x}\|^{-1}\|(\mathsf{R}_{\mathcal{E}} - \mathsf{R}_{\mathsf{A}\mathcal{U}})\mathsf{Q}\mathbf{x}\| \leq \gamma.$$

As the last expression in (8.19) is a decreasing function of ξ for $\xi \geq 0$, it follows that

$$\|\mathsf{Q}^T\mathsf{A}\mathsf{Q}\mathbf{x}\|^2 = \|\mathsf{A}\mathsf{Q}\mathbf{x}\|^2 \geq \lambda_{n-m}^2(1-\gamma^2) + \lambda_n^2\gamma^2.$$

The rest is an easy consequence of (8.17). □

The above theorem suggests that the preconditioning by the conjugate projector is efficient when $\mathcal{U}$ approximates the subspace spanned by the eigenvectors, which correspond to the smallest eigenvalues of A. If $\mathsf{U}^T\mathsf{E}$ is nonsingular and $\lambda_n < \lambda_{n-m}$, then $\gamma < 1$ and

$$\kappa(\mathsf{Q}^T\mathsf{A}\mathsf{Q}|\mathsf{A}\mathcal{V}) < \kappa(\mathsf{A}).$$

If the minimization problem arises from the discretization of elliptic partial differential equations, then U can be obtained using coarse basis functions of the hierarchical bases. For example, we can define the coarse space by the aggregation matrix

$$\mathsf{U} = \operatorname{diag}(\mathbf{e}_1, \ldots, \mathbf{e}_m), \quad \mathbf{e}_i = [1, \ldots, 1]^T, \quad i = 1, \ldots, m.$$

Experience shows that even the subspace with a small dimension can considerably improve the convergence rate. See Sect. 8.6.2 for a numerical example.

8.4 Conjugate Gradients for the Least Square Problems

Let us assume that A is SPS so that the cost function f is convex but not strictly convex, and let the unconstrained minimization problem (8.1) be well posed, i.e., $\mathbf{b} \in \operatorname{Im}\mathsf{A}$ by Proposition 2.1. If we start the conjugate gradient algorithm from arbitrary $\mathbf{x}^0 \in \operatorname{Im}\mathsf{A}$, then the gradient $\mathbf{g}^0$ and the Krylov space $\mathcal{K}^n(\mathsf{A}, \mathbf{g}^0)$ satisfy

$$\mathbf{g}^0 \in \operatorname{Im}\mathsf{A} \quad \text{and} \quad \mathcal{K}^n(\mathsf{A}, \mathbf{g}^0) \subseteq \operatorname{Im}\mathsf{A}.$$

Since the CG method picks the conjugate directions from $\mathcal{K}^n(\mathsf{A}, \mathbf{g}^0)$, it follows that the method works only on the range of A. Thus the algorithm generates the iterates $\mathbf{x}^k$ which converge to a unique solution $\widehat{\mathbf{x}}$ with the rate of convergence which we can describe by the bounds on the spectrum of the restriction $\mathsf{A}|\mathcal{K}^n(\mathsf{A}, \mathbf{g}^0)$. Observing that the eigenvalues of $\mathsf{A}|\mathcal{K}^n(\mathsf{A}, \mathbf{g}^0)$ are bounded from below by the least nonzero eigenvalue $\overline{\lambda}_{\min}$ of A, we get the error estimate

$$\|\mathbf{e}^k\|_{\mathsf{A}} \leq 2\left(\frac{\sqrt{\overline{\kappa}(\mathsf{A})}-1}{\sqrt{\overline{\kappa}(\mathsf{A})}+1}\right)^k \|\mathbf{e}^0\|_{\mathsf{A}}, \tag{8.20}$$

where $\overline{\kappa}(\mathsf{A})$ denotes the *regular spectral condition number* of A defined by

$$\overline{\kappa}(\mathsf{A}) = \kappa(\mathsf{A}|\mathrm{Im}\mathsf{A}) = \lambda_{\max}/\overline{\lambda}_{\min}.$$

Let P and $\mathsf{Q} = \mathsf{I} - \mathsf{P}$ denote the orthogonal projectors on $\mathrm{Im}\mathsf{A}$ and $\mathrm{Ker}\mathsf{A}$, respectively, so that

$$\overline{\mathbf{x}} = \mathsf{P}\overline{\mathbf{x}} + \mathsf{Q}\overline{\mathbf{x}}.$$

Since the restriction $\mathsf{A}|\mathrm{Im}\mathsf{A}$ is nonsingular, it follows that there is a unique solution $\overline{\mathbf{x}}_{\mathrm{LS}} \in \mathrm{Im}\mathsf{A}$ of (8.1), and by Proposition 2.1 any solution $\overline{\mathbf{x}}$ satisfies

$$\overline{\mathbf{x}} = \widehat{\mathbf{x}}_{\mathrm{LS}} + \mathbf{d}, \quad \mathbf{d} \in \mathrm{Ker}\mathsf{A}.$$

Thus if $\overline{\mathbf{x}}$ solves (8.1), then $\mathsf{P}\overline{\mathbf{x}} = \overline{\mathbf{x}}_{\mathrm{LS}}$ and

$$\mathsf{Q}\overline{\mathbf{x}} = \mathsf{Q}\mathbf{x}^0.$$

It follows that $\mathbf{x}_{\mathrm{LS}}$ is the least square solution of $\mathsf{A}\mathbf{x} = \mathbf{b}$, and to get it by the conjugate gradient algorithm, it is enough to take $\mathbf{x}^0 \in \mathrm{Im}\mathsf{A}$.

Let $\mathsf{A} \neq \mathsf{O}$ be only SPS, and let us admit that unconstrained minimization problem (3.1) is not well posed, i.e., $\mathbf{b} \notin \mathrm{Im}\mathsf{A}$ by Proposition 2.1. Then we can consider the *least square problem* (8.1), that appears in many applications. Here, we adapt the conjugate gradient algorithm (4.1) so that it generates in parallel the iterations for the solution of the least square problem (8.1) and projection of $\mathbf{b}$ on the range of A.

8.4.1 Equivalent Formulation and Krylov Spaces

First, recall that problem (8.1) has the same solution as

$$\min_{\mathbf{x}\in\mathrm{Im}\mathsf{A}} q(\mathbf{x}), \qquad q(\mathbf{x}) = \frac{1}{2}\mathbf{x}^T\mathsf{A}^2\mathbf{x} - (\mathsf{A}\mathbf{b})^T\mathbf{x} \tag{8.21}$$

or

$$\min\{\|\mathbf{x}\| : \mathbf{x} = \arg\min_{\mathbf{y}\in\mathbb{R}^n} \|\mathsf{A}\mathbf{y} - \mathbf{b}\|^2\}. \tag{8.22}$$

The solution $\widehat{\mathbf{x}} \in \mathrm{Im}\mathsf{A}$ of (8.21) is a unique vector which satisfies $\widehat{\mathbf{x}} \in \mathrm{Im}\mathsf{A}$ and the *normal equation*

$$\mathsf{A}^2\mathbf{x} = \mathsf{A}\mathbf{b}. \tag{8.23}$$

If we multiply both sides of (8.23) by the Moore–Penrose inverse $\mathsf{A}^\dagger$ on the left, we get that the solution $\widehat{\mathbf{x}}$ of (8.21) satisfies also

$$\mathsf{A}\mathbf{x} = \mathsf{Q}\mathbf{b}, \tag{8.24}$$

where Q is the orthogonal projector onto $\mathrm{Im}\mathsf{A}$. Since for any $\mathbf{x} \in \mathrm{Im}\mathsf{A}$

$$\mathbf{b}^T\mathbf{x} = (\mathsf{Q}\mathbf{b})^T\mathbf{x},$$

it follows that $\widehat{\mathbf{x}}$ is also a unique solution of (8.1) or

$$\min_{\mathbf{x}\in\mathrm{Im}\mathsf{A}} f_{\mathsf{Q}}(\mathbf{x}), \qquad f_{\mathsf{Q}}(\mathbf{x}) = \frac{1}{2}\mathbf{x}^T\mathsf{A}\mathbf{x} - (\mathsf{Q}\mathbf{b})^T\mathbf{x}. \tag{8.25}$$

We can approximate solution of (8.25) by the minimizers of f on $\mathcal{K}^k(\mathsf{A}, \mathsf{Q}\mathbf{b})$. Since $\mathcal{K}^k(\mathsf{A}, \mathsf{Q}\mathbf{b}) \subseteq \mathbb{R}^n$, we have $k \le n$. It follows that there is

$$d = d(\mathsf{A}, \mathsf{Q}\mathbf{b}) > 0$$

such that

$$\mathcal{K}^{d-1}(\mathsf{A}, \mathsf{Q}\mathbf{b}) \subsetneq \mathcal{K}^d(\mathsf{A}, \mathsf{Q}\mathbf{b}) = \mathcal{K}^{d+1}(\mathsf{A}, \mathsf{Q}\mathbf{b}) = \mathcal{K}^{d+2}(\mathsf{A}, \mathsf{Q}\mathbf{b}) = \dots .$$

The following lemma recalls the cyclic property of the invariant Krylov space $\mathcal{K}(\mathsf{A}, \mathsf{A}\mathbf{b}) = \mathcal{K}^d(\mathsf{A}, \mathsf{A}\mathbf{b})$ that we shall use to generate $\mathcal{K}(\mathsf{A}, \mathsf{Q}\mathbf{b})$.

Lemma 8.5. Let A denote an SPS matrix and $\mathbf{b} \in \mathbb{R}^n$. Then

$$\mathsf{Q}\mathbf{b} \in \mathcal{K}(\mathsf{A}, \mathsf{A}\mathbf{b}) \quad \text{and} \quad \mathcal{K}(\mathsf{A}, \mathsf{Q}\mathbf{b}) = \mathcal{K}(\mathsf{A}, \mathsf{A}\mathbf{b}) = \mathsf{A}\mathcal{K}(\mathsf{A}, \mathsf{Q}\mathbf{b}). \tag{8.26}$$

Proof. Let us denote

$$f_{\mathsf{A}}(\mathbf{x}) = \frac{1}{2}\mathbf{x}^T\mathsf{A}\mathbf{x} - (\mathsf{A}\mathbf{b})^T\mathbf{x}.$$

Since $\mathcal{K}(\mathsf{A}, \mathsf{A}\mathbf{b})$ is a subspace of $\mathrm{Im}\mathsf{A}$ and the restriction $f_{\mathsf{A}}|\mathcal{K}(\mathsf{A}, \mathsf{A}\mathbf{b})$ of f_{A} to $\mathcal{K}(\mathsf{A}, \mathsf{A}\mathbf{b})$ is coercive, it follows that there is

$$\mathbf{z} = \arg\min f_{\mathsf{A}}(\mathbf{x}) \quad \text{s.t.} \quad \mathbf{x} \in \mathcal{K}(\mathsf{A}, \mathsf{A}\mathbf{b}),$$

and $\mathbf{z}$ satisfies

$$\nabla f_{\mathsf{A}}(\mathbf{z})^T\mathbf{y} = (\mathsf{A}\mathbf{z} - \mathsf{A}\mathbf{b})^T\mathbf{y} = 0 \quad \text{for any} \quad \mathbf{y} \in \mathcal{K}(\mathsf{A}, \mathsf{A}\mathbf{b}).$$

Since $\mathsf{A}\mathbf{z}, \mathsf{A}\mathbf{b} \in \mathcal{K}(\mathsf{A}, \mathsf{A}\mathbf{b}) = \mathsf{A}\mathcal{K}(\mathsf{A}, \mathsf{A}\mathbf{b})$, we have $\mathsf{A}\mathbf{z} - \mathsf{A}\mathbf{b} \in \mathcal{K}(\mathsf{A}, \mathsf{A}\mathbf{b})$ and

$$\|\mathsf{A}\mathbf{z} - \mathsf{A}\mathbf{b}\|^2 = 0.$$

Since $\mathbf{z}, \mathsf{Q}\mathbf{b} \in \mathrm{Im}\mathsf{A}$, we have

$$\|\mathbf{z} - \mathsf{Q}\mathbf{b}\| = 0.$$

Thus $\mathsf{Q}\mathbf{b} \in \mathcal{K}(\mathsf{A}, \mathsf{A}\mathbf{b})$. The rest is easy. □

8.4.2 Algorithm cgSLS

Using Lemma 8.5 and the reasoning from the beginning of Sect. 8.4, we can apply the CG procedure to generate a conjugate basis of

$$\mathcal{K}(\mathsf{A}, \mathsf{A}\mathbf{b}) = \mathcal{K}(\mathsf{A}, \mathsf{Q}\mathbf{b})$$

and approximations $\mathbf{y}^k$ for the unique solution $\widehat{\mathbf{y}} = \mathsf{Q}\mathbf{b}$ of

$$\min_{\mathbf{y} \in \mathrm{Im}\mathsf{A}} f_{\mathsf{A}}(\mathbf{y}), \quad f_{\mathsf{A}}(\mathbf{y}) = \frac{1}{2}\mathbf{y}^T \mathsf{A}\mathbf{y} - (\mathsf{A}\mathbf{b})^T \mathbf{y}. \tag{8.27}$$

At the same time, we can use the conjugate basis to generate approximations $\mathbf{x}^k$ of the minimizer $\widehat{\mathbf{x}}$ of f on

$$\mathcal{K}(\mathsf{A}, \mathsf{A}\mathbf{b}) = \mathcal{K}(\mathsf{A}, \mathsf{Q}\mathbf{b}).$$

Since for any $\mathbf{x} \in \mathrm{Im}\mathsf{A}$, $f(\mathbf{x}) = f_{\mathsf{Q}}(\mathbf{x})$, we get the solution of (8.1) and (8.25). The resulting algorithm reads as follows.

Algorithm 8.3. Conjugate gradients for SPS least-squares problem (cgSLS)

Given an SPS matrix A of the order n and an n-vector $\mathbf{b}$.
Step 0. {*Initialization.*}
  Set $k = 0$, $\mathbf{x}^0 = \mathbf{o}$, $\mathbf{y}^0 = \mathbf{o}$, $\mathbf{g}^0 = -\mathbf{b}$, $\mathbf{g}_{\mathsf{A}}^0 = -\mathsf{A}\mathbf{b}$, $\mathbf{p}^1 = \mathbf{g}_{\mathsf{A}}^0$.
  while $\|\mathsf{A}\mathbf{x}^k - \mathbf{y}^k\| + \|\mathbf{g}_{\mathsf{A}}^k\|$ is not small
Step 1. {*Minimization of f and f_{A} on $\mathcal{K}^k(\mathsf{A}, \mathsf{A}\mathbf{b})$.*}
  $\alpha = (\mathbf{g}^k)^T \mathbf{p}^{k+1}/(\mathbf{p}^{k+1})^T \mathsf{A}\mathbf{p}^{k+1} = \arg\min_{\alpha \in \mathbb{R}} f(\mathbf{x}^k - \alpha\mathbf{p}^{k+1})$
  $\alpha_{\mathsf{A}} = (\mathbf{g}_{\mathsf{A}}^k)^T \mathbf{p}^{k+1}/(\mathbf{p}^{k+1})^T \mathsf{A}\mathbf{p}^{k+1} = \arg\min_{\alpha \in \mathbb{R}} f_{\mathsf{A}}(\mathbf{x}^k - \alpha\mathbf{p}^{k+1})$
  $\mathbf{x}^{k+1} = \mathbf{x}^k - \alpha\mathbf{p}^{k+1}$, $\mathbf{y}^{k+1} = \mathbf{y}^k - \alpha_{\mathsf{A}}\mathbf{p}^{k+1}$
Step 2. {*Update of the gradients of f and f_{A}.*}
  $\mathbf{g}^{k+1} = \mathbf{g}^k - \alpha\mathsf{A}\mathbf{p}^{k+1}$, $\mathbf{g}_{\mathsf{A}}^{k+1} = \mathbf{g}_{\mathsf{A}}^k - \alpha_{\mathsf{A}}\mathsf{A}\mathbf{p}^{k+1}$
Step 3. {*New conjugate direction.*}
  $k = k + 1$
  $\beta = (\mathbf{g}_{\mathsf{A}}^k)^T \mathsf{A}\mathbf{p}^k/(\mathbf{p}^k)^T \mathsf{A}\mathbf{p}^k$, $\mathbf{p}^{k+1} = \mathbf{g}_{\mathsf{A}}^k - \beta\mathbf{p}^k$
  end while
Step 4. {*Return (possibly inexact) solutions.*}
  $\widetilde{\mathbf{x}} = \mathbf{x}^k$, $\widetilde{\mathbf{y}} = \mathbf{y}^k$

Recall that $\mathbf{y}^k$ are just the standard CG iterates for solving (8.27). Because

$$\mathsf{A}\mathbf{x}^k = \mathbf{g}^k + \mathbf{b},$$

we can evaluate the test without additional multiplications by A. Thus, if we store $\mathsf{A}\mathbf{p}^{k+1}$, the most expensive operation, the multiplication by A, is executed only once in each step. The algorithm requires the storage for one generation of $\mathbf{x}, \mathbf{y}, \mathbf{g}, \mathbf{g}_{\mathsf{A}}, \mathbf{p}, \mathbf{b}$, and $\mathsf{A}\mathbf{p}$. One matrix–vector product and six scalar products dominate the iteration cost. Since $\widehat{\mathbf{y}} = \mathsf{Q}\mathbf{b}$,

$$\|\mathbf{y}^k - \widehat{\mathbf{y}}\| \le \|\mathsf{A}^\dagger\| \|\mathsf{A}\mathbf{y}^k - \mathsf{A}\mathbf{b}\| = \|\mathsf{A}^\dagger\| \|\mathbf{g}_\mathsf{A}^k\|,$$

and

$$\|\mathsf{A}\mathbf{x}^k - \mathsf{Q}\mathbf{b}\| \le \|\mathsf{A}\mathbf{x}^k - \mathbf{y}^k\| + \|\mathbf{y}^k - \widehat{\mathbf{y}}\| \le \|\mathsf{A}\mathbf{x}^k - \mathbf{y}^k\| + \|\mathsf{A}^\dagger\| \|g_\mathsf{A}^k\|,$$

the test guarantees a small approximation error of both $\mathbf{x}^k$ and $\mathbf{y}^k$.

The following proposition guarantees that the algorithm is well-defined and finds the solution in a finite number of steps.

Proposition 8.1. Let $\mathbf{x}^k$, $\mathbf{y}^k$, and $\mathbf{p}^k$ be generated by the cgSLS algorithm for the solution of (4.2); let

$$\mathbf{g}^k = \mathsf{A}\mathbf{x}^k - \mathbf{b} = \nabla f(\mathbf{x}^k), \quad \text{and} \quad \mathbf{g}_\mathsf{A}^k = \mathsf{A}\mathbf{y}^k - \mathsf{A}\mathbf{b} = \nabla f_\mathsf{A}(\mathbf{y}^k).$$

Then the following statements hold:
(i) If $\mathbf{g}_\mathsf{A}^k \ne \mathbf{o}$, then $\mathbf{x}^k$, $\mathbf{y}^k$, and $\mathbf{p}^{k+1}$ are well defined and

$$\|\mathbf{p}^{k+1}\| \ge \|\mathbf{g}_\mathsf{A}^k\|.$$

(ii) If $\mathbf{g}_\mathsf{A}^k = \mathbf{o}$, then $\mathbf{y}^k = \widehat{\mathbf{y}} = \mathsf{Q}\mathbf{b}$ and $\mathbf{x}^k = \widehat{\mathbf{x}} = \mathsf{A}^\dagger\mathbf{b}$.
(iii) cgSLS finds $\widehat{\mathbf{x}}$ and $\widehat{\mathbf{y}}$ in a finite number of steps.

Proof. (i) The statement is trivial for $i = 0$. If $k \ge 1$, then $\mathbf{y}^k$ minimizes f_A on $\mathcal{K}^k(\mathsf{A}, \mathsf{A}\mathbf{b})$, so $(\mathbf{g}_\mathsf{A}^k)^T\mathbf{y} = 0$ for any $\mathbf{y} \in \mathcal{K}^k(\mathsf{A}, \mathsf{A}\mathbf{b})$. Thus $(\mathbf{g}_\mathsf{A}^k)^T\mathbf{p}^k = 0$ and

$$\|\mathbf{p}^{k+1}\|^2 = \|\mathbf{g}_\mathsf{A}^k\|^2 + \beta^2\|\mathbf{p}^k\|^2 \ge \|\mathbf{g}_\mathsf{A}^k\|^2.$$

Since $\mathbf{p}^{k+1} \in \mathcal{K}^{k+1}(\mathsf{A}, \mathsf{A}\mathbf{b})$, it follows that

$$\mathbf{p}^{k+1} \in \mathrm{Im}\mathsf{A} \quad \text{and} \quad (\mathbf{p}^{k+1})^T\mathsf{A}\mathbf{p}^{k+1} > 0.$$

(ii) The first part of (ii) is rather trivial. Indeed, if $\mathbf{g}_\mathsf{A}^k = \mathbf{o}$, then $\mathsf{A}\mathbf{y}^k = \mathsf{A}\mathbf{b}$. Since $\mathbf{y}^k \in \mathcal{K}^k(\mathsf{A}, \mathsf{A}\mathbf{b})$ it follows that $\mathbf{y}^k = \widehat{\mathbf{y}} = \mathsf{Q}\mathbf{b}$.

To prove the second part, let $\mathbf{g}_\mathsf{A}^k = \mathbf{o}$ and notice that $\mathbf{y}^k = \mathsf{Q}\mathbf{b}$ implies

$$\mathsf{Q}\mathbf{b} \in \mathcal{K}^k(\mathsf{A}, \mathsf{A}\mathbf{b}) \quad \text{and} \quad \mathcal{K}^k(\mathsf{A}, \mathsf{A}\mathbf{b}) = \mathcal{K}^{k+1}(\mathsf{A}, \mathsf{Q}\mathbf{b}). \tag{8.28}$$

Because $\mathbf{g}^0 \ne \mathbf{o}, \dots \mathbf{g}^{k-1} \ne \mathbf{o}$, it follows that

$$\mathcal{K}^1(\mathsf{A}, \mathsf{A}\mathbf{b}) \subsetneq \dots \subsetneq \mathcal{K}^k(\mathsf{A}, \mathsf{A}\mathbf{b}),$$

so $\mathsf{A}\mathbf{b}, \dots, \mathsf{A}^k\mathbf{b}$ is a basis of $\mathcal{K}^k(\mathsf{A}, \mathsf{A}\mathbf{b})$. Observing that the mapping

$$\tau : \mathcal{K}^k(\mathsf{A}, \mathsf{Q}\mathbf{b}) \ni \mathbf{z} \mapsto \mathsf{A}\mathbf{z} \in \mathcal{K}^k(\mathsf{A}, \mathsf{A}\mathbf{b})$$

is isomorphism, we get that $\mathsf{Q}\mathbf{b}, \mathsf{A}\mathbf{b}, \dots \mathsf{A}^{k-1}\mathbf{b}$ is a basis of $\mathcal{K}^k(\mathsf{A}, \mathsf{Q}\mathbf{b})$, so by (8.28) $\mathsf{A}^k\mathbf{b} \in \mathcal{K}^k(\mathsf{A}, \mathsf{Q}\mathbf{b})$ and

$$\mathcal{K}^k(\mathsf{A}, \mathsf{Q}\mathbf{b}) = \mathcal{K}^k(\mathsf{A}, \mathsf{A}\mathbf{b}) = \mathcal{K}^{k+1}(\mathsf{A}, \mathsf{Q}\mathbf{b}). \tag{8.29}$$

It follows that k is the *grade* of $\mathsf{Q}\mathbf{b}$ with respect to A and

$$\mathcal{K}^k(\mathsf{A}, \mathsf{A}\mathbf{b}) = \mathcal{K}(\mathsf{A}, \mathsf{Q}\mathbf{b}).$$

To finish the proof, notice that

$$\mathbf{x}^k = \arg \min_{\mathbf{x} \in \mathcal{K}^k(\mathsf{A},\mathsf{A}\mathbf{b})} f(\mathbf{x}) = \arg \min_{\mathbf{x} \in \mathcal{K}(\mathsf{A},\mathsf{Q}\mathbf{b})} f(\mathbf{x}),$$

so that

$$(\mathsf{A}\mathbf{x}^k - \mathbf{b})^T \mathbf{z} = 0, \quad \mathbf{z} \in \mathcal{K}(\mathsf{A}, \mathsf{Q}\mathbf{b}).$$

Taking

$$\mathbf{z} = \mathsf{A}\mathbf{x}^k - \mathsf{Q}\mathbf{b} = \mathsf{Q}(\mathsf{A}\mathbf{x}^k - \mathbf{b}),$$

we get

$$(\mathsf{A}\mathbf{x}^k - \mathbf{b})^T \mathsf{Q}(\mathsf{A}\mathbf{x}^k - \mathbf{b}) = \|\mathsf{A}\mathbf{x}^k - \mathsf{Q}\mathbf{b}\|^2 = 0,$$

i.e., $\mathbf{x}^k = \widehat{\mathbf{x}} = \mathsf{A}^\dagger \mathbf{b}$.

(iii) Since $\mathbf{g}_\mathsf{A}^k$ is orthogonal to $\mathcal{K}^k(\mathsf{A}, \mathsf{A}\mathbf{b})$ and belongs to $\mathcal{K}^{k+1}(\mathsf{A}, \mathsf{A}\mathbf{b})$, it follows that that $\mathbf{g}_\mathsf{A}^k = \mathbf{o}$ in at most n steps. The rest follows by (ii). □

8.4.3 Convergence and Error Bounds

Let us show that it is possible to give bounds on the error of cgSLS iterations similar to the standard estimates developed in Sect. 8.1.2. Indeed, the iterates $\mathbf{y}^k$ are just the CG iterations on the range of A for the minimal norm solution of (8.27), so that we can use the error bound (8.20) in the form

$$\|\mathbf{e}_\mathsf{A}^k\|_\mathsf{A} \le 2 \left(\frac{\sqrt{\overline{\kappa}(\mathsf{A})} - 1}{\sqrt{\overline{\kappa}(\mathsf{A})} + 1} \right)^k \|\mathbf{e}_\mathsf{A}^0\|_\mathsf{A}, \tag{8.30}$$

where

$$\mathbf{e}_\mathsf{A}^k = \mathbf{y}^k - \widehat{\mathbf{y}}, \quad \mathbf{e}_\mathsf{A}^0 = \mathbf{y}^0 - \widehat{\mathbf{y}} = -\mathsf{Q}\mathbf{b}.$$

The following theorem gives the bound on the approximation error of the iterates $\mathbf{x}^k$ and the asymptotic convergence rate.

Theorem 8.7. *Let $\mathbf{x}^k$ denote the iterations generated by cgSLS for the least-squares problem (8.1) with the solution $\widehat{\mathbf{x}}$. Then*

$$\|\mathbf{e}^{2k-1}\|_\mathsf{A} \le 6 \left(\frac{\sqrt{\overline{\kappa}(\mathsf{A})} - 1}{\sqrt{\overline{\kappa}(\mathsf{A})} + 1} \right)^k \|\mathbf{e}^0\|_\mathsf{A} \tag{8.31}$$

where

$$\mathbf{e}^k = \mathbf{x}^k - \widehat{\mathbf{x}} \quad \text{and} \quad \mathbf{e}^0 = \mathbf{x}^0 - \widehat{\mathbf{x}} = -\widehat{\mathbf{x}}.$$

Proof. We shall adapt the analysis of CG developed in Sect. 8.1. Let us denote

$$\mathbf{e} = \mathbf{e}(\mathbf{x}) = \mathbf{x} - \widehat{\mathbf{x}}, \quad \mathbf{e}^0 = -\widehat{\mathbf{x}},$$

$$\mathbf{g} = \mathbf{g}(\mathbf{x}) = \mathsf{A}\mathbf{x} - \mathsf{Q}\mathbf{b} = \mathsf{A}\mathbf{e}(\mathbf{x}), \quad \mathbf{g}^k = \mathsf{A}\mathbf{x}^k - \mathsf{Q}\mathbf{b}, \quad \mathbf{e}^0_{\mathsf{A}} = -\mathsf{Q}\mathbf{b},$$

and recall that for any $\mathbf{x} \in \mathrm{Im}\mathsf{A}$

$$f(\mathbf{x}) - f(\widehat{\mathbf{x}}) = \frac{1}{2}\|\mathbf{x} - \widehat{\mathbf{x}}\|^2_{\mathsf{A}} = \frac{1}{2}\|\mathbf{e}\|^2_{\mathsf{A}}.$$

Since cgSLS looks for the iterations in $\mathcal{K}^k(\mathsf{A}, \mathsf{A}\mathbf{b})$, it follows that they can be written in the form

$$\mathbf{x}^k = \xi_1 \mathsf{A}\mathbf{b} + \ldots + \xi_i \mathsf{A}^k \mathbf{b}.$$

Taking into account that $\mathbf{e}^0 = -\widehat{\mathbf{x}}$ and $\mathsf{A}\mathbf{b} = \mathsf{A}^2\widehat{\mathbf{x}} = -\mathsf{A}^2\mathbf{e}^0$, we get

$$\mathbf{e}^k = \xi_1 \mathsf{A}\mathbf{b} + \cdots + \xi_k \mathsf{A}^k \mathbf{b} - \widehat{\mathbf{x}} = \mathbf{e}^0 - \xi_1 \mathsf{A}^2 \mathbf{e}^0 - \cdots - \xi_k \mathsf{A}^{k+1}\mathbf{e}^0 = p_{k+1}(\mathsf{A})\mathbf{e}^0,$$

where

$$p_{k+1}(\lambda) = 1 - \xi_1\lambda^2 - \xi_2\lambda^3 - \cdots - \xi_k\lambda^{k+1} = 1 - \lambda^2(\xi_1 + \xi_2\lambda + \cdots + \xi_k\lambda^{k-1}).$$

Since the iterates $\mathbf{x}^k$ minimize $\mathbf{e}^k$ on the Krylov space $\mathcal{K}^k(\mathsf{A}, \mathsf{A}\mathbf{b})$, we get the estimate

$$\|\mathbf{e}^k\|_{\mathsf{A}} \le \min_{p \in \mathcal{P}_0^{k+1}} \max_{\lambda \in \sigma(\mathsf{A})} |p(\lambda)| \|\mathbf{e}^0\| \le \min_{p \in \mathcal{P}_0^{k+1}} \max_{\lambda \in [\lambda_{\min}, \lambda_{\max}]} |p(\lambda)| \|\mathbf{e}^0\|, \tag{8.32}$$

where $\mathcal{P}_0^k$ denotes the set of all polynomials p of the degree k that satisfy $p(0) = 1$ and have zero coefficient of the linear term. Since for any polynomial $p \in \mathcal{P}_0^{k+1}$

$$\min_{p \in \mathcal{P}_0^{k+1}} \max_{\lambda \in [\lambda_{\min}, \lambda_{\max}]} |p(\lambda)| \|\mathbf{e}^0\| \le \max_{\lambda \in [\lambda_{\min}, \lambda_{\max}]} |p(\lambda)| \|\mathbf{e}^0\|,$$

it is enough to find a good bound on any polynomial $p \in \mathcal{P}_0^{k+1}$ with small values on $[\overline{\lambda}_{\min}, \lambda_{\max}]$

We shall use the extremal properties of the Chebyshev polynomials introduced in Sect. 8.1.2 to find a practical bound. Recall that the transformed (weighted and shifted) Chebyshev polynomial

$$t_k = t_k(\lambda)$$

minimizes $\max|1 - \lambda q(\lambda)|$ for $\lambda \in [\overline{\lambda}_{\min}, \lambda_{\max}]$ and $q \in \mathcal{Q}^{k-1}$, where $\mathcal{Q}^k$ denotes the set of all polynomials of the degree k. The transformed Chebyshev polynomials are related to the Chebyshev polynomials

$$T_k(\lambda) = \cos(k \arccos(\lambda))$$

by

$$t_k = T_k\left(\frac{2\lambda - \lambda_{\max} - \overline{\lambda}_{\min}}{\lambda_{\max} - \overline{\lambda}_{\min}}\right) / T_k\left(-\frac{\lambda_{\max} + \overline{\lambda}_{\min}}{\lambda_{\max} - \overline{\lambda}_{\min}}\right)$$

and satisfy

$$|t_k(\lambda)| \le 2\left(\frac{\sqrt{\overline{\kappa}(\mathsf{A})} - 1}{\sqrt{\overline{\kappa}(\mathsf{A})} + 1}\right)^k, \quad \lambda \in [\overline{\lambda}_{\min}, \lambda_{\max}]. \tag{8.33}$$

Though $t_k \notin \mathcal{P}_0^k$, we can use Chebyschev polynomials t_k to define

$$p_{2k}(\lambda) = t_k(\lambda)\,(2 - t_k(\lambda)), \quad k = 1, 2, \dots\ . \tag{8.34}$$

The polynomials p_{2k} are small as $|t_k(\lambda)| \le 1$ and $\lambda \in [\lambda_{\min}, \lambda_{\max}]$ imply

$$|p_{2k}(\lambda)| \le 3|t_k(\lambda)|, \quad \lambda \in [\lambda_{\min}, \lambda_{\max}]. \tag{8.35}$$

To see that $p_{2k} \in \mathcal{P}_0^{2k}$, recall that $t_k(0) = 1$, so that $p_{2k} = 1$, and, t_k can be written in the form

$$t_k(\lambda) = 1 - \lambda q_{k-1}(\lambda), \quad q_{k-1} \in \mathcal{Q}^{k-1}.$$

Thus

$$p_{2k}(\lambda) = (1 - \lambda\, q_{k-1}(\lambda))\,(1 + \lambda\, q_{k-1}(\lambda)) = 1 - \lambda^2 q_{k-1}^2(\lambda).$$

To get (8.31), it is now enough to substitute (8.33) into (8.35). □

8.5 Convergence in the Presence of Rounding Errors

The elegant mathematical theory presented above assumes the implementation of the conjugate gradient algorithm in exact arithmetic. It captures the performance of only a limited number of conjugate gradient iterations in computer arithmetics. Since we will use the conjugate gradient method mainly for a low-precision approximation of well-conditioned auxiliary problems, we shall base our exposition on this theory in what follows. However, it is still helpful to be aware of the possible effects of rounding errors accompanying any computer implementation of the conjugate gradient algorithm.

It has been known since the introduction of the CG method and the Lanczos method [199], which generates the same iterates, that, when used in finite precision arithmetic, the vectors generated by these algorithms can seriously violate their theoretical properties. In particular, it has been observed that the evaluated gradients can lose their orthogonality after as small number of iterations as twenty and that the conjugate directions can become nearly dependent. Despite these effects, it has been observed that the conjugate gradient method still converges in finite precision arithmetic, but the convergence is delayed [149, 151].

We can reduce the undesirable effects of rounding errors by reorthogonalization. A simple analysis reveals that the full reorthogonalization of the gradients is costly and requires large memory. A key to an efficient implementation of the reorthogonalization uses the observation that accumulation of the rounding errors has a regular pattern, namely, that large perturbations of the generated vectors belong to the space generated by the eigenvectors of A which are approximated well by the vectors from the current Krylov space. This observation has led to the efficient implementation of the conjugate gradient method based on the *selective orthogonalization* proposed by Parlett and Scott [225]. More details can be found in the comprehensive review paper by Meurant and Strakoš [215].

8.6 Numerical Experiments

Here, we illustrate the performance of the CG algorithm and the effect of preconditioning on the solution of an ill-conditioned benchmark and a class of well-conditioned problems. We proposed the latter to resemble the problems from the multigrid or domain decomposition methods applied to the elliptic partial differential equations. The cost functions $f_{\mathrm{L},h}$ and $f_{\mathrm{LW},h}$ introduced here are used in Sects. 9.5, 10.8, and 12.9 as benchmarks for the solution of constrained problems so that we can assess additional complexity arising from the implementation of various constraints and better understand our algorithms. Moreover, using the same cost functions in our benchmarks considerably simplifies their implementation.

8.6.1 Discretized Energy Cost Benchmark

Let $\Omega = (0,1) \times (0,1)$ denote an open domain with the boundary Γ and its part $\Gamma_u = \{0\} \times [0,1]$. Let $H^1(\Omega)$ denote the Sobolev space of the first order in the space $L^2(\Omega)$ of functions on Ω whose squares are integrable in the Lebesgue sense, let

$$\mathcal{V} = \{u \in H^1(\Omega) : \; u = 0 \text{ on } \Gamma_u\},$$

and let us define for any $u \in H^1(\Omega)$

$$f_{\mathrm{L}}(u) = \frac{1}{2} \int_{\Omega} \|\nabla u(x)\|^2 \mathrm{d}\Omega + \int_{\Omega} u \mathrm{d}\Omega.$$

Thus, we can define the continuous problem to find

$$\min_{u \in \mathcal{V}} f_{\mathrm{L}}(u). \tag{8.36}$$

Our ill-conditioned benchmark was obtained from (8.36) by the finite element discretization using a regular grid defined by the discretization parameter h and linear elements. The Dirichlet conditions were enhanced into the Hessian $\mathsf{A}_{\mathrm{L},h}$ of the discretized cost function $f_{\mathrm{L},h}$, so that $\mathsf{A}_{\mathrm{L},h} \in \mathbb{R}^{n \times n}$ is positive definite, $n = p(p-1)$, and $p = 1/h + 1$. Moreover, $\mathsf{A}_{\mathrm{L},h}$ is known to be ill-conditioned with the condition number $\kappa(\mathsf{A}_{\mathrm{L},h}) \approx h^{-2}$. We performed the computations with $h = 1/32$, corresponding to $n = 1056$ unknowns.

8.6.2 Basic CG and Preconditioning

We used the benchmark to compare the performance of CG, CG with SSOR preconditioning, and CG with preconditioning by the conjugate projector. To define the conjugate projector, we decomposed the domain into 4×4 squares with typically 8×8 variables aggregated by the matrix U with 16 columns.

The graph of the norm of the gradient (vertical axis) against the number of iterations for the basic CG algorithm (CG), the CG algorithm with SSOR preconditioning (CG–SSOR), and the CG algorithm with preconditioning by the conjugate projector (CG–CP) is in Fig. 8.2. We can see that though the performance of the CG algorithm is poor if the Hessian of the cost function is ill-conditioned, we can considerably improve the convergence rate by preconditioning.

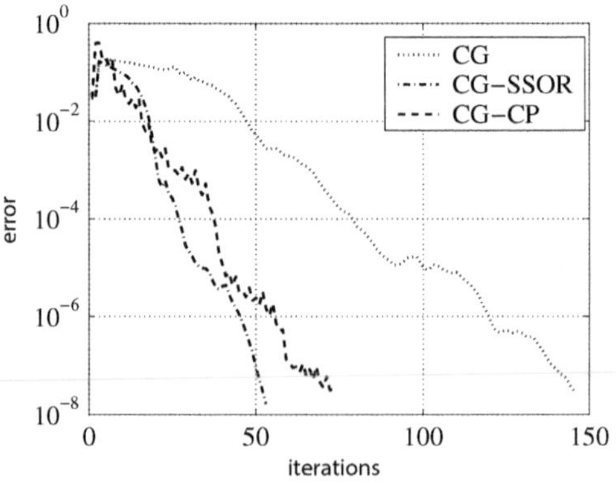

Fig. 8.2. Convergence of CG, CG–SSOR, and CG–CP algorithms

8.6.3 Numerical Demonstration of Optimality

To illustrate the concept of optimality, let us consider the class of problems to minimize

$$f_{\mathrm{LW},h}(\mathbf{x}) = \frac{1}{2}\mathbf{x}^T \mathsf{A}_{\mathrm{LW},h}\mathbf{x} - \mathbf{b}_{\mathrm{LW},h}^T \mathbf{x},$$

where

$$\mathsf{A}_{\mathrm{LW},h} = \mathsf{A}_{\mathrm{L},h} + 2\mathsf{I}, \quad [\mathbf{b}_{\mathrm{LW},h}]_i = -1, \quad i = 1, \dots, n, \quad n = 1/h + 1.$$

There is a mechanical interpretation of the problem that is associated with the expanding spring systems on Winkler's foundation. Using Gershgorin's theorem, it can be proved that the spectrum of the Hessian $\mathsf{A}_{\mathrm{LW},h}$ of $f_{\mathrm{LW},h}$ is located in the interval [4, 14], so that $\kappa(\mathsf{A}_{\mathrm{LW},h}) \le 5$.

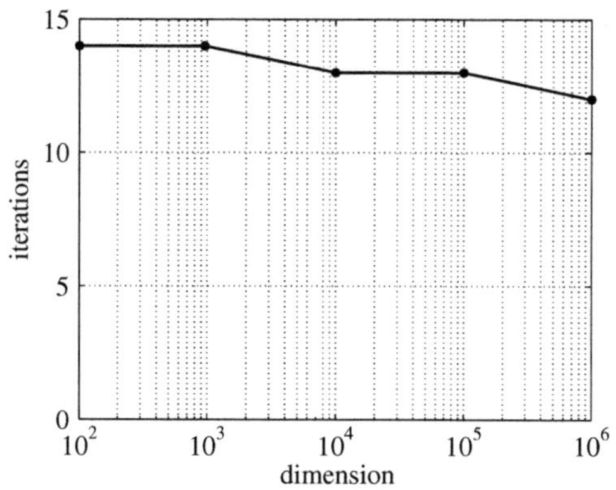

Fig. 8.3. Optimality of CG for a class of well-conditioned problems

In Fig. 8.3, we can see the numbers of CG iterations k_n (vertical axis) that were necessary to reduce the norm of the gradient by 10^{-6} for the problems with the dimension n ranging from 100 to 1000000. The dimension n on the horizontal axis is in the logarithmic scale. We can see that k_n varies mildly with varying n, in agreement with the theory developed in Sect. 8.1. Moreover, since the cost of the matrix–vector multiplications is, in our case, proportional to the dimension n of the matrix $\mathsf{A}_{\mathrm{LW},h}$, it follows that the cost of the solution is also proportional to n.

8.6.4 Performance of cgSLS

We tested the cgSLS algorithm on the least-squares solution of the Poisson equation with the inconsistent Neumann boundary condition given by

$$\begin{array}{rll} -\Delta u = 0 & \text{on } \Omega \ , \\ \frac{\partial u}{\partial n} = q & \text{on } \Gamma, \end{array} \tag{8.37}$$

where $\Omega = [0,1] \times [0,1]$ and

$$q(x) = \begin{cases} \sin(\pi x_1) & \text{for } x = (x_1, x_2) \in \mathbb{R}^2, \ 0 \le x_1 \le 1 \text{ and } \mathrm{x}_2 = 0, \\ 0 & \text{elsewhere.} \end{cases}$$

The function q is chosen so that it *does not satisfy* the compatibility condition

$$\int_\Gamma q \mathrm{d}\Gamma = 0.$$

For the discretization, we used the finite element method with linear elements defined on 100×100 grid.

We compared the performance of cgSLS with the performance of the standard CG algorithm applied to the nearest feasible problem (8.23) (results are labeled CGQ) and to the modified problem (8.24) (results are labeled CGLS). We resolved the problem with cgSLS, CGLS, and CGQ algorithms and drew the relative energetic errors in Fig. 8.4. The results confirm that cgSLS outperforms CGLS and that the performance of cgSLS copies that of CGQ with some delay. The latter observation is in agreement with the estimates provided by Theorem 8.7, which gives the same bounds on the asymptotic rate of convergence of cgSLS and CGQ but provides larger error bounds for the iterations of cgSLS in the early stage of computations.

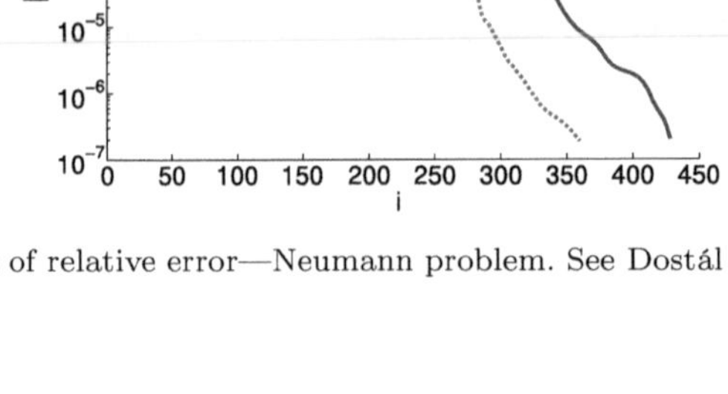

Fig. 8.4. Decrease of relative error—Neumann problem. See Dostál and Pospíšil [84]

8.7 Comments and References

The proof of (8.4) appeared in Daniel [46]. Due to the importance of the conjugate gradient method and the beauty of its analysis, there are many references concerning this subject. We refer an interested reader to the textbooks and research monographs by Saad [240], van der Vorst [259], Greenbaum [150], Hackbusch [156], Chen [33], and Axelsson [6] for more information. A comprehensive account of the development of the CG method up to 1989 is in the paper by Golub and O'Leary [146].

Preconditioning by conjugate projector presented in Sect. 8.3 was introduced by Dostál [60]. The same preconditioning with different analyses was presented independently by Marchuk and Kuznetsov [213] as the *conjugate gradients in subspace* or the *generalized conjugate gradient method* and by Nicolaides [219] as the *deflation method.* The preconditioning by the conjugate projector is an effective tool for preconditioning auxiliary linear problems arising in the active set strategy for solving problems with bound constraints.

Finding at each step the minimum over the subspace generated by all the previous search directions, the conjugate gradient method exploits all the information gathered during the earlier iterations. The practical application assumes long, uninterrupted sequences of the conjugate gradient iterations. This strategy also supports using yet another unique feature of the conjugate gradient method: its self-preconditioning capabilities described by van der Sluis and van der Vorst [247]. The latter property can also be described in terms of the preconditioning by the conjugate projector. Indeed, if Q_k denotes the conjugate projector onto the conjugate complement $\mathcal{V}$ of $\mathcal{U} = \mathrm{Span}\{\mathbf{p}_1, \ldots, \mathbf{p}_k\}$, then it is possible to give the bound on the rate of convergence of the conjugate gradient method starting from $\mathbf{x}_{k+1}$ in terms of the regular condition number $\overline{\kappa}_k = \overline{\kappa}(\mathsf{Q}_k^T\mathsf{A}\mathsf{Q}_k|\mathcal{V})$ of $\mathsf{Q}_k^T\mathsf{A}\mathsf{Q}_k|\mathcal{V}$ and observe that $\overline{\kappa}_k$ decreases with the increasing k.

A modification of CG for the least square problems exploiting the Krylov spaces associated with the image of A was proposed by Calvetti et al. [31]. The variant of CG which generates a conjugate basis of $\mathrm{Im}\mathsf{A}$ by the standard CG and uses it to get both the approximations of the solution of (8.1) and orthogonal projection $\mathsf{Q}\mathbf{b}$ of $\mathbf{b}$ on $\mathrm{Im}\mathsf{A}$ appeared with analysis in Dostál and Pospíšil [84].

For solving large problems, the basic CG algorithm is most efficient when combined with the preconditioning, which exploits additional information about A, often obtained by tracing its generation. Thus the *multigrid* (see, e.g., Hackbusch [155] or Krause and Margenov [191])-based preconditioners for solving problems arising from the discretization of elliptic partial differential equations exploit the information about the original continuous problems so efficiently that they can solve the discretized problems at a cost proportional to the number of unknowns.

The domain decomposition methods based on *FETI* (finite element tearing and interconnecting, see, e.g., Farhat et al. [125], or Toselli and Widlund [256]) share the asymptotically linear complexity with multigrid methods and enable effective exploitation of massively parallel facilities of modern supercomputers. Special preconditioners for singular or nearly singular systems arising in optimization were proposed, e.g., by Hager [160].

Notice that the gradient algorithms in Chap. 3 generate the iterates in the same Krylov space as the conjugate gradient method presented here so that they cannot be faster in exact arithmetic than CG. However, the new insights and numerical experiments (see, e.g., Zou and Magoules [271] or Oviedo et al. [223]) suggest that it is possible to exploit their robustness to develop competitive algorithms based on gradient methods.

9
SMALE for Equality Constraints

We shall now be interested in optimal algorithms for the problem

$$\min_{\mathbf{x}\in\Omega_E} f(\mathbf{x}), \tag{9.1}$$

where $f(\mathbf{x}) = \frac{1}{2}\mathbf{x}^T\mathsf{A}\mathbf{x} - \mathbf{x}^T\mathbf{b}$, $\mathbf{b}$ is a given column n-vector, A is an $n \times n$ SPD matrix, $\Omega_E = \{\mathbf{x} \in \mathbb{R}^n : \mathsf{B}\mathbf{x} = \mathbf{c}\}$, $\mathsf{B} \in \mathbb{R}^{m\times n}$, and $\mathbf{c} \in \mathrm{Im}\mathsf{B}$. Unlike Chap. 6, we assume that A is large but reasonably conditioned so that iterative methods can find approximate solutions to auxiliary problems at a reasonable cost. We assume that $\mathsf{B} \neq \mathsf{O}$ is not a full column rank matrix so that $\mathrm{Ker}\mathsf{B} \neq \{\mathbf{o}\}$, and we allow dependent rows of B. There are at least two reasons why we consider the constraint matrix B with dependent rows:

- For large problems, it may be expensive to identify dependent rows of B.
- The removal of dependent constraints can complicate the precision control of the removed constraints.

Here, we present two variants of the augmented Lagrangian method that accept the inexact solution of auxiliary unconstrained problems.

The *asymptotically exact augmented Lagrangian method* of Sect. 9.1 controls the precision of the solution of auxiliary problems by a user-defined forcing sequence. The method is a standard technique applicable to more general nonlinear problems.

In Sect. 9.2, we describe more sophisticated *semimonotonic augmented Lagrangian method* for equality constraints called SMALE. The unique theoretical results concerning SMALE include a bound on the number of iterations necessary to find an approximate solution with prescribed error and independent of the conditioning of constraints. The performance of SMALE can be improved by preconditioning that preserves the gap in the spectrum (see Sect. 9.3).

Z. Dostál, *Optimal Quadratic Programming and QCQP Algorithms with Case Studies*, Springer Optimization and Its Applications 23,
https://doi.org/10.1007/978-3-031-95167-1_9

9.1 Asymptotically Exact Augmented Lagrangian Method

The augmented Lagrangian method considered in Chap. 6 assumed that the minimization in Step 1 is carried out exactly. Since such iterations are expensive, there is a good chance to reduce the cost of the outer iterations without significantly increasing the number of iterations due to the approximate minimization, especially when we recall that the gradient is a robust ascent direction.

In this section, we analyze the convergence of the augmented Lagrangian algorithm controlling the precision of the solutions of the auxiliary problems in Step 1 by the gradient norm. We assume that the bounds are prescribed by the *forcing sequence* $\{\varepsilon_k\}$, where $\varepsilon_k > 0$ and $\lim_{k\to\infty} \varepsilon_k = 0$. The latter condition implies that the stopping criterion becomes more stringent with the increasing index of the outer iterations, so the minimization turns asymptotically exact. Taking into account the discussion of Sect. 6.3.3, we consider the step length $r_k = \varrho_k$.

9.1.1 Algorithm

The augmented Lagrangian algorithm with asymptotically exact solutions of auxiliary unconstrained problems differs from the exact algorithm only in Step 1. We first restrict our attention to the inexact version of the original augmented Lagrangian method, which reads as follows.

Algorithm 9.1. Asymptotically exact augmented Lagrangians

Given a symmetric positive definite matrix $\mathsf{A} \in \mathbb{R}^{n\times n}$, $\mathsf{B} \in \mathbb{R}^{m\times n}$, $\mathbf{b} \in \mathbb{R}^n$, *and* $\mathbf{c} \in \mathrm{Im}\mathsf{B}$.

Step 0. {*Initialization.*}
    Choose $\varepsilon_i > 0$ *so that* $\lim_{i\to\infty} \varepsilon_i = 0$, $\boldsymbol{\lambda}^0 \in \mathbb{R}^m$, $\varrho_i \geq \varrho > 0$
    for *k=0,1,2,...*
Step 1. {*Minimization with respect to* $\mathbf{x}$.}
    Choose $\mathbf{x}^k \in \mathbb{R}^n$ *so that* $\|\nabla_{\mathbf{x}} L(\mathbf{x}^k, \boldsymbol{\lambda}^k, \varrho_k)\| \leq \varepsilon_k$
Step 2. {*Updating the Lagrange multipliers.*}
    $\boldsymbol{\lambda}^{k+1} = \boldsymbol{\lambda}^k + \varrho_k(\mathsf{B}\mathbf{x}^k - \mathbf{c})$
    end for

We assume that the inexact solutions of the auxiliary problems in Step 1 of Algorithm 9.1 are implemented by a suitable iterative method, such as the conjugate gradient method introduced in Chap. 8. Thus, the algorithm solves approximately the auxiliary unconstrained problems in the inner loop while it generates the approximations of the Lagrange multipliers in the outer loop. Let us recall that the effective application of the conjugate gradient method assumes a favorable distribution of the spectrum of the Hessian matrix A. The letter can be achieved by a problem-dependent preconditioning discussed in Sect. 8.2 in combination with the gap-preserving strategy of Sect. 9.4.

9.1.2 Auxiliary Estimates

Our analysis of the augmented Lagrangian algorithm is based on the following lemma.

Lemma 9.1. *Let* $\mathsf{A}, \mathsf{B}, \mathbf{b}$, *and* $\mathbf{c}$ *be those of the definition of problem (9.1) with* $\mathsf{B} \neq \mathsf{O}$ *not necessarily a full rank matrix. For any vectors* $\mathbf{x} \in \mathbb{R}^n$ *and* $\boldsymbol{\lambda} \in \mathbb{R}^m$, *let us denote*

$$\widetilde{\boldsymbol{\lambda}} = \boldsymbol{\lambda} + \varrho(\mathsf{B}\mathbf{x} - \mathbf{c}),$$
$$\mathbf{g} = \nabla_{\mathbf{x}} L(\mathbf{x}, \boldsymbol{\lambda}, \varrho) = \mathsf{A}_\varrho \mathbf{x} - \mathbf{b} + \mathsf{B}^T \boldsymbol{\lambda} - \varrho \mathsf{B}^T \mathbf{c}.$$

Let $\boldsymbol{\lambda}_{\mathrm{LS}}$ *denote the vector of the least square Lagrange multipliers for problem (9.1), and let* $\overline{\beta}_{\min}$ *denote the least nonzero eigenvalue of* $\mathsf{B}\mathsf{A}^{-1}\mathsf{B}$.

Then for any $\boldsymbol{\lambda} \in \mathrm{Im}\mathsf{B}$

$$\|\widetilde{\boldsymbol{\lambda}} - \boldsymbol{\lambda}_{\mathrm{LS}}\| \le \frac{\|\mathsf{B}\mathsf{A}^{-1}\|}{\overline{\beta}_{\min} + \varrho^{-1}} \|\mathbf{g}\| + \frac{\varrho^{-1}}{\overline{\beta}_{\min} + \varrho^{-1}} \|\boldsymbol{\lambda} - \boldsymbol{\lambda}_{\mathrm{LS}}\|. \tag{9.2}$$

Proof. The definitions of $\widetilde{\boldsymbol{\lambda}}$ and $\mathbf{g}$ imply that

$$\begin{aligned} \mathsf{A}\mathbf{x} + \mathsf{B}^T \widetilde{\boldsymbol{\lambda}} &= \mathbf{b} + \mathbf{g}, \\ \mathsf{B}\mathbf{x} - \varrho^{-1}\widetilde{\boldsymbol{\lambda}} &= -\varrho^{-1}(\boldsymbol{\lambda} - \boldsymbol{\lambda}_{\mathrm{LS}}) - \varrho^{-1}\boldsymbol{\lambda}_{\mathrm{LS}} + \mathbf{c}, \end{aligned} \tag{9.3}$$

and the solution $\widehat{\mathbf{x}}$ and $\boldsymbol{\lambda}_{\mathrm{LS}}$ satisfy by the assumptions

$$\begin{aligned} \mathsf{A}\widehat{\mathbf{x}} + \mathsf{B}^T \boldsymbol{\lambda}_{\mathrm{LS}} &= \mathbf{b}, \\ \mathsf{B}\widehat{\mathbf{x}} - \varrho^{-1}\boldsymbol{\lambda}_{\mathrm{LS}} &= -\varrho^{-1}\boldsymbol{\lambda}_{\mathrm{LS}} + \mathbf{c}. \end{aligned} \tag{9.4}$$

Subtracting (9.4) from (9.3) and switching to the matrix notation, we get

$$\begin{bmatrix} \mathsf{A} & \mathsf{B}^T \\ \mathsf{B} & -\varrho^{-1}\mathsf{I} \end{bmatrix} \begin{bmatrix} \mathbf{x} - \widehat{\mathbf{x}} \\ \widetilde{\boldsymbol{\lambda}} - \boldsymbol{\lambda}_{\mathrm{LS}} \end{bmatrix} = \begin{bmatrix} \mathbf{g} \\ -\varrho^{-1}(\boldsymbol{\lambda} - \boldsymbol{\lambda}_{\mathrm{LS}}) \end{bmatrix}. \tag{9.5}$$

After multiplying the first block row of (9.5) by $-\mathsf{BA}^{-1}$, adding the result to the second block row, and simple manipulations, we get

$$\widetilde{\boldsymbol{\lambda}} - \boldsymbol{\lambda}_{\mathrm{LS}} = \mathsf{S}_\varrho^{-1}\mathsf{BA}^{-1}\mathbf{g} + \varrho^{-1}\mathsf{S}_\varrho^{-1}(\boldsymbol{\lambda} - \boldsymbol{\lambda}_{\mathrm{LS}}), \tag{9.6}$$

where

$$\mathsf{S}_\varrho = \mathsf{BA}^{-1}\mathsf{B}^T + \varrho^{-1}\mathsf{I}.$$

Noticing that $\boldsymbol{\lambda}_{\mathrm{LS}} - \boldsymbol{\lambda} \in \mathrm{Im}\mathsf{B}$ and taking norms, we get

$$\|\widetilde{\boldsymbol{\lambda}} - \boldsymbol{\lambda}\| \le \|\mathsf{S}_\varrho^{-1}|\mathrm{Im}\mathsf{B}\|(\|\mathsf{BA}^{-1}\|\|\mathbf{g}\| + \varrho^{-1}\|\boldsymbol{\lambda}_{\mathrm{LS}} - \boldsymbol{\lambda}\|). \tag{9.7}$$

To estimate the first factor on the right-hand side, notice that by (1.51) our task reduces to finding an upper bound for

$$\|(\varrho^{-1}\mathsf{I} + \mathsf{BA}^{-1}\mathsf{B}^T)^{-1}|\mathrm{Im}\mathsf{BA}^{-1}\mathsf{B}^{\mathrm{T}}\|.$$

Since $\mathrm{Im}\mathsf{B}^{\mathrm{T}}\mathsf{A}^{-1}\mathsf{B}$ is an invariant subspace of any matrix function of $\mathsf{B}^T\mathsf{A}^{-1}\mathsf{B}$, and the eigenvectors of $\mathsf{BA}^{-1}\mathsf{B}^T|\mathrm{Im}\mathsf{BA}^{-1}\mathsf{B}^{\mathrm{T}}$ are just the eigenvectors of $\mathsf{BA}^{-1}\mathsf{B}^T$ which correspond to the nonzero eigenvalues, it follows by (1.30) and (1.32) that

$$\|\mathsf{S}_\varrho^{-1}|\mathrm{Im}\mathsf{B}\| = \|(\varrho^{-1}\mathsf{I} + \mathsf{BA}^{-1}\mathsf{B}^{\mathrm{T}})^{-1}|\mathrm{Im}\mathsf{B}\| = 1/(\varrho^{-1} + \overline{\beta}_{\min}).$$

After substituting into (9.7), we get (9.2). □

To simplify applications of Lemma 9.1 for $\boldsymbol{\lambda}^0 \notin \mathrm{Im}\mathsf{B}$, let us formulate the following easy lemma.

Lemma 9.2. *Let* $\boldsymbol{\lambda}^0 \in \mathbb{R}^m$, *let*

$$\boldsymbol{\lambda}^{k+1} = \boldsymbol{\lambda}^k + \mathbf{u}^k, \quad \mathbf{u}^k \in \mathrm{Im}\mathsf{B}, \quad k = 0, 1, \ldots,$$

let $\boldsymbol{\lambda}_{\mathrm{LS}}$ *denote the vector of the least square Lagrange multipliers for (9.1), so that* $\boldsymbol{\lambda}_{\mathrm{LS}} \in \mathrm{Im}\mathsf{B}$, *let* P *denote the orthogonal projector onto* $\mathrm{Im}\mathsf{B}$, *and let*

$$\overline{\boldsymbol{\lambda}} = \boldsymbol{\lambda}_{\mathrm{LS}} + (\mathsf{I} - \mathsf{P})\boldsymbol{\lambda}^0. \tag{9.8}$$

Then

$$\boldsymbol{\lambda}^k - \overline{\boldsymbol{\lambda}} = \mathsf{P}\boldsymbol{\lambda}^k - \boldsymbol{\lambda}_{\mathrm{LS}}, \quad k = 0, 1, \ldots. \tag{9.9}$$

Proof. Since for $k = 0, 1, \ldots$

$$(\mathsf{I} - \mathsf{P})\boldsymbol{\lambda}^{k+1} = (\mathsf{I} - \mathsf{P})(\boldsymbol{\lambda}^k + \mathbf{u}^k) = (\mathsf{I} - \mathsf{P})\boldsymbol{\lambda}^k = \cdots = (\mathsf{I} - \mathsf{P})\boldsymbol{\lambda}^0,$$

we have

$$\boldsymbol{\lambda}^k - \overline{\boldsymbol{\lambda}} = \mathsf{P}\boldsymbol{\lambda}^k + (\mathsf{I} - \mathsf{P})\boldsymbol{\lambda}^k - \boldsymbol{\lambda}_{\mathrm{LS}} - (\mathsf{I} - \mathsf{P})\boldsymbol{\lambda}^0 = \mathsf{P}\boldsymbol{\lambda}^k - \boldsymbol{\lambda}_{\mathrm{LS}}.$$

□

9.1.3 Convergence Analysis

Now we can use Lemma 9.1 in the proof of convergence of Algorithm 9.1.

Theorem 9.1. *Let* $\mathbf{x}^k, \boldsymbol{\lambda}^k, k = 0, 1, \ldots$, *be generated by Algorithm 9.1 for the solution of* (9.1) *with given* $\boldsymbol{\lambda}^0 \in \mathbb{R}^m$, $\varrho > 0, \varrho_k \geq \varrho$, *and* $\varepsilon_k > 0$ *such that* $\lim_{k\to\infty} \varepsilon_k = 0$. *Let* $(\widehat{\mathbf{x}}, \boldsymbol{\lambda}_{\mathrm{LS}})$ *denote the least square KKT pair for (9.1). Let* P *denote the orthogonal projector on* $\mathrm{Im}\mathsf{B}$, *let* $\overline{\beta}_{\min}$ *denote the least nonzero eigenvalue of* $\mathsf{B}\mathsf{A}^{-1}\mathsf{B}^T$, *let* $\boldsymbol{\lambda}_{\min}$ *denote the least eigenvalue of* A, *and denote*

$$\overline{\boldsymbol{\lambda}} = \mathsf{P}\boldsymbol{\lambda}_{\mathrm{LS}} + (\mathsf{I} - \mathsf{P})\boldsymbol{\lambda}^0 = \boldsymbol{\lambda}_{\mathrm{LS}} + (\mathsf{I} - \mathsf{P})\boldsymbol{\lambda}^0.$$

Then

$$\lim_{i\to\infty} \mathbf{x}^k = \widehat{\mathbf{x}}, \qquad \lim_{i\to\infty} \boldsymbol{\lambda}^k = \overline{\boldsymbol{\lambda}},$$

and for any positive integers k, s, $k + s = i$,

$$\|\boldsymbol{\lambda}^{k+s} - \overline{\boldsymbol{\lambda}}\| \leq C\overline{\varepsilon}_k \frac{1}{1-\nu} + \nu^s C\overline{\varepsilon}_0 \frac{1}{1-\nu} + \nu^{k+s}\|\boldsymbol{\lambda}^0 - \overline{\boldsymbol{\lambda}}\|, \tag{9.10}$$

$$\|\mathbf{x}^i - \widehat{\mathbf{x}}\| \leq \lambda_{\min}^{-1}\|\mathsf{B}\|(\|\boldsymbol{\lambda}^i - \overline{\boldsymbol{\lambda}}\| + \varepsilon_i), \tag{9.11}$$

where

$$\overline{\varepsilon}_k = \max\{\varepsilon_k, \varepsilon_{k+1}, \ldots\},\ C = \frac{\|\mathsf{B}\mathsf{A}^{-1}\|}{\overline{\beta}_{\min} + \varrho^{-1}},\ \textit{and}\ \nu = \frac{\varrho^{-1}}{\overline{\beta}_{\min} + \varrho^{-1}} < 1. \tag{9.12}$$

Proof. First notice that by the assumptions

$$(\mathsf{A} + \varrho_k \mathsf{B}^T\mathsf{B})\mathbf{x}^k = \mathbf{b} + \varrho_k \mathsf{B}^T\mathbf{c} - \mathsf{B}^T\boldsymbol{\lambda}^k + \mathbf{g}^k, \quad \|\mathbf{g}^k\| \leq \varepsilon_k, \tag{9.13}$$

where $\mathbf{g}^k = \nabla_{\mathbf{x}} L(\mathbf{x}^k, \boldsymbol{\lambda}^k, \varrho_k)$, and observe that the update rule in Step 2 of Algorithm 9.1 and Lemma 9.2 with $\mathbf{u}^k = \varrho_k(\mathsf{B}\mathbf{x}^k - \mathbf{c})$ imply that

$$\boldsymbol{\lambda}^k - \overline{\boldsymbol{\lambda}} = \mathsf{P}\boldsymbol{\lambda}^k - \boldsymbol{\lambda}_{\mathrm{LS}},\ k = 0, 1, \ldots.$$

Since

$$\mathsf{P}\boldsymbol{\lambda}^{k+1} = \mathsf{P}\boldsymbol{\lambda}^k + \varrho_k(\mathsf{B}\mathbf{x}^k - \mathbf{c})$$

and $\mathsf{P}\boldsymbol{\lambda}^k \in \mathrm{Im}\mathsf{B}$, we can apply Lemma 9.1 with

$$\boldsymbol{\lambda} = \mathsf{P}\boldsymbol{\lambda}^k \quad \text{and} \quad \widetilde{\boldsymbol{\lambda}} = \mathsf{P}\boldsymbol{\lambda}^k + \varrho_k(\mathsf{B}\mathbf{x}^k - \mathbf{c}) = \mathsf{P}\boldsymbol{\lambda}^{k+1}$$

and use the assumptions to get

$$\begin{aligned}\|\boldsymbol{\lambda}^{k+1} - \overline{\boldsymbol{\lambda}}\| = \|\mathsf{P}\boldsymbol{\lambda}^{k+1} - \boldsymbol{\lambda}_{\mathrm{LS}}\| &\leq \frac{\|\mathsf{B}\mathsf{A}^{-1}\|}{\overline{\beta}_{\min} + \varrho_k^{-1}}\|\mathbf{g}^k\| + \frac{\varrho_k^{-1}}{\overline{\beta}_{\min} + \varrho_k^{-1}}\|\mathsf{P}\boldsymbol{\lambda}^k - \boldsymbol{\lambda}_{\mathrm{LS}}\| \\ &\leq C\varepsilon_k + \nu\|\boldsymbol{\lambda}^k - \overline{\boldsymbol{\lambda}}\|,\end{aligned}$$

where C and ν are defined by (9.12).

It follows that for any positive integer s and $k = 0, 1, \ldots$, we have

$$\begin{aligned}\|\boldsymbol{\lambda}^{k+s} - \overline{\boldsymbol{\lambda}}\| &\le C\varepsilon_{k+s-1} + \nu\|\boldsymbol{\lambda}^{k+s-1} - \overline{\boldsymbol{\lambda}}\| \\ &\le C(\varepsilon_{k+s-1} + \nu\varepsilon_{k+s-2} + \cdots + \nu^{s-1}\varepsilon_k) + \nu^s\|\boldsymbol{\lambda}^k - \overline{\boldsymbol{\lambda}}\| \\ &\le C\overline{\varepsilon}_k(1 + \nu + \nu^2 + \cdots + \nu^{s-1}) + \nu^s\|\boldsymbol{\lambda}^k - \overline{\boldsymbol{\lambda}}\| \\ &\le C\overline{\varepsilon}_k\frac{1}{1-\nu} + \nu^s\|\boldsymbol{\lambda}^k - \overline{\boldsymbol{\lambda}}\|.\end{aligned}$$

To prove (9.10), it is enough to use the above inequalities to bound the last term by

$$\|\boldsymbol{\lambda}^k - \overline{\boldsymbol{\lambda}}\| = \|\boldsymbol{\lambda}^{0+k} - \overline{\boldsymbol{\lambda}}\| \le C\overline{\varepsilon}_0\frac{1}{1-\nu} + \nu^k\|\boldsymbol{\lambda}^0 - \overline{\boldsymbol{\lambda}}\|.$$

Observing that any large integer may be expressed as a sum of two large integers, and that $\overline{\varepsilon}_k$ converges to zero, we conclude that $\boldsymbol{\lambda}^k$ converges to $\overline{\boldsymbol{\lambda}}$.

To prove the convergence of the primal variables, denote $\mathsf{A}_{\varrho_k} = \mathsf{A} + \varrho_k\mathsf{B}^T\mathsf{B}$ and observe that

$$\mathsf{A}_{\varrho_k}\widehat{\mathbf{x}} = \mathbf{b} + \varrho_k\mathsf{B}^T\mathbf{c} - \mathsf{B}^T\overline{\boldsymbol{\lambda}}.$$

After subtracting the last equation from (9.13) and simple manipulations, we get

$$\mathbf{x}^k - \widehat{\mathbf{x}} = \mathsf{A}_{\varrho_k}^{-1}\mathsf{B}^T(\overline{\boldsymbol{\lambda}} - \boldsymbol{\lambda}^k + \mathbf{g}^k).$$

Taking norms, using the properties of norms, the assumptions, and

$$\|\mathsf{A}_{\varrho_k}^{-1}\| \le \lambda_{\min}^{-1},$$

we get (9.11). It follows by assumptions that $\mathbf{x}^k$ converges to $\widehat{\mathbf{x}}$. □

The analysis of the asymptotically exact augmented Lagrangian algorithm for more general equality-constrained problems may be found in Chap. 2 of Bertsekas [15].

9.2 Semimonotonic Augmented Lagrangians (SMALE)

The analysis of the previous section reveals that it is possible to use an inexact solution of the auxiliary problems in Step 1 of the augmented Lagrangian algorithm. However, the terms related to the precision in the inequalities (9.10) and (9.11) indicate that the convergence can be slow if the precision control is relaxed. The price paid for the inexact minimization is an additional term in the convergence rate estimate. Moreover, this term depends on the distribution of the singular values of B.

Is it possible to propose an inexact algorithm with any reasonable kind of convergence *independent of the constraint matrix* B? A key to a positive answer is to view the augmented Lagrangian algorithm as a repeated application of the penalty method with the update of linear term.

9.2.1 SMALE Algorithm

The algorithm presented here implements the observations of Sect. 6.3. First, having a minimizer $\mathbf{x}^k$ of $L(\mathbf{x}, \boldsymbol{\lambda}^k, \varrho)$ subject to $\mathbf{x}$, we can define $\boldsymbol{\lambda}^{k+1}$ in such a way that $\mathbf{x}$ is also an unconstrained minimizer of $L(\mathbf{x}, \boldsymbol{\lambda}^{k+1}, 0)$. The new iterate $\mathbf{x}^{k+1}$ is a minimizer of $L(\mathbf{x}, \boldsymbol{\lambda}^{k+1}, \varrho)$. In this way, the exact augmented Lagrangian method generates increasing values of $L(\mathbf{x}^k, \boldsymbol{\lambda}^k, \varrho)$, see (6.21).

The algorithm presented here tries to implement these observations with inexact solutions to auxiliary problems. We introduce a balancing parameter M_k to control the precision of auxiliary problems and update M_k when the Lagrangian does not increase. We shall show that there is a relatively large lower bound on the balancing parameter M_k. Most importantly, this bound is independent of the conditioning of the constraint matrix B. The complete algorithm reads as follows.

Algorithm 9.2. Semimonotonic augmented Lagrangians (SMALE-M)

Given a symmetric positive definite matrix $\mathsf{A} \in \mathbb{R}^{n\times n}$, $\mathsf{B} \in \mathbb{R}^{m\times n}$, $\mathbf{b} \in \mathbb{R}^n$, and $\mathbf{c} \in \mathrm{Im}\mathsf{B}$.

Step 0. {*Initialization.*}
Choose $\eta > 0$, $0 < \beta < 1$, $M_0 > 0$, $\varrho > 0$, $\boldsymbol{\lambda}^0 \in \mathbb{R}^m$
for k=0,1,2,...

Step 1. {*Inner iteration with adaptive precision control.*}
Find $\mathbf{x}^k$ such that

$$\|\mathbf{g}(\mathbf{x}^k, \boldsymbol{\lambda}^k, \varrho)\| \le \min\{M_k\|\mathsf{B}\mathbf{x}^k - \mathbf{c}\|, \eta\}. \tag{9.14}$$

Step 2. {*Updating the Lagrange multipliers.*}

$$\boldsymbol{\lambda}^{k+1} = \boldsymbol{\lambda}^k + \varrho(\mathsf{B}\mathbf{x}^k - \mathbf{c}) \tag{9.15}$$

Step 3. {*Reduce M_k provided the increase of Lagrangian is insufficient.*}
if $k > 0$ *and*

$$L(\mathbf{x}^k, \boldsymbol{\lambda}^k, \varrho) < L(\mathbf{x}^{k-1}, \boldsymbol{\lambda}^{k-1}, \varrho) + \frac{\varrho}{2}\|\mathsf{B}\mathbf{x}^k - \mathbf{c}\|^2 \tag{9.16}$$

$M_{k+1} = \beta M_k$
else
$M_{k+1} = M_k$.
end if
end for

In Step 1 we can use any convergent algorithm for the minimization of the strictly convex quadratic function such as the preconditioned conjugate gradient method of Sect. 4.3. The proof that Algorithm 9.2 is well defined is based on the following lemma.

Lemma 9.3. *Let $M > 0$, $\boldsymbol{\lambda} \in \mathbb{R}^m$, and $\varrho \geq 0$ be given and let $\{\mathbf{y}^k\}$ denote any sequence that converges to the unique solution $\widehat{\mathbf{y}}$ of the problem*

$$\min_{\mathbf{y} \in \mathbb{R}^n} L(\mathbf{y}, \boldsymbol{\lambda}, \varrho). \tag{9.17}$$

Then $\{\mathbf{y}^k\}$ either converges to the solution $\widehat{\mathbf{x}}$ of problem (9.1), or there is an index k such that

$$\|\nabla L(\mathbf{y}^k, \boldsymbol{\lambda}, \varrho)\| \leq M \|\mathsf{B}\mathbf{y}^k - \mathbf{c}\|. \tag{9.18}$$

Proof. First observe that $\nabla L(\mathbf{y}^k, \boldsymbol{\lambda}, \varrho)$ converges to zero by the assumptions. Thus if (9.18) does not hold for any k, then we must have $\mathsf{B}\widehat{\mathbf{y}} = \mathbf{c}$. In this case, since $\widehat{\mathbf{y}}$ is the solution of (9.17), it follows that

$$\mathsf{A}\widehat{\mathbf{y}} - \mathbf{b} + \mathsf{B}^T\boldsymbol{\lambda} + \varrho\mathsf{B}^T(\mathsf{B}\widehat{\mathbf{y}} - \mathbf{c}) = \mathbf{o}. \tag{9.19}$$

After substituting $\mathsf{B}\widehat{\mathbf{y}} = \mathbf{c}$ into (9.19), we get

$$\mathsf{B}\widehat{\mathbf{y}} - \mathbf{b} + \mathsf{B}^T\boldsymbol{\lambda} = \mathbf{o}. \tag{9.20}$$

However, since (9.20) and $\mathsf{B}\widehat{\mathbf{y}} = \mathbf{c}$ are sufficient conditions for $\widehat{\mathbf{y}}$ to be the unique solution of (9.1), we have $\widehat{\mathbf{y}} = \widehat{\mathbf{x}}$. □

It follows that there is no hidden enforcement of the exact solution in (9.14) and consequently typically inexact solutions of the auxiliary unconstrained problems are obtained in Step 1.

9.2.2 Relations for Augmented Lagrangians

In this section, we establish the fundamental inequalities that relate the bound on the norm of the gradient $\mathbf{g}$ of the augmented Lagrangian L to the values of the augmented Lagrangian L. These inequalities are the essential ingredients in the proof of convergence of Algorithm 9.2.

Lemma 9.4. *Let $\mathsf{A}, \mathsf{B}, \mathbf{b}$, and $\mathbf{c}$ be those of problem (9.1),*

$$\mathbf{x} \in \mathbb{R}^n, \quad \boldsymbol{\lambda} \in \mathbb{R}^m, \quad \varrho > 0, \quad \eta > 0, \quad \text{and} \quad M > 0.$$

Let $\lambda_{\min}$ denote the least eigenvalue of A and

$$\widetilde{\boldsymbol{\lambda}} = \boldsymbol{\lambda} + \varrho(\mathsf{B}\mathbf{x} - \mathbf{c}).$$

(i) If

$$\|\mathbf{g}(\mathbf{x}, \boldsymbol{\lambda}, \varrho)\| \le M\|\mathsf{B}\mathbf{x} - \mathbf{c}\|, \tag{9.21}$$

then for any $\mathbf{y} \in \mathbb{R}^n$

$$L(\mathbf{y}, \widetilde{\boldsymbol{\lambda}}, \varrho) \ge L(\mathbf{x}, \boldsymbol{\lambda}, \varrho) + \frac{1}{2}\left(\varrho - \frac{M^2}{\lambda_{\min}}\right)\|\mathsf{B}\mathbf{x} - \mathbf{c}\|^2 + \frac{\varrho}{2}\|\mathsf{B}\mathbf{y} - \mathbf{c}\|^2. \tag{9.22}$$

(ii) If

$$\|\mathbf{g}(\mathbf{x}, \boldsymbol{\lambda}, \varrho)\| \le \eta, \tag{9.23}$$

then for any $\mathbf{y} \in \mathbb{R}^n$

$$L(\mathbf{y}, \widetilde{\boldsymbol{\lambda}}, \varrho) \ge L(\mathbf{x}, \boldsymbol{\lambda}, \varrho) + \frac{\varrho}{2}\|\mathsf{B}\mathbf{x} - \mathbf{c}\|^2 + \frac{\varrho}{2}\|\mathsf{B}\mathbf{y} - \mathbf{c}\|^2 - \frac{\eta^2}{2\lambda_{\min}}. \tag{9.24}$$

(iii) If (9.23) holds and $\mathbf{z}_0$ *is any vector such that* $\mathsf{B}\mathbf{z}_0 = \mathbf{c}$, *then*

$$L(\mathbf{x}, \boldsymbol{\lambda}, \varrho) \le f(\mathbf{z}_0) + \frac{\eta^2}{2\lambda_{\min}}. \tag{9.25}$$

Proof. Let us denote

$$\boldsymbol{\delta} = \mathbf{y} - \mathbf{x}, \quad \mathsf{A}_\varrho = \mathsf{A} + \varrho\mathsf{B}^T\mathsf{B}, \quad \mathbf{g} = \mathbf{g}(\mathbf{x}, \boldsymbol{\lambda}, \varrho), \quad \text{and} \quad \widetilde{\mathbf{g}} = \mathbf{g}(\mathbf{x}, \widetilde{\boldsymbol{\lambda}}, \varrho).$$

(i) Using

$$\begin{aligned} L(\mathbf{x}, \widetilde{\boldsymbol{\lambda}}, \varrho) &= L(\mathbf{x}, \boldsymbol{\lambda}, \varrho) + \varrho\|\mathsf{B}\mathbf{x} - \mathbf{c}\|^2, \\ \mathbf{g}(\mathbf{x}, \widetilde{\boldsymbol{\lambda}}, \varrho) &= \mathbf{g}(\mathbf{x}, \boldsymbol{\lambda}, \varrho) + \varrho\mathsf{B}^T(\mathsf{B}\mathbf{x} - \mathbf{c}), \end{aligned}$$

we get

$$\begin{aligned} L(\mathbf{y}, \widetilde{\boldsymbol{\lambda}}, \varrho) &= L(\mathbf{x}, \widetilde{\boldsymbol{\lambda}}, \varrho) + \boldsymbol{\delta}^T\widetilde{\mathbf{g}} + \frac{1}{2}\boldsymbol{\delta}^T\mathsf{A}_\varrho\boldsymbol{\delta} \\ &= L(\mathbf{x}, \boldsymbol{\lambda}, \varrho) + \boldsymbol{\delta}^T\mathbf{g} + \frac{1}{2}\boldsymbol{\delta}^T\mathsf{A}_\varrho\boldsymbol{\delta} + \varrho\boldsymbol{\delta}^T\mathsf{B}^T(\mathsf{B}\mathbf{x} - \mathbf{c}) + \varrho\|\mathsf{B}\mathbf{x} - \mathbf{c}\|^2 \\ &\ge L(\mathbf{x}, \boldsymbol{\lambda}, \varrho) + \boldsymbol{\delta}^T\mathbf{g} + \frac{\lambda_{\min}}{2}\|\boldsymbol{\delta}\|^2 + \varrho\boldsymbol{\delta}^T\mathsf{B}^T(\mathsf{B}\mathbf{x} - \mathbf{c}) + \frac{\varrho}{2}\|\mathsf{B}\boldsymbol{\delta}\|^2 \\ &\quad + \varrho\|\mathsf{B}\mathbf{x} - \mathbf{c}\|^2. \end{aligned}$$

Noticing that

$$\begin{aligned} \frac{\varrho}{2}\|\mathsf{B}\mathbf{y} - \mathbf{c}\|^2 &= \frac{\varrho}{2}\|\mathsf{B}\boldsymbol{\delta} + (\mathsf{B}\mathbf{x} - \mathbf{c})\|^2 \\ &= \varrho\boldsymbol{\delta}^T\mathsf{B}^T(\mathsf{B}\mathbf{x} - \mathbf{c}) + \frac{\varrho}{2}\|\mathsf{B}\boldsymbol{\delta}\|^2 + \frac{\varrho}{2}\|\mathsf{B}\mathbf{x} - \mathbf{c}\|^2, \end{aligned}$$

we get

$$L(\mathbf{y}, \widetilde{\boldsymbol{\lambda}}, \varrho) \geq L(\mathbf{x}, \boldsymbol{\lambda}, \varrho) + \boldsymbol{\delta}^T \mathbf{g} + \frac{\lambda_{\min}}{2}\|\boldsymbol{\delta}\|^2 + \frac{\varrho}{2}\|\mathsf{B}\mathbf{x} - \mathbf{c}\|^2 + \frac{\varrho}{2}\|\mathsf{B}\mathbf{y} - \mathbf{c}\|^2. \quad (9.26)$$

Assuming (9.21) and using simple manipulations, we get

$$\begin{aligned}
L(\mathbf{y}, \widetilde{\boldsymbol{\lambda}}, \varrho) &\geq L(\mathbf{x}, \boldsymbol{\lambda}, \varrho) - M\|\boldsymbol{\delta}\|\|\mathsf{B}\mathbf{x} - \mathbf{c}\| + \frac{\lambda_{\min}}{2}\|\boldsymbol{\delta}\|^2 \\
&\quad + \frac{\varrho}{2}\|\mathsf{B}\mathbf{x} - \mathbf{c}\|^2 + \frac{\varrho}{2}\|\mathsf{B}\mathbf{y} - \mathbf{c}\|^2 \\
&= L(\mathbf{x}, \boldsymbol{\lambda}, \varrho) + \left(\frac{\lambda_{\min}}{2}\|\boldsymbol{\delta}\|^2 - M\|\boldsymbol{\delta}\|\|\mathsf{B}\mathbf{x} - \mathbf{c}\| + \frac{M^2\|\mathsf{B}\mathbf{x} - \mathbf{c}\|^2}{2\lambda_{\min}}\right) \\
&\quad - \frac{M^2\|\mathsf{B}\mathbf{x} - \mathbf{c}\|^2}{2\lambda_{\min}} + \frac{\varrho}{2}\|\mathsf{B}\mathbf{x} - \mathbf{c}\|^2 + \frac{\varrho}{2}\|\mathsf{B}\mathbf{y} - \mathbf{c}\|^2 \\
&\geq L(\mathbf{x}, \boldsymbol{\lambda}, \varrho) + \frac{1}{2}\left(\varrho - \frac{M^2}{\lambda_{\min}}\right)\|\mathsf{B}\mathbf{x} - \mathbf{c}\|^2 + \frac{\varrho}{2}\|\mathsf{B}\mathbf{y} - \mathbf{c}\|^2,
\end{aligned}$$

which proves (i).

(ii) If we assume that (9.23) holds, then

$$\boldsymbol{\delta}^T \mathbf{g} \geq -\eta\|\boldsymbol{\delta}\|$$

and

$$\boldsymbol{\delta}^T \mathbf{g} + \frac{\lambda_{\min}}{2}\|\boldsymbol{\delta}\|^2 \geq \min_{\xi} \frac{\lambda_{\min}}{2}\xi^2 - \eta\xi = -\frac{\eta^2}{2\lambda_{\min}}. \quad (9.27)$$

Substituting the latter inequality into (9.26), we get (ii).

(iii) Finally, let $\widehat{\mathbf{y}}$ denote the solution of the auxiliary problem

$$\text{minimize } L(\mathbf{y}, \boldsymbol{\lambda}, \varrho) \ \text{ s.t. } \ \mathbf{y} \in \mathbb{R}^n, \quad (9.28)$$

$\mathsf{B}\mathbf{z}_0 = \mathbf{c}$, and $\widehat{\boldsymbol{\delta}} = \widehat{\mathbf{y}} - \mathbf{x}$. If (9.23) holds, then by (9.27)

$$0 \geq L(\widehat{\mathbf{y}}, \boldsymbol{\lambda}, \varrho) - L(\mathbf{x}, \boldsymbol{\lambda}, \varrho) = \widehat{\boldsymbol{\delta}}^T \mathbf{g} + \frac{1}{2}\widehat{\boldsymbol{\delta}}^T \mathsf{A}_\varrho \widehat{\boldsymbol{\delta}} \geq -\frac{\eta^2}{2\lambda_{\min}}.$$

Since $L(\widehat{\mathbf{y}}, \boldsymbol{\lambda}, \varrho) \leq L(\mathbf{z}_0, \boldsymbol{\lambda}, \varrho) = f(\mathbf{z}_0)$, we conclude that

$$\begin{aligned}
L(\mathbf{x}, \boldsymbol{\lambda}, \varrho) &\leq L(\mathbf{x}, \boldsymbol{\lambda}, \varrho) + (f(\mathbf{z}_0) - L(\widehat{\mathbf{y}}, \boldsymbol{\lambda}, \varrho)) \\
&= f(\mathbf{z}_0) + (L(\mathbf{x}, \boldsymbol{\lambda}, \varrho) - L(\widehat{\mathbf{y}}, \boldsymbol{\lambda}, \varrho)) \\
&\leq f(\mathbf{z}_0) + \frac{\eta^2}{2\lambda_{\min}}.
\end{aligned}$$

□

9.2.3 Convergence and Monotonicity

The analysis of SMALE is based on the following lemma.

Lemma 9.5. *Let* $\{\mathbf{x}^k\}, \{\boldsymbol{\lambda}^k\}$, *and* $\{M_k\}$ *be generated by Algorithm 9.2 for the solution of problem (9.1) with*

$$\eta > 0, \quad 0 < \beta < 1, \quad M_k > 0, \quad \varrho > 0, \quad \text{and} \quad \boldsymbol{\lambda}^0 \in \mathbb{R}^m.$$

Denote $\lambda_{\min} = \lambda_{\min}(\mathsf{A})$.

(i) If $k \geq 0$ *and*

$$0 < M_k^2 \leq \varrho \lambda_{\min}, \tag{9.29}$$

then

$$L(\mathbf{x}^{k+1}, \boldsymbol{\lambda}^{k+1}, \varrho) \geq L(\mathbf{x}^k, \boldsymbol{\lambda}^k, \varrho) + \frac{\varrho}{2}\|\mathsf{B}\mathbf{x}^{k+1} - \mathbf{c}\|^2. \tag{9.30}$$

(ii) For any $k \geq 0$

$$L(\mathbf{x}^{k+1}, \boldsymbol{\lambda}^{k+1}, \varrho) \geq L(\mathbf{x}^k, \boldsymbol{\lambda}^k, \varrho) + \frac{\varrho}{2}\|\mathsf{B}\mathbf{x}^k - \mathbf{c}\|^2 + \frac{\varrho}{2}\|\mathsf{B}\mathbf{x}^{k+1} - \mathbf{c}\|^2 - \frac{\eta^2}{2\lambda_{\min}}. \tag{9.31}$$

(iii) For any $k \geq 0$ *and* $\mathbf{z}_0$ *such that* $\mathsf{B}\mathbf{z}_0 = \mathbf{c}$

$$L(\mathbf{x}^k, \boldsymbol{\lambda}^k, \varrho) \leq f(\mathbf{z}_0) + \frac{\eta^2}{2\lambda_{\min}}. \tag{9.32}$$

Proof. In Lemma 9.4, let us substitute

$$\mathbf{x} = \mathbf{x}^k, \quad \boldsymbol{\lambda} = \boldsymbol{\lambda}^k, \quad M = M_k, \quad \text{and} \quad \mathbf{y} = \mathbf{x}^{k+1},$$

so that inequality (9.21) holds by (9.14), and by (9.15) $\widetilde{\boldsymbol{\lambda}} = \boldsymbol{\lambda}^{k+1}$. (i) If (9.29) holds, we get by (9.22)

$$L(\mathbf{x}^{k+1}, \boldsymbol{\lambda}^{k+1}, \varrho) \geq L(\mathbf{x}^k, \boldsymbol{\lambda}^k, \varrho) + \frac{\varrho}{2}\|\mathsf{B}\mathbf{x}^{k+1} - \mathbf{c}\|^2. \tag{9.33}$$

This proves (9.30).

(ii) If we notice that by the definition of Step 1 of Algorithm 9.2

$$\|\mathbf{g}(\mathbf{x}^k, \boldsymbol{\lambda}^k, \varrho)\| \leq \eta,$$

we can apply the same substitution as above to Lemma 9.4(ii) to get (9.31).

(iii) Similarly as above, the inequality (9.32) results from application of the substitution to Lemma 9.4(iii). □

Remark 9.1. Notice that by Lemma 9.5, in particular the condition (9.29), large ϱ can prevent the balancing parameter M_k from decreasing. Thus we can fix the balancing parameter M_k and replace the update of M_k in Step 3 by increasing ϱ. However, this approach can result in ill-conditioning of the augmented matrix, so we shall not develop it here.

Theorem 9.2. *Let $\{\mathbf{x}^k\}, \{\boldsymbol{\lambda}^k\}$, and $\{M_k\}$ be generated by Algorithm 9.2 for the solution of problem (9.1) with*

$$\eta > 0, \quad 0 < \beta < 1, \quad M_0 > 0, \quad \varrho > 0, \quad \text{and} \quad \boldsymbol{\lambda}^0 \in \mathbb{R}^m.$$

Let $\lambda_{\min}$ denote the least eigenvalue of the Hessian A of f and let $s_{\max} \geq 0$ denote the smallest integer such that

$$M_0^2 \leq \beta^{-2s} \varrho \lambda_{\min}. \tag{9.34}$$

(i) The sequence $\{M_k\}$ is decreasing, bounded away from zero, and

$$M_k \geq \beta^s M_0, \quad s \leq s_{\max}. \tag{9.35}$$

(ii) If $\mathbf{z}_0$ denotes any vector such that $\mathsf{B}\mathbf{z}_0 = \mathbf{c}$, then

$$\sum_{k=1}^{\infty} \frac{\varrho}{2} \|\mathsf{B}\mathbf{x}^k - \mathbf{c}\|^2 \leq f(\mathbf{z}_0) - L(\mathbf{x}^0, \boldsymbol{\lambda}^0, \varrho) + (1+s)\frac{\eta^2}{2\lambda_{\min}}. \tag{9.36}$$

(iii) The sequence $\{\mathbf{x}^k\}$ converges to the solution $\widehat{\mathbf{x}}$ of (9.1).
(iv) The sequence $\{\boldsymbol{\lambda}^k\}$ converges to the vector

$$\overline{\boldsymbol{\lambda}} = \boldsymbol{\lambda}_{\mathrm{LS}} + (\mathsf{I} - \mathsf{P})\boldsymbol{\lambda}^0,$$

where P is the orthogonal projector onto $\mathrm{Im}\mathsf{B}$, and $\boldsymbol{\lambda}_{\mathrm{LS}}$ is the least square Lagrange multiplier of (9.1).
(v) The sequence $\|\mathsf{B}\mathbf{x}^k - \mathbf{c}\|$ converges to zero R-linearly.

Proof. (i) Let $s \geq 0$ satisfy (9.35) and let $\mathcal{I}$ denote the set of all indices k_i such that

$$M_{k_i} = \beta M_{k_{i-1}}, \; i = 1, \ldots, s \leq s_{\max}, \quad M_{k_s} = M_{k_s+1} = \cdots .$$

We admit $\mathcal{I} = \emptyset$; in this case $M_k = M_0$, and we define $s = s_{\max} = 0$.
Using Lemma 9.5(i), we get that $M_k = M_{k+1}$ when $M_k^2 \leq \varrho\lambda_{\min}$. Since

$$M_{k_i} = \beta M_{k_{i-1}} \geq \beta^s M_0 \quad \text{for} \quad i \in \mathcal{I},$$

we conclude that

$$M_k \geq \beta^s M_0 \geq \beta^{s_{\max}} M_0.$$

Thus $\mathcal{I}$ has at most $s_{\max}$ elements and (9.35) holds.

(ii) By the definition of Step 3, for $k > 0$ either $k+1 \notin \mathcal{I}$ and

$$\frac{\varrho}{2}\|\mathsf{B}\mathbf{x}^k - \mathbf{c}\|^2 \le L(\mathbf{x}^k, \boldsymbol{\lambda}^k, \varrho) - L(\mathbf{x}^{k-1}, \boldsymbol{\lambda}^{k-1}, \varrho),$$

or $k+1 \in \mathcal{I}$ and by (9.31)

$$\begin{aligned}\frac{\varrho}{2}\|\mathsf{B}\mathbf{x}^k - \mathbf{c}\|^2 &\le \frac{\varrho}{2}\|\mathsf{B}\mathbf{x}^{k-1} - \mathbf{c}\|^2 + \frac{\varrho}{2}\|\mathsf{B}\mathbf{x}^k - \mathbf{c}\|^2 \\ &\le L(\mathbf{x}^k, \boldsymbol{\lambda}^k, \varrho) - L(\mathbf{x}^{k-1}, \boldsymbol{\lambda}^{k-1}, \varrho) + \frac{\eta^2}{2\lambda_{\min}}.\end{aligned}$$

Summing up the appropriate cases of the last two inequalities for $k = 1, \ldots, n$ and taking into account that $\mathcal{I}$ has at most $s_{\max}$ elements, we get

$$\sum_{k=1}^{n} \frac{\varrho}{2}\|\mathsf{B}\mathbf{x}^k - \mathbf{c}\|^2 \le L(\mathbf{x}^n, \boldsymbol{\lambda}^n, \varrho) - L(\mathbf{x}^0, \boldsymbol{\lambda}^0, \varrho) + s_{\max}\frac{\eta^2}{2\lambda_{\min}}. \tag{9.37}$$

To get (9.36), it is enough to replace $L(\mathbf{x}^n, \boldsymbol{\lambda}^n, \varrho)$ by the upper bound (9.32).
(iii) and (iv) Let us denote

$$\mathbf{g}^k = \mathbf{g}(\mathbf{x}^k, \boldsymbol{\lambda}^k, \varrho) = \mathsf{A}_\varrho\mathbf{x}^k + \mathsf{B}^T\boldsymbol{\lambda}^k - \mathbf{b} - \varrho_k\mathsf{B}^T\mathbf{c},$$

and let us assume that B is a full row rank matrix. Since the unique KKT pair $(\widehat{\mathbf{x}}, \widehat{\boldsymbol{\lambda}})$ is fully determined by

$$\begin{aligned}\mathsf{A}\widehat{\mathbf{x}} + \mathsf{B}^T\widehat{\boldsymbol{\lambda}} &= \mathbf{b}, \\ \mathsf{B}\widehat{\mathbf{x}} \qquad &= \mathbf{c},\end{aligned}$$

we can rewrite $\mathbf{g}^k$ as

$$\mathbf{g}^k = \mathsf{A}_\varrho(\mathbf{x}^k - \widehat{\mathbf{x}}) + \mathsf{B}^T(\boldsymbol{\lambda}^k - \widehat{\boldsymbol{\lambda}}). \tag{9.38}$$

The last equation together with

$$\mathsf{B}(\mathbf{x}^k - \widehat{\mathbf{x}}) = \mathsf{B}\mathbf{x}^k - \mathbf{c} \tag{9.39}$$

may be written in the matrix form as

$$\begin{pmatrix} \mathsf{A}_\varrho & \mathsf{B}^T \\ \mathsf{B} & 0 \end{pmatrix}\begin{pmatrix} \mathbf{x}^k - \widehat{\mathbf{x}} \\ \boldsymbol{\lambda}^k - \widehat{\boldsymbol{\lambda}} \end{pmatrix} = \begin{pmatrix} \mathbf{g}^k \\ \mathsf{B}\mathbf{x}^k - \mathbf{c} \end{pmatrix}. \tag{9.40}$$

Since $\|\mathsf{B}\mathbf{x}^k - \mathbf{c}\|$ converges to zero due to (9.36), $\|\mathbf{g}^k\| \le M\|\mathsf{B}\mathbf{x}^k - \mathbf{c}\|$, and the matrix of the system (9.40) is regular, we conclude, using Proposition 2.12, that $\mathbf{x}^k$ converges to $\widehat{\mathbf{x}}$ and $\boldsymbol{\lambda}^k$ converges to $\widehat{\boldsymbol{\lambda}}$. Since B is a full rank matrix, it follows that $\widehat{\boldsymbol{\lambda}} = \boldsymbol{\lambda}_{\mathrm{LS}} = \overline{\boldsymbol{\lambda}}$.

If B is not a full rank matrix, then the augmented matrix on the left-hand side of (9.40) is singular, but the solution $\widehat{\mathbf{x}}$ is still uniquely determined, as

$$\mathrm{Ker}\mathsf{A} \cap \mathrm{Ker}\mathsf{B} \subseteq \mathrm{Ker}\mathsf{A} = \{\mathbf{o}\}$$

by the assumptions. Since any KKT pair $(\widehat{\mathbf{x}}, \overline{\boldsymbol{\lambda}})$ satisfies

$$\begin{pmatrix} \mathsf{A}_\varrho & \mathsf{B}^T \\ \mathsf{B} & 0 \end{pmatrix} \begin{pmatrix} \mathbf{x}^k - \widehat{\mathbf{x}} \\ \boldsymbol{\lambda}^k - \overline{\boldsymbol{\lambda}} \end{pmatrix} = \begin{pmatrix} \mathbf{g}^k \\ \mathsf{B}\mathbf{x}^k - \mathbf{c} \end{pmatrix}, \tag{9.41}$$

we can use the same arguments as above and Proposition 2.12 to find out again that $\mathbf{x}^k$ converges to $\widehat{\mathbf{x}}$, but now we shall get only that $\mathsf{B}^T\boldsymbol{\lambda}^k$ converges to $\mathsf{B}^T\overline{\boldsymbol{\lambda}}$. However, using Lemma 9.2, we get

$$\boldsymbol{\lambda}^k - \overline{\boldsymbol{\lambda}} = \mathsf{P}\boldsymbol{\lambda}^k - \boldsymbol{\lambda}_{\mathrm{LS}},$$

so that in particular $\boldsymbol{\lambda}^k - \overline{\boldsymbol{\lambda}} \in \mathrm{Im}\mathsf{B}$. It follows by (1.40) that

$$\|\mathsf{B}^T(\boldsymbol{\lambda}^k - \overline{\boldsymbol{\lambda}})\| \geq \overline{\sigma}_{\min}\|\boldsymbol{\lambda}^k - \overline{\boldsymbol{\lambda}}\|.$$

Since the left-hand side converges to zero, we conclude that $\boldsymbol{\lambda}^k$ converges to $\overline{\boldsymbol{\lambda}}$, which completes the proof of (iii) and (iv).

The proof of (v) is technical and can be found in [107]. The result can also be considered as a special case of Theorem 12.7. □

9.2.4 Optimality of the Outer Loop

Theorem 9.2 suggests that for homogeneous constraints, it is possible to give a rate of convergence of the feasibility error that does not depend on the constraint matrix B. To present explicitly this qualitatively new feature of Algorithm 9.2, let $\mathcal{T}$ denote a set of indices and define for any $t \in \mathcal{T}$ a problem

$$\text{minimize } f_t(\mathbf{x}) \text{ s.t. } \mathbf{x} \in \Omega_t, \quad f_t(\mathbf{x}) = \frac{1}{2}\mathbf{x}^T\mathsf{A}_t\mathbf{x} - \mathbf{b}_t^T\mathbf{x}, \tag{9.42}$$

with

$$\Omega_t = \{\mathbf{x} \in \mathbb{R}^{n_t} : \mathsf{B}_t\mathbf{x} = \mathbf{o}\},$$

and

$$\mathsf{A}_t \in \mathbb{R}^{n_t \times n_t} \text{ SPD}, \ \mathsf{B}_t \in \mathbb{R}^{m_t \times n_t}, \text{ and } \mathbf{b}_t, \ \mathbf{x} \in \mathbb{R}^{n_t}.$$

Moreover, we assume

$$0 < a_{\min} \leq \lambda_{\min}(\mathsf{A}_t), \quad 0 < \|\mathbf{b}_t\| \leq b. \tag{9.43}$$

Our optimality result then reads as follows.

Theorem 9.3. *Let $\{\mathbf{x}_t^k\}$, $\{\boldsymbol{\lambda}_t^k\}$, and $\{M_{t,k}\}$ be generated by Algorithm 9.2 for (9.42) and* (9.43) *with*

$$\|\mathbf{b}_t\| \geq \eta_t > 0, \quad 0 < \beta < 1, \varrho > 0, \quad M_{t,0} = M_0 > 0, \quad \boldsymbol{\lambda}_t^0 = \mathbf{o}.$$

Let $0 < a_{\min}$ denote a given constant and let

$$a_{\min} \leq \lambda_{\min}(\mathsf{A}_t),$$

where $\lambda_{\min}(\mathsf{A}_t)$ denotes the smallest eigenvalue of A_t. Finally, let us denote

$$a = (2+s)/(a_{\min}\varrho),$$

where $s \geq 0$ is the smallest integer such that

$$M_0^2 \leq \beta^{-2s} \varrho a_{\min}.$$

Then for each $\varepsilon > 0$ there are the indices $k_t, t \in \mathcal{T}$, such that

$$k_t \leq a/\varepsilon^2 + 1$$

and $\mathbf{x}_t^{k_t}$ is an approximate solution of (9.42) satisfying

$$\|\mathbf{g}_t(\mathbf{x}_t^{k_t}, \boldsymbol{\lambda}_t^{k_t}, \varrho)\| \leq M_0 \varepsilon \|\mathbf{b}_t\| \quad \text{and} \quad \|\mathsf{B}_t \mathbf{x}_t^{k_t}\| \leq \varepsilon \|\mathbf{b}_t\|. \tag{9.44}$$

Proof. First, notice that for any index k

$$\frac{k\varrho}{2} \min_{i \in \{1,\ldots,k\}} \|\mathsf{B}_t \mathbf{x}_t^i\|^2 \leq \sum_{i=1}^{k} \frac{\varrho}{2} \|\mathsf{B}_t \mathbf{x}_t^i\|^2 \leq \sum_{i=1}^{\infty} \frac{\varrho}{2} \|\mathsf{B}_t \mathbf{x}_t^i\|^2. \tag{9.45}$$

Denoting by $L_t(\mathbf{x}, \boldsymbol{\lambda}, \varrho)$ the Lagrangian for problem (9.42), we get for any $\mathbf{x} \in \mathbb{R}^{n_t}$ and $\varrho \geq 0$

$$L_t(\mathbf{x}, \mathbf{o}, \varrho) = \frac{1}{2}\mathbf{x}^T(\mathsf{A}_t + \varrho \mathsf{B}_t^T \mathsf{B}_t)\mathbf{x} - \mathbf{b}_t^T \mathbf{x} \geq \frac{1}{2} a_{\min} \|\mathbf{x}\|^2 - \|\mathbf{b}_t\|\|\mathbf{x}\| \geq -\frac{\|\mathbf{b}_t\|^2}{2a_{\min}}.$$

If we substitute this inequality and $\mathbf{z} = \mathbf{z}_0^t = \mathbf{o}$ into (9.36) and use $\|\mathbf{b}_t\| \geq \eta_t$, we get for any $t \in \mathcal{T}$

$$\sum_{k=1}^{\infty} \frac{\varrho}{2} \|\mathsf{B}_t \mathbf{x}_t^k\|^2 \leq \frac{\|\mathbf{b}_t\|^2}{2a_{\min}} + \frac{(1+s)\eta_t^2}{2a_{\min}} \leq \frac{2+s}{2a_{\min}} \|\mathbf{b}_t\|^2 = \frac{a\varrho}{2} \|\mathbf{b}_t\|^2. \tag{9.46}$$

Combining the latter inequality with (9.45), we get

$$\min\{\|\mathsf{B}_t \mathbf{x}_t^i\|^2 : i = 1, \ldots, k\} \leq a\|\mathbf{b}_t\|^2/k. \tag{9.47}$$

Taking for j the smallest integer that satisfies $a/j \le \varepsilon^2$, so that

$$a/\varepsilon^2 \le j \le a/\varepsilon^2 + 1,$$

and denoting for any $t \in \mathcal{T}$ by $k_t \in \{1, \ldots, j\}$ the index which minimizes $\{\|\mathsf{B}_t \mathbf{x}_t^i\| : i = 1, \ldots, j\}$, we can use (9.47) to obtain

$$\|\mathsf{B}_t \mathbf{x}_t^{k_t}\|^2 = \min\{\|\mathsf{B}_t \mathbf{x}_t^i\|^2 : i = 1, \ldots, j\} \le a\|\mathbf{b}_t\|^2/j \le \varepsilon^2 \|\mathbf{b}_t\|^2.$$

Thus

$$\|\mathsf{B}_t \mathbf{x}_t^{k_t}\|^2 \le \varepsilon^2 \|\mathbf{b}_t\|^2,$$

and, using the definition of Step 1 of Algorithm 9.2, we get also the inequality

$$\|\mathbf{g}_t(\mathbf{x}_t^{k_t}, \boldsymbol{\lambda}_t^{k_t}, \varrho)\| \le M_{k_t} \|\mathsf{B}_t \mathbf{x}_t^{k_t}\| \le M_0 \varepsilon \|\mathbf{b}_t\|.$$

□

Let us recall that

$$\|\mathbf{g}_t(\mathbf{x}_t^{k_t}, \boldsymbol{\lambda}_t^{k_t+1}, 0)\| = \|\mathbf{g}_t(\mathbf{x}_t^{k_t}, \boldsymbol{\lambda}_t^{k_t}, \varrho)\|,$$

so that $(\mathbf{x}_t^{k_t}, \boldsymbol{\lambda}_t^{k_t+1})$ is an approximate KKT pair of problem (9.42). We used the assumption on the homogeneity of the constraints to find $\mathbf{z}_0^t$ such that $f(\mathbf{z}_0^t)$ is uniformly bounded, in this case by zero.

9.2.5 Optimality of SMALE with Conjugate Gradients

We shall need the following simple lemma to prove the optimality of the inner loop.

Lemma 9.6. *Let $\{\mathbf{x}^k\}, \{\boldsymbol{\lambda}^k\}$, and $\{M_k\}$ be generated by Algorithm 9.2 for problem (9.1) with $\mathbf{c} = \mathbf{o}$ and*

$$\eta > 0, \quad 1 > \beta > 0, \quad \varrho > 0, \quad M_0 > 0, \quad \text{and} \quad \boldsymbol{\lambda}^0 \in \mathbb{R}^m.$$

Then for any $k \ge 0$

$$L(\mathbf{x}^k, \boldsymbol{\lambda}^{k+1}, \varrho) - L(\mathbf{x}^{k+1}, \boldsymbol{\lambda}^{k+1}, \varrho) \le \frac{\eta^2}{2\lambda_{\min}} + \frac{\varrho}{2}\|\mathsf{B}\mathbf{x}^k\|^2. \tag{9.48}$$

Proof. Notice that by definition of the Lagrangian function and by the update rule (9.15)

$$L(\mathbf{x}^k, \boldsymbol{\lambda}^{k+1}, \varrho) = L(\mathbf{x}^k, \boldsymbol{\lambda}^k, \varrho) + \varrho\|\mathsf{B}\mathbf{x}^k\|^2.$$

After subtracting $L(\mathbf{x}^{k+1}, \boldsymbol{\lambda}^{k+1}, \varrho)$ from both sides and observing that by (9.31)

$$L(\mathbf{x}^k, \boldsymbol{\lambda}^k, \varrho) - L(\mathbf{x}^{k+1}, \boldsymbol{\lambda}^{k+1}, \varrho) \leq \frac{\eta^2}{2\lambda_{\min}} - \frac{\varrho}{2}\|\mathsf{B}\mathbf{x}^k\|^2,$$

we get

$$L(\mathbf{x}^k, \boldsymbol{\lambda}^{k+1}, \varrho) - L(\mathbf{x}^{k+1}, \boldsymbol{\lambda}^{k+1}, \varrho) \leq \frac{\eta^2}{2\lambda_{\min}} + \frac{\varrho}{2}\|\mathsf{B}\mathbf{x}^k\|^2.$$

□

We are ready to prove our main result concerning the inner loop.

Theorem 9.4. *Let* $\{\mathbf{x}_t^k\}$, $\{\boldsymbol{\lambda}_t^k\}$, *and* $\{M_{t,k}\}$ *be generated by Algorithm 9.2 for (9.42) and* (9.43) *with*

$$\|\mathbf{b}_t\| \geq \eta_t > 0, \quad 0 < \beta < 1, \quad \varrho > 0, \quad M_{t,0} = M_0 > 0, \quad \boldsymbol{\lambda}_t^0 = \mathbf{o}.$$

Assume that Step 1 is implemented by the conjugate gradient method, which generates the iterates $\mathbf{x}_t^{k,0}, \mathbf{x}_t^{k,1}, \ldots, \mathbf{x}_t^{k,l} = \mathbf{x}_t^k$ *starting from* $\mathbf{x}_t^{k,0} = \mathbf{x}_t^{k-1}$ *with* $\mathbf{x}_t^{-1} = \mathbf{o}$, *where* $l = l(k,t)$ *is the first index satisfying either*

$$\|\mathbf{g}(\mathbf{x}_t^{k,l}, \boldsymbol{\lambda}_t^k, \varrho)\| \leq M_{t,k}\|\mathsf{B}_t\mathbf{x}_t^{k,l}\| \tag{9.49}$$

or

$$\|\mathbf{g}(\mathbf{x}_t^{k,l}, \boldsymbol{\lambda}_t^k, \varrho)\| \leq \varepsilon M_{t,k}\|\mathbf{b}_t\|. \tag{9.50}$$

Finally, let the class of problems (9.42) satisfy

$$a_{\min} \leq \lambda_{\min}(\mathsf{A}_t) \leq \lambda_{\max}(\mathsf{A}_t) = \|\mathsf{A}_t\| \leq a_{\max} \quad \text{and} \quad \|\mathsf{B}_t\| \leq B_{\max}, \tag{9.51}$$

where $0 < a_{\min} < a_{\max}$ *and* $0 < B_{\max}$ *are given constants.*

Then Algorithm 9.2 generates an approximate solution $\mathbf{x}_t^{k_t}$ *of any problem (9.42) with (9.43) which satisfies*

$$\|\mathbf{g}_t(\mathbf{x}_t^{k_t}, \boldsymbol{\lambda}_t^{k_t}, \varrho)\| \leq M_0\varepsilon\|\mathbf{b}_t\| \quad \text{and} \quad \|\mathsf{B}_t\mathbf{x}_t^{k_t}\| \leq \varepsilon\|\mathbf{b}_t\| \tag{9.52}$$

at $O(1)$ *matrix–vector multiplications by the Hessian of the augmented Lagrangian* L_t *for (9.42).*

Proof. Let $t \in \mathcal{T}$ be fixed and let us denote by $L_t(\mathbf{x}, \boldsymbol{\lambda}, \varrho)$ the augmented Lagrangian for problem (9.42), so that for any $\mathbf{x} \in \mathbb{R}^p$ and $\varrho \geq 0$

$$\begin{aligned} L_t(\mathbf{x}, \mathbf{o}, \varrho) &= \frac{1}{2}\mathbf{x}^T(\mathsf{A}_t + \varrho\mathsf{B}_t^T\mathsf{B}_t)\mathbf{x} - \mathbf{b}_t^T\mathbf{x} \\ &\geq \frac{1}{2}a_{\min}\|\mathbf{x}\|^2 - \|\mathbf{b}_t\|\|\mathbf{x}\| \geq -\frac{\|\mathbf{b}_t\|^2}{2a_{\min}}. \end{aligned}$$

Applying the latter inequality to (9.36) with $\mathbf{z}_0 = \mathbf{o}$ and $\boldsymbol{\lambda}_t^0 = \mathbf{o}$, we get, using the assumption $\|\mathbf{b}_t\| \ge \eta_t$, that for any $k \ge 0$

$$\frac{\varrho}{2}\|\mathsf{B}_t\mathbf{x}_t^k\|^2 \le \sum_{i=1}^{\infty}\frac{\varrho}{2}\|\mathsf{B}_t\mathbf{x}_t^i\|^2 \le f(\mathbf{z}_0) - L(\mathbf{x}_t^0, \boldsymbol{\lambda}_t^0, \varrho) + (1+s)\frac{\eta_t^2}{2a_{\min}}$$
$$\le (2+s)\|\mathbf{b}_t\|^2/(2a_{\min}),$$

where $s \ge 0$ denotes the smallest integer such that

$$M_0^2 \le \beta^{-2s} a_{\min}\varrho.$$

Using (9.48), we get that there is a constant $C > 0$ independent of t such that

$$0 \le L_t(\mathbf{x}_t^k, \boldsymbol{\lambda}_t^{k+1}, \varrho) - L_t(\mathbf{x}_t^{k+1}, \boldsymbol{\lambda}_t^{k+1}, \varrho) \le \frac{\eta_t^2}{2a_{\min}} + \frac{\varrho}{2}\|\mathsf{B}_t\mathbf{x}_t^k\|^2 \le C,$$

and, since the minimizer $\overline{\mathbf{x}}_t^{k+1}$ of $L_t(., \boldsymbol{\lambda}_t^{k+1}, \varrho)$ satisfies (9.49) and is a possible choice for $\mathbf{x}_t^{k+1}$, also

$$0 \le L_t(\mathbf{x}_t^k, \boldsymbol{\lambda}_t^{k+1}, \varrho) - L_t(\overline{\mathbf{x}}_t^{k+1}, \boldsymbol{\lambda}_t^{k+1}, \varrho) \le C. \tag{9.53}$$

Now recall that

$$\|\mathbf{g}_t(\mathbf{x}_t^k, \boldsymbol{\lambda}_t^{k+1}, \varrho)\|^2_{\mathsf{A}_{t,\varrho}^{-1}} = 2(L_t(\mathbf{x}_t^k, \boldsymbol{\lambda}_t^{k+1}, \varrho) - L_t(\overline{\mathbf{x}}_t^{k+1}, \boldsymbol{\lambda}_t^{k+1}, \varrho)),$$

where

$$\mathsf{A}_{t,\varrho} = \mathsf{A}_t + \frac{\varrho}{2}\mathsf{B}_t^T\mathsf{B}_t.$$

Since

$$a_{\min} \le \lambda_{\min}(\mathsf{A}_t) \le \|\mathsf{A}_{t,\varrho}\| \le \|\mathsf{A}_t\| + \varrho\|\mathsf{B}_t\|^2 \le a_{\max} + \varrho B_{\max}^2,$$

we can also bound the spectral condition number $\kappa(\mathsf{A}_{t,\varrho})$ of $\mathsf{A}_{t,\varrho}$ by

$$K = (a_{\max} + \varrho B_{\max}^2)/a_{\min}.$$

Combining this bound with the estimate (8.4) which reads in our case

$$\|\mathbf{g}_t(\mathbf{x}_t^{k,m}, \boldsymbol{\lambda}_t^k, \varrho)\|_{\mathsf{A}_{t,\varrho}^{-1}} \le 2\left(\frac{\sqrt{K}-1}{\sqrt{K}+1}\right)^m \|\mathbf{g}_t(\mathbf{x}_t^{k,0}, \boldsymbol{\lambda}_t^k, \varrho)\|^2_{\mathsf{A}_{t,\varrho}^{-1}},$$

we get that we can generate a new iteration by a uniformly bounded number of CG steps. To finish the proof, combine this result with Theorem 9.3. □

We can observe optimality in the solution of more general classes of problems than those considered in Theorem 9.4 provided we can bound the number of iterations in the inner loop; see, e.g., Sect. 9.5.2.

9.3 Preconditioning Preserving Gap in the Spectrum

We have seen that the penalty method reduces the solution of the equality constrained problem (6.1) to the unconstrained problem (6.4). The resulting problem may be solved by a suitable direct method, such as the Cholesky decomposition, or by an iterative method, such as the CG method. If ϱ is large, then the Hessian matrix

$$\mathsf{A}_\varrho = \mathsf{A} + \varrho \mathsf{B}^T \mathsf{B}$$

of the cost function f_ϱ of the problem (6.4) is obviously ill-conditioned. Thus, the estimates based on the condition number do not guarantee fast convergence of the CG method, and the natural idea is to reduce the condition number of A_ϱ by a suitable preconditioning, see, e.g., Hager [157, 159].

Here, we consider an alternative approach exploiting the fast convergence of the CG method for problems with a gap in the spectrum. We use two results: the bounds on the rate of convergence independent of ϱ given by (8.6) and (8.7) and Lemma 1.9 on the distribution of the spectrum of A_ϱ. The method presented here is applicable for large ϱ provided we have an effective preconditioner M for A. To simplify our exposition, we assume that $\mathsf{M} = \mathsf{L}\mathsf{L}^T$, where L is a sparse lower triangular matrix.

To express briefly the effects of the preconditioning strategies presented in this section, let $k(\mathsf{W}, \varepsilon)$ denote the number of the conjugate gradient iterations that are necessary to reduce the residual of any system with the symmetric positive definite matrix W by ε, and let

$$\overline{k}(\mathsf{W}, \varepsilon) = \operatorname{int}(\frac{1}{2}\sqrt{\kappa(\mathsf{W})}\ln(2/\varepsilon) + 1) \tag{9.54}$$

denote the upper bound on $k(\mathsf{W}, \varepsilon)$ which may be easily obtained from (8.6).

Let us first assume that the rank m of the constraint matrix $\mathsf{B} \in \mathbb{R}^{p\times n}$ in the original problem (6.1) is small. Then it is possible to use L to redistribute the spectrum of the penalized matrix A_ϱ directly. In this case (1.59) and the estimate (8.6) of the rate of the conjugate gradient method for the case that the Hessian of the cost function has m isolated eigenvalues give the bound

$$k(L^{-1}\mathsf{A}_\varrho \mathsf{L}^{-T}, \varepsilon) \le \overline{k}(\mathsf{L}^{-1}\mathsf{A}\mathsf{L}^{-T}, \varepsilon) + m. \tag{9.55}$$

Such preconditioning can be implemented even without the factorization of the preconditioner $\mathsf{M} = \mathsf{L}\mathsf{L}^T$ as in Algorithm 8.1, provided we can solve efficiently the linear systems with the matrix M.

If m dominates in the expression on the right-hand side of (9.55), then the bound (9.55) can be improved at the cost of increased computational complexity. Since for any nonsingular matrix Q

$$\Omega_E = \{\mathbf{x} \in \mathbb{R}^n : \ \mathsf{QB}\mathbf{x} = \mathsf{Q}\mathbf{c}\},$$

we can choose the matrix Q in such a way that the rows of QBL^{-T} are orthonormal. Denoting $\overline{\mathsf{B}} = \mathsf{QB}$, we get that minimizer of the augmented cost function with the Hessian

$$\overline{\mathsf{A}}_\varrho = \mathsf{A} + \varrho \overline{\mathsf{B}}^T \overline{\mathsf{B}}$$

also approximates the solution of (6.1), but the spectrum $\sigma(\mathsf{L}^{-1}\overline{\mathsf{A}}_\varrho \mathsf{L}^{-T})$ of the preconditioned Hessian $\mathsf{L}^{-1}\overline{\mathsf{A}}_\varrho \mathsf{L}^{-T}$ satisfies by (1.58) and (1.59)

$$\sigma(\mathsf{L}^{-1}\overline{\mathsf{A}}_\varrho \mathsf{L}^{-T}) \subseteq [a_{\min}, a_{\max}] \cup [a_{\min} + \varrho, a_{\max} + \varrho],$$

where

$$a_{\min} = \lambda_1(\mathsf{L}^{-1}\mathsf{A}\mathsf{L}^{-T}) \ \text{ and } \ a_{\max} = \lambda_n(\mathsf{L}^{-1}\mathsf{A}\mathsf{L}^{-T}).$$

Since the spectrum is in two intervals of the same length, we can use (8.7) to get the bound

$$k(\mathsf{L}^{-1}\overline{\mathsf{A}}_\varrho \mathsf{L}^{-T}, \varepsilon) \le \min\{\overline{k}(\mathsf{L}^{-1}\mathsf{A}\mathsf{L}^{-T}, \varepsilon) + m, 2\overline{k}(\mathsf{L}^{-1}\mathsf{A}\mathsf{L}^{-T}, \varepsilon)\}, \tag{9.56}$$

which is optimal with respect to both ϱ and m. Results of some numerical experiments with this strategy are in [65]. This approach may be very effective when we have several problems (6.4) with the same matrix B.

Observe that $\mathsf{Q}^T\mathsf{Q}$ defines a scalar product on $\mathbb{R}^m$. The method can be efficient also in the case that the rows of QBL^{-T} are orthonormal only approximately [206]. If $\mathsf{A} = \mathsf{L}\mathsf{L}^T$ and QBL^{-T} are orthonormal, then

$$\sigma(\mathsf{L}^{-1}\overline{\mathsf{A}}_\varrho \mathsf{L}^{-T}) = \{1, \varrho\},$$

and the CG algorithm finds the solution in just two steps.

9.4 Implementation of Inexact Augmented Lagrangians

We shall complete the discussion of inexact augmented Lagrangian algorithms by a few hints concerning their implementation.

9.4.1 Stopping, Modification of Constraints, and Preconditioning

While implementing the inexact augmented Lagrangian algorithms of Sects. 9.1 and 9.2, we should add a stopping criterion should not only after Step 1 but also into the procedure which generates $\mathbf{x}^k$ in Step 1. We use in our experiments the stopping criterion

$$\|\nabla L(\mathbf{x}^k, \boldsymbol{\lambda}^k, \varrho)\| \le \varepsilon_g \|\mathbf{b}\| \quad \text{and} \quad \|\mathsf{B}\mathbf{x}^k - \mathbf{c}\| \le \varepsilon_f \|\mathbf{b}\|.$$

The relative precisions ε_f and ε_g should be judiciously determined. Our stopping criterion in the inner conjugate gradient loop of SMALE reads

$$\|\mathbf{g}(\mathbf{y}^i, \boldsymbol{\lambda}^k, \varrho)\| \le \min\{M\|\mathsf{B}\mathbf{y}^i - \mathbf{c}\|, \eta\} \;\text{ or }\; \|\mathbf{g}(\mathbf{y}^i, \boldsymbol{\lambda}^k, \varrho)\| \le \min\{\varepsilon_g, M\varepsilon_f\}\|\mathbf{b}\|,$$

so that the inner loop is interrupted when either the solution or a new iterate $\mathbf{x}^{k+1} = \mathbf{y}^i$ is found.

Before applying the algorithms we strongly recommend rescaling the equality constraints so that $\|\mathsf{A}\| \approx \|\mathsf{B}\|$. Considering the convergence rate estimate like (9.2), it is also helpful to orthonormalize or at least normalize the constraints. This approach has been successfully applied, e.g., in the FETI-DP-based solver for analysis of layered composites [195].

If the Hessian A is ill-conditioned and there is an approximation M of A that can be used as preconditioner, then we can use the preconditioning strategies introduced in the discussion on the implementation of the penalty method in Sect. 9.3. The construction of the matrix M is typically problem dependent. We refer interested readers to the books by Axelsson [6], Saad [240], van der Vorst [259], or Chen [33].

Sometimes, it is possible to exploit the structure of the problem for very efficient implementation of preconditioning. For example, it has been shown that it is possible to find multigrid preconditioners to the discretized Stokes problem so that SMALE can solve the latter with asymptotically linear complexity [206].

9.4.2 Initialization of Constants

Though all the inexact algorithms converge with

$$0 < \beta < 1, \;\; \eta > 0, \;\; \varrho > 0, \;\; M_0 > 0, \;\; \text{and} \;\; \boldsymbol{\lambda}^0 = \mathbf{o},$$

their choice affects the performance of the algorithms and should exploit available information. Here, we give a few hints and heuristics that can be useful for their efficient implementation.

SMALE uses the parameter β to increase the precision of the auxiliary problems' solution. Our experience indicates that $\beta = 0.5$ is a reasonable choice.

The parameter η is used only by SMALE algorithm 9.2. It helps to avoid outer iterations that do not invoke the inner CG iterations; we use $\eta = 0.1\|\mathbf{b}\|$.

The estimate (9.36) shows that a large value of the penalty parameter ϱ guarantees fast convergence of the outer loop. By the analysis of the penalty method in Sect. 6.2, it is possible to find the solution in one outer iteration. At the same time, the large value of the penalty parameter slows down the rate of convergence of the CG method in the inner loop, but the analysis of the CG method in Sect. 9.3 based on the effective condition number of $\mathsf{A}_\varrho = \mathsf{A} + \varrho \mathsf{B}^T\mathsf{B}$ indicates that the slowdown need not be severe when the number of constraints is small, or when the constraints are close to orthogonal. If neither is the case and at least crude estimates of $\|\mathsf{A}\|$ and $\|\mathsf{B}\|$ are available, a simple strategy can be based on the observation that

$$\lambda_{\min}(\mathsf{A}) \le \lambda_{\min}(\mathsf{A}_\varrho) \quad \text{and} \quad \|\mathsf{A}_\varrho\| \le \|\mathsf{A}\| + \varrho\|\mathsf{B}\|^2,$$

so that

$$\varrho\|\mathsf{B}\|^2 \le C\|\mathsf{A}\| \Rightarrow \kappa(\mathsf{A} + \varrho\mathsf{B}^T\mathsf{B}) \le (C+1)\kappa(\mathsf{A}).$$

If $\|\mathsf{A}\| \approx \|\mathsf{B}\|$, then $\varrho = 2$ seems to be a reasonable choice.

The parameter M balances the weight of the cost function and the constraints. If $\|\mathsf{A}\| \approx \|\mathsf{B}\|$, the $M = 1$ is a reasonable choice. If we use different relative precision ε_g for the norm of gradient and for the gradient and ε_f for the feasibility error, then a reasonable choice is

$$M = \varepsilon_g/\varepsilon_f.$$

If there is no better guess of the initial approximation of $\boldsymbol{\lambda}^0$, we use $\boldsymbol{\lambda}^0 = \mathbf{o}$. Recall that using $\boldsymbol{\lambda}^0 \in \mathrm{Im}\mathsf{B}$ results in $\boldsymbol{\lambda}^k$ converging to the least square Lagrange multiplier $\boldsymbol{\lambda}_{\mathrm{LS}}$.

9.4.3 Solution of More General Problems

If A is positive definite only on the kernel of B, we can use a suitable penalization to reduce such a problem to the convex one. Using Lemma 1.5, it follows that there is $\overline{\varrho} > 0$ such that $\mathsf{A} + \overline{\varrho}\mathsf{B}^T\mathsf{B}$ is positive definite, so that we can apply our SMALE algorithm to the equivalent penalized problem

$$\min_{\mathbf{x}\in\Omega_E} f_{\overline{\varrho}}(\mathbf{x}), \tag{9.57}$$

where

$$f_{\overline{\varrho}}(\mathbf{x}) = \mathbf{x}^T(\mathsf{A} + \overline{\varrho}\mathsf{B}^T\mathsf{B})\mathbf{x} - \mathbf{b}^T\mathbf{x}.$$

Alternatively, we can modify the inner loop of SMALE so that it increases the penalty parameter whenever the negative curvature is recognized. Let us point out that such modification does not guarantee the optimality of the modified algorithm.

9.5 Numerical Experiments

Here we illustrate the performance of the exact Uzawa algorithm, the exact augmented Lagrangian algorithm, and SMALE Algorithm 9.2 on minimization of the cost functions $f_{\mathrm{L},h}$ and $f_{\mathrm{LW},h}$ introduced in Sect. 8.6 subject to ill-conditioned multipoint constraints. Let us recall that we refer to Algorithm 6.2 as the Uzawa algorithm when $\varrho = 0$ and the augmented Lagrangian algorithm when $\varrho > 0$.

9.5.1 Uzawa, Exact Augmented Lagrangians, and SMALE

Let us start with the minimization of the quadratic function $f_{\mathrm{L},h}$ defined by the discretization parameter h (see Sect. 8.6.1) subject to the multipoint constraints which join the displacements of the node with the coordinates $(0, 1/3)$ with all the other nodes in the square $[h, 1/3] \times [1/3, 2/3]$. Let us recall that the Hessian $\mathsf{A}_{\mathrm{L},h}$ of $f_{\mathrm{L},h}$ is ill-conditioned with the spectral condition number $\kappa(\mathsf{A}_{\mathrm{L},h}) \approx h^{-2}$.

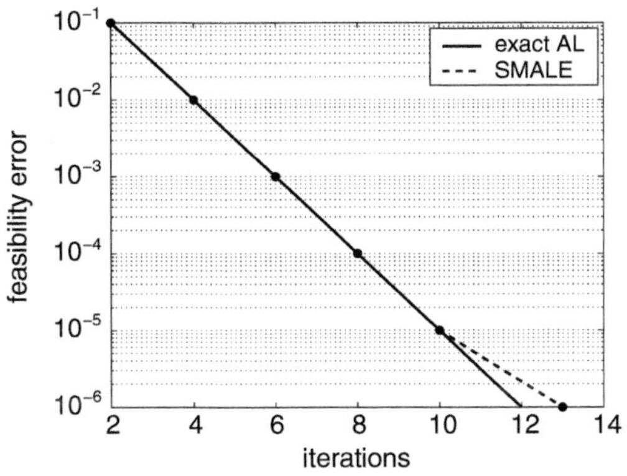

Fig. 9.1. Outer iterations of exact AL and SMALE algorithms

The graph of the relative feasibility error (vertical axis) against the numbers of outer iterations (horizontal axis) for exact augmented Lagrangians (exact AL) with $r_k = \varrho_k = 10$ and SMALE algorithm with $M_0 = 1$ and $\varrho_0 = 10$ is in Fig. 9.1. We obtained the results with $h = 1/33$, corresponding to $n = 1156$ unknowns and 131 multipliers. The inexact solution of auxiliary problems by SMALE has little effect on the convergence of outer iterations. The SMALE algorithm required 964 CG iterations to reach the final precision. We achieved the same result using the original Uzawa algorithm with the optimal steplength only after 3840 (!!!) iterations, each comprising direct solutions of auxiliary linear problems. We conclude that even moderate regularization improves the convergence of the outer loop, and the inexact solution of auxiliary problems need not slow down its convergence rate.

9.5.2 Numerical Demonstration of Optimality

To illustrate the optimality of SMALE for the solution of (9.1), let us consider the class of problems to minimize the quadratic function $f_{\mathrm{LW},h}$ (see Sect. 8.6.3) subject to the multipoint constraints defined above. The class of problems can be given a mechanical interpretation associated with the expanding and partly stiff spring systems on Winkler's foundation. The spectrum of the Hessian $\mathsf{A}_{\mathrm{LW},h}$ of $f_{\mathrm{LW},h}$ is located in the interval [4, 14]. Thus, the assumptions of Theorem 9.3 are satisfied, and the number of outer iterations is bounded. Moreover, the rows of $\mathsf{B} \in \mathbb{R}^{m \times n}$ have a simple pattern given by

$$\mathsf{B}_{i*} = [0, \dots, 0, 1, 0, \dots, 0, -1, 0, \dots, 0], \quad i = 1, \dots, m.$$

It can be checked that $\mathsf{B}^T\mathsf{B}$ can be expressed as the sum of a matrix with the norm not exceeding four and a matrix of rank two. Using the reasoning of Sect. 9.3, we get that the number of inner iterations is also bounded.

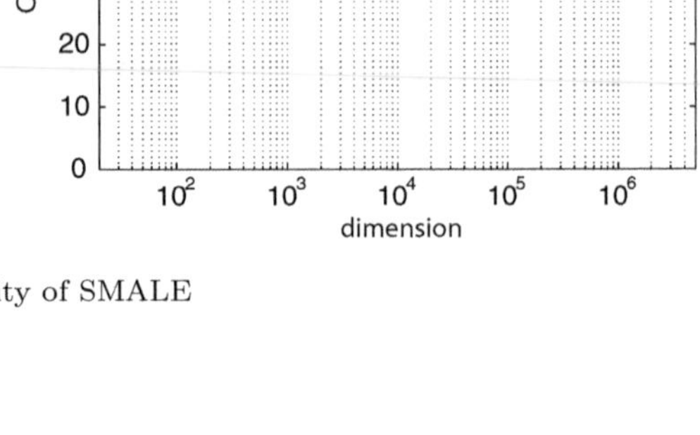

Fig. 9.2. Optimality of SMALE

In Fig. 9.2, on the vertical axis, we can see the numbers of the CG iterations k_n required to reduce the norm of the gradient and of the feasibility error to $10^{-6}\|\nabla f_{\mathrm{LW},h}(\mathbf{o})\|$ for the problems with the dimension n ranging from $n = 49$ to $n = 2,362,369$. The dimension n on the horizontal axis is in the logarithmic scale. We can see that k_n varies mildly with varying n, in agreement with Theorem 9.4 and the optimality of CG. Moreover, since the cost of the matrix–vector multiplications is in our case proportional to the dimension n of $\mathsf{A}_{\mathrm{LW},h}$, it follows that the cost of the solution is also proportional to n. The number of outer iterations ranged from seven to ten.

The purpose of the above numerical experiment was to illustrate the concept of optimality. For large practical applications, combining SMALE with a suitable preconditioning can be necessary. Application of SMALE with the multigrid preconditioning to development of, in a sense, an optimal algorithm for the solution of the discretized Stokes problem is in Lukáš and Dostál [206].

9.6 Comments and References

Several researchers exploited the penalty method for the solution of contact problems of elasticity [13, 154, 173, 177] and many other problems. Theoretical results concerning the penalty method (e.g., Dostál [61], or Sect. 8.1 of Kikuchi and Oden [180]) yield that the norm of the approximation error depends on the condition number of the Hessian of the cost function. The optimal feasibility estimate for the penalty methods (6.14) was used in the first scalable algorithms for variational inequalities [78, 79].

An early analysis of the inexact augmented Lagrangian method can be found in Hager [157, 159]. He obtained global convergence results for an algorithm of this type using inexact minimization to solve the auxiliary problems. In both papers, Hager compared the optimality error with the size of the feasibility error of the solution of the auxiliary problems, trying to balance these quantities throughout the whole process. This comparison was used to decide whether the penalty parameter will be increased [157] or the minimization of the auxiliary problem will be stopped [159]. When applied to the least squares estimate of the Lagrange multipliers, the convergence rate was free of any term related to the inexact minimization. Dostál et al. [90, 96] obtained similar results for the linear update combined with the update of the penalty parameter enforcing a priori prescribed reduction of feasibility error. Conn et al. [39] used the same strategy to solve more general bound and equality-constrained problems. A helpful discussion of the application of augmented Lagrangian methods can be found in Birgin and Martínez [19].

The SMALE algorithm was proposed by Dostál [67, 72]. The main new feature of SMALE was the update of the penalty parameter with aim to guarantee the increase of the Lagrangian. The original variant of SMALE enforced the convergence by increasing the penalty parameter; however, this approach deteriorated the conditioning of the Hessian of augmented Lagrangian. The experience showed better results by enforcing the convergence by reducing the parameter M_k and keeping the conditioning fixed. We sometimes refer to SMALE-ϱ or SMALE-M.

The most attractive feature of SMALE is a uniform bound on the number of iterations necessary to achieve prescribed precision. The bound results from a global analysis; we can hardly obtain it by analyzing one step of the algorithm. The first optimality results are in Dostál et al. [107]. These authors also proved that the resulting SMALE-M algorithm enjoys the R-linear convergence for any $\varrho > 0$.

Let us point out that the optimality results for the SMALE algorithm refer to the type of convergence known from the classical analysis of infinite series but seldom exploited in numerical analysis. We shall call it the *sum bounding convergence of the second order* as it uses the bound on the sum of the squares of errors. The sum-bounding convergence is an alternative characteristic of convergence. For example, if the algorithm generates a "bad" (here far from feasible) iteration, the gap between the current sum and the upper bound is essentially reduced. Since we can control the upper bound by the penalty parameter, the sum bounding convergence explains the fast convergence of the outer loop of SMALE observed in our numerical experiments [206].

The preconditioning preserving the gap in the spectrum appeared in the Axelsson and Lindskög [8]; see also Dostál [65].

Since solving equality-constrained quadratic programming problems with equality constraints can be reduced to solving linear systems with saddle point matrices, there is a vast amount of literature on the solution and preconditioning of such systems or the related Schur complements. See, e.g., Benzi et al. [14] or Axelsson [7].

10 MPRGP for Bound Constraints

Let us consider again the *bound constrained problem* to find

$$\min_{\mathbf{x} \in \Omega_B} f(\mathbf{x}) \tag{10.1}$$

with $\Omega_B = \{\mathbf{x} \in \mathbb{R}^n : \ \mathbf{x} \geq \boldsymbol{\ell}\}$, $f(\mathbf{x}) = \frac{1}{2}\mathbf{x}^T\mathsf{A}\mathbf{x} - \mathbf{x}^T\mathbf{b}$, $\boldsymbol{\ell}$ and $\mathbf{b}$ given column n-vectors, and A an $n \times n$ SPD or SPS matrix. To include unconstrained components of $\mathbf{x}$, we admit $\ell_i = -\infty$. Unlike Chap. 7, we are interested in large, possibly sparse problems with a well-conditioned A. Such problems can also appear as auxiliary problems in algorithms for solving bound and equality-constrained problems.

The starting point of our development is the Polyak algorithm introduced in Chap. 7. We present two essential improvements to cope with the drawbacks of the Polyak algorithm. We first replace the original expansion step with the free gradient projection step, which can add many indices to the active set and guarantees a sufficient decrease in the cost function. The second modification balances the norms of the free and chopped gradients to ensure that the reduced free gradient is always sufficiently large in the CG steps. Recall that the conjugate gradient step efficiently reduces only the free gradient and is inefficient when the chopped gradient norm dominates the KKT conditions' error.

The modified algorithm enjoys the *R-linear rate of convergence* of both the cost function and norm of projected gradient. Moreover, it has the *finite termination property* even for dual degenerate QP problems. The performance of MPRGP can be improved by a problem-dependent *preconditioning in face.* We show that *preconditioning by conjugate projector* (deflation) can improve the convergence rate. We also show how to adapt MPRGP to solve problems with SPS Hessian. The performance of the algorithms, including scalability, is illustrated by solving academic benchmarks and particle simulation.

Z. Dostál, *Optimal Quadratic Programming and QCQP Algorithms with Case Studies*, Springer Optimization and Its Applications 23,
https://doi.org/10.1007/978-3-031-95167-1_10

10.1 Quantitative Refinement of KKT Conditions

If the dimension n of the problem (10.1) is large, it can be too ambitious to look for a solution that satisfies the KKT conditions exactly. A natural idea is to consider the weaker condition

$$\|\mathbf{g}^P(\mathbf{x})\| \le \varepsilon, \tag{10.2}$$

but require that $\mathbf{x} \in \Omega_B$ is satisfied exactly. Recall that $\mathbf{g}^P$ is the projected gradient introduced in Sect. 7.1. Thus, the typical "solution" returned by an iterative solver is just a feasible $\mathbf{x}$ that satisfies the condition (10.2) with a small ε. The following lemma shows that (10.2) also provides a computable bound on the solution error.

Lemma 10.1. *Let $\widehat{\mathbf{x}}$ be the solution of (10.1) with a positive definite A and let $\mathbf{g}^P = \mathbf{g}^P(\mathbf{x})$ denote the projected gradient at $\mathbf{x} \in \Omega_B$. Then*

$$\|\mathbf{x} - \widehat{\mathbf{x}}\|_{\mathsf{A}}^2 \le 2(f(\mathbf{x}) - f(\widehat{\mathbf{x}})) \le \|\mathbf{g}^P\|_{\mathsf{A}^{-1}}^2 \le \lambda_{\min}^{-1}\|\mathbf{g}^P\|^2, \tag{10.3}$$

where $\lambda_{\min}$ denotes the smallest eigenvalue of A.

Proof. Let $\widehat{\mathcal{A}}, \widehat{\mathcal{F}}$, and $\widehat{\mathbf{g}}$ denote the active set, free set, and the gradient in the solution, respectively. Since

$$[\mathbf{x} - \widehat{\mathbf{x}}]_{\widehat{\mathcal{A}}} \ge \mathbf{o}_{\widehat{\mathcal{A}}}, \quad \widehat{\mathbf{g}}_{\widehat{\mathcal{F}}} = \mathbf{o}_{\widehat{\mathcal{F}}}, \quad \text{and} \quad \widehat{\mathbf{g}} \ge \mathbf{o},$$

we get

$$\begin{aligned} f(\mathbf{x}) - f(\widehat{\mathbf{x}}) &= \widehat{\mathbf{g}}^T(\mathbf{x} - \widehat{\mathbf{x}}) + \frac{1}{2}(\mathbf{x} - \widehat{\mathbf{x}})^T\mathsf{A}(\mathbf{x} - \widehat{\mathbf{x}}) \\ &= \widehat{\mathbf{g}}_{\widehat{\mathcal{A}}}^T[\mathbf{x} - \widehat{\mathbf{x}}]_{\widehat{\mathcal{A}}} + \frac{1}{2}\|\mathbf{x} - \widehat{\mathbf{x}}\|_{\mathsf{A}}^2 \ge \frac{1}{2}\|\mathbf{x} - \widehat{\mathbf{x}}\|_{\mathsf{A}}^2. \end{aligned}$$

This proves the left inequality of (10.3).

To prove the middle inequality, let $\mathcal{A} = \mathcal{A}(\mathbf{x})$ and $\mathcal{F} = \mathcal{F}(\mathbf{x})$ denote the active set and the free set of $\mathbf{x} \in \Omega_B$, respectively. Since

$$\mathbf{g}_{\mathcal{F}}^P = \mathbf{g}_{\mathcal{F}}, \quad [\widehat{\mathbf{x}} - \mathbf{x}]_{\mathcal{A}} \ge \mathbf{o}_{\mathcal{A}}, \quad \mathbf{g} = (\mathbf{g} - \mathbf{g}^P) + \mathbf{g}^P, \quad \textbf{and} \quad \mathbf{g} - \mathbf{g}^P \ge \mathbf{o},$$

we get

$$\begin{aligned} 0 \ge 2(f(\widehat{\mathbf{x}}) - f(\mathbf{x})) &= \|\widehat{\mathbf{x}} - \mathbf{x}\|_{\mathsf{A}}^2 + 2\mathbf{g}^T(\widehat{\mathbf{x}} - \mathbf{x}) \\ &= \|\widehat{\mathbf{x}} - \mathbf{x}\|_{\mathsf{A}}^2 + 2(\mathbf{g} - \mathbf{g}^P)^T(\widehat{\mathbf{x}} - \mathbf{x}) + 2(\mathbf{g}^P)^T(\widehat{\mathbf{x}} - \mathbf{x}) \\ &= \|\widehat{\mathbf{x}} - \mathbf{x}\|_{\mathsf{A}}^2 + 2[\mathbf{g} - \mathbf{g}^P]_{\mathcal{A}}^T[\widehat{\mathbf{x}} - \mathbf{x}]_{\mathcal{A}} + 2(\mathbf{g}^P)^T(\widehat{\mathbf{x}} - \mathbf{x}) \\ &\ge \|\widehat{\mathbf{x}} - \mathbf{x}\|_{\mathsf{A}}^2 + 2(\mathbf{g}^P)^T(\widehat{\mathbf{x}} - \mathbf{x}) \\ &\ge 2\left(\min_{\mathbf{y} \in \mathbb{R}^n} \frac{1}{2}\mathbf{y}^T\mathsf{A}\mathbf{y} + (\mathbf{g}^P)^T\mathbf{y}\right) = -(\mathbf{g}^P)^T\mathsf{A}^{-1}\mathbf{g}^P = -\|\mathbf{g}^P\|_{\mathsf{A}^{-1}}^2. \end{aligned}$$

We used (2.10) in the last step. Using simple manipulations and (1.30), we get the middle inequality and the right inequality of (10.3). □

Remark 10.1. By the above lemma, small $\mathbf{g}^P(\mathbf{x})$ guarantees that the solution $\widehat{\mathbf{x}}$ is near to $\mathbf{x}$, but not the other way round. The projected gradient is not a continuous mapping and can be large near the solution as in Fig. 10.1!

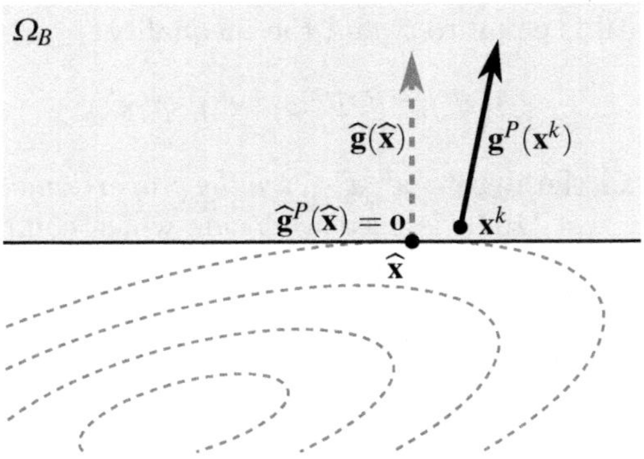

Fig. 10.1. Large projected gradient near the solution

Alternatively, we can use test based on the *α-reduced gradient* $\widetilde{\mathbf{g}}_\alpha(\mathbf{x})$ defined in Sect. 5.1 component-wise by

$$[\widetilde{\mathbf{g}}_\alpha(\mathbf{x})]_i = \min\left\{g_i, \frac{x_i - \ell_i}{\alpha}\right\}. \tag{10.4}$$

Indeed, $\widehat{\mathbf{x}}$ solves (10.1) if and only if $\widetilde{\mathbf{g}}_\alpha(\widehat{\mathbf{x}})$ vanishes and $\widetilde{\mathbf{g}}_\alpha(\mathbf{x})$ is a continuous function of $\mathbf{x}$. Moreover, since $\widetilde{\mathbf{g}}_\alpha$ represents the feasible increment of the gradient projection with step α, it cannot be large near the solution and

$$\|\widetilde{\mathbf{g}}_\alpha(\mathbf{x})\| \leq \|\mathbf{g}^P(\mathbf{x})\|. \tag{10.5}$$

The reduced gradient is most useful for more general optimization problems discussed in the next chapter.

Recall that $\mathbf{g}^P$ is the sum of chopped and free gradients (see (7.4))

$$\mathbf{g}^P(\mathbf{x}) = \boldsymbol{\beta}(\mathbf{x}) + \boldsymbol{\varphi}(\mathbf{x}). \tag{10.6}$$

We can decompose similarly the reduced gradient

$$\widetilde{\mathbf{g}}_\alpha(\mathbf{x}) = \boldsymbol{\beta}(\mathbf{x}) + \widetilde{\boldsymbol{\varphi}}_\alpha(\mathbf{x}), \quad \widetilde{\boldsymbol{\varphi}}_\alpha(\mathbf{x}) = [\widetilde{\mathbf{g}}_\alpha(\mathbf{x})]_{\mathcal{F}(\mathbf{x})}. \tag{10.7}$$

10.2 MPRGP Schema

The algorithm that we propose here exploits user-defined constants $\Gamma > 0$ and $\overline{\alpha} \in (0, 2\|\mathsf{A}\|^{-1})$, a test used to decide when to leave the face, and three types of steps. Let us denote

$$\widehat{\alpha} = \min\{\overline{\alpha}, 2\|\mathsf{A}\|^{-1} - \overline{\alpha}\} \in (0, \|\mathsf{A}^{-1}). \tag{10.8}$$

If the steplength is equal to $\overline{\alpha}$ and the inequality

$$\|\boldsymbol{\beta}(\mathbf{x}^k)\|^2 \leq \Gamma^2 \widetilde{\varphi}_{\widehat{\alpha}}(\mathbf{x}^k)^T \varphi(\mathbf{x}^k) \tag{10.9}$$

holds, then we call the iterate $\mathbf{x}^k$ $\Gamma-$*strictly proportional* or briefly *strictly proportional.* The test (10.9) is used to decide which components of $\mathbf{g}^P(\mathbf{x}^k)$ should be reduced in the next step. Notice that the test (10.9) compares the potential decrease of the cost function in the direction of the two components of the gradient with the same step length.

The *conjugate gradient step* is defined as in Polyak's algorithm by

$$\mathbf{x}^{k+1} = \mathbf{x}^k - \alpha_{cg}\mathbf{p}^{k+1}, \quad \alpha_{cg} = \mathbf{b}^T\mathbf{p}^{k+1} / \left(\mathbf{p}^{k+1}\right)^T \mathsf{A}\mathbf{p}^{k+1}, \tag{10.10}$$

where $\mathbf{p}^{k+1}$ is the conjugate gradient direction constructed recurrently. The recurrence starts (or restarts) with $\mathbf{p}^{s+1} = \varphi(\mathbf{x}^s)$ whenever $\mathbf{x}^s$ is strictly proportional and

$$\mathbf{x}^k - \alpha_{cg}\varphi \in \Omega_B.$$

If $\mathbf{x}^k$ is generated by the conjugate gradient step, then

$$\mathbf{p}^{k+1} = \varphi(\mathbf{x}^k) - \beta\mathbf{p}^k, \quad \beta = \frac{\varphi(\mathbf{x}^k)^T \mathsf{A}\boldsymbol{p}^k}{(\mathbf{p}^k)^T \mathsf{A}\mathbf{p}^k}. \tag{10.11}$$

The coefficient α_{cg} is chosen so that

$$f(\mathbf{x}^{s+m}) = \min\{f(\mathbf{x}) : \mathbf{x} \in \mathbf{x}^s + \mathrm{Span}\{\mathbf{p}^{s+1}, \dots, \mathbf{p}^{s+m}\}.$$

We can check directly that

$$f(\mathbf{x}^{k+1}) \leq f\left(\mathbf{x}^k - \alpha_{cg}\varphi(\mathbf{x}^k)\right) = f(\mathbf{x}^k) - \frac{1}{2}\frac{\|\varphi(\mathbf{x}^k)\|^4}{\varphi(\mathbf{x}^k)^T \mathsf{A}\varphi(\mathbf{x}^k)}. \tag{10.12}$$

The conjugate gradient steps carry out the minimization in the working face $\mathcal{W}_{\mathcal{I}}$ given by $\mathcal{I} = \mathcal{A}(\mathbf{x}^s)$. We shall use in our proofs that by Theorem 4.1

$$f(\mathbf{x}^{s+m}) = \min\{f(\mathbf{x}^s + \mathbf{y}) : \ \mathbf{y} \in \mathrm{Span}\{\varphi(\mathbf{x}^s), \dots, \varphi(\mathbf{x}^{s+m-1})\}\}. \tag{10.13}$$

The *expansion step* is defined by

$$\mathbf{x}^{k+1} = P_{\Omega_B}(\mathbf{x}^k - \overline{\alpha}\boldsymbol{\varphi}(\mathbf{x}^k)) = \max\{\boldsymbol{\ell}, \mathbf{x}^k - \overline{\alpha}\boldsymbol{\varphi}(\mathbf{x}^k)\}. \tag{10.14}$$

The expansion step is carried out when the CG step is not feasible. This step typically adds indices to the current working set. To describe the expansion step in a form suitable for analysis, let us introduce the *reduced free gradient* $\widetilde{\boldsymbol{\varphi}}_\alpha(\mathbf{x})$ with the entries

$$\widetilde{\varphi}_i = \widetilde{\varphi}_i(\mathbf{x}, \alpha) = \min\{(x_i - \ell_i)/\alpha, \varphi_i\}, \quad i \in \mathcal{N} = \{1, \dots, n\}. \tag{10.15}$$

It is defined for any $\mathbf{x} \in \Omega_B$ and $\alpha > 0$. Thus

$$P_{\Omega_B}(\mathbf{x} - \alpha\boldsymbol{\varphi}(\mathbf{x})) = \mathbf{x} - \alpha\widetilde{\boldsymbol{\varphi}}_\alpha(\mathbf{x}). \tag{10.16}$$

The *proportioning step* is defined by

$$\mathbf{x}^{k+1} = \mathbf{x}^k - \alpha_{cg}\boldsymbol{\beta}(\mathbf{x}^k) \tag{10.17}$$

with the steplength α_{cg} that minimizes $f\left(\mathbf{x}^k - \alpha\boldsymbol{\beta}(\mathbf{x}^k)\right)$. It is invoked when $\mathbf{x}^k$ is not strictly proportional. It has been shown in Sect. 4.1 that the CG steplength α_{cg} that minimizes $f(\mathbf{x} - \alpha\mathbf{d})$ for a given $\mathbf{d}$ and $\mathbf{x}$ can be evaluated using the gradient $\mathbf{g} = \mathbf{g}(\mathbf{x}) = \nabla f(\mathbf{x})$ at $\mathbf{x}$ by

$$\alpha_{cg} = \alpha_{cg}(\mathbf{d}) = \frac{\mathbf{d}^T\mathbf{g}}{\mathbf{d}^T\mathsf{A}\mathbf{d}}. \tag{10.18}$$

The proportioning step removes the indices that violate the KKT conditions from the active set. Note that if $\mathbf{x}^k \in \Omega_B$, then

$$\mathbf{x}^{k+1} = \mathbf{x}^k - \alpha_{cg}\boldsymbol{\beta}(\mathbf{x}^k) \in \Omega_B, \quad \alpha_{cg} = \frac{\|\boldsymbol{\beta}\|^2}{\boldsymbol{\beta}^T\mathsf{A}\boldsymbol{\beta}}.$$

Now, we are ready to define the algorithm in a convenient form for analysis, postponing the discussion about implementation to the next section.

Algorithm 10.1. Modified proportioning with reduced gradient projections (MPRGP schema)

Given a symmetric positive definite matrix $\mathsf{A} \in \mathbb{R}^{n\times n}$ *and* n*-vectors* $\mathbf{b}, \boldsymbol{\ell}$.
Choose $\mathbf{x}^0 \in \Omega_B$, $\overline{\alpha} \in (0, 2\|\mathsf{A}\|^{-1})$, *and* $\Gamma > 0$. *Set* $k = 0$.
For $k \geq 0$ *and* $\mathbf{x}^k$ *known, choose* $\mathbf{x}^{k+1}$ *by the following rules:*

(i) If $\mathbf{g}^P(\mathbf{x}^k) = \mathbf{o}$, *set* $\mathbf{x}^{k+1} = \mathbf{x}^k$.
(ii) *If* $\mathbf{x}^k$ *is strictly proportional and* $\mathbf{g}^P(\mathbf{x}^k) \neq \mathbf{o}$, *try to generate* $\mathbf{x}^{k+1}$ *by the conjugate gradient step. If* $\mathbf{x}^{k+1} \in \Omega_B$, *then accept it, else generate* $\mathbf{x}^{k+1}$ *by the expansion step.*
(iii) *If* $\mathbf{x}^k$ *is not strictly proportional, define* $\mathbf{x}^{k+1}$ *by proportioning.*

We call our algorithm modified proportioning to distinguish it from the earlier algorithms introduced independently by Friedlander and Martínez with their collaborators [18, 51, 138–140] and Dostál [62, 63]. These earlier algorithms applied the proportioning step when

$$\|\boldsymbol{\beta}(\mathbf{x}^k)\| \leq \Gamma\|\boldsymbol{\varphi}(\mathbf{x}^k)\|.$$

10.3 Rate of Convergence

Now, we are ready to prove the convergence of MPRGP. We first prove its R-linear convergence rate in the energy norm. The proof uses the gradient information associated with the three types of MPRGP steps and does not consider the information from the previous steps. Then, we give a similar result for the Euclidean norm of the projected gradient using the gradient information from two subsequent iterations. This result is nontrivial as the projected gradient is not a continuous mapping and can be large near the solution. Recall that $\widetilde{\mathbf{g}}_\alpha(\mathbf{x}^k)$ and $\widetilde{\boldsymbol{\varphi}}_\alpha(\mathbf{x}^k)$ denote the α-reduced gradient and α-reduced free gradient defined by (5.4). In the proof, we shall use Lemma 5.3 which states that for $\overline{\alpha} \in (0, 2\|\mathsf{A}\|^{-1})$ and $\widehat{\alpha} = \min\{\overline{\alpha}, 2 - \overline{\alpha}\}$, we have

$$f(P_{\Omega_B}(\mathbf{x} - \overline{\alpha}\boldsymbol{\varphi}(\mathbf{x}) \leq f(\mathbf{x}) - \frac{\widehat{\alpha}}{2}\widetilde{\boldsymbol{\varphi}}_{\widehat{\alpha}}(\mathbf{x})^T\boldsymbol{\varphi}(\mathbf{x}). \tag{10.19}$$

10.3.1 Convergence in Energy Norm

The following theorem guarantees the R-linear rate of convergence of MPRGP in terms of bounds on the spectrum of the Hessian A for $\overline{\alpha} \in (0, 2\|\mathsf{A}\|^{-1})$.

Theorem 10.1. *Let $\{\mathbf{x}^k\}$ be generated by Algorithm 10.1 with $\mathbf{x}^0 \in \Omega_B$, $\Gamma > 0$, and $\overline{\alpha} \in (0, 2\|\mathsf{A}\|^{-1}]$. Then*

$$f(\mathbf{x}^{k+1}) - f(\widehat{\mathbf{x}}) \leq \eta_\Gamma(f(\mathbf{x}^k) - f(\widehat{\mathbf{x}})), \tag{10.20}$$

where $\widehat{\mathbf{x}}$ denotes the unique solution of (10.1),

$$\eta_\Gamma = 1 - \frac{\widehat{\alpha}\lambda_{\min}}{2 + 2\widehat{\Gamma}^2}, \qquad \widehat{\Gamma} = \max\{\Gamma, \Gamma^{-1}\}, \tag{10.21}$$

$$\widehat{\alpha} = \min\{\overline{\alpha}, 2\|\mathsf{A}\|^{-1} - \overline{\alpha}\}, \tag{10.22}$$

and $\lambda_{\min}$ denotes the smallest eigenvalue of A.

The error in the A-norm is bounded by

$$\|\mathbf{x}^k - \widehat{\mathbf{x}}\|_{\mathsf{A}}^2 \leq 2\eta_\Gamma^k(f(\mathbf{x}^0) - f(\widehat{\mathbf{x}})). \tag{10.23}$$

Proof. We shall examine the three steps separately. The main idea is to compare the decrease of the cost function f along the free gradient path with the known decrease of f along the projected gradient path. Recall that by Theorem 5.1 for $\alpha \in (0, 2\|\mathsf{A}^{-1})$, $\widehat{\alpha} = \min\{\alpha, 2\|\mathsf{A}\|^{-1} - \alpha\}$,

$$f(P_{\Omega_B}(\mathbf{x} - \alpha\mathbf{g}(\mathbf{x})) - f(\widehat{\mathbf{x}}) \leq \eta_f(f(\mathbf{x}) - f(\widehat{\mathbf{x}})), \quad \eta_f = 1 - \widehat{\alpha}\lambda_{\min}. \tag{10.24}$$

We shall use the representation of the projection

$$P_{\Omega_B}(\mathbf{x} - \alpha\mathbf{g}(\mathbf{x})) = \mathbf{x} - \alpha\widetilde{\mathbf{g}}_\alpha(\mathbf{x}) = \mathbf{x} - \alpha\big(\widetilde{\varphi}_\alpha(\mathbf{x}) + \boldsymbol{\beta}(\mathbf{x})\big), \quad \alpha \geq 0,$$

and the Taylor expansion

$$f(P_{\Omega_B}(\mathbf{x} - \alpha\mathbf{g}(\mathbf{x})) \geq f(\mathbf{x}) - \alpha\left(\widetilde{\varphi}_\alpha(\mathbf{x})^T\varphi(\mathbf{x}) + \|\boldsymbol{\beta}(\mathbf{x})\|^2\right). \tag{10.25}$$

The *expansion step* is used when $\mathbf{x}^k$ is strictly proportional, i.e.

$$\|\boldsymbol{\beta}(\mathbf{x}^k)\|^2 \leq \Gamma^2\widetilde{\varphi}_{\widehat{\alpha}}(\mathbf{x}^k)^T\varphi(\mathbf{x}^k). \tag{10.26}$$

After substituting (10.26) into (10.25) with $\alpha = \widehat{\alpha}$, we get

$$f(P_{\Omega_B}(\mathbf{x}^k - \widehat{\alpha}\mathbf{g}(\mathbf{x}^k))) \geq f(\mathbf{x}^k) - \widehat{\alpha}(1 + \Gamma^2)\widetilde{\varphi}_{\widehat{\alpha}}(\mathbf{x}^k)^T\varphi(\mathbf{x}^k). \tag{10.27}$$

Thus for $\mathbf{x}^{k+1}$ generated by the expansion step, we get by elementary algebra and application of Lemma 5.3 and (10.27) that for $0 < \overline{\alpha} < 2\|\mathsf{A}\|^{-1}$

$$\begin{aligned} f(\mathbf{x}^{k+1}) = f\big(P_{\Omega_B}(\mathbf{x}^k - \overline{\alpha}\varphi(\mathbf{x}^k))\big) &\overset{L.5.3}{\leq} f(\mathbf{x}^k) - \frac{\widehat{\alpha}}{2}\widetilde{\varphi}_{\widehat{\alpha}}(\mathbf{x}^k)^T\varphi(\mathbf{x}^k) \qquad (10.28)\\ &= \frac{1}{2 + 2\Gamma^2}\big((1 + 2\Gamma^2)f(\mathbf{x}^k) + f(\mathbf{x}^k) - \widehat{\alpha}(1 + \Gamma^2)\widetilde{\varphi}_{\widehat{\alpha}}(\mathbf{x}^k)^T\varphi(\mathbf{x}^k)\big)\\ &\overset{(10.27)}{\leq} \frac{1}{2 + 2\Gamma^2}\Big(f\big(P_\Omega\left(\mathbf{x}^k - \widehat{\alpha}\mathbf{g}(\mathbf{x}^k)\right)\big) + (1 + 2\Gamma^2)f(\mathbf{x}^k)\Big). \end{aligned}$$

Inserting $-f(\widehat{\mathbf{x}}) + f(\widehat{\mathbf{x}})$ into the last term and then substituting (10.24) with $\mathbf{x} = \mathbf{x}^k$ and $\alpha = \widehat{\alpha}$ into the result, we get

$$\begin{aligned} f(\mathbf{x}^{k+1}) &\leq \frac{\eta_f + 1 + 2\Gamma^2}{2 + 2\Gamma^2}f(\mathbf{x}^k) + \frac{1 - \eta_f}{2 + 2\Gamma^2}f(\widehat{\mathbf{x}})\\ &= \frac{\eta_f + 1 + 2\Gamma^2}{2 + 2\Gamma^2}f(\mathbf{x}^k) + \frac{-\eta_f - 1 - 2\Gamma^2 + 2 + 2\Gamma^2}{2 + 2\Gamma^2}f(\widehat{\mathbf{x}})\\ &= \frac{\eta_f + 1 + 2\Gamma^2}{2 + 2\Gamma^2}\big(f(\mathbf{x}^k) - f(\widehat{\mathbf{x}})\big) + f(\widehat{\mathbf{x}}). \end{aligned}$$

Thus

$$f(\mathbf{x}^{k+1}) - f(\widehat{\mathbf{x}}) \le \frac{\eta_f + 1 + 2\Gamma^2}{2 + 2\Gamma^2}(f(\mathbf{x}^k) - f(\widehat{\mathbf{x}}))$$
$$\overset{(10.24)}{=} \left(1 - \frac{\widehat{\alpha}\lambda_{\min}}{2 + 2\Gamma^2}\right)(f(\mathbf{x}^k) - f(\widehat{\mathbf{x}})). \qquad (10.29)$$

If $\mathbf{x}^{k+1}$ is generated by the *CG step* (10.10), then by (10.12)

$$f(\mathbf{x}^{k+1}) \overset{(10.10)}{\le} f(\mathbf{x}^k - \alpha_{cg}\varphi(\mathbf{x}^k)) \overset{(10.12)}{=} f(\mathbf{x}^k) - \frac{1}{2}\frac{\|\varphi(\mathbf{x}^k)\|^4}{\varphi(\mathbf{x}^k)^T \mathsf{A}\varphi(\mathbf{x}^k)}$$
$$\le f(\mathbf{x}^k) - \frac{1}{2}\|\mathsf{A}\|^{-1}\|\varphi(\mathbf{x}^k)\|^2.$$

The CG step is assumed to be feasible. Using $\widehat{\alpha} \le \|\mathsf{A}\|^{-1}$, we get

$$f(\mathbf{x}^{k+1}) \le f(\mathbf{x}^k) - \frac{1}{2}\|\mathsf{A}\|^{-1}\|\varphi(\mathbf{x}^k)\|^2 \le f(\mathbf{x}^k) - \frac{\widehat{\alpha}}{2}\widetilde{\varphi}_{\widehat{\alpha}}(\mathbf{x}^k)^T\varphi(\mathbf{x}^k). \qquad (10.30)$$

We have got the same inequality as (10.28) obtained for the expansion step. The CG step is also carried out when $\mathbf{x}^k$ is strictly proportional. Thus, we can use the above procedure to show that (10.29) is valid for the CG steps.

Let us finally assume that $\mathbf{x}^{k+1}$ is generated by the *proportioning step* (10.17) so that the iterate $\mathbf{x}^k$ is not strictly proportional, i.e.,

$$\widetilde{\varphi}_{\widehat{\alpha}}(\mathbf{x}^k)^T\varphi(\mathbf{x}^k) < \Gamma^{-2}\|\beta(\mathbf{x}^k)\|^2. \qquad (10.31)$$

After substituting (10.31) into (10.25) with $\alpha = \widehat{\alpha}$, we get

$$f(P_{\Omega_B}(\mathbf{x}^k - \widehat{\alpha}\mathbf{g}(\mathbf{x}^k))) \ge f(\mathbf{x}^k) - \widehat{\alpha}(1 + \Gamma^{-2})\|\beta(\mathbf{x}^k)\|^2. \qquad (10.32)$$

Before we shall use (10.31), notice that

$$f(\mathbf{x}^{k+1}) = f(\mathbf{x}^k - \alpha_{cg}\beta(\mathbf{x}^k)) = f(\mathbf{x}^k) - \frac{1}{2}\frac{\|\beta(\mathbf{x}^k)\|^4}{\beta(\mathbf{x}^k)^T\mathsf{A}\beta(\mathbf{x}^k)}$$
$$\le f(\mathbf{x}^k) - \frac{1}{2}\|\mathsf{A}\|^{-1}\|\beta(\mathbf{x}^k)\|^2.$$

Recall that α_{cg} is defined by (10.18). Since $\widehat{\alpha} \le \|\mathsf{A}\|^{-1}$, we get

$$f(\mathbf{x}^{k+1}) \le f(\mathbf{x}^k) - \frac{\widehat{\alpha}}{2}\|\beta(\mathbf{x}^k)\|^2. \qquad (10.33)$$

Using simple manipulations and (10.32), we get

$$f(\mathbf{x}^{k+1}) = f(\mathbf{x}^k - \alpha_{cg}\beta(\mathbf{x}^k)) \overset{(10.33)}{\le} f(\mathbf{x}^k) - \frac{\widehat{\alpha}}{2}\|\beta(\mathbf{x}^k)\|^2$$
$$= f(\mathbf{x}^k) - \frac{\widehat{\alpha}}{2 + 2\Gamma^{-2}}(1 + \Gamma^{-2})\|\beta(\mathbf{x}^k)\|^2$$
$$\overset{(10.32)}{\le} f\left(P_{\Omega_B}(\mathbf{x}^k - \frac{\widehat{\alpha}\lambda_{\min}}{2 + 2\Gamma^{-2}}\mathbf{g}(\mathbf{x}^k))\right)$$

and

$$f(\mathbf{x}^{k+1}) - f(\widehat{\mathbf{x}}) \le f\left(P_{\Omega_B}\left(\mathbf{x}^k - \frac{\widehat{\alpha}}{2+2\Gamma^{-2}}\mathbf{g}(\mathbf{x}^k)\right)\right) - f(\widehat{\mathbf{x}})$$
$$\overset{T.5.1}{\le} \left(1 - \frac{\widehat{\alpha}\lambda_{\min}}{2+2\Gamma^{-1}}\right)\left(f(\mathbf{x}^k) - f(\widehat{\mathbf{x}})\right).$$

Denoting $\widehat{\Gamma} = \max\{\Gamma, \Gamma^{-1}\}$ and using Theorem 5.1, we can give a common bound on the effect of all three steps of the MPRGP algorithm as

$$f(\mathbf{x}^{k+1}) - f(\widehat{\mathbf{x}}) \le f\Big(P_{\Omega_B}(\mathbf{x}^k - \frac{\widehat{\alpha}}{2+2\widehat{\Gamma}^2}\mathbf{g}(\mathbf{x}^k))\Big) - f(\widehat{\mathbf{x}})$$
$$\le \eta_{\widehat{\Gamma}}(f(\mathbf{x}^k) - f(\widehat{\mathbf{x}})), \quad \eta_{\widehat{\Gamma}} = 1 - \frac{\widehat{\alpha}\lambda_{\min}}{2+2\widehat{\Gamma}^2}.$$

To get the error bound (10.23), notice that by Lemma 10.1

$$\|\mathbf{x}^k - \widehat{\mathbf{x}}\|_{\mathsf{A}}^2 \le 2(f(\mathbf{x}^k) - f(\widehat{\mathbf{x}})) \le 2\eta_{\Gamma}^k(f(\mathbf{x}^0) - f(\widehat{\mathbf{x}})). \quad \square \tag{10.34}$$

Theorem 10.1 gives the best bound on the rate of convergence for the values $\Gamma = \widehat{\Gamma} = 1$ in agreement with the heuristics that we should leave the face when the chopped gradient dominates the violation of the Karush–Kuhn–Tucker conditions. The formula for the best bound η_{Γ}^{opt} which corresponds to $\Gamma = 1$ and $\overline{\alpha} = \|\mathsf{A}\|^{-1}$ reads

$$\eta_{\Gamma}^{opt} = 1 - \kappa(\mathsf{A})^{-1}/4, \tag{10.35}$$

where $\kappa(\mathsf{A})$ denotes the spectral condition number of A.

10.3.2 Rate of Convergence of Gradients

To use the MPRGP algorithm in the inner loop of other algorithms, we must be able to *recognize* when we are near the solution. However, there is a catch— though the algorithm generates iterates that are near the solution, Theorem 10.1 does not guarantee that corresponding projected gradients in these iterates are small. Recall that the projected gradient is not continuous and can be large near the solution, as shown in Fig. 10.1.

Good luck; the norms of projected gradients generated by MPRGP converge to zero. The proof uses the following lemma by Kučera [197] valid for any three iterations.

Lemma 10.2. *Let Ω and f denote a convex set and convex quadratic function, respectively. Let $\widehat{\mathbf{y}}$ denote the minimizer of f subject to $\mathbf{y} \in \Omega$, and let $\mathbf{y}^0$, $\mathbf{y}^1$, and $\mathbf{y}^2$ belong to Ω.*

If there is $\eta \in (0,1)$ such that

$$f(\mathbf{y}^2) - f(\widehat{\mathbf{x}}) \leq \eta\big(f(\mathbf{y}^1) - f(\widehat{\mathbf{x}})\big) \leq \eta^2\big(f(\mathbf{y}^0) - f(\widehat{\mathbf{x}})\big), \tag{10.36}$$

then

$$f(\mathbf{y}^1) - f(\mathbf{y}^2) \leq \frac{1+\eta}{1-\eta}\eta \left(f(\mathbf{y}^0) - f(\mathbf{y}^1)\right).$$

Proof. First, let us denote

$$\widehat{\mathbf{x}} = \arg \min_{\mathbf{x}\in\Omega}(\mathbf{x})$$

and apply repeatedly (10.36) to get

$$\begin{aligned} f(\mathbf{y}^0) - f(\mathbf{y}^1) &= \big(f(\mathbf{y}^0) - f(\widehat{\mathbf{x}})\big) - \big(f(\mathbf{y}^1) - f(\widehat{\mathbf{x}})\big) \geq (1-\eta)\left(f(\mathbf{y}^0) - f(\widehat{\mathbf{x}})\right) \\ &\geq \frac{1-\eta}{\eta}\left(f(\mathbf{y}^1) - f(\widehat{\mathbf{x}})\right) \end{aligned}$$

and

$$\begin{aligned} f(\mathbf{y}^1) - f(\mathbf{y}^2) &= 2\left(\frac{f(\mathbf{y}^1) + f(\mathbf{y}^2)}{2} - f(\mathbf{y}^2)\right) \leq 2\left(\frac{f(\mathbf{y}^1) + f(\mathbf{y}^2)}{2} - f(\widehat{\mathbf{x}})\right) \\ &\leq (1+\eta)\left(f(\mathbf{y}^1) - f(\widehat{\mathbf{x}})\right). \end{aligned}$$

Thus

$$\frac{1}{1+\eta}\left(f(\mathbf{y}^1) - f(\mathbf{y}^2)\right) \leq f(\mathbf{y}^1) - f(\widehat{\mathbf{x}}) \leq \frac{\eta}{1-\eta}\left(f(\mathbf{y}^0) - f(\mathbf{y}^1)\right).$$

This completes the proof. □

Now, we are ready to prove the main result of this section.

Theorem 10.2. *Let $\{\mathbf{x}^k\}$ denote the iterates generated by the MPRGP algorithm for the solution of (10.1) under the assumptions of Theorem 10.1 with*

$$\overline{\alpha} \in (0, 2\|\mathsf{A}\|^{-1}) \quad \text{and} \quad \Gamma > 0.$$

Then there is k_0 such that for any $k \geq k_0$

$$\|\mathbf{g}^P(\mathbf{x}^k)\|^2 \leq a_1 \eta_\Gamma^k \big(f(\mathbf{x}^0) - f(\widehat{\mathbf{x}})\big), \quad a_1 = \frac{2(1+\eta_\Gamma)}{\widehat{\alpha}(1-\eta_\Gamma)}, \tag{10.37}$$

where $\widehat{\alpha}$ and η_Γ are defined in Theorem 10.1, and $\widehat{\mathbf{x}}$ solves (10.1).

Proof. Let $\{\mathbf{x}^k\}$ be generated by the MPRGP algorithm with the parameters $\Gamma > 0$ and $\overline{\alpha} \in (0, 2\|\mathsf{A}\|^{-1})$, so that

$$\widehat{\alpha} \in (0, \|\mathsf{A}\|^{-1}) \quad \text{and} \quad \widehat{\alpha}\|\mathsf{A}\| < 1.$$

We shall apply Lemma 10.2 to $\mathbf{x}^{k-1}$, $\mathbf{x}^k$, and to the auxiliary iteration $\hat{\boldsymbol{x}}^{k+1}$. To specify $\hat{\boldsymbol{x}}^{k+1}$, we shall use that there is $k_0 > 0$ such that for $k \geq k_0$, the binding set $\widehat{\mathcal{B}}$ of the solution $\widehat{\mathbf{x}}$ is a subset of the binding set $\mathcal{B}_k$ of $\mathbf{x}^k$ (for the proof see Lemma 10.4), so that

$$\widehat{\mathcal{B}} = \mathcal{B}(\widehat{\mathbf{x}}) \subseteq \mathcal{B}(\mathbf{x}^k) = \mathcal{B}_k, \quad k \geq k_0 \geq 0.$$

It follows that for $k \geq k_0$,

$$f(\widehat{\mathbf{x}}) = \min_{\mathbf{x}\in\Omega_B} f(\mathbf{x}) = \min_{\mathbf{x}\in\Omega_{\widehat{\mathcal{B}}}} f(\mathbf{x}), \qquad \Omega_{\widehat{\mathcal{B}}} = \{\mathbf{x} \in \mathbb{R}^n : \mathbf{x}_{\widehat{\mathcal{B}}} \geq \ell_{\widehat{\mathcal{B}}}\}.$$

Let us denote the iterates associated with $\Omega_{\widehat{\mathcal{B}}}$ by hat and notice that $\hat{\mathbf{x}}^k = \mathbf{x}^k$ are feasible iterates for solving (10.1), $f(\hat{\mathbf{x}}^k) = f(\mathbf{x}^k)$, as well as $\mathbf{g}(\hat{\mathbf{x}}^k) = \mathbf{g}(\mathbf{x}^k)$. Using the definitions and (10.5), we get that for $k \geq k_0$

$$\|\widetilde{\mathbf{g}}_\alpha(\hat{\mathbf{x}}^k)\| = \|\mathbf{g}^P(\hat{\mathbf{x}}^k)\| \geq \|\mathbf{g}^P(\mathbf{x}^k)\| \geq \|\widetilde{\mathbf{g}}_\alpha(\mathbf{x}^k)\|,$$

so that it is enough to get bound on the first term of the above relations. We shall find it using the auxiliary iteration

$$\hat{\boldsymbol{x}}^{k+1} = P_{\Omega_{\widehat{\mathcal{B}}}}(\mathbf{x}^k - \alpha\mathbf{g}(\mathbf{x}^k)), \quad \alpha = \|\mathsf{A}\|^{-1}.$$

Using definitions and basic linear algebra, we get

$$\begin{aligned}
f(\hat{\boldsymbol{x}}^{k+1}) &= f\left(P_{\Omega_{\widehat{\mathcal{B}}}}(\mathbf{x}^k - \alpha\mathbf{g}\left(\mathbf{x}^k\right))\right)\\
&= f(\mathbf{x}^k) - \alpha\mathbf{g}^T(\mathbf{x}^k)\,\widetilde{\mathbf{g}}_\alpha(\hat{\mathbf{x}}^k) + \frac{\alpha^2}{2}\widetilde{\mathbf{g}}_\alpha^T(\hat{\mathbf{x}}^k)\mathsf{A}\widetilde{\mathbf{g}}_\alpha(\hat{\mathbf{x}}^k)\\
&\leq f(\mathbf{x}^k) - \alpha\|\widetilde{\mathbf{g}}_\alpha(\hat{\mathbf{x}}^k)\|^2 + \frac{\alpha^2\|\mathsf{A}\|}{2}\|\widetilde{\mathbf{g}}_\alpha(\hat{\mathbf{x}}^k)\|^2 = f(\mathbf{x}^k) - \frac{\alpha}{2}\|\widetilde{\mathbf{g}}_\alpha(\hat{\mathbf{x}}^k)\|^2.
\end{aligned}$$

It follows that

$$f(\mathbf{x}^k) - f\left(\hat{\boldsymbol{x}}^{k+1}\right) \geq \frac{\alpha}{2}\|\widetilde{\mathbf{g}}_\alpha(\hat{\mathbf{x}}^k)\|^2 \geq \frac{\alpha}{2}\|\mathbf{g}^P(\mathbf{x}^k)\|^2 \geq \frac{\alpha}{2}\|\widetilde{\mathbf{g}}_\alpha(\mathbf{x}^k)\|^2, \quad (10.38)$$

and by Theorem 5.1

$$f\left(\hat{\boldsymbol{x}}^{k+1}\right) - f(\widehat{\mathbf{x}}) = f\left(P_{\Omega_{\widehat{\mathcal{B}}}}(\mathbf{x}^k - \alpha\mathbf{g}\left(\mathbf{x}^k\right))\right) - f(\widehat{\mathbf{x}}) \leq \eta_\Gamma\big(f(\mathbf{x}^k) - f(\widehat{\mathbf{x}})\big).$$

Let us denote

$$\mathbf{y}^0 = \mathbf{x}^{k-1}, \quad \mathbf{y}^1 = \mathbf{x}^k, \quad \mathbf{y}^2 = \hat{\mathbf{x}}^{k+1} = P_{\Omega_{\widehat{\mathcal{B}}}}(\mathbf{x}^k - \alpha\mathbf{g}\left(\mathbf{x}^k\right))$$

and recall that by the assumptions

$$f\left(\mathbf{x}^k\right) - f(\widehat{\mathbf{x}}) \le \eta_\Gamma\big(f(\mathbf{x}^{k-1}) - f(\widehat{\mathbf{x}})\big), \quad k = 1, 2, \ldots .$$

Using Lemma 10.2 and the latter inequalities with $\eta = \eta_\Gamma$, we get

$$\begin{aligned} f\left(\mathbf{x}^k\right) - f\left(\hat{\mathbf{x}}^{k+1}\right) &\le \frac{1+\eta_\Gamma}{1-\eta_\Gamma}\eta_\Gamma\left(f(\mathbf{x}^{k-1}) - f(\mathbf{x}^k)\right) \\ &\le \frac{1+\eta_\Gamma}{1-\eta_\Gamma}\eta_\Gamma\left(f(\mathbf{x}^{k-1}) - f(\widehat{\mathbf{x}})\right) \\ &\le \frac{1+\eta_\Gamma}{1-\eta_\Gamma}\eta_\Gamma^k\left(f(\mathbf{x}^0) - f(\widehat{\mathbf{x}})\right). \end{aligned}$$

If we combine the above inequalities with (10.38), we get

$$\|\mathbf{g}^P(\mathbf{x}^k)\|^2 \le \frac{2(1+\eta_{\Gamma)}}{\overline{\alpha}(1-\eta_\Gamma)}\eta_\Gamma^k\left(f(\mathbf{x}^0) - f(\widehat{\mathbf{x}})\right).$$

This completes the proof. □

10.3.3 Optimality

Theorems 5.3 and 10.2 give the bounds on the rates of convergence of the iterates and corresponding projected gradients that depend only on the bounds on the spectrum but do not depend on the constraint vector $\boldsymbol{\ell}$. It follows that if we have a class of bound-constrained problems with the spectrum of the Hessian of the cost function in an a priori fixed interval, then the MPRGP algorithm's convergence rate can be bounded uniformly for the whole class. To present explicitly this feature of the MPRGP algorithm, let $\mathcal{T}$ denote any set of indices and assume that for any $t \in \mathcal{T}$ there is defined a problem

$$\text{minimize} \;\; f_t(\mathbf{x}) \;\; \text{s.t.} \;\; \mathbf{x} \in \Omega_{B_t} \tag{10.39}$$

with $\Omega_{B_t} = \{\mathbf{x} \in \mathbb{R}^{n_t} : \; \mathbf{x} \ge \boldsymbol{\ell}_t\}$, $f_t(\mathbf{x}) = \frac{1}{2}\mathbf{x}^T\mathsf{A}_t\mathbf{x} - \mathbf{b}_t^T\mathbf{x}$, $\mathsf{A}_t \in \mathbb{R}^{n_t \times n_t}$ SPD, and $\boldsymbol{\ell}_t \in \mathbb{R}^{n_t}$. Our optimality result then reads as follows.

Theorem 10.3. *Let $a_{\max} > a_{\min} > 0$ denote given constants and let $\{\mathbf{x}_t^k\}$ be generated by the MPRGP algorithm for solving of problem (10.39) with*

$$0 < \overline{\alpha} \le 2a_{\max}^{-1} \;\; \text{and} \;\; \Gamma > 0$$

starting from $\mathbf{x}_t^0 = \max\{\mathbf{o}, \boldsymbol{\ell}_t\}$. Let the class of problems (10.39) satisfy

$$0 < a_{\min} \le \lambda_{\min}(\mathsf{A}_t) \le \lambda_{\max}(\mathsf{A}_t) \le a_{\max},$$

where $\lambda_{\min}(\mathsf{A}_t)$ and $\lambda_{\max}(\mathsf{A}_t)$ denote the extreme eigenvalues of A_t.

Then there are integers $\overline{k}$ and $\overline{\ell}$ such that for any $t \in \mathcal{T}$ and $\varepsilon > 0$

$$\|\mathbf{g}_t^P(\mathbf{x}_t^{\overline{k}})\| \le \varepsilon \|\mathbf{g}_t^P(\mathbf{x}_t^0)\| \quad \text{and} \quad f_t(\mathbf{x}_t^{\overline{\ell}}) - f_t(\widehat{\mathbf{x}}_t) \le \varepsilon(f_t(\mathbf{x}_t^0) - f(\widehat{\mathbf{x}}_t)).$$

Proof. First denote

$$\eta_\Gamma^t = 1 - \frac{\widehat{\alpha}\lambda_{\min}^t}{2 + 2\widehat{\Gamma}^2}, \qquad \overline{\eta}_\Gamma = 1 - \frac{\widehat{\alpha}a_{\min}}{2 + 2\widehat{\Gamma}^2}$$

$$a_1^t = a_1 = \frac{2(1 + \eta_\Gamma^t)}{\widehat{\alpha}(1 - \eta_\Gamma^t)} \qquad \overline{a}_1 = \frac{2(1 + \overline{\eta}_\Gamma)}{\widehat{\alpha}(1 - \overline{\eta}_\Gamma)},$$

where $\widehat{\Gamma} = \max\{\Gamma, \Gamma^{-1}\}$, so that

$$\eta_\Gamma^t \le \overline{\eta}_\Gamma < 1 \quad \text{and} \quad a_1^t \le \overline{a}_1.$$

Combining these estimates with Theorem 10.2 and inequality (10.3), we get for any $k \ge 1$

$$\|\mathbf{g}_t^P(\mathbf{x}^{k+1})\|^2 \le \overline{a}_1 \overline{\eta}_\Gamma^k (f_t(\mathbf{x}_t^0) - f_t(\widehat{\mathbf{x}}_t)) \le \frac{\overline{a}_1}{a_{\min}} \overline{\eta}_\Gamma^k \|\mathbf{g}_t^P(\mathbf{x}_t^0)\|^2.$$

Similarly, using Theorem 5.3, we get

$$f_t(\mathbf{x}_t^k) - f_t(\widehat{\mathbf{x}}_t) \le \overline{\eta}_\Gamma^k (f_t(\mathbf{x}_t^0) - f(\widehat{\mathbf{x}}_t)).$$

To finish the proof, it is enough to take $\overline{k}$ and $\overline{\ell}$ so that

$$\frac{\overline{a}_1}{a_{\min}} \overline{\eta}_\Gamma^{\overline{k}-1} \le \varepsilon \qquad \text{and} \qquad \overline{\eta}_\Gamma^{\overline{\ell}} \le \varepsilon. \qquad \square$$

10.4 Finite Termination

We shall prove the finite termination property of MPRGP for problems with *sufficiently large* Γ or the *regular* solution, i.e., the vector of Lagrange multipliers $\widehat{\lambda}$ of the solution satisfies the *strict complementarity condition* $\widehat{\lambda}_i > 0$ for $i \in \mathcal{A}(\widehat{\mathbf{x}})$.

Let us first prove that MPRGP generates the *iterations which have active components in the binding set of the solution.* The proof uses simple geometrical observations. For example, examining Fig. 10.2, it is easy to see that the free sets of the iterates $\mathbf{x}^k$ soon contain the free set of the solution $\widehat{\mathbf{x}}$. The formal analysis is the subject of the following identification lemma.

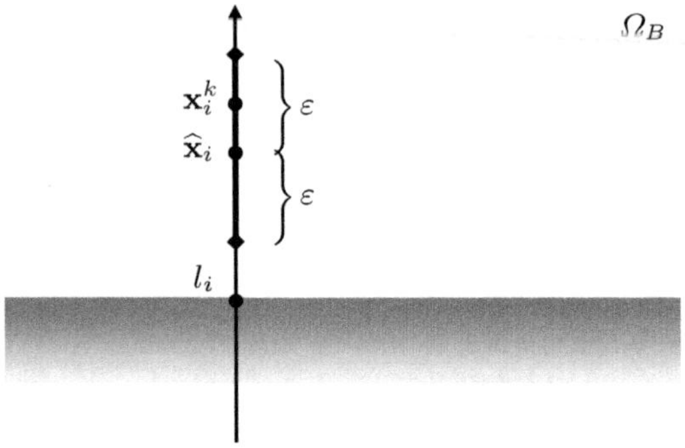

Fig. 10.2. Identification of the free set of the solution

Lemma 10.3. *Let $\{\mathbf{x}^k\}$ be generated by Algorithm 10.1 with $\mathbf{x}^0 \in \Omega_B$, $\Gamma > 0$, and $\overline{\alpha} \in (0, 2\|\mathsf{A}\|^{-1}]$. Then there is k_0 such that for $k \geq k_0$*

$$\mathcal{F}(\widehat{\mathbf{x}}) \subseteq \mathcal{F}(\mathbf{x}^k) \quad \text{and} \quad \widehat{\mathcal{B}} \subseteq \mathcal{B}(\mathbf{x}^k), \tag{10.40}$$

where $\widehat{\mathcal{B}} = \mathcal{B}(\widehat{\mathbf{x}})$ and $\widetilde{\varphi}(\mathbf{x}^k) = \widetilde{\varphi}_{\overline{\alpha}}(\mathbf{x}^k)$ is defined by (10.15).

Proof. Since (10.40) is trivially satisfied when there is $k = k_0$ such that $\mathbf{x}^k = \widehat{\mathbf{x}}$, we shall assume in what follows that $\mathbf{x}^k \neq \widehat{\mathbf{x}}$ for any $k \geq 0$. Let us denote $x_i^k = [\mathbf{x}^k]_i$ and $\widehat{x}_i = [\widehat{\mathbf{x}}]_i$, $i = 1, \ldots, n$.

Let us first assume that $\mathcal{F}(\widehat{\mathbf{x}}) \neq \emptyset$ and $\widehat{\mathcal{B}} \neq \emptyset$, so that we can define

$$\varepsilon = \min\{\widehat{x}_i - \ell_i : \ i \in \mathcal{F}(\widehat{\mathbf{x}})\} > 0 \quad \text{and} \quad \delta = \min\{g_i(\widehat{\mathbf{x}}) : \ i \in \widehat{\mathcal{B}}\} > 0.$$

Since by Theorem 10.1 $\{\mathbf{x}^k\}$ converges to $\widehat{\mathbf{x}}$, there is k_0 such that for any $k \geq k_0$

$$g_i(\mathbf{x}^k) \leq \frac{\varepsilon}{4\overline{\alpha}} \text{ for } i \in \mathcal{F}(\widehat{\mathbf{x}}) \tag{10.41}$$

$$x_i^k \geq \ell_i + \frac{\varepsilon}{2} \text{ for } i \in \mathcal{F}(\widehat{\mathbf{x}}) \tag{10.42}$$

$$x_i^k \leq \ell_i + \frac{\overline{\alpha}\delta}{8} \text{ for } i \in \widehat{\mathcal{B}} \tag{10.43}$$

$$g_i(\mathbf{x}^k) \geq \frac{\delta}{2} \text{ for } i \in \widehat{\mathcal{B}}. \tag{10.44}$$

In particular, for $k \geq k_0$, the first inclusion of (10.40) follows from (10.42).

Since there can be only a finite number of consecutive steps of MPRGP without the expansion, it follows that there is $k \geq k_0$ such that $\mathbf{x}^{k+1}$ is generated by the expansion step (10.14). For $i \in \mathcal{B}(\widehat{\mathbf{x}})$, we get by (10.43) and (10.44)

$$x_i^k - \overline{\alpha} g_i(\mathbf{x}^k) \leq \ell_i + \frac{\overline{\alpha}\delta}{8} - \frac{\overline{\alpha}\delta}{2} < \ell_i.$$

Thus

$$x_i^{k+1} = \max\{\ell_i, x_i^k - \overline{\alpha} g_i(\mathbf{x}^k)\} = \ell_i.$$

Moreover, for large k, the index i can never be released from the active set due to (10.44).

□

Proving the finite termination property of MPRGP for the problems with regular solutions is now easy. If the solution of problem (10.1) is dual degenerate, i.e., if there is i such that $\widehat{x}_i = \ell_i$ and $g_i(\widehat{\mathbf{x}}) = 0$, then we can use the estimate from Sect. 7.4.1 to get the finite termination property for large balancing parameter Γ. We shall formulate these observations in the following proposition.

Proposition 10.1. *Let $\{\mathbf{x}^k\}$ be generated by Algorithm 10.1 with*

$$\mathbf{x}^0 \in \Omega_B, \ \Gamma > 0, \ \text{and } \overline{\alpha} \in (0, 2\|\mathsf{A}\|^{-1}).$$

If $\Gamma \geq \sqrt{\kappa}(\mathsf{A})$ or the solution $\widehat{\mathbf{x}}$ is regular, i.e.,

$$\widehat{x}_i = \ell_i \text{ implies } g_i(\widehat{\mathbf{x}}) > 0, \tag{10.45}$$

then there is $k \geq 0$ such that $\mathbf{x}^k = \widehat{\mathbf{x}}$.

Proof. If

$$\Gamma \geq \sqrt{\kappa}(\mathsf{A}),$$

then MPRGP never returns to the face with non-proportional iterate by Lemma 7.4. Since the proportional iterate is the only one that reduces the active set and the algorithm changes the active set after at most n iterates, it follows that it must generate after at most n^2 iterates either the solution or non-proportional iterate. Since the number of faces is finite, MPRGP finds the solution in a finite number of steps.

If $\widehat{\mathbf{x}}$ satisfies the condition of strict complementarity, then

$$\mathcal{A}(\widehat{\mathbf{x}}) = \mathcal{B}(\widehat{\mathbf{x}}),$$

and, by Lemma 10.3, there is $k_0 \geq 0$ such that for $k \geq k_0$ we have

$$\mathcal{F}(\mathbf{x}^k) = \mathcal{F}(\widehat{\mathbf{x}}) \quad \text{and} \quad \mathcal{B}(\mathbf{x}^k) = \mathcal{B}(\widehat{\mathbf{x}}).$$

Thus, for $k \geq k_0$, all $\mathbf{x}^k$ that satisfy $\widehat{\mathbf{x}} \neq \mathbf{x}^{k-1}$ are generated by the conjugate gradient steps and, by the finite termination property of the CG, there is $k \leq k_0 + n$ such that $\mathbf{x}^k = \widehat{\mathbf{x}}$. □

Remark 10.2. For the regular solution, Moré and Toraldo proved the finite termination property of an algorithm combining conjugate gradients with Euclidean projections in their seminal paper [217]. See also the looking ahead Polyak algorithm discussed in Sect. 7.4.1.

10.5 Implementation of MPRGP and Optional Modifications

Here we describe Algorithm 10.1 in the form convenient for implementation. We also include some modifications that can improve its performance.

10.5.1 Expansion Step with Feasible Half-Step

To improve the efficiency of the expansion step, we can use the trial conjugate gradient direction $\mathbf{p}^k$ generated before the expansion step is invoked. We propose to generate first the auxiliary vectors

$$\mathbf{x}^{k+\frac{1}{2}} = \mathbf{x}^k - \alpha_f \mathbf{p}^k \quad \text{and} \quad \mathbf{g}^{k+\frac{1}{2}} = \mathbf{g}^k - \alpha_f \mathsf{A}\mathbf{p}^k,$$

where the feasible steplength α_f for $\mathbf{p}^k$ is defined by

$$\alpha_f = \max\{\alpha : \ \mathbf{x}^k - \alpha\mathbf{p}^k \in \Omega_B\} = \min_{i=1,\ldots,n} \{(x_i^k - \ell_i)/p_i^k, p_i^k > 0\},$$

and then define

$$\mathbf{x}^{k+1} = P_{\Omega_B}\left(\mathbf{x}^{k+\frac{1}{2}} - \overline{\alpha}\varphi(\mathbf{x}^{k+\frac{1}{2}})\right).$$

We illustrate the half-step in Fig. 10.3. Such modification does not require any additional matrix–vector multiplication and the monotonicity of MPRGP is preserved as

$$f(\mathbf{x}^{k+\frac{1}{2}}) - f(\mathbf{x}^k) \leq 0$$

and

$$f(\mathbf{x}^{k+1}) - f(\widehat{\mathbf{x}}) = \left((f(\mathbf{x}^{k+\frac{1}{2}}) - f(\mathbf{x}^k)\right) + f(\mathbf{x}^k) - f(\widehat{\mathbf{x}})$$
$$< (f(\mathbf{x}^k) - f(\widehat{\mathbf{x}})).$$

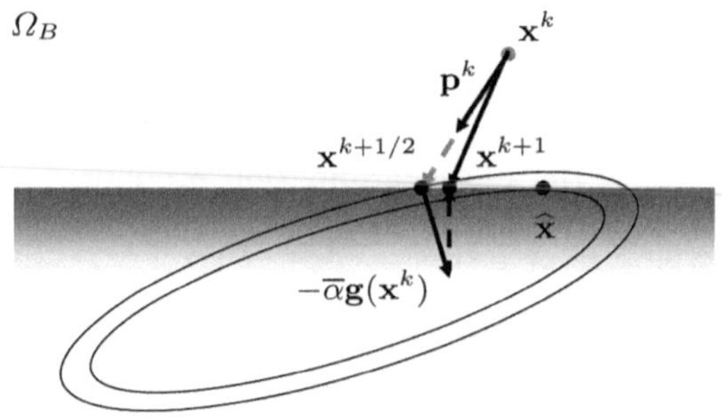

Fig. 10.3. Feasible half-step

10.5.2 MPRGP Algorithm

Now, we are ready to give the details of the implementation of the MPRGP algorithm, which is described in the form suitable for analysis as Algorithm 10.1. To preserve readability, we do not distinguish generations of variables by indices unless it is convenient for further reference.

Algorithm 10.2. Modified proportioning with reduced gradient projections (MPRGP)

Given an SPD matrix A *of the order* n*,* n*-vectors* $\mathbf{b}, \boldsymbol{\ell}, \Omega_B = \{\mathbf{x}:\ \mathbf{x} \geq \boldsymbol{\ell}\}$, $\mathbf{x}^0 \in \Omega_B$.

Step 0. {*Initialization.*}
  Choose $\Gamma > 0$, $\overline{\alpha} \in (0, 2\|\mathsf{A}\|^{-1}]$, *set* $k = 0$, $\mathbf{g} = \mathsf{A}\mathbf{x}^0 - \mathbf{b}$, $\mathbf{p} = \boldsymbol{\varphi}(\mathbf{x}^0)$
  while $\|\mathbf{g}^P(\mathbf{x}^k)\|$ *is not small*
  if $\|\boldsymbol{\beta}(\mathbf{x}^k)\|^2 \leq \Gamma^2 \widetilde{\boldsymbol{\varphi}}_{\widehat{\alpha}}(\mathbf{x}^k)^T \boldsymbol{\varphi}(\mathbf{x}^k)$

Step 1. {*Proportional* $\mathbf{x}^k$*. Trial conjugate gradient step.*}
  $\alpha_{cg} = \mathbf{g}^T\mathbf{p}/\mathbf{p}^T\mathsf{A}\mathbf{p}$, $\mathbf{y} = \mathbf{x}^k - \alpha_{cg}\mathbf{p}$
  $\alpha_f = \max\{\alpha:\ \mathbf{x}^k - \alpha\mathbf{p} \in \Omega_B\} = \min\{(x_i^k - \ell_i)/p_i:\ p_i > 0\}$
  if $\alpha_{cg} \leq \alpha_f$

Step 2. {*Conjugate gradient step.*}
  $\mathbf{x}^{k+1} = \mathbf{y}, \mathbf{g} = \mathbf{g} - \alpha_{cg}\mathsf{A}\mathbf{p}$,
  $\beta = \boldsymbol{\varphi}(\mathbf{y})^T\mathsf{A}\mathbf{p}/\mathbf{p}^T\mathsf{A}\mathbf{p}$, $\mathbf{p} = \boldsymbol{\varphi}(\mathbf{y}) - \beta\mathbf{p}$
  else

Step 3. {*Expansion step.*}
  $\mathbf{x}^{k+\frac{1}{2}} = \mathbf{x}^k - \alpha_f\mathbf{p}$, $\mathbf{g} = \mathbf{g} - \alpha_f\mathsf{A}\mathbf{p}$
  $\mathbf{x}^{k+1} = P_{\Omega_B}(\mathbf{x}^{k+\frac{1}{2}} - \overline{\alpha}\boldsymbol{\varphi}(\mathbf{x}^{k+\frac{1}{2}}))$
  $\mathbf{g} = \mathsf{A}\mathbf{x}^{k+1} - \mathbf{b}$, $\mathbf{p} = \boldsymbol{\varphi}(\mathbf{x}^{k+1})$
  end if
  else

Step 4. {*Proportioning step.*}
  $\mathbf{d} = \boldsymbol{\beta}(\mathbf{x}^k), \alpha_{cg} = \mathbf{g}^T\mathbf{d}/\mathbf{d}^T\mathsf{A}\mathbf{d}$
  $\mathbf{x}^{k+1} = \mathbf{x}^k - \alpha_{cg}\mathbf{d}, \mathbf{g} = \mathbf{g} - \alpha_{cg}\mathsf{A}\mathbf{d}$, $\mathbf{p} = \boldsymbol{\varphi}(\mathbf{x}^{k+1})$
  end if
  $k = k + 1$
  end while

Step 5. {*Return (possibly inexact) solution.*}
  $\widetilde{\mathbf{x}} = \mathbf{x}^k$

To simplify the notation, we denote by $\widetilde{\boldsymbol{\varphi}}(\mathbf{x}) = \widetilde{\boldsymbol{\varphi}}_{\overline{\alpha}}(\mathbf{x})$ the reduced free gradient defined by (10.15). Let us mention that $\widetilde{\boldsymbol{\varphi}}_{\widehat{\alpha}}(\mathbf{x}) \geq \widetilde{\boldsymbol{\varphi}}_{\overline{\alpha}}(\mathbf{x})$, so that we can simplify the proportioning test by using $\widetilde{\boldsymbol{\varphi}}_{\overline{\alpha}}(\mathbf{x})$. Recall that the condition $0 < \alpha < 2\|\mathsf{A}\|^{-1}$ must be satisfied!

10.5.3 Choice of Parameters

Our experience indicates that MPRGP is not sensitive to Γ as long as $\Gamma \approx 1$. Since $\Gamma = 1$ minimizes the upper bound on the convergence rate and guarantees that the CG steps reduce directly the larger of the two components of the projected gradient, we can expect good efficiency with this value.

The choice of $\overline{\alpha}$ requires an estimate of $\|\mathsf{A}\|$. If the entries of A are available, we can use $\|\mathsf{A}\| \leq \|\mathsf{A}\|_\infty$ to define $\overline{\alpha} = 2\|\mathsf{A}\|_\infty^{-1}$ which guarantees convergence. If this is not the case, or if $\|\mathsf{A}\|_\infty$ gives a poor upper bound on $\|\mathsf{A}\|$, then we can carry out a few, e.g., five, iterations of the following power method.

Though these methods typically give only a lower bound A on the norm of $\|\mathsf{A}\|$, the choice like $\overline{\alpha} = 1.8A^{-1}$ is often sufficient in practice. The decrease of f can be achieved more reliably by initializing

$$\overline{\alpha} = 2\|\mathbf{b}\|^2(\mathbf{b}^T\mathsf{A}\mathbf{b})^{-1}$$

and by inserting the following piece of code into the expansion step:

Algorithm 10.3. Power method for the estimate of $\|\mathsf{A}\|$

Given a symmetric positive definite matrix $\mathsf{A} \in \mathbb{R}^{n\times n}$, *returns* $A \approx \|\mathsf{A}\|$

Step 0. *Choose* $\mathbf{x} \in \mathbb{R}^n$ *such that* $\mathbf{x} \neq \mathbf{o}$, $n_{\text{it}} \geq 1$
Step 1. **for** $i = 1, 2, \ldots, n_{\text{it}}$
  $\mathbf{y} = \mathsf{A}\mathbf{x}$, $\mathbf{x} = \|\mathbf{y}\|^{-1}\mathbf{y}$
  end for
Step 2. $A = \|\mathsf{A}\mathbf{x}\|$

The modified algorithm can outperform that with $\overline{\alpha} = \|\mathsf{A}\|^{-1}$; the longer steps in an early stage of computations can be effective for identification of the solution's active set. We observed a good performance with $\overline{\alpha}$ close to, but not greater than $2\|\mathsf{A}\|^{-1}$, near α_{opt} which minimizes the coefficient η_E of the Euclidean contraction (3.1).

Algorithm 10.4. Modification of the steplength of the expansion step

A piece of code to be inserted at the end of the expansion step of Algorithm 10.2.

if $f(P_{\Omega_B}(\mathbf{x}^{k+1})) > f(\mathbf{x}^k)$
  $\overline{\alpha} = \overline{\alpha}/2$ *and repeat the expansion step*
end if

10.5.4 Proportioning with Newton Step (PRGPN)

The MPRGP algorithm described above was proposed for solving large minimization problems with reasonably conditioned Hessian so that the auxiliary linear problems could be solved effectively by the conjugate gradient method. This strategy need not be the most efficient for solving smaller problems with some hundreds or thousands of variables that admit direct solution of auxiliary linear systems

$$\mathsf{A}_{\mathcal{F}\mathcal{F}}\mathbf{x}_{\mathcal{F}} = \mathbf{b}_{\mathcal{F}}, \quad \mathcal{F} \subseteq \{1, \ldots, n\}.$$

A natural idea for how to improve the efficiency of MPRGP for smaller problems is to replace the conjugate gradient step with the *Newton step*

$$\mathbf{x}^{k+1} = \mathbf{x}^k - \mathbf{d}^N\left(\mathbf{x}^k\right), \tag{10.46}$$

where $\mathbf{d}^N = \mathbf{d}^N(\mathbf{x}^k)$ is defined by the components

$$\mathbf{d}^N_{\mathcal{F}} = \mathsf{A}^{-1}_{\mathcal{F}\mathcal{F}}\mathbf{g}^k_{\mathcal{F}}, \quad \mathbf{d}^N_{\mathcal{A}} = \mathbf{o}, \quad \mathbf{g}^k = \mathsf{A}\mathbf{x}^k - \mathbf{b}.$$

The full Newton step generates $\mathbf{x}^{k+1}$ which minimizes f on the manifold defined by the free variables, so if the Newton step does not find the solution, i.e., $\mathbf{x}^k - \mathbf{d}^N \neq \widehat{\mathbf{x}}$, then it is necessarily followed by the proportioning step. Notice that the Newton step reduces the cost function by

$$f\left(\mathbf{x}^k - \mathbf{d}^N\right) - f\left(\mathbf{x}^k\right) = -(\mathbf{g}^k)^T\mathbf{d}^N_{\mathcal{F}} + \frac{1}{2}\left(\mathbf{d}^N_{\mathcal{F}}\right)^T \mathsf{A}\mathbf{d}^N_{\mathcal{F}} = -\frac{1}{2}\left(\mathbf{g}^k_{\mathcal{F}}\right)^T \mathsf{A}^{-1}_{\mathcal{F}\mathcal{F}}\mathbf{g}^k_{\mathcal{F}},$$

which is typically a much larger decrease than that resulting from the conjugate gradient step. If we replace the CG step of the MPRGP algorithm with the Newton step, we get the algorithm with the same worst-case bounds and finite termination property as the original MPRGP algorithm, but with typically better performance in terms of the number of iterations.

We suggest two more modifications. The algorithm with the Newton steps uses the same expansion step and user-defined constants $\Gamma > 0$ and $\overline{\alpha} \in (0, 2\|\mathsf{A}\|^{-1})$ as the MPRGP algorithm, however, the proportioning test can be simplified—the reduced free gradient does not represent a full potential decrease of the cost function. We suggest to use the test proposed for the predecessors of the MPRGP algorithm by Friedlander and Martínez [63] or Dostál [138]: if

$$\|\boldsymbol{\beta}(\mathbf{x}^k)\| \leq \Gamma\|\boldsymbol{\varphi}(\mathbf{x}^k)\| \tag{10.47}$$

holds, then we call the iterate $\mathbf{x}^k$ *proportional*. The test (10.9) is again used to decide which components of $\mathbf{g}^P(\mathbf{x}^k)$ should be reduced in the next step. To reflect the modifications in the acronym of the algorithm, we shall call it *proportioning* with reduced gradient and Newton step (PRGPN).

The *expansion step* is defined by

$$\mathbf{x}^{k+1} = P_{\Omega_B}(\mathbf{x}^k - \overline{\alpha}\boldsymbol{\varphi}(\mathbf{x}^k)) = \max\{\boldsymbol{\ell}, \mathbf{x}^k - \overline{\alpha}\boldsymbol{\varphi}(\mathbf{x}^k)\} \tag{10.48}$$

as in the original MPRGP algorithm. It is carried out when the Newton step is not feasible.

We can also use the *proportioning step* of the original proportioning algorithm [138] defined by

$$\mathbf{x}^{k+1} = \mathbf{x}^k - \alpha_{cg}\boldsymbol{\beta}(\mathbf{x}^k) \tag{10.49}$$

with the steplength α_{cg} that minimizes $f\left(\mathbf{x}^k - \alpha\boldsymbol{\beta}(\mathbf{x}^k)\right)$. If A has nonnegative inverse, we can replace $\boldsymbol{\beta}$ by the *Newton release direction* $\boldsymbol{\beta}^N$ defined by

$$\boldsymbol{\beta}^N_{\mathcal{B}}\left(\boldsymbol{x}^k\right) = \mathsf{A}^{-1}_{\mathcal{B}\mathcal{B}}\boldsymbol{\beta}^N_{\mathcal{B}}\left(\boldsymbol{x}^k\right), \quad \boldsymbol{\beta}^N_{\mathcal{F}}\left(\boldsymbol{x}^k\right) = \mathbf{o}. \tag{10.50}$$

An important case of such matrices represents M-matrices (see, e.g., Pool and Boullion [230] or Axelsson [6]).

Let us define the modified algorithm in a form convenient for analysis, referring an interested reader to Sect. 10.5 for more implementation details.

Algorithm 10.5. Proportioning with reduced gradient projections and Newton step (PRGPN schema)

Given a symmetric positive definite matrix $\mathsf{A} \in \mathbb{R}^{n\times n}$ *and n-vectors* $\mathbf{b}, \boldsymbol{\ell}$.
Choose $\mathbf{x}^0 \in \Omega_B$, $\overline{\alpha} \in (0, 2\|\mathsf{A}\|^{-1})$, *and* $\Gamma > 0$. *Set* $k = 0$.
For $k \geq 0$ *and* $\mathbf{x}^k$ *known, choose* $\mathbf{x}^{k+1}$ *by the following rules:*

(i) If $\mathbf{g}^P(\mathbf{x}^k) = \mathbf{o}$, *set* $\mathbf{x}^{k+1} = \mathbf{x}^k$.
(ii) *If* $\mathbf{x}^k$ *is strictly proportional and* $\mathbf{g}^P(\mathbf{x}^k) \neq \mathbf{o}$, *try to generate* $\mathbf{x}^{k+1}$ *by the Newton step. If* $\mathbf{x}^{k+1} \in \Omega_B$, *then accept it, else generate* $\mathbf{x}^{k+1}$ *by the expansion step, possibly with the feasible half-step of Sect 10.5.1.*
(iii) *If* $\mathbf{x}^k$ *is not strictly proportional, define* $\mathbf{x}^{k+1}$ *by proportioning, possibly with a positive approximation of the Newton direction.*

10.6 Preconditioning

The performance of the conjugate gradient-based methods can be improved by the preconditioning described in Sect. 8.2. However, applying preconditioning requires some care, as the *preconditioning transforms the variables, turning the bound constraints into more general inequality constraints.* In this section, we present two strategies that preserve the bound constraints.

10.6.1 Preconditioning in Face

Probably the most straightforward preconditioning strategy that preserves the bound constraints is the preconditioning applied to the diagonal block $\mathsf{A}_{\mathcal{F}\mathcal{F}}$ of the Hessian matrix A in the conjugate gradient loop which minimizes the cost function f in the face defined by a free set $\mathcal{F}$. Such preconditioning requires that we can define for each diagonal block $\mathsf{A}_{\mathcal{F}\mathcal{F}}$ a regular matrix $\mathsf{M}(\mathcal{F})$ which satisfies the following two conditions:

- $\mathsf{M}(\mathcal{F})$ approximates $\mathsf{A}_{\mathcal{F}\mathcal{F}}$ so that the convergence of the conjugate gradients method is significantly accelerated.
- The solution of the system
$$\mathsf{M}(\mathcal{F})\mathbf{x} = \mathbf{y}$$
can be obtained easily.

The preconditioners $\mathsf{M}(\mathcal{F})$ can be generated, e.g., by any of the methods described in Sect. 8.2.

Though the preconditioning can considerably improve the performance of the algorithm, *preconditioning in the face does not result in the improved bound on the rate of convergence.* The reason is that such preconditioning affects only the feasible conjugate gradient step, leaving the expansion and the proportioning steps without any preconditioning.

In probably the first application of preconditioning to the solution of bound-constrained problems [222], O'Leary considered two simple methods that can be used to obtain the preconditioner for $\mathsf{A}_{\mathcal{F}\mathcal{F}}$ from the preconditioner M which approximates A, namely,

$$\mathsf{M}(\mathcal{F}) = \mathsf{M}_{\mathcal{F}\mathcal{F}} \quad \text{and} \quad \mathsf{M}(\mathcal{F}) = \mathrm{L}_{\mathcal{F}\mathcal{F}}\mathrm{L}_{\mathcal{F}\mathcal{F}}^T,$$

where L denotes the factor of the Cholesky factorization $\mathsf{M} = \mathsf{L}\mathsf{L}^T$ It can be proved that whichever preconditioning method we use, the convergence bound for the CG algorithm applied to the subproblems is at least as good as that of the conjugate gradient method applied to the original matrix [222].

To describe the MPRGP algorithm with the preconditioning in the face, let us assume that we are given the preconditioner $\mathsf{M}(\mathcal{F})$ for each set of indices $\mathcal{F}$, and let us denote $\mathcal{F}_k = \mathcal{F}(\mathbf{x}^k)$ and $\mathcal{A}_k = \mathcal{A}(\mathbf{x}^k)$ for each vector $\mathbf{x}^k \in \Omega_B$. To simplify the description of the algorithm, let M_k denote the preconditioner corresponding to the face $\mathcal{F}_k$ padded with zeros so that

$$[\mathsf{M}_k]_{\mathcal{F}\mathcal{F}} = \mathsf{M}(\mathcal{F}_k), \quad [\mathsf{M}_k]_{\mathcal{A}\mathcal{A}} = \mathsf{O}, \quad [\mathsf{M}_k]_{\mathcal{A}\mathcal{F}} = [\mathsf{M}_k]_{\mathcal{F}\mathcal{A}}^T = \mathsf{O},$$

and recall that $\mathsf{M}_k^\dagger$ denotes the Moore–Penrose generalized inverse of M_k which is defined by

$$[\mathsf{M}_k^\dagger]_{\mathcal{F}\mathcal{F}} = \mathsf{M}(\mathcal{F}_k)^{-1}, \quad [\mathsf{M}_k^\dagger]_{\mathcal{A}\mathcal{A}} = \mathsf{O}, \quad [\mathsf{M}_k^\dagger]_{\mathcal{A}\mathcal{F}} = [\mathsf{M}_k^\dagger]_{\mathcal{F}\mathcal{A}}^T = \mathsf{O}.$$

In particular, it follows that

$$\mathsf{M}_k^\dagger \mathbf{g}(\mathbf{x}^k) = \mathsf{M}_k^\dagger \boldsymbol{\varphi}(\mathbf{x}^k).$$

The MPRGP algorithm with preconditioning in face reads as follows.

Algorithm 10.6. MPRGP with preconditioning in face

Given a symmetric positive definite matrix A *of the order* n, n*-vectors* $\mathbf{b}$, $\boldsymbol{\ell}$, $\Omega_B = \{\mathbf{x} \in \mathbb{R}^n : \mathbf{x} \geq \boldsymbol{\ell}\}$; *choose* $\mathbf{x}^0 \in \Omega_B$, $\Gamma > 0$, $\overline{\alpha} \in (0, 2\|\mathsf{A}\|^{-1}]$, *and the rule which assigns to each* $\mathbf{x}^k \in \Omega_B$ *the preconditioner* M_k *which is SPD in the face defined by* $\mathcal{F}(\mathbf{x}^k)$.

Step 0. {*Initialization.*}
  Set $k = 0$, $\mathbf{g} = \mathsf{A}\mathbf{x}^0 - \mathbf{b}$, $\mathbf{z} = \mathsf{M}_0^\dagger \mathbf{g}$, $\mathbf{p} = \mathbf{z}$
  while $\|\mathbf{g}^P(\mathbf{x}^k)\|$ *is not small*
  if $\|\boldsymbol{\beta}(\mathbf{x}^k)\|^2 \leq \Gamma^2 \widetilde{\boldsymbol{\varphi}}(\mathbf{x}^k)^T \boldsymbol{\varphi}(\mathbf{x}^k)$
Step 1. {*Proportional* $\mathbf{x}^k$. *Trial conjugate gradient step.*}
  $\alpha_{cg} = \mathbf{z}^T \mathbf{g} / \mathbf{p}^T \mathsf{A}\mathbf{p}$, $\mathbf{y} = \mathbf{x}^k - \alpha_{cg}\mathbf{p}$
  $\alpha_f = \max\{\alpha : \mathbf{x}^k - \alpha\mathbf{p} \in \Omega_B\} = \min\{(x_i^k - \ell_i)/p_i : p_i > 0\}$
  if $\alpha_{cg} \leq \alpha_f$
Step 2. {*Conjugate gradient step.*}
  $\mathbf{x}^{k+1} = \mathbf{y}$, $\mathbf{g} = \mathbf{g} - \alpha_{cg}\mathsf{A}\mathbf{p}$, $\mathbf{z} = \mathsf{M}_k^\dagger \mathbf{g}$
  $\beta = \mathbf{z}^T \mathsf{A}\mathbf{p} / \mathbf{p}^T \mathsf{A}\mathbf{p}$, $\mathbf{p} = \mathbf{z} - \beta\mathbf{p}$
  else
Step 3. {*Expansion step.*}
  $\mathbf{x}^{k+\frac{1}{2}} = \mathbf{x}^k - \alpha_f \mathbf{p}$, $\mathbf{g} = \mathbf{g} - \alpha_f \mathsf{A}\mathbf{p}$
  $\mathbf{x}^{k+1} = P_{\Omega_B}\left(\mathbf{x}^{k+\frac{1}{2}} - \overline{\alpha}\boldsymbol{\varphi}(\mathbf{x}^{k+\frac{1}{2}})\right)$
  $\mathbf{g} = \mathsf{A}\mathbf{x}^{k+1} - \mathbf{b}$, $\mathbf{z} = \mathsf{M}_{k+1}^\dagger \mathbf{g}$, $\mathbf{p} = \mathbf{z}$
  end if
  else
Step 4. {*Proportioning step.*}
  $\mathbf{d} = \boldsymbol{\beta}(\mathbf{x}^k)$, $\alpha_{cg} = \mathbf{g}^T \mathbf{d} / \mathbf{d}^T \mathsf{A}\mathbf{d}$
  $\mathbf{x}^{k+1} = \mathbf{x}^k - \alpha_{cg}\mathbf{d}, \mathbf{g} = \mathbf{g} - \alpha_{cg}\mathsf{A}\mathbf{d}$, $\mathbf{z} = \mathsf{M}_{k+1}^\dagger \mathbf{g}$, $\mathbf{p} = \mathbf{z}$
  end if
  $k = k + 1$
  end while
Step 5. {*Return (possibly inexact) solution.*}
  $\widetilde{\mathbf{x}} = \mathbf{x}^k$

A reasonable conditioning of the matrix A can often be achieved by a suitable formulation of the problem. A significant potential is in exploiting duality and specific features of the problem. For example, exploiting the duality with domain decomposition, we solved some QP problems with hundreds of millions of nodal variables (see, e.g., Chap. 13).

10.6.2 Preconditioning by Conjugate Projector

Let $1 \le m < n$ and let the vector of bounds satisfy

$$\ell_{m+1} = -\infty, \ldots, \ell_n = -\infty,$$

so that problem (10.1) is only partially constrained, and the feasible set can be described by

$$\Omega_B = \{\mathbf{x} \in \mathbb{R}^n : \mathbf{x}_{\mathcal{I}} \ge \boldsymbol{\ell}_{\mathcal{I}}\}, \quad \mathcal{I} = \{1, \ldots, m\}. \tag{10.51}$$

Here, we show that such partially constrained problems can be preconditioned by the conjugate projector of Sect. 8.3 and that *it is possible to give an improved bound on the rate of convergence of the preconditioned problem.*

Let us assume that $\mathcal{U}$ is the subspace spanned by the full column rank matrix $\mathsf{U} \in \mathbb{R}^{n \times p}$ of the form

$$\mathsf{U} = \begin{bmatrix} \mathsf{O} \\ \mathsf{V} \end{bmatrix}, \quad \mathsf{V} \in \mathbb{R}^{(n-m) \times p}.$$

As in Sect. 8.3.2, we decompose our partially constrained problem using the conjugate projectors

$$\mathsf{P} = \mathsf{U}(\mathsf{U}^T\mathsf{A}\mathsf{U})^{-1}\mathsf{U}^T\mathsf{A} \tag{10.52}$$

and $\mathsf{Q} = \mathsf{I} - \mathsf{P}$ onto $\mathcal{U}$ and $\mathcal{V} = \mathrm{Im}\mathsf{Q}$, respectively. Due to our special choice of U, we get that for any $\mathbf{x} \in \mathbb{R}^n$

$$[\mathsf{Q}\mathbf{x}]_{\mathcal{I}} = \mathbf{x}_{\mathcal{I}},$$

and that for any $\mathbf{y} \in \mathcal{U}$ and $\mathbf{z} \in \mathcal{V}$, $\mathbf{y} + \mathbf{z} \in \Omega_B$ if and only if $\mathbf{z} \in \Omega_B$. Using (8.13), (8.14), and the observations of Sect. 8.3.3, we thus get

$$\begin{aligned}
\min_{\mathbf{x} \in \Omega_B} f(\mathbf{x}) &= \min_{\substack{\mathbf{y} \in \mathcal{U},\ \mathbf{z} \in \mathcal{V} \\ \mathbf{y}+\mathbf{z} \in \Omega_B}} f(\mathbf{y} + \mathbf{z}) = \min_{\mathbf{y} \in \mathcal{U}} f(\mathbf{y}) + \min_{\mathbf{z} \in \mathcal{V} \cap \Omega_B} f(\mathbf{z}) \\
&= f(\mathbf{x}^0) + \min_{\mathbf{z} \in \mathcal{V} \cap \Omega_B} f(\mathbf{z}) = f(\mathbf{x}^0) + \min_{\substack{\mathbf{z} \in \mathsf{A}\mathcal{V} \\ \mathbf{z}_{\mathcal{I}} \ge \boldsymbol{\ell}_{\mathcal{I}}}} \frac{1}{2}\mathbf{z}^T\mathsf{Q}^T\mathsf{A}\mathsf{Q}\mathbf{z} - \mathbf{b}^T\mathsf{Q}\mathbf{z} \\
&= f(\mathbf{x}^0) + \min_{\substack{\mathbf{z} \in \mathsf{A}\mathcal{V} \\ \mathbf{z}_{\mathcal{I}} \ge \boldsymbol{\ell}_{\mathcal{I}}}} \frac{1}{2}\mathbf{z}^T\mathsf{Q}^T\mathsf{A}\mathsf{Q}\mathbf{z} + (\mathbf{g}^0)^T\mathbf{z},
\end{aligned}$$

where $\mathbf{x}^0 = \mathsf{PA}^{-1}\mathbf{b}$ and $\mathbf{g}^0 = -\mathsf{Q}^T\mathbf{b}$. We have thus reduced our bound constrained problem (10.1) with the feasible set (10.51) to the problem

$$\min_{\substack{\mathbf{z}\in\mathsf{A}\mathcal{V}\\ \mathbf{z}_\mathcal{I}\geq\boldsymbol{\ell}_\mathcal{I}}} \frac{1}{2}\mathbf{z}^T\mathsf{Q}^T\mathsf{AQ}\mathbf{z} + (\mathbf{g}^0)^T\mathbf{z}. \tag{10.53}$$

The following lemma shows that the MPRGP algorithm can solve the above problem.

Lemma 10.4. *Let* $\mathbf{z}^1, \mathbf{z}^2, \dots$ *be generated by the MPRGP algorithm for the problem*

$$\min_{\mathbf{z}_\mathcal{I}\geq\boldsymbol{\ell}_\mathcal{I}} \frac{1}{2}\mathbf{z}^T\mathsf{Q}^T\mathsf{AQ}\mathbf{z} + (\mathbf{g}^0)^T\mathbf{z} \tag{10.54}$$

starting from $\mathbf{z}^0 = P_{\Omega_B}(\mathbf{g}^0)$*. Then* $\mathbf{z}^k \in \mathsf{A}\mathcal{V}$, $k = 0, 1, 2, \dots$.

Proof. First observe that since $\mathsf{A}\mathcal{V}$ is orthogonal to $\mathcal{U}$ and $\dim \mathsf{A}\mathcal{V} = \dim \mathcal{V}$, it follows that $\mathsf{A}\mathcal{V}$ is the orthogonal complement of $\mathcal{U}$. Thus, $\mathsf{A}\mathcal{V}$ is not only an invariant subspace of Q, but it is also an invariant subspace of P_{Ω_B}. Moreover, it also follows that $\mathsf{A}\mathcal{V}$ contains the set $\mathcal{V}_0 \subseteq \mathbb{R}^n$ of all the vectors of $\mathbb{R}^m$ padded with zeros,

$$\mathcal{V}_0 = \{\mathbf{x} \in \mathbb{R}^n : \ \mathbf{x}_\mathcal{J} = \mathbf{o}, \ \mathcal{J} = \{m+1, \dots, n\}\}.$$

More formally,

$$P_{\Omega_B}(\mathsf{A}\mathcal{V}) \subseteq \mathsf{A}\mathcal{V} \quad \text{and} \quad \mathcal{V}_0 \subseteq \mathsf{A}\mathcal{V}. \tag{10.55}$$

Let us now recall that by (8.14) $\mathbf{g}^0 \in \mathrm{Im}\mathsf{Q}^T$ and by (8.16) $\mathrm{Im}\mathsf{Q}^T = \mathsf{A}\mathcal{V}$, so that $\mathbf{g}^0 \in \mathsf{A}\mathcal{V}$. Using the definition of $\mathbf{z}^0$ and (10.55), we have $\mathbf{z}^0 \in \mathsf{A}\mathcal{V}$.

To finish the proof by induction, let us assume that $\mathbf{z}^k \in \mathsf{A}\mathcal{V}$. Since

$$\mathbf{g}^k = \mathsf{Q}^T\mathsf{AQ}\mathbf{z}^k - \mathsf{Q}^T\mathbf{b} = \mathsf{AQ}\mathbf{z}^k + \mathbf{g}^0,$$

we have $\mathbf{g}^k \in \mathsf{A}\mathcal{V}$. We shall use this simple observation to examine separately the three possible steps of the MPRGP algorithm of Sect. 10.2 that can be used to generate $\mathbf{z}^{k+1}$.

Let us first assume that $\mathbf{z}^{k+1}$ is generated by the proportioning step. Then

$$\mathbf{z}^{k+1} = \mathbf{z}^k - \alpha_{cg}\boldsymbol{\beta}(\mathbf{z}^k).$$

Using the definition of the chopped gradient, we can check that $\boldsymbol{\beta}(\mathbf{z}^k) \in \mathcal{V}_0$. Since $\mathcal{V}_0 \subseteq \mathsf{A}\mathcal{V}$, $\mathsf{A}\mathcal{V}$ is a subspace of $\mathbb{R}^n$, and $\mathbf{z}^k \in \mathsf{A}\mathcal{V}$ by the assumptions, this proves that $\mathbf{z}^{k+1} \in \mathsf{A}\mathcal{V}$ when it is generated by the proportioning step.

Before examining the other two steps, observe that $\boldsymbol{\varphi}(\mathbf{z}^k) - \mathbf{g}^k \in \mathcal{V}_0$, so that

$$\boldsymbol{\varphi}(\mathbf{z}^k) = (\boldsymbol{\varphi}(\mathbf{z}^k) - \mathbf{g}^k) + \mathbf{g}^k \in \mathsf{A}\mathcal{V}.$$

Thus

$$\mathbf{z}^k - \alpha\boldsymbol{\varphi}(\mathbf{z}^k) \in \mathsf{A}\mathcal{V}$$

for any $\alpha \in \mathbb{R}$. Using the first inclusion of (10.55), we get that

$$P_{\Omega_B}(\mathbf{z}^k - \overline{\alpha}\varphi(\mathbf{z}^k)) \in \mathsf{A}\mathcal{V}$$

for any $\overline{\alpha}$ of Algorithm 10.2. This proves that $\mathbf{z}^{k+1} \in \mathsf{A}\mathcal{V}$ for $\mathbf{z}^{k+1}$ generated by the expansion step. To finish the proof, observe that the conjugate direction $\mathbf{p}^k$ is either equal to $\varphi(\mathbf{z}^k)$, or it is defined by the recurrence (see (7.15)) $\mathbf{p}^{k+1} = \varphi(\mathbf{z}^k) - \beta\mathbf{p}^k$ starting from the restart $\mathbf{p}^{s+1} = \varphi(\mathbf{z}^s)$. In any case, $\mathbf{p}^k \in \mathsf{A}\mathcal{V}$. Since we assume that $\mathbf{z}^k \in \mathsf{A}\mathcal{V}$ and the iterate $\mathbf{z}^{k+1}$ generated by the conjugate gradient step is a linear combination of $\mathbf{z}^k$ and $\mathbf{p}^k$, this completes the proof. □

It follows that we can obtain the correction $\widehat{\mathbf{z}}$ which solves the auxiliary problem by the standard MPRGP algorithm. Since the iterations are reduced to the subspace, the *projector preconditions all three types of steps* and we can give an improved bound on the convergence rate. The solution $\widehat{\mathbf{x}}$ of the bound constrained problem (10.1) with the feasible set (10.51) can be expressed by $\widehat{\mathbf{x}} = \mathbf{x}^0 + \widehat{\mathbf{z}}$. For the reader's convenience, we give the complete algorithm for solving the preconditioned problem (10.54).

To describe the improved bound on the rate of convergence, let us denote, as in Sect. 8.3.4, the gap

$$\gamma = \|\mathsf{R}_{\mathsf{A}\mathcal{U}} - \mathsf{R}_{\mathcal{E}}\|$$

between $\mathsf{A}\mathcal{U}$ and the m-dimensional subspace $\mathcal{E}$ spanned by the eigenvectors corresponding to the m smallest eigenvalues

$$\lambda_{n-m+1} \geq \cdots \geq \lambda_{\min}$$

of A, so that the smallest nonzero eigenvalue $\overline{\lambda}_{\min}$ of $\mathsf{Q}^T\mathsf{A}\mathsf{Q}$ satisfies by Theorem 8.6

$$\overline{\lambda}_{\min} \geq \sqrt{(1-\gamma^2)\lambda_{n-m}^2 + \gamma^2\lambda_{\min}^2} \geq \lambda_{\min}. \tag{10.56}$$

Recall that by (8.17) and $\mathsf{A}\mathsf{Q} = \mathsf{Q}^T\mathsf{A}\mathsf{Q}$

$$\|\mathsf{A}\mathsf{Q}\| \leq \|\mathsf{A}\|.$$

Theorem 10.4. *Let $\{\mathbf{z}^k\}$ denote the sequence generated by Algorithm 10.7 for problem (10.54) with $\overline{\alpha} \in (0, 2\|\mathsf{A}\mathsf{Q}\|^{-1}]$ and $\Gamma > 0$ starting from $\mathbf{z}^0 = P_{\Omega_B}(\mathbf{g}^0)$. Let us denote*

$$f_{0,\mathsf{Q}}(\mathbf{z}) = \frac{1}{2}\mathbf{z}^T\mathsf{Q}^T\mathsf{A}\mathsf{Q}\mathbf{z} + (\mathbf{g}^0)^T\mathbf{z}.$$

Then

$$f_{0,\mathsf{Q}}(\mathbf{z}^{k+1}) - f_{0,\mathsf{Q}}(\widehat{\mathbf{z}}) \leq \eta_\Gamma(f_{0,\mathsf{Q}}(\mathbf{z}^k) - f_{0,\mathsf{Q}}(\widehat{\mathbf{z}})), \tag{10.57}$$

Algorithm 10.7. MPRGP projection preconditioning

Given a symmetric positive definite matrix A *of the order* n *and* $\mathbf{b}$, $\boldsymbol{\ell} \in \mathbb{R}^n$; *choose a full column rank matrix* $\mathsf{U} \in \mathbb{R}^{m \times n}$, $\mathbf{g}^0 = -\mathsf{Q}^T\mathbf{b}$, $\mathbf{x}^0 = \mathsf{P}\mathsf{A}^{-1}\mathbf{b}$, $\mathbf{z}^0 = P_{\Omega_B}(\mathbf{g}^0)$, $\Gamma > 0$, *and* $\overline{\alpha} \in (0, 2\|\mathsf{AQ}\|^{-1}]$, *where* P *is defined by (10.52) and* $\mathsf{Q} = \mathsf{I} - \mathsf{P}$.

Step 0. {*Initialization.*}
  Set $k = 0$, $\mathbf{g} = \mathsf{AQ}\mathbf{z}^0 + \mathbf{g}^0$, $\mathbf{p} = \varphi(\mathbf{z}^0)$
  while $\|\mathbf{g}^P(\mathbf{z}^k)\|$ *is not small*
  if $\|\boldsymbol{\beta}(\mathbf{z}^k)\|^2 \le \Gamma^2 \widetilde{\boldsymbol{\varphi}}(\mathbf{z}^k)^T \boldsymbol{\varphi}(\mathbf{z}^k)$
Step 1. {*Proportional* $\mathbf{z}^k$. *Trial conjugate gradient step.*}
  $\alpha_{cg} = \mathbf{g}^T\mathbf{p}/\mathbf{p}^T\mathsf{AQ}\mathbf{p}$, $\mathbf{y} = \mathbf{z}^k - \alpha_{cg}\mathbf{p}$
  $\alpha_f = \max\{\alpha : \mathbf{z}^k - \alpha\mathbf{p} \in \Omega_B\} = \min\{(z_i^k - \ell_i)/p_i : p_i > 0\}$
  if $\alpha_{cg} \le \alpha_f$
Step 2. {*Conjugate gradient step.*}
  $\mathbf{z}^{k+1} = \mathbf{y}$, $\mathbf{g} = \mathbf{g} - \alpha_{cg}\mathsf{AQ}\mathbf{p}$
  $\beta = \varphi(\mathbf{y})^T\mathsf{AQ}\mathbf{p}/\mathbf{p}^T\mathsf{AQ}\mathbf{p}$, $\mathbf{p} = \varphi(\mathbf{y}) - \beta\mathbf{p}$
  else
Step 3. {*Expansion step.*}
  $\mathbf{z}^{k+\frac{1}{2}} = \mathbf{z}^k - \alpha_f\mathbf{p}$, $\mathbf{g} = \mathbf{g} - \alpha_f\mathsf{AQ}\mathbf{p}$
  $\mathbf{z}^{k+1} = P_{\Omega_B}(\mathbf{z}^{k+\frac{1}{2}} - \overline{\alpha}\varphi(\mathbf{z}^{k+\frac{1}{2}}))$
  $\mathbf{g} = \mathsf{AQ}\mathbf{z}^{k+1} + \mathbf{g}^0$, $\mathbf{p} = \varphi(\mathbf{z}^{k+1})$
  end if
  else
Step 4. {*Proportioning step.*}
  $\mathbf{d} = \boldsymbol{\beta}(\mathbf{z}^k), \alpha_{cg} = \mathbf{g}^T\mathbf{d}/\mathbf{d}^T\mathsf{AQ}\mathbf{d}$
  $\mathbf{z}^{k+1} = \mathbf{z}^k - \alpha_{cg}\mathbf{d}$, $\mathbf{g} = \mathbf{g} - \alpha_{cg}\mathsf{AQ}\mathbf{d}$, $\mathbf{p} = \varphi(\mathbf{z}^{k+1})$
  end if
  $k = k + 1$
  end while
Step 5. {*Return (possibly inexact) solution.*}
  $\widetilde{\mathbf{x}} = \mathbf{z}^k + \mathbf{x}^0$

where

$$\eta_\Gamma = 1 - \frac{\widehat{\alpha}\overline{\lambda}_{\min}}{\vartheta + \vartheta\widehat{\Gamma}^2}, \qquad \widehat{\Gamma} = \max\{\Gamma, \Gamma^{-1}\}, \tag{10.58}$$

$$\vartheta = 2\max\{\overline{\alpha}\|\mathsf{A}\|, 1\}, \qquad \widehat{\alpha} = \min\{\overline{\alpha}, 2\|\mathsf{A}\|^{-1} - \overline{\alpha}\}, \tag{10.59}$$

and $\overline{\lambda}_{\min}$ *denotes the least nonzero eigenvalue of* $\mathsf{Q}^T\mathsf{AQ}$ *which satisfies* (10.56).

Proof. It is enough to combine Theorem 10.1 with the bounds given by Theorem 8.6. □

The efficiency of preconditioning by conjugate projector depends on the choice of the matrix U whose columns span the subspace which should approximate an invariant subspace spanned by the eigenvectors which correspond to small eigenvalues of A. For the minimization problems arising from the discretization of variational inequalities, U is typically obtained by aggregating variables using geometrical information or from the coarse discretization, as in the multigrid methods. A numerical example is in the next section. For references on related topics, see Sect. 10.9.

10.7 MPRGP for Problems with SPS Hessian

Let us adapt the MPRGP algorithm for solving

$$\min_{\mathbf{x}\in\Omega_B} f(\mathbf{x}), \quad \Omega = \{\mathbf{x}\in\mathbb{R}^n : \mathbf{x}\geq\boldsymbol{\ell}\}, \tag{10.60}$$

where $f(\mathbf{x}) = \frac{1}{2}\mathbf{x}^T\mathsf{A}\mathbf{x} - \mathbf{x}^T\mathbf{b}$, $\mathsf{A}\in\mathbb{R}^{n\times n}$ denotes an SPS matrix, $\mathbf{b},\mathbf{x}\in\mathbb{R}^n$, and $\boldsymbol{\ell} = [\ell_i]$, $\ell_i\in\mathbb{R}\cup\{-\infty\}$. We are especially interested in large, well-defined problems. Recall that by Proposition 2.16, a solution to (10.60) exists if

$$\mathbf{d}^T\mathbf{b}\leq 0 \quad \text{for } \mathbf{d}\in\mathcal{C}\cap\mathrm{Ker}\mathsf{A},$$

where $\mathcal{C}$ denotes the cone of recession directions. If $\boldsymbol{\ell}\in\mathbb{R}^n$, then $\mathcal{C}=\mathbb{R}^n_+$.

The MPRGPS algorithm described above differs from the standard MPRGP by the tests in the CG and proportioning steps. The algorithm generates iterations with a nonincreasing cost function as the expansion step does not need to reduce the cost function. The following simple proposition guarantees that MPRGPS generates iterates with the small norm of the projected gradient.

Proposition 10.2. *Let problem* (10.60) *has a solution and let* $\{\mathbf{x}^k\}$ *be generated by the MPRGPS algorithm for solving* (10.60). *Then for any* $\varepsilon>0$ *there is* $k\geq 0$ *such that*

$$\|\mathbf{g}^P(\mathbf{x}^k)\|\leq\varepsilon. \tag{10.61}$$

Proof: If MPRGPS does not find a solution in a finite number of steps, it generates at most n consecutive conjugate gradient iterations, followed by the expansion or proportioning steps. Thus, if MPRGPS does not generate a solution in the finite number of steps, then the index set $\mathcal{K}_{\mathrm{dis}}$ of disproportional iterations is infinite. Notice that $\{f(\mathbf{x}^k)\}$ is only non-increasing as the expansion step does not guarantee a decrease of the cost function.

Let $\mathcal{K}_{\mathrm{dis}}$ be infinite, assume $\overline{\alpha}\in(0,2\|\mathsf{A}^{-1})$, and denote

$$\widehat{\alpha} = \min\{\overline{\alpha}, 2\|\mathsf{A}^{-1}\| - \overline{\alpha}\} \quad\text{and}\quad \boldsymbol{\beta}^k = \boldsymbol{\beta}^k(\mathbf{x}^k).$$

Let $\overline{\mathbf{x}}$ denote a solution of (10.60). Then, the proportioning step satisfies

$$f(\mathbf{x}^k-\overline{\alpha}\boldsymbol{\beta}^k) - f(\mathbf{x}^k) = -\overline{\alpha}\|\boldsymbol{\beta}^k\|^2 + \frac{\overline{\alpha}^2}{2}(\boldsymbol{\beta}^k)^T\mathsf{A}\boldsymbol{\beta}^k \leq -\frac{1}{2}\widehat{\alpha}\|\boldsymbol{\beta}^k\|^2. \tag{10.62}$$

Algorithm 10.8. Modified proportioning for problems with SPS Hessian (MPRGPS)

Given an SPS matrix A *of the order* n, n*-vectors* $\mathbf{b}, \boldsymbol{\ell}, \Omega_B = \{\mathbf{x} : \mathbf{x} \geq \boldsymbol{\ell}\}$, $\mathbf{x}^0 \in \Omega_B$.

Step 0. {*Initialization.*}
  Choose $\Gamma > 0$, $\overline{\alpha} \in (0, 2\|\mathsf{A}\|^{-1})$, *set* $k = 0$, $\mathbf{g} = \mathsf{A}\mathbf{x}^0 - \mathbf{b}$, $\mathbf{p} = \boldsymbol{\varphi}(\mathbf{x}^0)$
  while $\|\mathbf{g}^P(\mathbf{x}^k)\|$ *is not small*
   if $\|\boldsymbol{\beta}(\mathbf{x}^k)\|^2 \leq \Gamma^2 \widetilde{\boldsymbol{\varphi}}(\mathbf{x}^k)^T \boldsymbol{\varphi}(\mathbf{x}^k)$
Step 1. {*Proportional* $\mathbf{x}^k$. *Trial conjugate gradient step.*}
   if $\mathsf{A}\mathbf{p} \neq \mathbf{o}$ **then** $\alpha_{cg} = \mathbf{g}^T\mathbf{p}/\mathbf{p}^T\mathsf{A}\mathbf{p}$ **else** $\alpha_{cg} = \infty$,
   $\alpha_f = \max\{\alpha : \mathbf{x}^k - \alpha\mathbf{p} \in \Omega_B\} = \min\{(x_i^k - \ell_i)/p_i : p_i > 0\}$
   if $\max\{\alpha_{cg}, \alpha_f\} = \infty$ **then** *stop (no solution)*
   if $\alpha_{cg} \leq \alpha_f$
Step 2. {*Conjugate gradient step.*}
   $\mathbf{x}^{k+1} = \mathbf{x}^k - \alpha_{cg}\mathbf{p}$, $\mathbf{g} = \mathbf{g} - \alpha_{cg}\mathsf{A}\mathbf{p}$,
   $\beta = \boldsymbol{\varphi}(\mathbf{y})^T\mathsf{A}\mathbf{p}/\mathbf{p}^T\mathsf{A}\mathbf{p}$, $\mathbf{p} = \boldsymbol{\varphi}(\mathbf{y}) - \beta\mathbf{p}$
   else
Step 3. {*Expansion step.*}
   $\mathbf{x}^{k+\frac{1}{2}} = \mathbf{x}^k - \alpha_f\mathbf{p}$, $\mathbf{g} = \mathbf{g} - \alpha_f\mathsf{A}\mathbf{p}$
   $\mathbf{x}^{k+1} = P_{\Omega_B}(\mathbf{x}^{k+\frac{1}{2}} - \overline{\alpha}\boldsymbol{\varphi}(\mathbf{x}^{k+\frac{1}{2}}))$
   $\mathbf{g} = \mathsf{A}\mathbf{x}^{k+1} - \mathbf{b}$, $\mathbf{p} = \boldsymbol{\varphi}(\mathbf{x}^{k+1})$
   end if
  else
Step 4. {*Proportioning step.*}
   $\mathbf{d} = \boldsymbol{\beta}(\mathbf{x}^k)$
   if $\mathsf{A}\mathbf{d} \neq \mathbf{o}$ **then** $\alpha_{cg} = \mathbf{g}^T\mathbf{d}/\mathbf{d}^T\mathsf{A}\mathbf{d}$ **else** *stop (no solution)*
   $\mathbf{x}^{k+1} = \mathbf{x}^k - \alpha_{cg}\mathbf{d}$, $\mathbf{g} = \mathbf{g} - \alpha_{cg}\mathsf{A}\mathbf{d}$, $\mathbf{p} = \boldsymbol{\varphi}(\mathbf{x}^{k+1})$
  end if
  $k = k + 1$
  end while
Step 5. {*Return (possibly inexact) solution.*}
  $\widetilde{\mathbf{x}} = \mathbf{x}^k$

It follows that

$$\begin{aligned}
\sum_{k\in\mathcal{K}_{\mathrm{dis}}} \|\boldsymbol{\beta}^k\|^2 &\leq 2\|\mathsf{A}\| \left(\sum_{k\in\mathcal{K}_{\mathrm{dis}}} \left(f(\mathbf{x}^k) - f(x^k - \overline{\alpha}\beta^k)\right) \right) \\
&= 2\|\mathsf{A}\| \left(\sum_{k\in\mathcal{K}_{\mathrm{dis}}} \left(f(\mathbf{x}^k) - f(\mathbf{x}^{k+1})\right) \right) \\
&\leq 2\|\mathsf{A}\| \left(\sum_{k=0}^{\infty} \left(f(\mathbf{x}^k) - f(\mathbf{x}^{k+1})\right) \right) \\
&= 2\|\mathsf{A}\| \lim_{k\to\infty} \left(f(\mathbf{x}^0) - f(\mathbf{x}^k)\right) \leq 2\|\mathsf{A}\| \left(f(\mathbf{x}^0) - f(\overline{\mathbf{x}})\right).
\end{aligned} \tag{10.63}$$

This proves

$$\lim_{k \in \mathcal{K}_{\text{dis}}} \|\boldsymbol{\beta}(\mathbf{x}^k)\| = 0.$$

Since the iterates $\mathbf{x}^k$ with $k \in \mathcal{K}_{\text{dis}}$ are disproportional, we have

$$\|\mathbf{g}^P(\mathbf{x}^k)\|^2 \leq (1 + \Gamma^{-2})\|\boldsymbol{\beta}(\mathbf{x}^k)\|^2.$$

which completes the proof of (10.61). □

10.8 Numerical Experiments

Here we illustrate the performance of some CG-based algorithms for the bound-constrained problem (10.1) on minimization of the cost functions $f_{\mathrm{L},h}$ and $f_{\mathrm{LW},h}$ introduced in Sect. 8.6 subject to bound constraints. All the computations are carried out with $\Gamma = 1$ and $\mathbf{x}^0 = \mathbf{o}$.

10.8.1 Polyak, MPRGP, and Preconditioned MPRGP

Let us first compare the performance of the algorithms presented in this chapter on minimization and of the energy function $f_{\mathrm{L},h}$ introduced in Sect. 8.6.2. The function is defined on the unit square $\Omega = [0,1] \times [0,1]$ and is specified by the parameter $h = 1/32$ the homogeneous boundary condition prescribed on $\Gamma_u = 0 \times [0,1]$. The boundary obstacle ℓ is defined by the upper part of the circle with the radius $R = 1$ and the center $S = (1, 0.5, -1.3)$. The boundary obstacle is placed under $\Gamma_c = 1 \times [0,1]$. The cost function and its mechanical interpretation are described in more detail in Sect. 8.6.2; its solution is in Fig. 10.4. Recall that the Hessian $\mathsf{A}_{\mathrm{L},h}$ of $f_{\mathrm{L},h}$ is ill-conditioned with $\kappa(\mathsf{A}_{\mathrm{L},h}) \approx h^{-2}$.

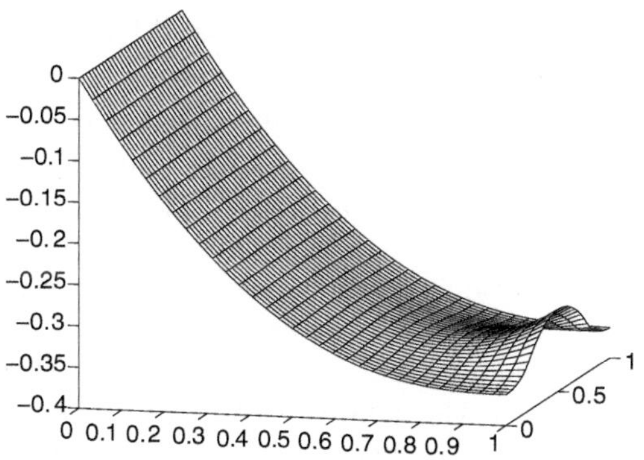

Fig. 10.4. Convergence of Polyak, MPRGP, and MPRGP–CP algorithms

The graph of the norm of the projected gradient (vertical axis) against the numbers of matrix–vector multiplications (horizontal axis) for Algorithm 7.2 (Polyak), Algorithm 10.2 (MPRGP), and MPRGP with preconditioning by the conjugate projector (MPRGP–CP) is in Fig. 10.5. We used $h = 1/32$, which corresponds to $n = 1056$ unknowns. The conjugate projector was defined by aggregating variables in the squares with 8×8 variables as in Sect. 8.6.2, so the matrix U has 16 columns.

We can see not only that the MPRGP algorithm outperforms Polyak's algorithm but also that the performance of MPRGP can be considerably improved by preconditioning. The difference between the Polyak and basic MPRGP algorithms is small due to the choice of $\boldsymbol{\ell}$, which makes identifying the active set easy; most iterations of both algorithms were CG steps. The picture can completely change for different $\boldsymbol{\ell}$ as documented in Dostál and Schöberl [85].

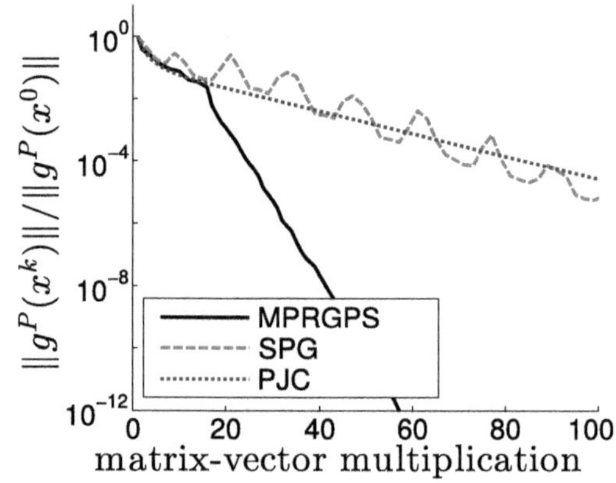

Fig. 10.5. Convergence of Polyak, MPRGP, and MPRGP–CP algorithms

10.8.2 Numerical Demonstration of Optimality

To illustrate the concept of optimality, let us consider the class of problems to minimize the quadratic function $f_{\mathrm{LW},h}$ (Sect. 8.6.3) subject to the bound constraints defined by the obstacle as above. The class of problems can be given a mechanical interpretation associated with the expanding spring systems on Winkler's foundation. The spectrum of the Hessian $\mathsf{A}_{\mathrm{LW},h}$ of $f_{\mathrm{LW},h}$ is located in the interval [4, 14]. Moreover, $\boldsymbol{\ell} \le \mathbf{o}$, so that the assumptions of Theorem 10.3 are satisfied.

In Fig. 10.6, we can see the numbers of the CG iterations k_n (vertical axis) that were necessary to reduce the norm of the projected gradient by 10^{-6} for the problems with the dimension n ranging from 100 to 1000000. The dimension n on the horizontal axis is in the logarithmic scale. We can

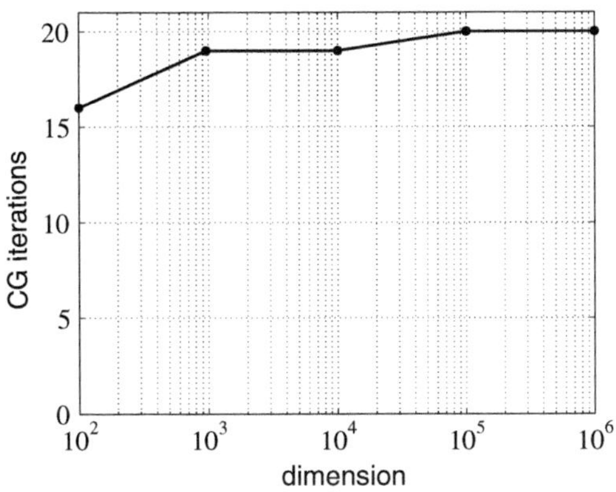

Fig. 10.6. Scalability of MPRGP algorithm

see that k_n varies mildly with varying n, in agreement with Theorem 10.3. Moreover, since the cost of the matrix–vector multiplications is in our case proportional to the dimension n of the matrix $\mathsf{A}_{\mathrm{LW},h}$, it follows that the cost of the solution is also proportional to n.

10.8.3 Numerical Experiments with SPS Hessian

We briefly review the results of two numerical experiments described in detail in Dostál and Pospíšil [84]. Our first benchmark is a 2D contact problem of elasticity. The goal is to find the displacement of a homogeneous cylinder pressed into the rigid corner by the gravity force as in Fig. 10.7. The discretized problem has the dimension $n = 660$, the regular condition number of the Hessian $\overline{\kappa}(\mathsf{A}) = 434.31$, and the three-dimensional kernel of the rigid body modes. The algorithm effectively identified the small contact interface and reduced the projected gradient by 10^4 after 258 iterations.

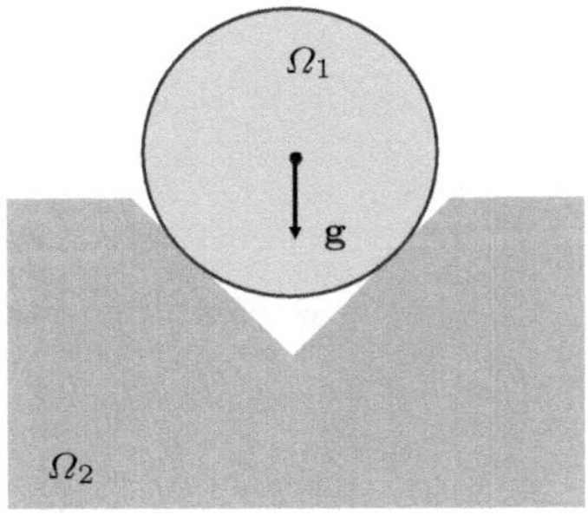

Fig. 10.7. Elastic cylinder pressed into the rigid corner by the gravitation

Our second benchmark is a multi-body particle simulation, a variant of the problems solved by Heyn et al. [165]. The spherical particles are scattered into a box with obstacles represented by nine walls as in Fig. 10.8, where we depicted only the bottom side of the box and the obstacles. The problem has the dimension $n = 1458$, the dimension of the kernel 729, and the regular condition number $\overline{\kappa}(\mathsf{A}) = 24.3$.

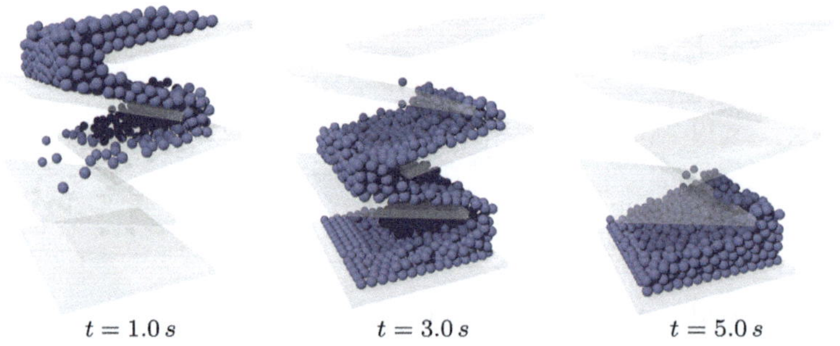

Fig. 10.8. Multibody particle simulation (courtesy of L. Pospíšil)

We compared the performance of MPRGPS with that of PBBF (projected Barzilain-Borwein algorithm with fallback) described in Sect. 11.7.1 and the projected Jacobi method PJC [255]. We chose the time $t = 3.0$ s and step-size $\Delta t = 6.25 \times 10^{-4}$ s. The decrease of $\|\mathbf{g}^P\|$ is in Fig. 10.9. MPRGPS used 50 iterations to achieve the relative precision $\epsilon = 10^{-12}$ with 8% of all bound constraints active.

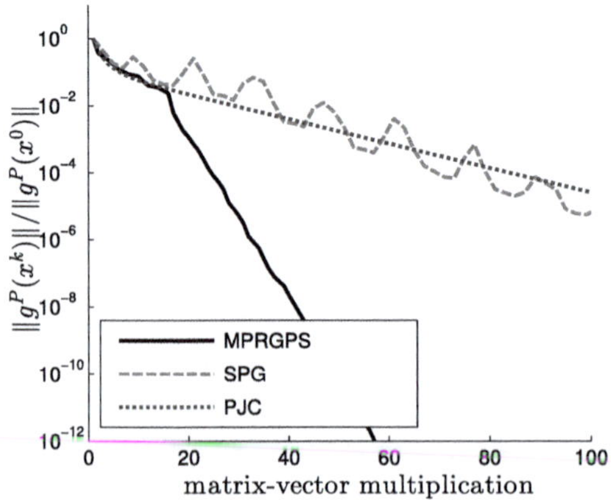

Fig. 10.9. The decrease of $\|\mathbf{g}^P\|$, $t = 3.0$ s

10.9 Comments and References

An essential step toward solving large bound-constrained problems was the development of algorithms with a rigorous convergence theory. Based on the results by Calamai and Moré [30], Moré and Toraldo [217] proposed the algorithm that uses conjugate gradients, but its convergence is driven by the gradient projections (see, e.g., Nocedal and Wright [220]). As in earlier algorithms, the steplength is found by possibly expensive backtracking. Despite the iterative basis of their algorithm, the authors proved that their algorithm preserves the finite termination property provided the solution satisfies the strict complementarity condition.

The effective theoretically supported strategies for adaptive precision control of the solution of auxiliary linear problems were presented independently by Friedlander and Martínez with their collaborators [18, 138–140], and Dostál [62, 63]. Comprehensive experiments and tests of heuristics can be found in Diniz-Ehrhardt et al. [52]. The research was not limited to the convex problems; see also Diniz-Ehrhardt et al. [51] or Mohy-ud-Din and Robinson [216].

A common drawback of all the strategies mentioned above is possible backtracking in search of the gradient projection step length and the lack of results on the convergence rate. A key to further progress provided the results by Schöberl [244], [245] on the decrease of the cost function along the projected-gradient path (see also Sect 10.3.1). Dostál [66] observed that these results can be plugged into the proportioning algorithm to preserve the convergence rate. See also Dostál and Schöberl [85]. The results have been extended for overrelaxation in Dostál et al. [103] and Dostál [73]. The preconditioning in the face appeared in O'Leary [222]. Preconditioning by a conjugate projector for bound-constrained problems appeared in the Ph.D. Thesis by Domorádová [54] and in Domorádová and Dostál [55]. The generalization of some results to problems with SPS Hessian and numerical experiments in the last section appeared in Pospíšil's Ph.D. Thesis [231] and Dostál and Pospíšil [83].

11
MPGP and PBBF for Separable QCQP

Let us consider more general *QCQP problems* (quadratic constraints quadratic program) to find

$$\min_{\mathbf{x}\in\Omega_S} f(\mathbf{x}), \quad f(\mathbf{x}) = \frac{1}{2}\mathbf{x}^T\mathsf{A}\mathbf{x} - \mathbf{x}^T\mathbf{b}, \tag{11.1}$$

where A is a large $n \times n$ SPD matrix, $\mathbf{b} \in \mathbb{R}^n$,

- $\Omega_S = \{\mathbf{x} \in \mathbb{R}^n : h_i(\mathbf{x}_i) \le 0,\ i = 1, \dots, s\}$, h_i convex quadratic functions,
- $\mathbf{x}_i = \mathbf{x}_{\mathcal{S}_i},\ \mathcal{S}_i \subseteq \mathcal{N},\ \mathcal{S}_i \cap \mathcal{S}_j = \emptyset$ for $i \ne j$, $\mathbf{x}_i \in \mathbb{R}^{n_i}$, and $n_1 + \dots + n_s = n$.

In particular, we consider *ellipses*, defined by SPD matrices $\mathsf{H}_i \in \mathbb{R}^{2\times 2}$ and

$$m_i = 2, \quad h_i(\mathbf{x}_i) = (\mathbf{x}_i - \mathbf{y}_i)^T\mathsf{H}_i(\mathbf{x}_i - \mathbf{y}_i) - c_i, \quad \mathbf{x}_i, \mathbf{y}_i \in \mathbb{R}^2, \quad c_i > 0;$$

and *spheres*

$$m_i \ge 2, \quad h_i(\mathbf{x}_i) = \|\mathbf{x}_i - \mathbf{y}_i\|^2 - c_i, \quad \mathbf{x}_i, \mathbf{y}_i \in \mathbb{R}^{m_i}, \quad c_i > 0.$$

We also admit bound constraints defined by $h_i(x_i) = \ell_i - x_i$, $\ell_i \in \mathbb{R}$.

There are two new features of Ω_S as compared with the bound constraints:

- If $h_i(\mathbf{x}_i) = h_i(\widehat{\mathbf{x}}_i)$, then it is possible that $\mathbf{x}_i \ne \widehat{\mathbf{x}}_i$
- If the generalized chopped gradient (see (11.8)) $\boldsymbol{\beta}_i(\mathbf{x}) \ne \mathbf{o}$, then for sufficiently large $\alpha > 0$, it can happen that $\mathbf{x}_i - \alpha\boldsymbol{\beta}_i(\mathbf{x}) \notin \Omega_i$.

We propose a variant of the active set strategy that we coined *MPGP* (Modified Proportioning with Gradient Projections), which enhances the above observations. The alternative PBBF (Projected Barzilai–Borwein with Fallback) algorithm combines the BB method with fixed step length gradient steps and uses the fallback to enforce the convergence.

Z. Dostál, *Optimal Quadratic Programming and QCQP Algorithms with Case Studies*, Springer Optimization and Its Applications 23,
https://doi.org/10.1007/978-3-031-95167-1_11

11.1 Projected Gradient and KKT Conditions

By Proposition 2.23, the solution to problem (11.1) always exists and is necessarily unique. As Ω_S satisfies ACQ, the unique solution $\widehat{\mathbf{x}}$ of (11.1) is fully determined by the KKT conditions (2.125), so that there is $\boldsymbol{\lambda} \in \mathbb{R}^s$ such that

$$\widehat{g}_i + \nabla h_i(\widehat{\mathbf{x}}_i)\lambda_i = \mathbf{o}, \; h_i(\widehat{\mathbf{x}}_i)\lambda_i = 0, \; \lambda_i \geq 0, \; h_i(\widehat{\mathbf{x}}_i) \leq 0, \; i = 1, \ldots, s, \quad (11.2)$$

where we use the notation

$$\widehat{\mathbf{g}} = \mathbf{g}(\widehat{\mathbf{x}}) = [\widehat{g}_i].$$

We begin with some notations that will help us to link the violation of the KKT conditions (11.2) to the solution error. Let $\mathscr{S}$ denote the set of all indices of the constraints so that

$$\mathscr{S} = \{1, 2, \ldots, s\}. \quad (11.3)$$

For any $\mathbf{x} \in \mathbb{R}^n$, we define the *active set* of $\mathbf{x}$ by

$$\mathscr{A}(\mathbf{x}) = \{i \in \mathscr{S} : h_i(\mathbf{x}_i) = 0\}. \quad (11.4)$$

Its complement

$$\mathscr{F}(\mathbf{x}) = \{i \in \mathscr{S} : h_i(\mathbf{x}_i) \neq 0\} \quad (11.5)$$

is called a *free set.*

For $\mathbf{x} \in \Omega_S$, we define the outer unit normal $\mathbf{n}$ by

$$\mathbf{n}_i = \mathbf{n}_i(\mathbf{x}) = \begin{cases} \|\nabla h_i(\mathbf{x}_i)\|^{-1}\nabla h_i(\mathbf{x}_i) & \text{for} \quad i \in \mathscr{A}(\mathbf{x}), \\ \mathbf{o} & \text{for} \quad i \in \mathscr{F}(\mathbf{x}). \end{cases} \quad (11.6)$$

The components of the gradient that violate the KKT conditions (11.2) in the free and active set are called a *free gradient* $\boldsymbol{\varphi}$ and a *chopped gradient* $\boldsymbol{\beta}$, respectively. The free gradient is defined by

$$\boldsymbol{\varphi}_i(\mathbf{x}) = \mathbf{g}_i(\mathbf{x}) \text{ for } i \in \mathscr{F}(\mathbf{x}), \quad \boldsymbol{\varphi}_i(\mathbf{x}) = \mathbf{o} \text{ for } i \in \mathscr{A}(\mathbf{x}), \quad (11.7)$$

the chopped gradient is defined by

$$\begin{aligned} &\boldsymbol{\beta}_i(\mathbf{x}) = \mathbf{o} \text{ for } i \in \mathscr{F}(\mathbf{x}), \\ &\boldsymbol{\beta}_i(\mathbf{x}) = \mathbf{g}_i(\mathbf{x}) - (\mathbf{n}_i^T\mathbf{g}_i)^-\mathbf{n}_i \text{ for } i \in \mathscr{A}(\mathbf{x}), \end{aligned} \quad (11.8)$$

where we use the notation

$$\mathbf{g}_i = \mathbf{g}_i(\mathbf{x}) \;\text{ and }\; (\mathbf{n}_i^T\mathbf{g}_i)^- = \min\{\mathbf{n}_i^T\mathbf{g}_i, 0\}. \quad (11.9)$$

Thus the KKT conditions (11.2) are satisfied if and only if the *projected gradient* (see Fig. 11.1)

$$\mathbf{g}^P(\mathbf{x}) = \boldsymbol{\varphi}(\mathbf{x}) + \boldsymbol{\beta}(\mathbf{x}) \tag{11.10}$$

is equal to zero.

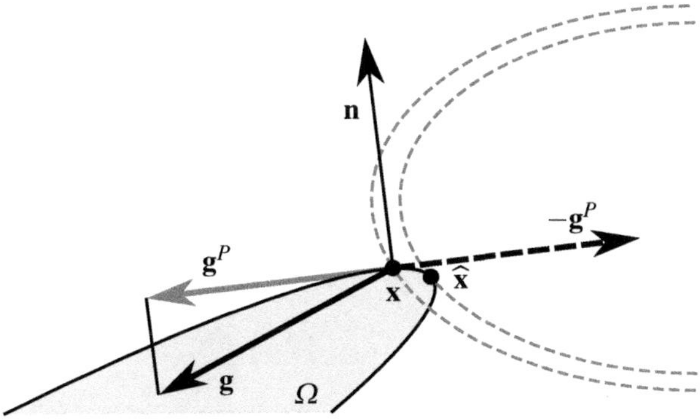

Fig. 11.1. Projected gradient near the solution

If $h_i(x_i)$ defines the bound constraint $x_i \geq \ell_i$, then $n_i = -1$ and the projected gradient is defined as in Sect. 7.1 by

$$g_i^P = g_i^- \quad \text{for} \quad i \in \mathscr{A}(\mathbf{x}), \quad g_i^P = g_i \quad \text{for} \quad i \in \mathscr{F}(\mathbf{x}).$$

Since

$$\mathbf{g}_i^P(\mathbf{x}) = \mathbf{g}_i(\mathbf{x}), \quad i \in \mathscr{F}(\mathbf{x}), \tag{11.11}$$

and for any $i \in \mathscr{A}(\mathbf{x})$

$$\begin{aligned}\|\mathbf{g}_i^P\|^2 =& \|\boldsymbol{\beta}_i\|^2 = \left(\mathbf{g}_i - (\mathbf{n}_i^T\mathbf{g}_i)^-\mathbf{n}_i\right)^T\left(\mathbf{g}_i - (\mathbf{n}_i^T\mathbf{g}_i)^-\mathbf{n}_i\right)\\ =& \|\mathbf{g}_i\|^2 - \left((\mathbf{n}_i^T\mathbf{g}_i)^-\right)^2 = \mathbf{g}_i^T\mathbf{g}_i^P,\end{aligned}$$

we have

$$\|\mathbf{g}^P\|^2 = \mathbf{g}^T\mathbf{g}^P \leq \|\mathbf{g}\|\ \|\mathbf{g}^P\| \tag{11.12}$$

and

$$\|\mathbf{g}^P\| \leq \|\mathbf{g}\|. \tag{11.13}$$

We need yet another simple property of the projected gradient.

Lemma 11.1. *Let* $\mathbf{x}, \mathbf{y} \in \Omega_S$ *and* $\mathbf{g} = \nabla f(\mathbf{x})$. *Then*

$$\mathbf{g}^T(\mathbf{y} - \mathbf{x}) \geq (\mathbf{g}^P)^T(\mathbf{y} - \mathbf{x}). \tag{11.14}$$

Proof. First, observe that

$$\mathbf{g}^T(\mathbf{y} - \mathbf{x}) = (\mathbf{g} - \mathbf{g}^P)^T(\mathbf{y} - \mathbf{x}) + (\mathbf{g}^P)^T(\mathbf{y} - \mathbf{x}). \tag{11.15}$$

Using the definition of projected gradient and $\mathbf{n}_i = \mathbf{n}_i(\mathbf{x})$, we get

$$(\mathbf{g} - \mathbf{g}^P)^T(\mathbf{y} - \mathbf{x}) = \sum_{i \in \mathscr{S}} (\mathbf{g}_i - \mathbf{g}_i^P)^T(\mathbf{y}_i - \mathbf{x}_i) \overset{(11.8)}{=} \sum_{i \in \mathscr{A}(\mathbf{x})} (\mathbf{n}_i^T \mathbf{g}_i)^- \mathbf{n}_i^T(\mathbf{y}_i - \mathbf{x}_i).$$

To finish the proof, observe that, due to the convexity of Ω_i, for $i \in \mathscr{A}(\mathbf{x})$

$$\mathbf{n}_i^T(\mathbf{y}_i - \mathbf{x}_i) \leq 0.$$

□

The following lemma is a quantitative refinement of the KKT conditions.

Lemma 11.2. *Let* $\widehat{\mathbf{x}}$ *be the solution of (11.1) and let* $\mathbf{g}^P(\mathbf{x})$ *denote the projected gradient at* $\mathbf{x} \in \Omega_S$. *Then*

$$\|\mathbf{x} - \widehat{\mathbf{x}}\|_{\mathsf{A}}^2 \leq 2\big(f(\mathbf{x}) - f(\widehat{\mathbf{x}})\big) \leq \|\mathbf{g}^P(\mathbf{x})\|_{\mathsf{A}^{-1}}^2 \leq \lambda_{\min}(\mathsf{A})^{-1}\|\mathbf{g}^P(\mathbf{x})\|^2. \tag{11.16}$$

Proof. Let $\widehat{\mathscr{A}}$, $\widehat{\mathscr{F}}$, and $\widehat{\mathbf{g}}$ denote the active set, free set, and the gradient at the solution, respectively. Observe that if $i \in \widehat{\mathscr{A}}$ and $\mathbf{x}_i \in \Omega_i$, then, using the convexity of h_i,

$$\left(\nabla h_i(\widehat{\mathbf{x}}_i)\right)^T(\mathbf{x}_i - \widehat{\mathbf{x}}_i) \leq h_i(\mathbf{x}_i) - h_i(\widehat{\mathbf{x}}_i) = h_i(\mathbf{x}_i) \leq 0. \tag{11.17}$$

It follows by the KKT conditions (11.2) and $\widehat{\mathbf{g}}_{\widehat{\mathscr{F}}} = \mathbf{o}_{\widehat{\mathscr{F}}}$ that

$$\widehat{\mathbf{g}}^T(\mathbf{x} - \widehat{\mathbf{x}}) = \sum_{i \in \widehat{\mathscr{A}}} \widehat{\mathbf{g}}_i^T(\mathbf{x}_i - \widehat{\mathbf{x}}_i) = \sum_{i \in \widehat{\mathscr{A}}} -\lambda_i\left(\nabla h_i(\widehat{\mathbf{x}}_i)\right)^T(\mathbf{x}_i - \widehat{\mathbf{x}}_i) \geq 0. \tag{11.18}$$

Thus, for any $\mathbf{x} \in \Omega_S$,

$$f(\mathbf{x}) - f(\widehat{\mathbf{x}}) = \widehat{\mathbf{g}}^T(\mathbf{x} - \widehat{\mathbf{x}}) + \frac{1}{2}(\mathbf{x} - \widehat{\mathbf{x}})^T\mathsf{A}(\mathbf{x} - \widehat{\mathbf{x}}) \geq \frac{1}{2}\|\mathbf{x} - \widehat{\mathbf{x}}\|_{\mathsf{A}}^2. \tag{11.19}$$

To prove the second inequality, we can use Lemma 11.1 and simple manipulations to get for any $\mathbf{x} \in \Omega_S$

$$\begin{aligned} 0 &\geq 2\big(f(\widehat{\mathbf{x}}) - f(\mathbf{x})\big) = \|\widehat{\mathbf{x}} - \mathbf{x}\|_{\mathsf{A}}^2 + 2\mathbf{g}^T(\widehat{\mathbf{x}} - \mathbf{x}) \\ &\overset{(11.14)}{\geq} \|\widehat{\mathbf{x}} - \mathbf{x}\|_{\mathsf{A}}^2 + 2\left(\mathbf{g}^P\right)^T(\widehat{\mathbf{x}} - \mathbf{x}) \\ &\geq 2 \min_{\mathbf{y}\in\mathbb{R}^n} \left(\frac{1}{2}\mathbf{y}^T\mathsf{A}\mathbf{y} + \left(\mathbf{g}^P\right)^T\mathbf{y}\right) = -(\mathbf{g}^P)^T\mathsf{A}^{-1}\mathbf{g}^P. \end{aligned}$$

The right inequality of (11.16) now follows easily. □

11.2 Alternative Error Measures

The projected gradient $\mathbf{g}^P$ is a natural error measure as it provides information about the cost function in a feasible neighborhood. However, experience shows that the projected gradient can be very sensitive to the curvature of the boundary.

The sensitivity of the projected gradient to the curvature of the feasible domain is documented in Fig. 11.2, where we can observe stagnation of the norm of the projected gradient in the iterations for the solution of a variant of the two clumped beams benchmark with orthotropic friction resolved in [27]. The feasible region was defined by the elliptic constraints with semi-major and semi-minor axes a and b satisfying $a/b = 10$. Let us point out that the active set was well approximated in the early stage of computations and stayed fixed after iteration 154, so the discontinuity of the projected gradient did not cause stagnation.

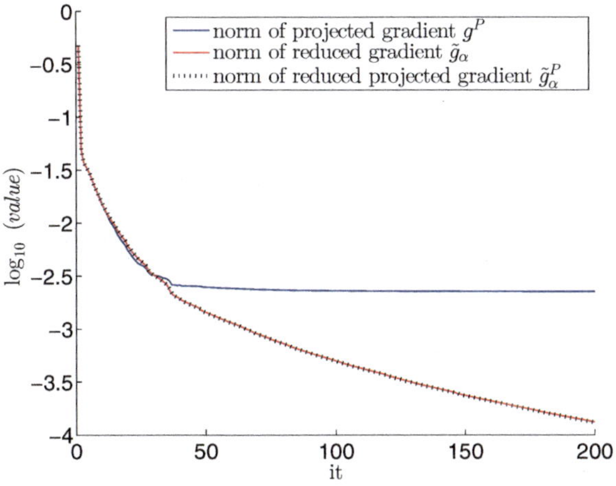

Fig. 11.2. Stagnation of $\|\mathbf{g}^P(\mathbf{x}^k)\|$ in convergent algorithm

The explanation is in Fig. 11.1 showing that the projected gradient can be large near the solution if the solution is on a strongly curved part of the boundary. In the following subsection, we introduce an alternative error measure that can be used as the stopping criterion in the algorithms for the minimization of a quadratic function subject to constraints with a strong curvature. This observation is most important for implementing the stopping criterion in the inner loops of algorithms for problems with separable inequality and linear equality constraints.

11.2.1 Reduced Gradient

If the curvature of the feasible region boundary near the solution is strong, and the norm of the gradient is large, which is typical for the elliptic constraints near a vertex or for the spherical constraints with a small radius, as in Fig. 11.1, then it is necessary to use more robust error measures, such as the *α-reduced gradient* (or briefly *reduced gradient*) $\widetilde{\mathbf{g}}_\alpha = \widetilde{\mathbf{g}}_\alpha(\mathbf{x})$, which is defined for any $\mathbf{x} \in \Omega_S$ and $\alpha > 0$ by

$$\widetilde{\mathbf{g}}_\alpha = \frac{1}{\alpha}\left(\mathbf{x} - P_{\Omega_S}(\mathbf{x} - \alpha\mathbf{g})\right). \tag{11.20}$$

The reduced gradient $\widetilde{\mathbf{g}}_\alpha$ is a continuous function of $\mathbf{x}$ and is not sensitive to α. The following lemma is an alternative to Lemma 11.2.

Lemma 11.3. *Let* $\widehat{\mathbf{x}}$ *be the solution of (11.1), let* $\widetilde{\mathbf{g}}_\alpha = \widetilde{\mathbf{g}}_\alpha(\mathbf{x})$ *denote the reduced gradient at* $\mathbf{x} \in \Omega_S$*, and let*

$$\alpha \in (0, 2(\lambda_{\min}(\mathsf{A}) + \|\mathsf{A}\|)^{-1}.$$

Then

$$\|\mathbf{x} - \widehat{\mathbf{x}}\| \le \lambda_{\min}(\mathsf{A})^{-1}\|\widetilde{\mathbf{g}}_\alpha\|. \tag{11.21}$$

Proof. Using the properties of norm and Corollary 2.7, we get for any $\alpha > 0$

$$\begin{aligned}
\|\mathbf{x} - \widehat{\mathbf{x}}\| \le& \|P_{\Omega_S}(\mathbf{x} - \alpha\mathbf{g}) - \mathbf{x}\| + \|P_{\Omega_S}(\mathbf{x} - \alpha\mathbf{g}) - \widehat{\mathbf{x}}\| \\
\overset{(5.7),(11.20)}{\le}& \alpha\|\widetilde{\mathbf{g}}_\alpha\| + \|P_{\Omega_S}(\mathbf{x} - \alpha\mathbf{g}) - P_{\Omega_S}\left(\widehat{\mathbf{x}} - \alpha\mathbf{g}(\widehat{\mathbf{x}})\right)\| \\
\overset{(2.16)}{\le}& \alpha\|\widetilde{\mathbf{g}}_\alpha\| + \|\mathbf{x} - \widehat{\mathbf{x}} - \alpha\mathsf{A}(\mathbf{x} - \widehat{\mathbf{x}})\| \\
\le& \alpha\|\widetilde{\mathbf{g}}_\alpha\| + \|\mathsf{I} - \alpha\mathsf{A}\|\|\mathbf{x} - \widehat{\mathbf{x}}\| \\
\le& \alpha\|\widetilde{\mathbf{g}}_\alpha\| + \max\{|1 - \alpha\lambda_{\min}(\mathsf{A})|, |1 - \alpha\lambda_{\max}(\mathsf{A})|\}\|\mathbf{x} - \widehat{\mathbf{x}}\|.
\end{aligned}$$

If α satisfies the assumption, then

$$1 - \alpha\lambda_{\min} = |1 - \alpha\lambda_{\min}| \ge |1 - \alpha\lambda_{\max}|$$

and

$$\|\mathbf{x} - \widehat{\mathbf{x}}\| \le \alpha \|\widetilde{\mathbf{g}}_\alpha\| + \left(1 - \alpha\lambda_{\min}(\mathsf{A})\right) \|\mathbf{x} - \widehat{\mathbf{x}}\|.$$

After simple manipulations, we get (11.21). □

We shall also need the inequalities formulated in the following lemma.

Lemma 11.4. *Let* $\mathbf{x} \in \Omega_S$, $\alpha > 0$, *and let us denote* $\mathbf{g} = \mathbf{g}(\mathbf{x}) = \mathsf{A}\mathbf{x} - \mathbf{b}$, $\widetilde{\mathbf{g}}_\alpha = \widetilde{\mathbf{g}}_\alpha(\mathbf{x})$, *and* $\mathbf{g}^P = \mathbf{g}^P(\mathbf{x})$. *Then*

$$\|\widetilde{\mathbf{g}}_\alpha\|^2 \le \mathbf{g}^T \widetilde{\mathbf{g}}_\alpha \le \mathbf{g}^T \mathbf{g}^P = \|\mathbf{g}^P\|^2 \le \|\mathbf{g}\|^2. \tag{11.22}$$

Proof. Recall that by Proposition 2.6 for any $\mathbf{x} \in \Omega_S$ and $\mathbf{y} \in \mathbb{R}^n$

$$\left(\mathbf{x} - P_{\Omega_S}(\mathbf{y})\right)^T \left(\mathbf{y} - P_{\Omega_S}(\mathbf{y})\right) \le 0, \tag{11.23}$$

and by the definition (11.20) for any $\alpha > 0$

$$\alpha\widetilde{\mathbf{g}}_\alpha = \mathbf{x} - P_{\Omega_S}(\mathbf{x} - \alpha\mathbf{g}) \quad \text{and} \quad P_{\Omega_S}(\mathbf{x} - \alpha\mathbf{g}) = \mathbf{x} - \alpha\widetilde{\mathbf{g}}_\alpha.$$

Denoting $\mathbf{y} = \mathbf{x} - \alpha\mathbf{g}$ and using (11.23), we get

$$\alpha\widetilde{\mathbf{g}}_\alpha^T(-\alpha\mathbf{g} + \alpha\widetilde{\mathbf{g}}_\alpha) = \left(\mathbf{x} - (\mathbf{x} - \alpha\widetilde{\mathbf{g}}_\alpha)\right)^T \left(\mathbf{x} - \alpha\mathbf{g} - (\mathbf{x} - \alpha\widetilde{\mathbf{g}}_\alpha)\right) \le 0. \tag{11.24}$$

After simple manipulations, we get the left inequality of (11.22).

To prove the rest, notice that $\mathbf{x} - \mathbf{g}^P$ is the projection of $\mathbf{x} - \mathbf{g}$ to the set defined by separable linear constraints

$$\hat{\Omega}(\mathbf{x}) = \hat{\Omega} = \hat{\Omega}_1 \times \cdots \times \hat{\Omega}_s, \tag{11.25}$$

where

$$\begin{aligned} &\hat{\Omega}_i = \{\mathbf{y}_i \in \mathbb{R}^{\ell_i} : \hat{h}_i(\mathbf{y}_i) \le 0\}, && \\ &\hat{h}_i(\mathbf{y}_i) = (\mathbf{y}_i - \mathbf{x}_i)^T \nabla h_i(\mathbf{x}_i) && \text{for } i \in \mathscr{A}(\mathbf{x}), \\ &\hat{h}_i(\mathbf{y}_i) = -1 && \text{for } i \in \mathscr{F}(\mathbf{x}). \end{aligned} \tag{11.26}$$

Since

$$\mathbf{x} - \alpha\widetilde{\mathbf{g}}_\alpha = P_{\Omega_S}(\mathbf{x} - \alpha\mathbf{g}), \quad \mathbf{y} = \mathbf{x} - \alpha\mathbf{g}^P = P_{\hat{\Omega}}(\mathbf{x} - \alpha\mathbf{g}),$$

and $\Omega_S \subseteq \hat{\Omega}$ implies

$$\|\mathbf{y} - P_{\hat{\Omega}}(\mathbf{y})\| \le \|\mathbf{y} - P_{\Omega_S}(\mathbf{y})\|,$$

we have $\mathbf{x} - \alpha\widetilde{\mathbf{g}}_\alpha \in \Omega_S$ and

$$\begin{aligned} \alpha^2\|\mathbf{g} - \mathbf{g}^P\|^2 &= \alpha^2\|(\mathbf{x} - \alpha\mathbf{g}) - (\mathbf{x} - \alpha\mathbf{g}^P)\|^2 \\ &= \alpha^2\|\mathbf{y} - P_{\hat{\Omega}}(\mathbf{y})\|^2 \le \|\mathbf{y} - P_{\Omega_S}(\mathbf{y})\|^2 \\ &= \alpha^2\|(\mathbf{x} - \alpha\mathbf{g}) - (\mathbf{x} - \alpha\widetilde{\mathbf{g}}_\alpha)\|^2 = \alpha^2\|\mathbf{g} - \widetilde{\mathbf{g}}_\alpha\|^2. \end{aligned}$$

It follows that

$$-\mathbf{g}^T\mathbf{g}^P = -2\mathbf{g}^T\mathbf{g}^P + \|\mathbf{g}^P\|^2 \le -2\mathbf{g}^T\widetilde{\mathbf{g}}_\alpha + \|\widetilde{\mathbf{g}}_\alpha\|^2 \le -\mathbf{g}^T\widetilde{\mathbf{g}}_\alpha. \tag{11.27}$$

The rest follows by (11.12) and (11.13). □

11.3 MPGP Scheme

The MPGP algorithm exploits a user-defined constant $\Gamma > 0$, the steplength $\overline{\alpha} \in (0, 2\|\mathsf{A}\|^{-1})$, a test which is used to decide when to change the face, and two types of steps.

If for a given $\Gamma > 0$ the inequality

$$\|\boldsymbol{\beta}(\mathbf{x}^k)\| \le \Gamma\|\boldsymbol{\varphi}(\mathbf{x}^k)\| \tag{11.28}$$

holds, then we call the iterate $\mathbf{x}^k$ *proportional*. Test (11.28) decides if the algorithm should continue in the current working set by conjugate gradient steps or change the working set by gradient projection step. If $\Gamma = 1$, then the iterate is proportional if the free gradient dominates the violation of the KKT conditions.

The *conjugate gradient step* is exactly that of the Polyak or MPRGP algorithm. It is defined by

$$\mathbf{x}^{k+1} = \mathbf{x}^k - \alpha_{cg}\mathbf{p}^{k+1}, \quad \alpha_{cg} = \mathbf{b}^T\mathbf{p}^{k+1}/\left(\mathbf{p}^{k+1}\right)^T\mathsf{A}\mathbf{p}^{k+1}, \tag{11.29}$$

where $\mathbf{p}^{k+1}$ is the conjugate gradient (see Chap. 8) constructed recurrently. The recurrence starts (or restarts) with $\mathbf{p}^{k+1} = \boldsymbol{\varphi}(\mathbf{x}^k)$ whenever a proportional $\mathbf{x}^k$ is generated by the gradient projection step. If $\mathbf{x}^k$ is generated by the conjugate gradient step, then $\mathbf{p}^{k+1}$ is given by

$$\mathbf{p}^{k+1} = \boldsymbol{\varphi}(\mathbf{x}^k) - \beta\mathbf{p}^k, \quad \beta = \boldsymbol{\varphi}(\mathbf{x}^k)^T\mathsf{A}\mathbf{p}^k/(\mathbf{p}^k)^T\mathsf{A}\mathbf{p}^k. \tag{11.30}$$

The coefficient α_{cg} is chosen so that

$$\begin{aligned} f(\mathbf{x}^{k+1}) &= \min\{f(\mathbf{x}^k - \alpha\mathbf{p}^{k+1}) : \alpha \in \mathbb{R}\} \\ &= \min\{f(\mathbf{x}) : \mathbf{x} \in \mathbf{x}^r + \operatorname{Span}\{\mathbf{p}^{r+1}, \ldots, \mathbf{p}^{k+1}\}. \end{aligned}$$

It can be checked directly that

$$f(\mathbf{x}^{k+1}) = f\left(\mathbf{x}^k - \alpha_{cg}\boldsymbol{\varphi}(\mathbf{x}^k)\right) = f(\mathbf{x}^k) - \frac{1}{2}\frac{\|\boldsymbol{\varphi}(\mathbf{x}^k)\|^4}{\boldsymbol{\varphi}(\mathbf{x}^k)^T\mathsf{A}\boldsymbol{\varphi}(\mathbf{x}^k)}. \tag{11.31}$$

The CG steps are used to speed up the minimization in the working face

$$\mathscr{W}_{\mathscr{J}} = \{\mathbf{x} : h_i(\mathbf{x}_i) = 0, \quad i \in \mathscr{J}\}, \quad \mathscr{J} = \mathscr{A}(\mathbf{x}^k).$$

The *gradient projection step* is defined by the gradient projection

$$\mathbf{x}^{k+1} = P_{\Omega_S}\left(\mathbf{x}^k - \alpha \mathbf{g}(\mathbf{x}^k)\right) = \mathbf{x}^k - \alpha \widetilde{\mathbf{g}}_\alpha \tag{11.32}$$

with a fixed step length $\alpha > 0$. This step can both add and remove the indices from the current working set.

Now, we are ready to describe the basic algorithm in a form that is convenient for the analysis. More details about the implementation and the choice of parameters can be found in Sect. 11.6.2.

Algorithm 11.1. Modified proportioning with gradient projections (MPGP schema)

Given an SPD matrix $\mathsf{A} \in \mathbb{R}^{n\times n}$, *and* an n-*vector* $\mathbf{b} \in \mathbb{R}^n$.
Choose $\mathbf{x}^0 \in \Omega_B$, $\overline{\alpha} \in (0, 2\|\mathsf{A}\|^{-1})$, *and* $\Gamma > 0$. *Set* $k = 0$.
For $k \geq 0$ *and* $\mathbf{x}^k$ *known, choose* $\mathbf{x}^{k+1}$ *by the following rules:*
(i) If $\mathbf{g}^P(\mathbf{x}^k) = \mathbf{o}$, *set* $\mathbf{x}^{k+1} = \mathbf{x}^k$.
(ii) If $\mathbf{x}^k$ *is proportional and* $\mathbf{g}^P(\mathbf{x}^k) \neq \mathbf{o}$, *try to generate* $\mathbf{x}^{k+1}$ *by the CG step. If* $\mathbf{x}^{k+1} \in \Omega_S$, *then accept it, else set* $\mathbf{x}^{k+1} = P_{\Omega_S}\left(\mathbf{x}^k - \overline{\alpha}\mathbf{g}(\mathbf{x}^k)\right)$.
(iii) If $\mathbf{x}^k$ *is not proportional, define* $\mathbf{x}^{k+1}$ *by the gradient projection step.*

Remark 11.1. The gradient projection (ii) can be replaced by the expansion step

$$\mathbf{x}^{k+1} = P_\Omega\left(\mathbf{x}^k - \overline{\alpha}\varphi(\mathbf{x}^k)\right), \tag{11.33}$$

but this choice can slow down the convergence of active variables.

11.4 Rate of Convergence

Let us examine the bounds on the difference between the cost function value at the solution and the current iterate. The main result reads as follows.

Theorem 11.1. *Let* $\{\mathbf{x}^k\}$ *be generated by Algorithm 11.1 with* $\mathbf{x}^0 \in \Omega_S$, $\Gamma > 0$, *and* $\overline{\alpha} \in (0, 2\|\mathsf{A}\|^{-1})$. *Then*

$$f(\mathbf{x}^{k+1}) - f(\widehat{\mathbf{x}}) \leq \eta_\Gamma(f(\mathbf{x}^k) - f(\widehat{\mathbf{x}})), \tag{11.34}$$

where $\widehat{\mathbf{x}}$ *denotes the unique solution of* (10.1),

$$\eta_\Gamma = 1 - \frac{\widehat{\alpha}\lambda_{\min}}{2+2\Gamma^2} \leq 1 - \frac{1}{2+2\Gamma^2}\kappa(\mathsf{A})^{-1}, \tag{11.35}$$

$$\widehat{\alpha} = \min\{\overline{\alpha}, 2\|\mathsf{A}\|^{-1} - \overline{\alpha}\}, \tag{11.36}$$

and $\lambda_{\min}$ *and* $\kappa(\mathsf{A})$ *denote the smallest eigenvalue and condition number of* A, *respectively.*

The error in the A*-norm is bounded by*

$$\|\mathbf{x}^k - \widehat{\mathbf{x}}\|_{\mathsf{A}}^2 \leq 2\eta_\Gamma^k(f(\mathbf{x}^0) - f(\widehat{\mathbf{x}})). \tag{11.37}$$

Proof. If $\mathbf{x}^{k+1}$ is generated by the gradient projection step, then the (11.34) is satisfied by Proposition 5.1.

To estimate the effect of a feasible CG step, let us start with the estimate

$$\begin{aligned} f(\mathbf{x}^{k+1}) &= f\left(\mathbf{x}^k - \alpha_{cg}\varphi(\mathbf{x}^k)\right) = f(\mathbf{x}^k) - \frac{1}{2}\frac{\|\varphi(\mathbf{x}^k)\|^4}{\varphi(\mathbf{x}^k)^T\mathsf{A}\varphi(\mathbf{x}^k)} \\ &\leq f(\mathbf{x}^k) - \frac{1}{2}\|\mathsf{A}\|^{-1}\|\varphi(\mathbf{x}^k)\|^2. \end{aligned} \tag{11.38}$$

We shall compare (11.38) with the effect of the gradient projection applied to the $\mathbf{x}^k$ which satisfies

$$\|\boldsymbol{\beta}(\mathbf{x}^k)\| \leq \Gamma\|\varphi(\mathbf{x}^k)\|.$$

For $\alpha > 0$, $\mathbf{x}^k - \alpha\varphi(\mathbf{x}^k) \in \Omega_S$, and $\|\widetilde{\boldsymbol{\beta}}_\alpha\| \leq \|\boldsymbol{\beta}\|$, we get

$$\begin{aligned} f\left(P_{\Omega_S}\left(\mathbf{x}^k - \alpha\mathbf{g}(\mathbf{x}^k)\right)\right) &= f(\mathbf{x}^k) - \alpha\widetilde{\mathbf{g}}_\alpha(\mathbf{x}^k)\mathbf{g}(\mathbf{x}^k) + \frac{\alpha^2}{2}\widetilde{\mathbf{g}}_\alpha(\mathbf{x}^k)^T\mathsf{A}\widetilde{\mathbf{g}}_\alpha(\mathbf{x}^k) \\ &\geq f(\mathbf{x}^k) - \alpha\left(\|\varphi(\mathbf{x}^k)\|^2 + \|\widetilde{\boldsymbol{\beta}}_\alpha(\mathbf{x}^k)\|^2\right) \\ &\geq f(\mathbf{x}^k) - \alpha\left(1+\Gamma^2\right)\|\varphi(\mathbf{x}^k)\|^2. \end{aligned}$$

Substituting

$$\alpha = \frac{1}{2+2\Gamma^2}\|\mathsf{A}\|^{-1}$$

into the above inequalities, we get

$$f(\mathbf{x}^{k+1}) \leq f(\mathbf{x}^k) - \frac{1}{2}\|\mathsf{A}\|^{-1}\|\varphi(\mathbf{x}^k)\|^2 \leq f\left(P_{\Omega_S}\left(\mathbf{x}^k - \alpha\mathbf{g}(\mathbf{x}^k)\right)\right) \tag{11.39}$$

and

$$f(\mathbf{x}^{k+1}) - f(\widehat{\mathbf{x}}) \leq f\left(P_{\Omega_S}\left(\mathbf{x}^k - \alpha\mathbf{g}(\mathbf{x}^k)\right)\right) - f(\widehat{\mathbf{x}}). \tag{11.40}$$

Using Theorem 5.1, we get

$$f(\mathbf{x}^{k+1}) - f(\widehat{\mathbf{x}}) \leq f\left(P_{\Omega_S}\left(\mathbf{x}^k - \alpha\mathbf{g}(\mathbf{x}^k)\right)\right) - f(\widehat{\mathbf{x}}) \leq \eta_\Gamma\left(f(\mathbf{x}^k) - f(\widehat{\mathbf{x}})\right)$$

with

$$\eta_\Gamma = 1 - \frac{\lambda_{\min}}{2+2\Gamma^2}\|\mathsf{A}\|^{-1} = 1 - \frac{1}{2+2\Gamma^2}\kappa(\mathsf{A})^{-1}.$$

Notice that $\|\mathsf{A}\|^{-1}$ is the maximum value of $\widehat{\alpha}$ in the estimate (10.22), so the estimate depends on the parameter α of the gradient projection step only. □

11.5 Bound on the Norm of Gradients

To solve the QCQP problems with separable constraints effectively, we must be able to *recognize* when we are near the solution. However, as in the previous chapter, there is a catch: the projected gradient is not continuous and can be large near the solution, as shown in Fig. 10.1. We shall return to the problems with the sensitivity of projected gradient in the next chapter on applying the MPGP algorithm in the inner loops of other algorithms. Here, we give the convergence results that support practical computations. We shall use the following simple relation.

Lemma 11.5. *Let $\mathbf{x}^k$ denote an iterate generated by the MPGP algorithm under the assumptions of Theorem 11.1, let the constraints in the problem (11.2) are separable linear, and let $\mathbf{g}^P(\mathbf{x}^k)$ and $\mathbf{g}_\alpha(\mathbf{x}^k)$ denote the projected gradient and reduced gradient, respectively.*

Then

$$\mathbf{g}^P(\mathbf{x}^k) = \widetilde{\mathbf{g}}_\alpha(\mathbf{x}^k). \tag{11.41}$$

Proof. Let us denote

$$\mathbf{n} = \mathbf{n}(\mathbf{x}^k),\ \mathbf{g} = \mathbf{g}(\mathbf{x}^k),\ \alpha > 0,\ \widetilde{\mathbf{g}}_\alpha = \widetilde{\mathbf{g}}_\alpha(\mathbf{x}^k),\ \text{and } \mathbf{n}_i^T\mathbf{g}_i < 0,\ i \in \mathscr{A}(\mathbf{x}^k).$$

Using standard linear algebra and linearity of the constraints, we get

$$\mathbf{x}_i^k - P_{\Omega_i}(\mathbf{x}_i^k - \alpha\mathbf{g}_i) = \alpha\widetilde{\mathbf{g}}_i^\alpha = \alpha\mathbf{g}_i - \alpha(\mathbf{n}_i^T\mathbf{g}_i)^-\mathbf{n}_i = \alpha\mathbf{g}_i^P.$$

Thus $\alpha\widetilde{\mathbf{g}}_i^\alpha = \alpha\mathbf{g}_i^P$. If $i \in \mathscr{F}(\mathbf{x}^k)$ or $\mathbf{n}_i^T\mathbf{g}_i \geq 0$, then obviously

$$\mathbf{g}_i^P = \widetilde{\mathbf{g}}_i^\alpha = \mathbf{g}_i. \tag{11.42}$$

□

Now, we are ready to prove the main result of this section.

Theorem 11.2. *Let $\{\mathbf{x}^k\}$ denote the iterates generated by the MPGP algorithm for problem* (11.1) *with the assumptions of Theorem 11.1 with*

$$\alpha \in (0, 2\|\mathsf{A}\|^{-1}) \quad \text{and} \quad \Gamma > 0. \tag{11.43}$$

Then for any $k \geq 1$

$$\|\widetilde{\mathbf{g}}_\alpha(\mathbf{x}^k)\|^2 \leq a_1 \eta_\Gamma^k (f(\mathbf{x}^0) - f(\mathbf{x})), \quad a_1 = 2\|\mathsf{A}\| \frac{1+\eta_\Gamma}{1-\eta_\Gamma}, \tag{11.44}$$

where η_Γ *is defined in Theorem 11.1. If the constraints are linear, then*

$$\|\mathbf{g}^P(\mathbf{x}^k)\|^2 \leq a_1 \eta_\Gamma^k (f(\mathbf{x}^0) - f(\mathbf{x}). \tag{11.45}$$

Proof. Let us first recall that by the assumptions

$$\begin{aligned} f(\mathbf{x}^k) - f(\widehat{\mathbf{x}}) &\leq \eta_\Gamma (f(\mathbf{x}^{k-1}) - f(\widehat{\mathbf{x}})) \leq \eta_\Gamma^2 (f(\mathbf{x}^{k-2}) - f(\widehat{\mathbf{x}})) \\ &\leq \cdots \leq \eta_\Gamma^k (f(\mathbf{x}^0) - f(\widehat{\mathbf{x}})). \end{aligned} \tag{11.46}$$

Let us denote

$$\mathbf{y}^0 = \mathbf{x}^{k-1}, \ \mathbf{y}^1 = \mathbf{x}^k, \quad \text{and} \ \ \mathbf{y}^2 = P_\Omega(\mathbf{x}^k - \alpha \mathbf{g}(\mathbf{x}^k)). \tag{11.47}$$

Using (11.46), we get

$$(f(\mathbf{y}^2) - f(\widehat{\mathbf{x}})) \leq \eta_\Gamma (f(\mathbf{y}^1) - f(\widehat{\mathbf{x}})) \leq \eta_\Gamma^2 (f(\mathbf{y}^0) - f(\widehat{\mathbf{x}})). \tag{11.48}$$

It follows that the assumptions of Lemma 10.2 are satisfied with $\eta = \eta_\Gamma$ and

$$\begin{aligned} f(\mathbf{y}^1) - f(\mathbf{y}^2) \leq & \frac{1+\eta_\Gamma}{1-\eta_\Gamma} \eta_\Gamma \left(f(\mathbf{y}^0) - f(\mathbf{y}^1)\right) = \frac{1+\eta_\Gamma}{1-\eta_\Gamma} \eta_\Gamma \left(f(\mathbf{x}^{k-1}) - f(\mathbf{x}^k)\right) \\ \leq & \frac{1+\eta_\Gamma}{1-\eta_\Gamma} \eta_\Gamma \left(f(\mathbf{x}^{k-1}) - f(\widehat{\mathbf{x}})\right) \leq \frac{1+\eta_\Gamma}{1-\eta_\Gamma} \eta_\Gamma^k \left(f(\mathbf{x}^0) - f(\widehat{\mathbf{x}})\right). \end{aligned}$$

Finally, we get for $\alpha = \|\mathsf{A}\|^{-1}$

$$\begin{aligned} f(\mathbf{y}^1) - f(\mathbf{y}^2) =& f(\mathbf{y}^1) - f(P_\Omega(\mathbf{y}^1 - \alpha \mathbf{g}(\mathbf{y}^1))) \\ \overset{\text{Taylor}}{=} & \alpha \widetilde{\mathbf{g}}_\alpha^T(\mathbf{y}^1) \widetilde{\mathbf{g}}(\mathbf{y}^1) - \frac{\alpha^2}{2} \widetilde{\mathbf{g}}_\alpha^T(\mathbf{y}^1) \mathsf{A} \widetilde{\mathbf{g}}_\alpha(\mathbf{y}^1) & (11.49) \\ \geq & \alpha \widetilde{\mathbf{g}}_\alpha^T(\mathbf{y}^1) \mathbf{g}(\mathbf{y}^1) - \frac{\alpha^2}{2} \|\mathsf{A}\| \|\widetilde{\mathbf{g}}_\alpha(\mathbf{y}^1)\|^2 & (11.50) \\ \geq & (\alpha - \frac{\alpha^2}{2}\|\mathsf{A}\|)\, \widetilde{\mathbf{g}}_\alpha^T(\mathbf{y}^1) \mathbf{g}(\mathbf{y}^1) \overset{(11.22)}{\geq} \frac{\alpha}{2} \|\widetilde{\mathbf{g}}_\alpha\|^2. & (11.51) \end{aligned}$$

After substituting $\mathbf{y}^1 = \mathbf{x}^{k-1}$, $\mathbf{y}^2 = \mathbf{x}^k$, we get (11.44). Similarly, if the constraints are separable linear, we get

$$f(\mathbf{y}^1) - f(\mathbf{y}^2) \overset{(11.41)}{=} \frac{\alpha}{2} \left(\mathbf{g}^P(\mathbf{y}^1)\right)^T \mathbf{g}(\mathbf{y}^1) \overset{(11.22)}{\geq} \frac{\alpha}{2} \|\mathbf{g}^P(\mathbf{y}^1)\|^2 \tag{11.52}$$

and

$$\frac{\alpha}{2}\|\mathbf{g}^P(\mathbf{x}^k)\|^2 \le f(\mathbf{x}^{k+1}) - f(\mathbf{x}^k) \le \frac{1+\eta_\Gamma}{1-\eta_\Gamma}\eta_\Gamma^k \left(f(\mathbf{x}^0) - f(\widehat{\mathbf{x}})\right). \quad (11.53)$$

□

Theorem 11.2 indicates that the projected gradient can be used in the stopping criterion, provided the boundary's curvature is modest. For the mechanics problems, the projected gradient norm is related to unbalanced contact forces. The norms of all three gradients provide bounds on the solution error (see lemmas 11.2, 11.3, and 12.5).

11.6 Implementation

In this section, we describe Algorithm 11.1 in a convenient form for implementation. We also include some modifications that can improve its performance.

11.6.1 Projection Step with Feasible Half-Step

To improve the efficiency of the projection step, we can use the trial conjugate gradient direction $\mathbf{p}^k$ generated before the projection step is invoked. We propose first to generate

$$\mathbf{x}^{k+\frac{1}{2}} = \mathbf{x}^k - \alpha_f \mathbf{p}^k \quad \text{and} \quad \mathbf{g}^{k+\frac{1}{2}} = \mathbf{g}^k - \alpha_f \mathsf{A}\mathbf{p}^k, \quad (11.54)$$

where the feasible step length α_f for $\mathbf{p}^k$ is defined by

$$\alpha_f = \max\{\alpha : \ \mathbf{x}^k - \alpha\mathbf{p}^k \in \Omega_S\}, \quad (11.55)$$

and then define

$$\mathbf{x}^{k+1} = P_{\Omega_S}\left(\mathbf{x}^{k+\frac{1}{2}} - \alpha\mathbf{g}(\mathbf{x}^{k+\frac{1}{2}})\right). \quad (11.56)$$

The half-step is illustrated in Fig. 11.3. It does not require any additional matrix–vector multiplication and estimate (11.34) remains valid as

$$\begin{aligned} f(\mathbf{x}^{k+1}) - f(\widehat{\mathbf{x}}) \le & \eta_\Gamma \left((f(\mathbf{x}^{k+\frac{1}{2}}) - f(\mathbf{x}^k)) + f(\mathbf{x}^k) - f(\widehat{\mathbf{x}}) \right) \\ \le & \eta_\Gamma \left(f(\mathbf{x}^k) - f(\widehat{\mathbf{x}}) \right). \end{aligned}$$

Since our analysis is based on the worst-case analysis, implementing the feasible half-step does not improve the estimate of the rate of convergence (11.34), but it does improve the real performance of the algorithm.

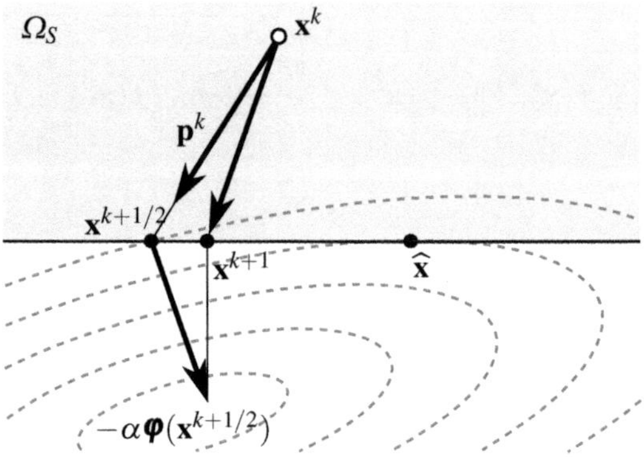

Fig. 11.3. Feasible half-step

Algorithm 11.2. Modified proportioning with a feasible half-step (MPGP)

Given an SPD matrix $\mathsf{A} \in \mathbb{R}^{n\times n}$, $\mathbf{b} \in \mathbb{R}^n$, and Ω_S.

Step 0. {*Initialization of parameters.*}
Choose $\mathbf{x}_0,\ \mathbf{x}_1 \in \Omega_S,\ \overline{\alpha} \in (0, 2\|\mathsf{A}\|^{-1})$
set $k = 0,\ \mathbf{g} = \mathsf{A}\mathbf{x}^0 - \mathbf{b},\ \mathbf{p} = \varphi(\mathbf{x}^0)$
while $\|\mathbf{g}^P(\mathbf{x}^k)\|$ *is not small*
if $\|\boldsymbol{\beta}(\mathbf{x}^k)\|^2 \le \Gamma^2 \widetilde{\varphi}(\mathbf{x}^k)^T \varphi(\mathbf{x}^k)$
Step 1. {*Proportional* $\mathbf{x}^k$. *Trial conjugate gradient step.*}
$\alpha_{cg} = \mathbf{g}^T\mathbf{p}/\mathbf{p}^T\mathsf{A}\mathbf{p}$
$\alpha_f = \max\{\alpha :\ \mathbf{x}^k - \alpha\mathbf{p} \in \Omega_S\}$
if $\alpha_{cg} \le \alpha_f$
Step 2. {*Conjugate gradient step.*}
$\mathbf{x}^{k+1} = \mathbf{x}^k - \alpha_{cg}\mathbf{p},\ \mathbf{g} = \mathbf{g} - \alpha_{cg}\mathsf{A}\mathbf{p},$
$\beta = \varphi(\mathbf{y})^T\mathsf{A}\mathbf{p}/\mathbf{p}^T\mathsf{A}\mathbf{p},\ \mathbf{p} = \varphi(\mathbf{y}) - \beta\mathbf{p}$
else
Step 3. {*Gradient projection step with half-step.*}
$\mathbf{x}^{k+\frac{1}{2}} = \mathbf{x}^k - \alpha_f\mathbf{p},\ \mathbf{g} = \mathbf{g} - \alpha_f\mathsf{A}\mathbf{p}$
$\mathbf{x}^{k+1} = P_{\Omega_B}(\mathbf{x}^{k+\frac{1}{2}} - \overline{\alpha}\mathbf{g}(\mathbf{x}^{k+\frac{1}{2}}))$
$\mathbf{g} = \mathsf{A}\mathbf{x}^{k+1} - \mathbf{b},\ \mathbf{p} = \varphi(\mathbf{x}^{k+1})$
end if
else
Step 4. {*Gradient proportioning step.*}
$\mathbf{x}^{k+1} = P_{\Omega_S}(\mathbf{x}^k - \alpha\mathbf{g}), \mathbf{g} = \mathsf{A}\mathbf{x}^{k+1} - \mathbf{b},\ \mathbf{p} = \varphi(\mathbf{x}^{k+1})$
end if
$k = k + 1$
end while
Step 5. {*Return (possibly inexact) solution.*}
$\widetilde{\mathbf{x}} = \mathbf{x}^k$

11.6.2 MPGP Algorithm in More Detail

Now, we are ready to give the details of the implementation of the MPGP algorithm, which was briefly described in a form suitable for analysis as Algorithm 11.1. To preserve readability, we do not distinguish the generations of variables by indices unless it is convenient for further reference.

Our experience indicates that the performance of MPGP is not sensitive to Γ as long as $\Gamma \approx 1$. Recall that $\Gamma = 1$ corresponds to $\delta = 1/4$. We can use the suggestions described in Sect. 10.5.3 to choose $\overline{\alpha}$.

11.7 Spectral Projected Gradients with Fallback

We have described above how to combine the conjugate gradient method of Chap. 8 with the gradient projection steps to develop the MPGP algorithm for solving (11.1) with the R-linear rate of convergence in terms of the bounds on the spectrum of the Hessian A of the cost function f. Here, we introduce an alternative approach using the Barzilai-Borwein (spectral gradient) method introduced in Sect. 3.3 and the gradient projections. Combining the fall-back strategy and the fixed step-length gradient projection enforces the convergence. Recall that using the fixed step-length gradient makes the step cost proportional to one or two matrix-vector multiplications.

11.7.1 Algorithm PBBF

The Barzilai-Borwein (BB, spectral) gradient method can be adapted for solving problems with separable constraints. The first choice is replacing the gradient step with the BB steplength by its gradient projection. However, there is a catch. Though the spectral gradient method has been proved to converge for unconstrained problems, Dai and Fletcher [44] found that it can cycle when applied to bound constrained problems. Hence, a globalization strategy is necessary to guarantee the convergence of the method. Moreover, replacing the BB-step with something like the sufficient decrease step results in slow convergence.

The algorithm PBBF (projected Barzilai-Borwain with fallback) presented here combines the spectral projected gradients with a fallback comprising the gradient projection step applied to the best of the last few approximations. We use a popular globalization strategy associated with the non-monotone line search schemes developed by Grippo et al. [152] for Newton's method.

The algorithm uses the same $\overline{\alpha} \in (0, 2\|\mathsf{A}\|^{-1})$ and $\Gamma > 0$ as the MPGP algorithm described above. Starting from $\mathbf{x}_0, \mathbf{x}_1 \in \Omega_S$, the algorithm generates the projected BB iterations

$$\mathbf{x} = P_{\Omega_S}\left(\mathbf{x}^k - \alpha_{BB}\nabla f(\mathbf{x}^k)\right)$$

until there is either the improvement of the cost function or there are K consecutive iterations without improvement. In the first case, an additional fixed step guarantees a sufficient decrease in the cost function. Otherwise, the next iteration is defined by the fixed step-length gradient projection step applied to the best of the last K iterations.

The algorithm is formulated to simplify the proof of its convergence. In particular, it is unnecessary to generate the copies of $\mathbf{x}^k_{\min}$.

The procedure reads as follows.

Algorithm 11.3. PBBF—projected Barzilai-Borwein method with Fallback

Given an SPD matrix $\mathsf{A} \in \mathbb{R}^{n\times n}$, $\mathbf{b} \in \mathbb{R}^n$, and Ω_S.

Step 0. {*Initialization of parameters.*}
Choose $\mathbf{x}^0, \mathbf{x}^1 \in \Omega_S$ $\overline{\alpha} \in (0, 2\|\mathsf{A}\|^{-1})$, $K \in \mathbb{N}$
set $k = 1$, $\hat{k} = 1$, $\mathbf{x}^1_{\min} = \mathbf{x}^1$
while $\|\mathbf{g}^P(\mathbf{x}^k)\|$ *is not small*

Step 1. {*Trial projected spectral gradient step* }
$\mathbf{s} = \mathbf{x}^k - \mathbf{x}^{k-1}$, $\alpha_{BB} = \mathbf{s}^T\mathbf{s}/\mathbf{s}^T\mathsf{A}\mathbf{s}$
$\mathbf{x} = P_\Omega\left(\mathbf{x}^k - \alpha_{BB}\nabla f(\mathbf{x}^k)\right)$
if $\hat{k} < K$ and $f(\mathbf{x}) \geq f(\mathbf{x}^k)$

Step 2. {*Accept and copy minimizer*}
$\mathbf{x}^{k+1} = \mathbf{x}$, $\mathbf{x}^{k+1}_{\min} = \mathbf{x}^k_{\min}$, $\hat{k} = \hat{k} + 1$
end if
if $\hat{k} < K$ *and* $f(\mathbf{x}) < f(\mathbf{x}^k_{\min})$

Step 3. {*Accept and update minimizer*}
$\mathbf{x}^{k+1} = \mathbf{x}$, $\mathbf{x}^{k+1}_{\min} = P_\Omega\left(\mathbf{x} - \overline{\alpha}\nabla f(\mathbf{x})\right)$, $\hat{k} = \hat{k} + 1$
end if
if $\hat{k} = K$

Step 4. {*Apply fall-back*}
$\mathbf{x}^{k+1} = P_\Omega(\mathbf{x}^k_{\min} - \overline{\alpha}\nabla f(\mathbf{x}^k_{\min})$
$\mathbf{x}^{k+1}_{\min} = \mathbf{x}^{k+1}$, $\hat{k} = 0$
end if
$k = k + 1$, $\hat{k} = \hat{k} + 1$
end while

Step 5. {*Return (possibly inexact) solution.*}
$\widetilde{\mathbf{x}} = \mathbf{x}^k$

11.7.2 Rate of Convergence

The following theorem guarantees the R-linear convergence.

Theorem 11.3. *Let* $\widehat{\mathbf{x}}$ *denote the unique solution of* (11.1)*, let* $\lambda_{\min}$ *denote the smallest eigenvalue of* A*, and let* $\{\mathbf{x}^k\}$ *be generated by PBBF algorithm 11.3 with* $\mathbf{x}^0, \mathbf{x}^1 \in \Omega_S$*,* $\overline{\alpha} \in (0, 2\|A\|^{-1})$*, and* $K > 0$.

If $\overline{\alpha} \in (0, \|\mathsf{A}\|^{-1}])$ *or* Ω_S *is a product of infinite half-closed intervals, ellipses, spheres, or other subsymmetric sets and* $\overline{\alpha} \in (0, 2\|\mathsf{A}\|^{-1}])$*, then*

$$f\left(\mathbf{x}_{\min}^{k+K}\right) - f(\widehat{\mathbf{x}}_{\min}) \le \eta_f(\overline{\alpha})\big(f(\mathbf{x}_{\min}^k) - f(\widehat{\mathbf{x}})\big), \tag{11.57}$$

where $\eta(\overline{\alpha})$ *is defined by*

$$\eta_f(\overline{\alpha}) = 1 - \widehat{\alpha}\lambda_{\min} \quad \text{and} \quad \widehat{\alpha} = \min\{\overline{\alpha}, 2\|\mathsf{A}^{-1}\| - \overline{\alpha}\}.$$

Proof. The PBBF algorithm generates the projected Barzilai-Borwein iterates $\mathbf{x}^k$ in Step 2, the fixed step-length gradient projection steps applied to the BB-iterates in Step 3, or the *fallback iterates* obtained by the gradient projection steps from the iterates minimizing the cost function in the last $K-1$ steps. Let us call the iterates generated in Step 3 or Step 4 the *minimizing iterates.* To simplify the analysis, our PBBF algorithm also generates the iterates $\mathbf{x}_{\min}^k$ corresponding to the minimizing iterates and their copies so that $\{f\left(\mathbf{x}_{\min}^k\right)\}$ is non-increasing and

$$\mathbf{x}_{\min}^k = \mathbf{x}_{\min}^{k+1} = \cdots \mathbf{x}_{\min}^{k+\hat{k}}$$

implies that $\hat{k} < K$.

The algorithm generates a trial iterations $\mathbf{x}$ to get the minimizing iterations. If $f(\mathbf{x}) \le f(\mathbf{x}_{\min}^k)$, then the next iterate is assigned in Step 3 by

$$\mathbf{x}_{\min}^{k+1} = \mathbf{x}^{k+1} = P_{\Omega_S}\left(\mathbf{x} - \overline{\alpha}\nabla f(\mathbf{x})\right). \tag{11.58}$$

If $f(\mathbf{x}) > f(\mathbf{x}_{\min}^k)$ and $\mathbf{x}_{\min}^k = \mathbf{x}_{\min}^{k-1} = \cdots \mathbf{x}_{\min}^{k-K+2}$, then

$$\mathbf{x}_{\min}^{k+1} = \mathbf{x}^{k+1} = P_{\Omega_S}\left(\mathbf{x}_{\min}^k - \overline{\alpha}\nabla f(\mathbf{x}_{\min}^k)\right). \tag{11.59}$$

It follows that either

$$f\left(\mathbf{x}_{\min}^{k+1}\right) - f(\widehat{\mathbf{x}}_{\min}) \le \eta_f(\alpha)\big(f(\mathbf{x}) - f(\widehat{\mathbf{x}})\big) \le \eta_f(\alpha)\big(f(\mathbf{x}_{\min}^k) - f(\widehat{\mathbf{x}})\big)$$

or

$$f\left(\mathbf{x}_{\min}^{k+1}\right) - f(\widehat{\mathbf{x}}_{\min}) \le \eta_f(\overline{\alpha})\big(f(\mathbf{x}_{\min}^k) - f(\widehat{\mathbf{x}})\big).$$

To finish the proof, recall that a minimizing iterate is present in any K-tuple of consecutive iterations. □

Remark 11.2. It follows by Theorem 11.3 that

$$f(\mathbf{x}_{\min}^{1+kK}) - f(\widehat{\mathbf{x}}) \leq \eta_f(\overline{\alpha})^k \big(f(\mathbf{x}_{\min}^1) - f(\widehat{\mathbf{x}})\big).$$

If we have a class of problems (11.1) with the spectrum of the Hessian of f in a given positive interval (a, A) and $\varepsilon > 0$, we can set $\overline{\alpha} = A^{-1}$ to get that there is $k_0 > 0$ such that for $k \geq k_0$ and iterations for any problem

$$\eta_f(\overline{\alpha})^k \leq (1 - \widehat{\alpha}a)^k \leq \varepsilon$$

and

$$\|\mathbf{x}^{kK} - \widehat{\mathbf{x}}\|_{\mathsf{A}}^2 \leq \eta_f(\overline{\alpha})^k \big(f(\mathbf{x}_{\min}^1) - f(\widehat{\mathbf{x}})\big) \leq \varepsilon \left(f(\mathbf{x}^1) - f(\widehat{\mathbf{x}})\right).$$

11.8 Numerical Experiments

Let us illustrate the performance of the MPGP algorithm by solving a 3D elasticity problem to find the stress and deformation of two elastic cantilever beams in mutual contact with orthotropic friction. Here, we use a variant of MPGP called the MwPGP algorithm, which differs from MPGP by using a *week proportioning*. Recall that the iterate is *weakly proportional* with the parameter $\delta \in (0, 1)$ if

$$\delta \|\widetilde{\boldsymbol{g}}_{\alpha}^{P}(\mathbf{x})\|^2 \leq \|\boldsymbol{\varphi}(\mathbf{x})\|^2.$$

The performance of MwPGP is compared with the variant of the spectral gradient method. The problem depicted in Fig. 11.4 is described in detail and resolved in Pospíšil and Dostál [232]. Here, we only give the essential characteristics of the problem to illustrate the performance of the MwPGP and PBBF algorithms. The solution of large contact problems with friction and the demonstration of the scalability of MPGP is in Chap. 14.

We consider three cases—a problem without friction (bound constraints), with isotropic Tresca friction (bound and circular constraints), and with orthotropic Tresca friction (bound and elliptic constraints). The bound constraints arise in the dual formulation of the non-penetration conditions; the other constraints are related to the friction in the dual formulation of equilibrium conditions. The feasible region corresponding to the orthotropic friction comprises ellipses with the principal axes in two orthogonal directions Ψ_1 and Ψ_2. The ratio of the length of the major axis to that of the minor axis equal to 10. The geometry, prescribed boundary conditions, and material properties are in Fig. 11.4.

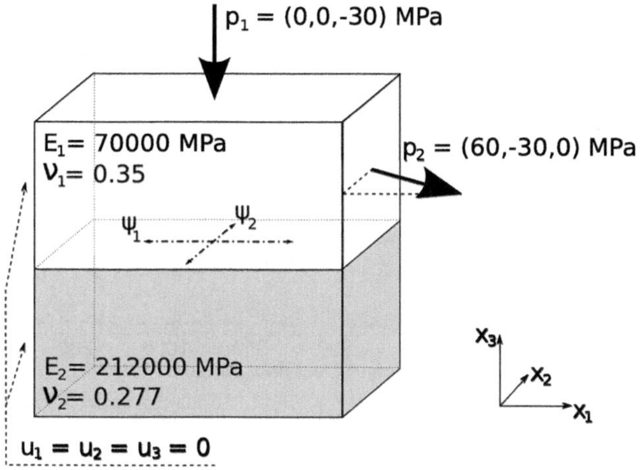

Fig. 11.4. Model specification

The problem was discretized by the finite element method using the tetrahedral elements in regular grids and reduced by duality to the interface of the blocks. Using various discretization parameters, we got the problems with 20% bound constraints, 60% with quadratic constraints, and dual dimensions ranging from 135 to 1305. The projection onto the ellipses was implemented by the algorithm described in [162].

We resolved the benchmark with three algorithms:

- PBBF with $K \in \{0, 5, 100\}$,
- the SPG (spectral projected gradient) method [20]) with the same choice of K and the GLL condition introduced by Grippo et al. [152],
- our recently developed MwPGP (modified weak proportioning with gradient projections) algorithm [27]) (inner loop of SMALSE-Mw, see Algorithm 12.2).

We considered six discretizations that induced dual problems of the dimension $n \in \{135, 273, 459, 693, 975, 1305\}$. The spectrum of the Hessian of the dual energy function was uniformly bounded. We used the relative stopping tolerance $\varepsilon = 10^{-4}$ of the norm of the reduced projected gradient, i.e.

$$\|\mathbf{g}^P(\mathbf{x}^k)\| \leq \varepsilon \|\mathbf{g}^P(\mathbf{x}^0).\|$$

To compare the algorithms, we present in Fig. 11.5, Fig. 11.6, and Fig. 11.7 the graphs show the matrix-vector multiplications for each benchmark and n. The results indicate a good agreement with the theory and scalability of both PBBF and MwPGP.

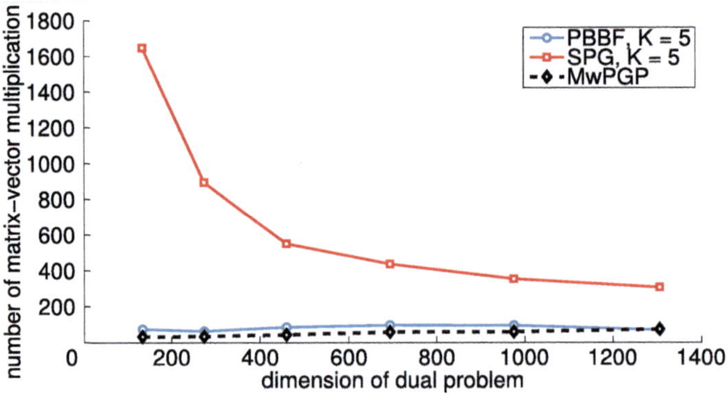

Fig. 11.5. Number of Hessian multiplications—bound constraints, $n_h = 14$

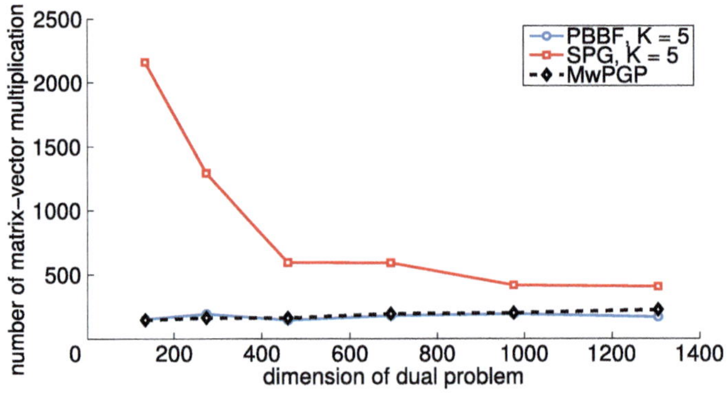

Fig. 11.6. Number of Hessian multiplications, bound and circular constraints, $n_h = 14$

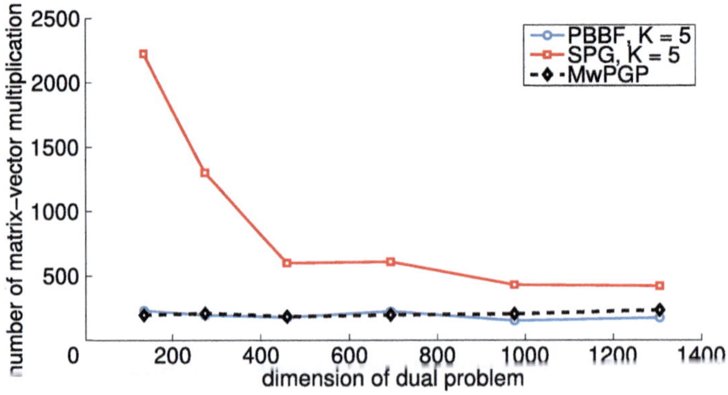

Fig. 11.7. Number of Hessian multiplications, bound and elliptic constraints, $n_h = 14$

11.9 Comments

The MPGP algorithm presented in this section is a variant of the algorithms that combine conjugate gradients with Euclidean projections. Such algorithms were first developed to solve the bound-constrained QP problems. See Chap. 10 for more information.

It seems that Kučera [197] was the first who adapted the MPRGP algorithm for solving separable QCQP problems and proved the R-linear convergence of his KPRGP algorithm for the step length $\alpha \in (0, \|\mathsf{A}\|^{-1})$. Later, he formulated the lemma on three consequent iterations and used it to prove the R-linear convergence of projected gradients for the step length $\alpha \in (0, \|\mathsf{A}\|^{-1}]$ [82]. Recall that the proof of convergence of the projected gradient is essential for the precision control of the inner loop of the algorithms for separable inequality and linear equality constraints presented in the next section.

The MPGP algorithm presented here appeared in Dostál and Kozubek [81]. The estimate of the rate of convergence presented in Theorem 11.1 is better than that in [82] and guarantees the R-linear convergence rate of the projected gradient for $\alpha \in (0, 2\|\mathsf{A}\|^{-1})$. The proof uses estimates from Bouchala et al. Vodstrčil [26, 27]. We observed that the performance of MPGP can sometimes be improved using some heuristic modifications, mainly using longer step lengths or unfeasible steps, such as the conjugate gradient or Barzialai–Borwein step length.

The presentation of the Barzilai–Borwain method with fallback is based on Pospíšil and Dostál [232]. See also Sect. 5.9. The performance of MPGP can also be improved by *preconditioning* described in Sect. 10.6.

12
Solvers for Separable and Equality QP/QCQP Problems

We finally combine the results of our previous investigations to develop efficient algorithms for minimizing strictly convex quadratic functions subject to *convex separable inequality* and *linear equality* constraints

$$\min_{\mathbf{x}\in\Omega_{SE}} f(\mathbf{x}), \qquad f(\mathbf{x}) = \frac{1}{2}\mathbf{x}^T\mathsf{A}\mathbf{x} - \mathbf{x}^T\mathbf{b}, \tag{12.1}$$

where A is a large $n \times n$ SPD matrix, $\mathbf{b} \in \mathbb{R}^n$, and

- $\Omega_{SE} = \{\mathbf{x} \in \mathbb{R}^n : \mathsf{B}\mathbf{x} = \mathbf{o} \text{ and } \mathbf{x} \in \Omega_S\}$, $\mathsf{B} \in \mathbb{R}^{m\times n}$,
- $\Omega_S = \{\mathbf{x} \in \mathbb{R}^n : h_i(\mathbf{x}_i) \le 0,\ i = 1, \dots, s\}$, h_i are convex functions,
- $\mathbf{x}_i = \mathbf{x}_{\mathcal{S}_i},\ \mathcal{S}_i \subseteq \mathcal{N},\ \mathcal{S}_i \cap \mathcal{S}_j = \emptyset$ for $i \ne j$, $\mathbf{x}_i \in \mathbb{R}^{n_i}$, and $n_1 + \cdots + n_s = n$.

To include the possibility that not all components of $\mathbf{x}$ are constrained, we admit $h_i(\mathbf{x}_i) = -1$. We are especially interested in problems including bound, spherical, and elliptic inequality constraints. We assume $\Omega_{SE} \ne \emptyset$ and admit dependent rows of B. We assume that $\mathsf{B} \ne \mathsf{O}$ and $\operatorname{Ker}\mathsf{B} \ne \{\mathbf{o}\}$. Moreover, we assume that the constraints satisfy the Abadie constraint qualification introduced in Sect. 2.8.3 so that the tangent cone characterizes the solution. However, we shall give some convergence results without this assumption. Observe that we can reduce some more general QP or QCQP problems to (12.1) by duality, a suitable shift of variables, or a modification of f. If A is regular only on $\operatorname{Ker}\mathsf{B}$, we use the regularization as in Sect. 9.4.3.

The main idea of the algorithms we develop here is to treat both sets of constraints separately. We restrict our attention to the SMALSE-M algorithm (Semi-Monotonic Augmented Lagrangian Algorithm for Separable and Equality constraints) and its variants: SMALSE-Mw for separable constraints with a strong curvature of the boundary of the feasible set and SMALBE for bound and equality constraints. We prove the convergence results that support the solving of large problems.

Z. Dostál, *Optimal Quadratic Programming and QCQP Algorithms with Case Studies*, Springer Optimization and Its Applications 23,
https://doi.org/10.1007/978-3-031-95167-1_12

12.1 KKT Conditions

Since Ω_{SE} is closed and convex and f is assumed to be strictly convex, the solution of (12.1) exists and is necessarily unique by Proposition 2.23. The solvability of (12.1) can be formulated by means of the augmented Lagrangian

$$L(\mathbf{x}, \boldsymbol{\lambda}_E, \boldsymbol{\mu}, \rho) = \frac{1}{2}\mathbf{x}^T\mathsf{A}\mathbf{x} - \mathbf{x}^T\mathbf{b} + \mathbf{x}^T\mathsf{B}^T\boldsymbol{\lambda}_E + \frac{\rho}{2}\|\mathsf{B}\mathbf{x}\|^2 + \sum_{i=1}^{s}\mu_i h_i(\mathbf{x}_i),$$

the gradient of which reads

$$\nabla_{\mathbf{x}} L(\mathbf{x}, \boldsymbol{\lambda}_E, \boldsymbol{\mu}, \rho) = (\mathsf{A} + \rho\mathsf{B}^T\mathsf{B})\mathbf{x} - \mathbf{b} + \mathsf{B}^T\boldsymbol{\lambda}_E + \sum_{i=1}^{s}\mu_i\nabla h_i(\mathbf{x}_i). \tag{12.2}$$

Using (2.134) and the Abadie constraint qualification, we get that a feasible vector $\mathbf{x} \in \Omega_{SE}$ is a solution of (12.1) if and only if there are $\boldsymbol{\lambda}_E \in \mathbb{R}^m$ and $\mu_1, \dots, \mu_s$ such that

$$\nabla_{\mathbf{x}} L(\mathbf{x}, \boldsymbol{\lambda}_E, \boldsymbol{\mu}, \rho) = \mathbf{o}, \quad \mu_i \geq 0, \quad \text{and} \quad \mu_i h_i(\mathbf{x}_i) = 0, \quad i = 1, \dots, s. \tag{12.3}$$

Having effective algorithms for solving QCQP problems with separable constraints, it is convenient to use explicitly the Lagrange multipliers only for the equality constraints, i.e., to use

$$L(\mathbf{x}, \boldsymbol{\lambda}_E, \rho) = \frac{1}{2}\mathbf{x}^T\mathsf{A}\mathbf{x} - \mathbf{x}^T\mathbf{b} + \mathbf{x}^T\mathsf{B}^T\boldsymbol{\lambda}_E + \frac{\rho}{2}\|\mathsf{B}\mathbf{x}\|^2. \tag{12.4}$$

Denoting by $\mathbf{g} = \mathbf{g}(\mathbf{x}, \boldsymbol{\lambda}_E, \rho)$ the gradient of the reduced augmented Lagrangian, we get

$$\mathbf{g} = \mathbf{g}(\mathbf{x}, \boldsymbol{\lambda}_E, \rho) = (\mathsf{A} + \rho\mathsf{B}^T\mathsf{B})\mathbf{x} - \mathbf{b} + \mathsf{B}^T\boldsymbol{\lambda}_E, \tag{12.5}$$

and $\mathbf{x} \in \Omega_{SE}$ solves (12.1) if and only if there is $\boldsymbol{\lambda}_E \in \mathbb{R}^m$ such that

$$\mathbf{g}^P(\mathbf{x}, \boldsymbol{\lambda}_E, \rho) = \mathbf{o}, \tag{12.6}$$

where $\mathbf{g}^P$ is the projected gradient of Sect. 11.1 for the auxiliary problem

$$\min_{\mathbf{x}\in\Omega_S} L(\mathbf{x}, \boldsymbol{\lambda}_E, \rho), \quad \Omega_S = \{\mathbf{x} \in \mathbb{R}^n : \mathbf{h}(\mathbf{x}) \leq \mathbf{o}\}. \tag{12.7}$$

However, the condition (12.6) is sensitive to the curvature of the boundary, so we shall discuss some more robust alternatives in the next section.

12.2 Penalty and Exact Method of Multipliers

Probably the most simple way how to exploit the algorithms of the previous chapters to the solution of (12.1) is to enhance the equality constraints into the objective function f by adding the penalty term introduced in Sect. 6.2. Thus the solution of (12.1) can be approximated by the solution of

$$\min_{\mathbf{x}\in\Omega_S} f_\rho(\mathbf{x}), \qquad f_\rho(\mathbf{x}) = f(\mathbf{x}) + \frac{\rho}{2}\|\mathsf{B}\mathbf{x}\|^2. \tag{12.8}$$

We illustrate the effect of the penalty term in Fig. 12.1. Notice that $\widehat{\mathbf{x}}_\rho$ is in Ω_S and typically only near Ω_{SE}.

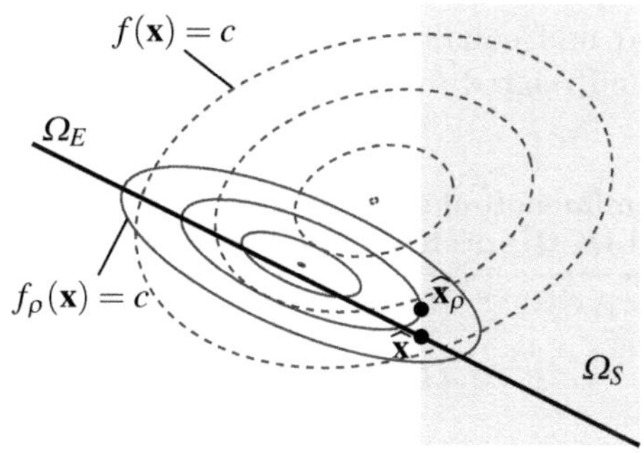

Fig. 12.1. The effect of the quadratic penalty

Because of its simplicity and intuitive appeal, the penalty method is often used in computations. However, a good approximation of the solution may require a very large penalty parameter, which can complicate computer implementation. The remedy can use the observation from Sect. 6.2.2 that the solution $\widehat{\mathbf{x}}$ to (12.1) solves also

$$\min_{\mathbf{x}\in\Omega_S} L(\mathbf{x}, \boldsymbol{\lambda}_E, \rho) \tag{12.9}$$

with a suitable $\boldsymbol{\lambda}_E \in \mathbb{R}^m$. The point is that having the exact solution $\widehat{\mathbf{x}}_\rho$ of (12.9) with ρ and any $\boldsymbol{\lambda}_E \in \mathbb{R}^m$, we can modify the linear term of L so that the minimum of the modified cost function *without the penalization term* with respect to $\mathbf{x} \in \Omega_S$ is achieved again at $\widehat{\mathbf{x}}_\rho$. The formula follows from

$$\min_{\mathbf{x}\in\Omega_S} L(\mathbf{x}, \boldsymbol{\lambda}_E, \rho) = \min_{\mathbf{x}\in\Omega_S} L(\mathbf{x}, \boldsymbol{\lambda}_E + \rho\mathsf{B}\widehat{\mathbf{x}}_\rho, 0). \tag{12.10}$$

Then we can find a better approximation by looking for the minimizer of $L(\mathbf{x}, \boldsymbol{\lambda} + \rho\mathsf{B}\widehat{\mathbf{x}}_\rho, \rho)$. We shall carry out the minimization only approximately as in previous chapters.

12.3 SMALSE-M/SMALBE-M Algorithm

The SMALSE-M algorithm (semi-monotonic augmented Lagrangians for separable and equality constraints with the precision of auxiliary problems controlled by parameter M) is a modification of the exact augmented Lagrangian that accepts inexact solutions of auxiliary problems with separable inequality constraints. The SMALSE-M algorithm uses a fixed regularization parameter ϱ and the parameter M_k for the adaptive precision control of auxiliary problems. Since we use only the multipliers for the equality constraints, we denote $\boldsymbol{\lambda} = \boldsymbol{\lambda}_E$.

If we apply SMALSE-M to problems with bound constraints, we can prove stronger results concerning convergence. We call SMALSE-M applied to the bound and equality constraints SMALBE-M (semi-monotonic augmented Lagrangians for bound and equality constraints) to simplify the reference to these applications and related results.

Algorithm 12.1. Semimonotonic augmented Lagrangians for separable and equality constrained QCQP problems (SMALSE-M/SMALBE-M)

Given an SPD matrix $\mathsf{A} \in \mathbb{R}^{n\times n}$, $\mathsf{B} \in \mathbb{R}^{m\times n}$, *n-vector* $\mathbf{b}$, *constraints* $\mathbf{h}$.

Step 0. {*Initialization.*}

Choose $\eta > 0,\ 0 < \beta < 1, M_{-1} = 0, M_0 > 0,\ \varrho > 0,\ \boldsymbol{\lambda}^0 \in \mathbb{R}^m$

for $k = 0, 1, 2, \dots$

Step 1. {*Inner iteration with adaptive precision control.*}

Find $\mathbf{x}^k \in \Omega_S$ *such that*

$$\|\mathbf{g}^P(\mathbf{x}^k, \boldsymbol{\lambda}^k, \varrho)\| \le \min\{M_k\|\mathsf{B}\mathbf{x}^k\|, \eta\} \tag{12.11}$$

Step 2. {*Updating the Lagrange multipliers.*}

$$\boldsymbol{\lambda}^{k+1} = \boldsymbol{\lambda}^k + \varrho\mathsf{B}\mathbf{x}^k \tag{12.12}$$

Step 3. {*Update M if the increase of Lagrangian is insufficient.*}

if $M_k = M_{k-1}$ *and*

$$L(\mathbf{x}^k, \boldsymbol{\lambda}^k, \varrho) < L(\mathbf{x}^{k-1}, \boldsymbol{\lambda}^{k-1}, \varrho) + \frac{\varrho}{2}\|\mathsf{B}\mathbf{x}^k\|^2 \tag{12.13}$$

$M_{k+1} = \beta M_k$

else

$M_{k+1} = M_k$

end if

end for

Remark 12.1. If we apply SMALSE-M to problems with strong curvature, we can have problems to recognize solution (see Fig. 11.1). The reason is the sensitivity of the projected gradient on the curvature of the boundary. The effective resolution of these problems is discussed in Sect. 12.5.

In Step 1, we can use any algorithm for minimizing strictly convex quadratic functions subject to separable constraints, provided it guarantees the convergence of projected gradients to zero. Our optimality theory requires that the algorithm in the inner loop has a rate of convergence in terms of bounds on the spectrum of A, such as the MPGP Algorithm 11.2 or MPRGP Algorithm 10.2. A *stopping criterion* should either follow Step 1 or be enhanced in Step 1. A natural choice is the *relative precision*

$$\|\mathbf{g}^P(\mathbf{x}^k)\| \le \varepsilon_g \|\mathbf{b}\| \quad \text{and} \quad \|\mathsf{B}\mathbf{x}^k\| \le \varepsilon_e \|\mathbf{b}\| \tag{12.14}$$

prescribed by small parameters $\varepsilon_g > 0$ and $\varepsilon_e > 0$.

The following lemma shows that Algorithm 12.1 is well defined, i.e., any algorithm for the solution of auxiliary problems in Step 1 that guarantees the convergence of projected gradients to zero generates either $\mathbf{x}^k$ which satisfies (12.11) or the iterates converging to the solution of (12.1).

Lemma 12.1. *Let $M > 0$, $\boldsymbol{\lambda} \in \mathbb{R}^m$, $\eta > 0$, and $\rho \ge 0$ be given. Assume that $\{\mathbf{y}^k\} \in \Omega_S$ denote any sequence such that*

$$\widehat{\mathbf{y}} = \lim_{k \to \infty} \mathbf{y}^k = \arg\min_{\mathbf{y} \in \Omega_S} L(\mathbf{y}, \boldsymbol{\lambda}, \rho), \tag{12.15}$$

and let $\mathbf{g}^P(\mathbf{y}^k, \boldsymbol{\lambda}, \rho)$ converges to zero vector. Then $\{\mathbf{y}^k\}$ either converges to the unique solution $\widehat{\mathbf{x}}$ of problem (12.1), or there is an index k such that

$$\|\mathbf{g}^P(\mathbf{y}^k, \boldsymbol{\lambda}, \rho)\| \le \min\{M\|\mathsf{B}\mathbf{y}^k\|, \eta\}. \tag{12.16}$$

Proof. If (12.16) does not hold for any k, then

$$\|\mathbf{g}^P(\mathbf{y}^k, \boldsymbol{\lambda}, \rho)\| > M\|\mathsf{B}\mathbf{y}^k\| \tag{12.17}$$

for any k. Since $\mathbf{g}^P(\mathbf{y}^k, \boldsymbol{\lambda}, \rho)$ converges to zero vector by the assumption, it follows that $\|\mathsf{B}\mathbf{y}^k\|$ converges to zero. Thus $\mathsf{B}\widehat{\mathbf{y}} = \mathbf{o}$ and

$$\mathbf{g}^P(\widehat{\mathbf{y}}, \boldsymbol{\lambda}, \rho) = \mathbf{o}. \tag{12.18}$$

It follows that $\widehat{\mathbf{y}}$ satisfies the KKT conditions (12.3) and $\widehat{\mathbf{y}} = \widehat{\mathbf{x}}$. □

Remark 12.2. We can keep $M_k = M_0$ and replace the constant regularization parameter ϱ by ϱ_k. In this case, we can update ϱ_k in Step 3 by

$$\rho_{k+1} = \rho_k / \beta. \tag{12.19}$$

We do not elaborate this option here.

12.4 Convergence and Optimality of SMALSE

The analysis of the convergence and optimality of SMALSE comprises several steps. We shall start with inequalities satisfied by the increments of augmented Lagrangian.

12.4.1 Inequalities Involving the Augmented Lagrangian

In this section, we establish fundamental inequalities that relate the bound on the norm of the projected gradient $\mathbf{g}^P$ of the augmented Lagrangian L to the values of L. These inequalities will be vital in proving convergence and other analyses concerning Algorithm 12.1.

Lemma 12.2. *Let* $\mathbf{x}, \mathbf{y} \in \Omega_S$, $\boldsymbol{\lambda} \in \mathbb{R}^m$, $\rho > 0$, $\eta > 0$, *and* $M > 0$. *Let* $\lambda_{\min}$ *denote the least eigenvalue of* A *and* $\widetilde{\boldsymbol{\lambda}} = \boldsymbol{\lambda} + \rho \mathsf{B}\mathbf{x}$.
(i) If

$$\|\mathbf{g}^P(\mathbf{x}, \boldsymbol{\lambda}, \rho)\| \leq M\|\mathsf{B}\mathbf{x}\|, \tag{12.20}$$

then

$$L(\mathbf{y}, \widetilde{\boldsymbol{\lambda}}, \rho) \geq L(\mathbf{x}, \boldsymbol{\lambda}, \rho) + \frac{1}{2}\left(\rho - \frac{M^2}{\lambda_{\min}}\right)\|\mathsf{B}\mathbf{x}\|^2 + \frac{\rho}{2}\|\mathsf{B}\mathbf{y}\|^2. \tag{12.21}$$

(ii) If

$$\|\mathbf{g}^P(\mathbf{x}, \boldsymbol{\lambda}, \rho)\| \leq \eta, \tag{12.22}$$

then

$$L(\mathbf{y}, \widetilde{\boldsymbol{\lambda}}, \rho) \geq L(\mathbf{x}, \boldsymbol{\lambda}, \rho) + \frac{\rho}{2}\|\mathsf{B}\mathbf{x}\|^2 + \frac{\rho}{2}\|\mathsf{B}\mathbf{y}\|^2 - \frac{\eta^2}{2\lambda_{\min}}. \tag{12.23}$$

(iii) If $\mathbf{z}_0 \in \Omega_{SE}$ *and* (12.22), *then*

$$L(\mathbf{x}, \boldsymbol{\lambda}, \rho) \leq f(\mathbf{z}_0) + \frac{\eta^2}{2\lambda_{\min}}. \tag{12.24}$$

Proof. Let us denote $\boldsymbol{\delta} = \mathbf{y} - \mathbf{x}$ and $\mathsf{A}_\rho = \mathsf{A} + \rho\mathsf{B}^T\mathsf{B}$ and recall that by the assumptions and Lemma 11.1

$$\mathbf{g}^T(\mathbf{y} - \mathbf{x}) \geq (\mathbf{g}^P)^T(\mathbf{y} - \mathbf{x}). \tag{12.25}$$

Using

$$L(\mathbf{x}, \widetilde{\boldsymbol{\lambda}}, \rho) = L(\mathbf{x}, \boldsymbol{\lambda}, \rho) + \rho\|\mathsf{B}\mathbf{x}\|^2 \quad \text{and} \quad \mathbf{g}(\mathbf{x}, \widetilde{\boldsymbol{\lambda}}, \rho) = \mathbf{g}(\mathbf{x}, \boldsymbol{\lambda}, \rho) + \rho\mathsf{B}^T\mathsf{B}\mathbf{x},$$

we get

$$\begin{aligned}
L(\mathbf{y}, \widetilde{\boldsymbol{\lambda}}, \rho) =& L(\mathbf{x}, \widetilde{\boldsymbol{\lambda}}, \rho) + \boldsymbol{\delta}^T \mathbf{g}(\mathbf{x}, \widetilde{\boldsymbol{\lambda}}, \rho) + \frac{1}{2}\boldsymbol{\delta}^T \mathsf{A}_\rho \boldsymbol{\delta} \\
=& L(\mathbf{x}, \boldsymbol{\lambda}, \rho) + \boldsymbol{\delta}^T \mathbf{g}(\mathbf{x}, \boldsymbol{\lambda}, \rho) + \frac{1}{2}\boldsymbol{\delta}^T \mathsf{A}_\rho \boldsymbol{\delta} + \rho \boldsymbol{\delta}^T \mathsf{B}^T \mathsf{B}\mathbf{x} + \rho \|\mathsf{B}\mathbf{x}\|^2 \\
\geq& L(\mathbf{x}, \boldsymbol{\lambda}, \rho) + \boldsymbol{\delta}^T \mathbf{g}^P(\mathbf{x}, \boldsymbol{\lambda}, \rho) + \frac{1}{2}\boldsymbol{\delta}^T \mathsf{A}_\rho \boldsymbol{\delta} + \rho \boldsymbol{\delta}^T \mathsf{B}^T \mathsf{B}\mathbf{x} + \rho \|\mathsf{B}\mathbf{x}\|^2 \\
\geq& L(\mathbf{x}, \boldsymbol{\lambda}, \rho) + \boldsymbol{\delta}^T \mathbf{g}^P(\mathbf{x}, \boldsymbol{\lambda}, \rho) + \frac{\lambda_{\min}}{2}\|\boldsymbol{\delta}\|^2 + \frac{\rho}{2}\|\mathsf{B}\boldsymbol{\delta}\|^2 + \rho \boldsymbol{\delta}^T \mathsf{B}^T \mathsf{B}\mathbf{x} \\
&+ \rho \|\mathsf{B}\mathbf{x}\|^2.
\end{aligned}$$

Noticing that

$$\frac{\rho}{2}\|\mathsf{B}\mathbf{y}\|^2 = \frac{\rho}{2}\|\mathsf{B}(\boldsymbol{\delta} + \mathbf{x})\|^2 = \rho \boldsymbol{\delta}^T \mathsf{B}^T \mathsf{B}\mathbf{x} + \frac{\rho}{2}\|\mathsf{B}\boldsymbol{\delta}\|^2 + \frac{\rho}{2}\|\mathsf{B}\mathbf{x}\|^2, \tag{12.26}$$

we get

$$\begin{aligned}
L(\mathbf{y}, \widetilde{\boldsymbol{\lambda}}, \rho) \geq& L(\mathbf{x}, \boldsymbol{\lambda}, \rho) + \boldsymbol{\delta}^T \mathbf{g}^P(\mathbf{x}, \boldsymbol{\lambda}, \rho) \\
&+ \frac{\lambda_{\min}}{2}\|\boldsymbol{\delta}\|^2 + \frac{\rho}{2}\|\mathsf{B}\mathbf{x}\|^2 + \frac{\rho}{2}\|\mathsf{B}\mathbf{y}\|^2.
\end{aligned} \tag{12.27}$$

Using (12.20) and simple manipulations then yields

$$\begin{aligned}
L(\mathbf{y}, \widetilde{\boldsymbol{\lambda}}, \rho) \geq& L(\mathbf{x}, \boldsymbol{\lambda}, \rho) - M\|\boldsymbol{\delta}\|\|\mathsf{B}\mathbf{x}\| + \frac{\lambda_{\min}}{2}\|\boldsymbol{\delta}\|^2 + \frac{\rho}{2}\|\mathsf{B}\mathbf{x}\|^2 + \frac{\rho}{2}\|\mathsf{B}\mathbf{y}\|^2 \\
=& L(\mathbf{x}, \boldsymbol{\lambda}, \rho) + \left(\frac{\lambda_{\min}}{2}\|\boldsymbol{\delta}\|^2 - M\|\boldsymbol{\delta}\|\|\mathsf{B}\mathbf{x}\| + \frac{M^2\|\mathsf{B}\mathbf{x}\|^2}{2\lambda_{\min}}\right) \\
&- \frac{M^2\|\mathsf{B}\mathbf{x}\|^2}{2\lambda_{\min}} + \frac{\rho}{2}\|\mathsf{B}\mathbf{x}\|^2 + \frac{\rho}{2}\|\mathsf{B}\mathbf{y}\|^2 \\
\geq& L(\mathbf{x}, \boldsymbol{\lambda}, \rho) + \frac{1}{2}\left(\rho - \frac{M^2}{\lambda_{\min}}\right)\|\mathsf{B}\mathbf{x}\|^2 + \frac{\rho}{2}\|\mathsf{B}\mathbf{y}\|^2.
\end{aligned}$$

This proves (i).
(ii) If we assume that (12.22) holds, then by (12.27)

$$\begin{aligned}
L(\mathbf{y}, \widetilde{\boldsymbol{\lambda}}, \rho) \geq& L(\mathbf{x}, \boldsymbol{\lambda}, \rho) - \|\boldsymbol{\delta}\|\eta + \frac{\lambda_{\min}}{2}\|\boldsymbol{\delta}\|^2 + \frac{\rho}{2}\|\mathsf{B}\mathbf{x}\|^2 + \frac{\rho}{2}\|\mathsf{B}\mathbf{y}\|^2 \\
\geq& L(\mathbf{x}, \boldsymbol{\lambda}, \rho) + \frac{\rho}{2}\|\mathsf{B}\mathbf{x}\|^2 + \frac{\rho}{2}\|\mathsf{B}\mathbf{y}\|^2 - \frac{\eta^2}{2\lambda_{\min}}.
\end{aligned} \tag{12.28}$$

This proves (ii).
(iii) Let $\widehat{\mathbf{z}}$ denote the solution of the auxiliary problem

$$\text{minimize} \quad L(\mathbf{z}, \boldsymbol{\lambda}, \rho) \quad \text{s.t.} \quad \mathbf{z} \geq \boldsymbol{\ell}, \tag{12.29}$$

let $\mathbf{z}_0 \in \Omega_{SE}$ so that $\mathsf{B}\mathbf{z}_0 = \mathbf{o}$, and let $\widehat{\boldsymbol{\delta}} = \widehat{\mathbf{z}} - \mathbf{x}$. If (12.22) holds, then

$$\begin{aligned} 0 \geq & L(\widehat{\mathbf{z}}, \boldsymbol{\lambda}, \rho) - L(\mathbf{x}, \boldsymbol{\lambda}, \rho) = \widehat{\boldsymbol{\delta}}^T \mathbf{g}(\mathbf{x}, \boldsymbol{\lambda}, \rho) + \frac{1}{2}\widehat{\boldsymbol{\delta}}^T \mathsf{A}_\rho \widehat{\boldsymbol{\delta}} \\ \geq & \widehat{\boldsymbol{\delta}}^T \mathbf{g}^P(\mathbf{x}, \boldsymbol{\lambda}, \rho) + \frac{1}{2}\widehat{\boldsymbol{\delta}}^T \mathsf{A}_\rho \widehat{\boldsymbol{\delta}} \geq -\|\widehat{\boldsymbol{\delta}}\|\eta + \frac{1}{2}\lambda_{\min}\|\widehat{\boldsymbol{\delta}}\|^2 \geq -\frac{\eta^2}{2\lambda_{\min}}. \end{aligned} \tag{12.30}$$

Since $L(\widehat{\mathbf{z}}, \boldsymbol{\lambda}, \rho) \leq L(\mathbf{z}_0, \boldsymbol{\lambda}, \rho) = f(\mathbf{z}_0)$, we conclude that

$$L(\mathbf{x}, \boldsymbol{\lambda}, \rho) \leq L(\mathbf{x}, \boldsymbol{\lambda}, \rho) - L(\widehat{\mathbf{z}}, \boldsymbol{\lambda}, \rho) + f(\mathbf{z}_0) \leq f(\mathbf{z}_0) + \frac{\eta^2}{2\lambda_{\min}}. \tag{12.31}$$

□

12.4.2 Monotonicity and Feasibility

Now, we shall translate the results on the relations satisfied by the augmented Lagrangian into the relations concerning the iterates generated by SMALSE-M.

Lemma 12.3. *Let $\{\mathbf{x}^k\}, \{\boldsymbol{\lambda}^k\}$, and M_k be generated by Algorithm 12.1 for solving (12.1) with $\eta > 0$, $0 < \beta < 1$, $M_0 > 0$, ρ, and $\boldsymbol{\lambda}^0 \in \mathbb{R}^m$. Let $\lambda_{\min}$ denote the least eigenvalue of the Hessian A of the quadratic function f.*
(i) If $k > 0$, $M_k = M_{k+1}$, and

$$M_k \leq \sqrt{\lambda_{\min}\rho}, \tag{12.32}$$

then

$$L(\mathbf{x}^{k+1}, \boldsymbol{\lambda}^{k+1}, \rho) \geq L(\mathbf{x}^k, \boldsymbol{\lambda}^k, \rho) + \frac{\rho}{2}\|\mathsf{B}\mathbf{x}^{k+1}\|^2. \tag{12.33}$$

(ii) For any $k \geq 0$

$$\begin{aligned} L(\mathbf{x}^{k+1}, \boldsymbol{\lambda}^{k+1}, \rho) \geq L(\mathbf{x}^k, \boldsymbol{\lambda}^k, \rho) &+ \frac{\rho}{2}\|\mathsf{B}\mathbf{x}^k\|^2 \\ &+ \frac{\rho}{2}\|\mathsf{B}\mathbf{x}^{k+1}\|^2 - \frac{\eta^2}{2\lambda_{\min}}. \end{aligned} \tag{12.34}$$

(iii) For any $k \geq 0$ and $\mathbf{z}_0 \in \Omega_{SE}$

$$L(\mathbf{x}^k, \boldsymbol{\lambda}^k, \rho) \leq f(\mathbf{z}_0) + \frac{\eta^2}{2\lambda_{\min}}. \tag{12.35}$$

Proof. In Lemma 12.2, let us substitute $\mathbf{x} = \mathbf{x}^k$, $\boldsymbol{\lambda} = \boldsymbol{\lambda}^k$, $M = M_k$, and $\mathbf{y} = \mathbf{x}^{k+1}$, so that inequalities (12.20) and (12.22) hold by (12.11) and $\widetilde{\boldsymbol{\lambda}} = \boldsymbol{\lambda}^{k+1}$. Thus we get Lemma 12.3 from Lemma 12.2.

If (12.32) holds, we can use (12.21) to get (12.33). Similarly, inequalities (12.34) and (12.35) can be obtained by the substitution into Lemma 12.2 (ii) and (iii), respectively. □

Theorem 12.1. *Let* $\{\mathbf{x}^k\}, \{\boldsymbol{\lambda}^k\}$, *and* $\{M_k\}$ *be generated by Algorithm 12.1 for the solution of (12.1) with* $\eta > 0$, $0 < \beta < 1$, $M_0 > 0$, $\rho > 0$, *and* $\boldsymbol{\lambda}^0 \in \mathbb{R}^m$. *Let* $\lambda_{\min}$ *denote the least eigenvalue of the Hessian* A *of the cost function* f, *and let* $p \geq 0$ *denote the smallest integer such that* $\rho \geq (\beta^p M_0)^2/\lambda_{\min}$. *Then the following statements hold.*
(i) There is k_0 *such that*

$$\min\{M_0, \beta\sqrt{\rho\lambda_{\min}}\} \leq M_{k_0} = M_{k_0+1} = M_{k_0+2} = \cdots \tag{12.36}$$

(ii) If $\mathbf{z}_0 \in \Omega_{SE}$, *then*

$$\frac{\rho}{2}\sum_{k=1}^{\infty} \|\mathsf{B}\mathbf{x}^k\|^2 \leq f(\mathbf{z}_0) - L(\mathbf{x}^0, \boldsymbol{\lambda}^0, \rho) + (1+p)\frac{\eta^2}{2\lambda_{\min}}. \tag{12.37}$$

(iii)

$$\lim_{k\to 0} \mathbf{g}^P(\mathbf{x}^k, \boldsymbol{\lambda}^k, \rho) = \mathbf{o} \quad \text{and} \quad \lim_{k\to 0} \mathsf{B}\mathbf{x}^k = \mathbf{o}. \tag{12.38}$$

Proof. Let $p \geq 0$ denote the smallest integer such that $\rho \geq (\beta^p M_0)^2/\lambda_{\min}$ and let $\mathscr{I} \subseteq \{1, 2, \dots\}$ denote a possibly empty set of the indices k_i such that $M_{k_i} < M_{k_i - 1}$. Using Lemma 12.3(i), $M_{k_i} = \beta M_{k_i-1} = \beta^i M_0$ for $k_i \in \mathscr{I}$, and $\rho \geq (\beta^p M_0)^2/\lambda_{\min}$, we conclude that there is no k such that $M_k < \beta^p M_0$. Thus $\mathscr{I}$ has at most p elements and (12.36) holds. By the definition of Step 3, if $k > 0$, then either $k \notin \mathscr{I}$ and

$$\frac{\rho}{2}\|\mathsf{B}\mathbf{x}^k\|^2 \leq L(\mathbf{x}^k, \boldsymbol{\lambda}^k, \rho) - L(\mathbf{x}^{k-1}, \boldsymbol{\lambda}^{k-1}, \rho), \tag{12.39}$$

or $k \in \mathscr{I}$ and by (12.34)

$$\begin{aligned} \frac{\rho}{2}\|\mathsf{B}\mathbf{x}^k\|^2 \leq& \frac{\rho}{2}\|\mathsf{B}\mathbf{x}^{k-1}\|^2 + \frac{\rho}{2}\|\mathsf{B}\mathbf{x}^k\|^2 \\ \leq& L(\mathbf{x}^k, \boldsymbol{\lambda}^k, \rho) - L(\mathbf{x}^{k-1}, \boldsymbol{\lambda}^{k-1}, \rho) + \frac{\eta^2}{2\lambda_{\min}}. \end{aligned} \tag{12.40}$$

Summing up the appropriate cases of the last two inequalities for $k = 1, \dots, j$ and taking into account that $\mathscr{I}$ has at most p elements, we get

$$\sum_{k=1}^{j} \frac{\rho}{2}\|\mathsf{B}\mathbf{x}^k\|^2 \leq L(\mathbf{x}^j, \boldsymbol{\lambda}^j, \rho) - L(\mathbf{x}^0, \boldsymbol{\lambda}^0, \rho) + p\frac{\eta^2}{2\lambda_{\min}}. \tag{12.41}$$

To get (12.37), replace $L(\mathbf{x}^j, \boldsymbol{\lambda}^j, \rho)$ by the upper bound (12.35). The relations (12.38) are easy corollaries of (12.37) and the definition of Step 1. □

12.4.3 Boundedness

The first step toward the proof of convergence of SMALSE-M is to show that $\mathbf{x}^k$ are bounded.

Proposition 12.1. *Let $\{\mathbf{x}^k\}$ and $\{\boldsymbol{\lambda}^k\}$ be generated by Algorithm 12.1 for the solution of (12.1) with $\eta > 0,\ 0 < \beta < 1,\ M_0 > 0,\ \rho > 0$, and $\boldsymbol{\lambda}^0 \in \mathbb{R}^m$. For each $i \in \{1, \dots, s\}$, let $\mathscr{I}_i$ denote the indices of the components of $\mathbf{x}$ associated with the argument of the constraint function h_i, so that $\mathbf{x}_i = [\mathbf{x}]_{\mathscr{I}_i}$, and let us define*

$$\widetilde{\mathscr{A}}(\mathbf{x}) = \cup_{i \in \mathscr{A}(\mathbf{x})} \mathscr{I}_i, \quad \widetilde{\mathscr{F}}(\mathbf{x}) = \mathscr{N} \setminus \widetilde{\mathscr{A}}(\mathbf{x}), \quad \mathbf{x}_{\mathscr{F}} = \mathbf{x}_{\widetilde{\mathscr{F}}(\mathbf{x})}, \quad \mathbf{x}_{\mathscr{A}} = \mathbf{x}_{\widetilde{\mathscr{A}}(\mathbf{x})}.$$

Let the boundary of Ω_S is bounded, i.e., there is $C > 0$ such that for any $\mathbf{x} \in \Omega_S$

$$\|\mathbf{x}_{\mathscr{A}(\mathbf{x})}\|^2 \le C. \tag{12.42}$$

Then $\{\mathbf{x}^k\}$ is bounded and there is a bounded sequence such that

$$L(\mathbf{x}^k, \boldsymbol{\lambda}^k, \varrho) = L(\mathbf{x}^k, \hat{\boldsymbol{\lambda}}^k, \varrho).$$

Proof. Since there is only a finite number of different subsets $\mathscr{F}$ of the set of all indices $\mathscr{N} = \{1, \dots, n\}$ and $\{\mathbf{x}^k\}$ is bounded if and only if $\{\mathbf{x}^k_{\mathscr{F}}\}$ is bounded for any $\mathscr{F} \subseteq \mathscr{N}$, we can restrict our attention to the analysis of the infinite subsequences $\{\mathbf{x}^k_{\mathscr{F}} : \widetilde{\mathscr{F}}(\mathbf{x}^k) = \mathscr{F}\}$ defined by the nonempty subsets $\mathscr{F}$ of $\mathscr{N}$.

Let $\mathscr{F} \subseteq \mathscr{N}$, $\mathscr{F} \ne \emptyset$, let $\mathscr{K} = \{k : \widetilde{\mathscr{F}}(\mathbf{x}^k) = \mathscr{F}\}$ be infinite, and denote

$$\mathscr{A} = \mathscr{N} \setminus \mathscr{F}, \quad \mathsf{H} = \mathsf{A} + \rho_k \mathsf{B}^T \mathsf{B}. \tag{12.43}$$

We get

$$\mathbf{g}^k = \mathbf{g}(\mathbf{x}^k, \boldsymbol{\lambda}^k, \rho) = \mathsf{H}\mathbf{x}^k + \mathsf{B}^T \boldsymbol{\lambda}^k - \mathbf{b} \tag{12.44}$$

and

$$\begin{bmatrix} \mathsf{H}_{\mathscr{F}\mathscr{F}} & \mathsf{B}^T_{*\mathscr{F}} \\ \mathsf{B}_{*\mathscr{F}} & \mathsf{O} \end{bmatrix} \begin{bmatrix} \mathbf{x}^k_{\mathscr{F}} \\ \boldsymbol{\lambda}^k \end{bmatrix} = \begin{bmatrix} \mathbf{g}^k_{\mathscr{F}} + \mathbf{b}_{\mathscr{F}} - \mathsf{H}_{\mathscr{F}A} \mathbf{x}^k_{\mathscr{A}} \\ \mathsf{B}_{*\mathscr{F}} \mathbf{x}^k_{\mathscr{F}} \end{bmatrix}. \tag{12.45}$$

Since for $k \in \mathscr{K}$

$$\mathsf{B}_{*\mathscr{F}} \mathbf{x}^k_{\mathscr{F}} = \mathsf{B}\mathbf{x}^k - \mathsf{B}_{*\mathscr{A}} \mathbf{x}_{\mathscr{A}}, \quad \|\mathbf{g}^k_{\mathscr{F}}\| = \|\mathbf{g}_{\mathscr{F}}(\mathbf{x}^k, \boldsymbol{\lambda}^k, \rho_k)\| \le \|\mathbf{g}^P(\mathbf{x}^k \boldsymbol{\lambda}^k, \rho_k)\|,$$

and both $\|\mathbf{g}^P(\mathbf{x}^k, \boldsymbol{\lambda}^k, \rho)\|$ and $\|\mathsf{B}\mathbf{x}^k\|$ converge to zero by the definition of $\mathbf{x}^k$ in Step 1 of Algorithm 12.1 and (12.37), the right-hand side of (12.45) is bounded. Since $\mathsf{H}_{\mathscr{F}F}$ is nonsingular, it is easy to check that the matrix of the system (12.45) is nonsingular when $\mathsf{B}_{*\mathscr{F}}$ is a full row rank matrix. It simply follows that both $\{\mathbf{x}^k\}$ and $\{\boldsymbol{\lambda}^k\}$ are bounded provided the matrix of (12.45) is nonsingular.

If $\mathsf{B}_{*\mathscr{F}}$ is not a full row rank matrix, then its rank r satisfies $r < m$, and by the RSVD formula (1.36) there are matrices

$$\mathsf{U} \in \mathbb{R}^{m\times r},\ \mathsf{V} \in \mathbb{R}^{n\times r},\ \Sigma = \operatorname{diag}(\sigma_1, \dots, \sigma_r),\ \sigma_i > 0,\ \mathsf{U}^T\mathsf{U} = \mathsf{I},\ \mathsf{V}^T\mathsf{V} = \mathsf{I},$$

such that $\mathsf{B}_{*\mathscr{F}} = \mathsf{U}\Sigma\mathsf{V}^T$. Thus $\hat{\mathsf{B}}_{*\mathscr{F}} = \Sigma\mathsf{V}^T$ has full rank and satisfies

$$\|\hat{\mathsf{B}}_{*\mathscr{F}}\mathbf{x}_{\mathscr{F}}\| = \|\Sigma\mathsf{V}^T\mathbf{x}_{\mathscr{F}}\| = \|\mathsf{U}\Sigma\mathsf{V}^T\mathbf{x}_{\mathscr{F}}\| = \|\mathsf{B}_{*\mathscr{F}}\mathbf{x}_{\mathscr{F}}\| \tag{12.46}$$

for any vector $\mathbf{x}$. Let us assign to any $\boldsymbol{\lambda} \in \mathbb{R}^m$ the vector $\hat{\boldsymbol{\lambda}} = \mathsf{U}^T\boldsymbol{\lambda}$, so that

$$\hat{\mathsf{B}}_{*\mathscr{F}}^T\hat{\boldsymbol{\lambda}} = \mathsf{V}\Sigma\mathsf{U}^T\boldsymbol{\lambda} = \mathsf{B}_{*\mathscr{F}}^T\boldsymbol{\lambda}. \tag{12.47}$$

Using the latter identity and (12.45), we get the system

$$\begin{bmatrix} \mathsf{H}_{\mathscr{F}F} & \hat{\mathsf{B}}_{*\mathscr{F}}^T \\ \hat{\mathsf{B}}_{*\mathscr{F}} & \mathsf{O} \end{bmatrix} \begin{bmatrix} \mathbf{x}_{\mathscr{F}}^k \\ \hat{\boldsymbol{\lambda}}^k \end{bmatrix} = \begin{bmatrix} \mathbf{g}_{\mathscr{F}}^k + \mathbf{b}_{\mathscr{F}} - \mathsf{H}_{\mathscr{F}A}\mathbf{x}_{\mathscr{A}} \\ \hat{\mathsf{B}}_{*\mathscr{F}}\mathbf{x}_{\mathscr{F}}^k \end{bmatrix} \tag{12.48}$$

with a nonsingular matrix. The right-hand side of (12.48) being bounded due to $\|\hat{\mathsf{B}}_{*\mathscr{F}}\mathbf{x}_{\mathscr{F}}^k\| = \|\mathsf{B}_{*\mathscr{F}}\mathbf{x}_{\mathscr{F}}^k\|$, we conclude that the sets $\{\mathbf{x}_{\mathscr{F}}^k : \mathscr{F}(\mathbf{x}^k) = \mathscr{F}\}$ and $\{\hat{\boldsymbol{\lambda}}^k\}$ are bounded. See also Dostál and Kučera [82] or Dostál and Kozubek [81]. □

12.4.4 Convergence

Now, we are ready to prove the main convergence results of this chapter. We shall call the solution $\widehat{\mathbf{x}}$ *regular* if $\mathsf{B}_{*\mathscr{F}}$ is a full row rank matrix, and *range regular* if $\operatorname{Im}\mathsf{B} = \operatorname{Im}\mathsf{B}_{*\mathscr{F}}$, where $\mathcal{F} = \mathcal{F}(\widehat{\mathbf{x}})$. The regular or range regular solutions of (12.1) satisfy the Abadie constraint qualification.

Theorem 12.2. *Let $\{\mathbf{x}^k\}$, $\{\boldsymbol{\lambda}^k\}$ be generated by Algorithm 12.1 for the solution of (12.1) with*

$$\eta > 0,\ \beta > 1,\ M > 0, \rho > 0, \text{and } \boldsymbol{\lambda}^0 \in \mathbb{R}^m. \tag{12.49}$$

Then, the following statements hold.
(i) The sequence $\{\mathbf{x}^k\}$ converges to the solution $\widehat{\mathbf{x}}$ of (12.1).
(ii) If the solution $\widehat{\mathbf{x}}$ of (12.1) is regular, then $\{\boldsymbol{\lambda}^k\}$ converges to a unique vector $\widehat{\boldsymbol{\lambda}}$ of Lagrange multipliers for the equality constraints of (12.1).
(iii) If the solution $\widehat{\mathbf{x}}$ of (12.1) is range regular, then $\{\boldsymbol{\lambda}^k\}$ converges to

$$\overline{\boldsymbol{\lambda}} = \boldsymbol{\lambda}_{\mathrm{LS}} + (\mathsf{I} - \mathsf{P})\boldsymbol{\lambda}^0, \tag{12.50}$$

where P *is the orthogonal projector onto* $\operatorname{Im}\mathsf{B} = \operatorname{Im}\mathsf{B}_{*\mathscr{F}}$ *and* $\boldsymbol{\lambda}_{\mathrm{LS}}$ *is the least square multiplier for the equality constraints of the solution of (12.1).*

Proof. (i) Let $\{\mathbf{x}^k\}$, $\{\boldsymbol{\lambda}^k\}$, and $\{\hat{\boldsymbol{\lambda}}^k\}$ denote the sequences of Proposition 12.1 so that

$$L(\mathbf{x}^k, \boldsymbol{\lambda}^k, \varrho) = L(\mathbf{x}^k, \hat{\boldsymbol{\lambda}}^k, \varrho). \tag{12.51}$$

Since both $\{\mathbf{x}^k\}$ and $\{\hat{\boldsymbol{\lambda}}^k\}$ are bounded, there is a cluster point $(\overline{\mathbf{x}}, \overline{\boldsymbol{\lambda}})$ of $(\mathbf{x}^k, \hat{\boldsymbol{\lambda}}^k)$ and $\mathscr{K} \subseteq \mathbb{N}$ such that

$$\lim_{k\to\infty} \{\mathbf{x}^k\}_{k\in\mathscr{K}} = \overline{\mathbf{x}} \;\text{ and }\; \lim_{k\to\infty} \{\hat{\boldsymbol{\lambda}}^k\}_{k\in\mathscr{K}} = \overline{\boldsymbol{\lambda}}. \tag{12.52}$$

Moreover, using (12.38) and the definition of Step 1 of SMALSE-M,, we get

$$\mathsf{B}\overline{\mathbf{x}} = \mathbf{o} \;\text{ and }\; \mathbf{g}^P(\overline{\mathbf{x}}, \overline{\boldsymbol{\lambda}}, \varrho) = \mathbf{o}. \tag{12.53}$$

Since the solution is unique and $\mathbf{x}^k \in \Omega_S$, it follows that $\{\mathbf{x}^k\}$ converges to $\overline{\mathbf{x}} = \widehat{\mathbf{x}}$.
(ii) Let us denote

$$\mathscr{F} = \mathscr{F}(\widehat{\mathbf{x}}) \text{ and } \mathsf{H} = \mathsf{A} + \rho\mathsf{B}^T\mathsf{B}, \tag{12.54}$$

so by the assumptions there is a unique Lagrange multiplier $\widehat{\boldsymbol{\lambda}}$ that solves

$$[\mathsf{H}\widehat{\mathbf{x}} - \mathbf{b} + \mathsf{B}^T\widehat{\boldsymbol{\lambda}}]_{\mathscr{F}} = \mathbf{o}. \tag{12.55}$$

Since $\{\mathbf{x}^k\}$ converges to $\widehat{\mathbf{x}}$, there is k_1 such that $\mathscr{F} \subseteq \mathscr{F}\{\mathbf{x}^k\}$ for $k \geq k_1$ and

$$\mathbf{g}_{\mathscr{F}}(\mathbf{x}^k, \boldsymbol{\lambda}^k, \rho) = \mathsf{H}_{\mathscr{F}*}\mathbf{x}^k - \mathbf{b}_{\mathscr{F}} + \mathsf{B}_{*\mathscr{F}}^T\boldsymbol{\lambda}^k \tag{12.56}$$

converges to zero. It follows that the sequence

$$\mathsf{B}_{*\mathscr{F}}^T\boldsymbol{\lambda}^k = \mathbf{b}_{\mathscr{F}} - \mathsf{H}_{\mathscr{F}*}\mathbf{x}^k + \mathbf{g}_{\mathscr{F}}(\mathbf{x}^k, \boldsymbol{\lambda}^k, \rho) \tag{12.57}$$

is bounded. Since the smallest nonzero singular value $\overline{\sigma}_{\min}$ of $\mathsf{B}_{*\mathscr{F}}$ satisfies

$$\overline{\sigma}_{\min}\|\boldsymbol{\lambda}^k\| \leq \|\mathsf{B}_{*\mathscr{F}}^T\boldsymbol{\lambda}^k\|, \tag{12.58}$$

it follows that there is a cluster point $\overline{\boldsymbol{\lambda}}$ of $\boldsymbol{\lambda}^k$. Moreover, $\overline{\boldsymbol{\lambda}}$ satisfies

$$[\mathsf{H}\widehat{\mathbf{x}} - \mathbf{b} + \mathsf{B}^T\overline{\boldsymbol{\lambda}}]_{\mathscr{F}} = \mathbf{o}. \tag{12.59}$$

After comparing the last relation with (12.55) and using the assumptions, we conclude that $\overline{\boldsymbol{\lambda}} = \widehat{\boldsymbol{\lambda}}$ and $\boldsymbol{\lambda}^k$ converges to $\widehat{\boldsymbol{\lambda}}$.
(iii) Let the solution $\widehat{\mathbf{x}}$ of (12.1) be range regular, $\mathscr{F} = \mathscr{F}(\widehat{\mathbf{x}})$, let $\overline{\boldsymbol{\lambda}}$ denote a vector of Lagrange multipliers for (12.1), and let $\mathsf{Q} = \mathsf{I} - \mathsf{P}$ denote the orthogonal projector onto $\operatorname{Ker}\mathsf{B}^T = \operatorname{Ker}\mathsf{B}_{*\mathscr{F}}^T$. Using (1.39) and

$$\mathsf{P}+\mathsf{Q}=\mathsf{I},\quad \mathsf{B}^T\mathsf{Q}=\mathsf{O}, \tag{12.60}$$

we get

$$\|\mathsf{B}_{*\mathscr{F}}^T(\boldsymbol{\lambda}^k-\overline{\boldsymbol{\lambda}})\| = \|\mathsf{B}_{*\mathscr{F}}^T(\mathsf{P}+\mathsf{Q})(\boldsymbol{\lambda}^k-\overline{\boldsymbol{\lambda}})\| = \|\mathsf{B}_{*\mathscr{F}}^T(\mathsf{P}\boldsymbol{\lambda}^k-\mathsf{P}\overline{\boldsymbol{\lambda}})\| \tag{12.61}$$

$$\geq \overline{\sigma}_{\min}\|\mathsf{P}\boldsymbol{\lambda}^k-\mathsf{P}\overline{\boldsymbol{\lambda}}\|. \tag{12.62}$$

Using the arguments from the proof of (ii), we get that the left-hand side of the above relation converges to zero, so $\mathsf{P}\boldsymbol{\lambda}^k$ converges to $\mathsf{P}\overline{\boldsymbol{\lambda}}$. Since

$$\boldsymbol{\lambda}^k=\boldsymbol{\lambda}^0+\rho\mathsf{B}\mathbf{x}^0+\cdots+\rho\mathsf{B}\mathbf{x}^k \tag{12.63}$$

with $\mathsf{B}\mathbf{x}^k\in\mathrm{Im}\mathsf{B}$, we get

$$\boldsymbol{\lambda}^k=(\mathsf{P}+\mathsf{Q})\boldsymbol{\lambda}^k=\mathsf{Q}\boldsymbol{\lambda}^0+\mathsf{P}\boldsymbol{\lambda}^k. \tag{12.64}$$

Since $\overline{\boldsymbol{\lambda}}=\boldsymbol{\lambda}_{\mathrm{LS}}+\mathsf{Q}\boldsymbol{\lambda}^0$ is a multiplier for (12.1) and $\mathsf{P}\overline{\boldsymbol{\lambda}}=\boldsymbol{\lambda}_{\mathrm{LS}}$, we get

$$\|\boldsymbol{\lambda}^k-\overline{\boldsymbol{\lambda}}\|=\|\mathsf{Q}\boldsymbol{\lambda}^0+\mathsf{P}\boldsymbol{\lambda}^k-\left(\boldsymbol{\lambda}_{\mathrm{LS}}+\mathsf{Q}\boldsymbol{\lambda}^0\right)\|=\|\mathsf{P}\boldsymbol{\lambda}^k-\mathsf{P}\overline{\boldsymbol{\lambda}}\|. \tag{12.65}$$

Since the right-hand side converges to zero, we conclude that $\boldsymbol{\lambda}^k$ converges to $\overline{\boldsymbol{\lambda}}$, which completes the proof of (iii). □

12.4.5 Optimality of the Outer Loop

Theorem 12.1 suggests that it is possible to give an independent of B upper bound on the number of outer iterations of Algorithm 12.1 (SMALSE-M) that are necessary to achieve a prescribed feasibility error for a class of problems like (12.1). To present explicitly this feature of SMALSE-M, let $\mathscr{T}$ denote any set of indices and let for any $t\in\mathscr{T}$ be defined the problem

$$\text{minimize}\ \ f_t(\mathbf{x})\ \ \text{s.t.}\ \ \mathbf{x}\in\Omega_{SE}^t,\quad f_t(\mathbf{x})=\frac{1}{2}\mathbf{x}^T\mathsf{A}_t\mathbf{x}-\mathbf{b}_t^T\mathbf{x}, \tag{12.66}$$

where

$$\Omega_{SE}^t=\{\mathbf{x}\in\mathbb{R}^{n_t}:\mathsf{B}_t\mathbf{x}=\mathbf{o}\ \text{ and }\ \mathbf{x}\in\Omega_S^t\},\quad \Omega_S^t=\{h_i^t(\mathbf{x}_i)\leq 0,\ i\in\mathscr{I}^t\},$$

h_i^t defines the bound, spherical, or elliptic constraints, $\mathsf{A}_t\in\mathbb{R}^{n_t\times n_t}$ denotes an SPD matrix, $\mathsf{B}_t\in\mathbb{R}^{m_t\times n_t}$, and $\mathbf{b}_t\in\mathbb{R}^{n_t}$. Our optimality result reads as follows.

Theorem 12.3. *Let $\{\mathbf{x}_t^k\},\{\boldsymbol{\lambda}_t^k\}$, and $\{M_{t,k}\}$ be generated by Algorithm 12.1 for (12.66) with*

$$0<\eta_t\leq\|\mathbf{b}_t\|,\quad 0<\beta<1,\quad M_{t,0}=M_0>0,\quad \rho>0,\quad \text{and}\quad \boldsymbol{\lambda}_t^0=\mathbf{o}.$$

Let $\mathbf{o} \in \Omega_{SE}^t$ *and let there be an* $a_{\min} > 0$ *such that the least eigenvalue* $\lambda_{\min}(\mathsf{A}_t)$ *of the Hessian* A_t *of the quadratic function* f_t *satisfies*

$$\lambda_{\min}(\mathsf{A}_t) \geq a_{\min}, \quad t \in \mathscr{T}.$$

Then for each $\varepsilon > 0$ *there are indices* k_t, $t \in \mathscr{T}$, *such that*

$$k_t \leq a/\varepsilon^2 + 1 \tag{12.67}$$

and $\mathbf{x}_t^{k_t}$ *is an approximate solution of (12.66) satisfying*

$$\|\mathsf{B}_t \mathbf{x}_t^{k_t}\| \leq \varepsilon \|\mathbf{b}_t\|. \tag{12.68}$$

Proof. First, notice that for any index j

$$\frac{\rho j}{2} \min\{\|\mathsf{B}_t \mathbf{x}_t^i\|^2 : \; i = 1, \ldots, j\} \leq \sum_{i=1}^{j} \frac{\rho}{2} \|\mathsf{B}_t \mathbf{x}_t^i\|^2 \leq \sum_{i=1}^{\infty} \frac{\rho}{2} \|\mathsf{B}_t \mathbf{x}_t^i\|^2. \tag{12.69}$$

Denoting by $L_t(\mathbf{x}, \boldsymbol{\lambda}, \rho)$ the augmented Lagrangian for problem (12.66), we get for any $\mathbf{x} \in \mathbb{R}^{n_t}$ and $\rho \geq 0$

$$L_t(\mathbf{x}, \mathbf{o}, \rho) = \frac{1}{2} \mathbf{x}^T (\mathsf{A}_t + \rho \mathsf{B}_t^T \mathsf{B}_t) \mathbf{x} - \mathbf{b}_t^T \mathbf{x} \geq \frac{1}{2} a_{\min} \|\mathbf{x}\|^2 - \|\mathbf{b}_t\| \|\mathbf{x}\| \geq -\frac{\|\mathbf{b}_t\|^2}{2a_{\min}}.$$

If we substitute this inequality and $\mathbf{z}_0 = \mathbf{o}$ into (12.37) and use the assumption $\|\mathbf{b}_t\| \geq \eta_t$, we get

$$\sum_{i=1}^{\infty} \frac{\rho}{2} \|\mathsf{B}_t \mathbf{x}_t^i\|^2 \leq \frac{\|\mathbf{b}_t\|^2}{2a_{\min}} + (1+p) \frac{\eta^2}{2a_{\min}} \leq \frac{(2+p)\|\mathbf{b}_t\|^2}{2a_{\min}}, \tag{12.70}$$

where $p \geq 0$ denotes the smallest integer such that $\rho \geq \beta^{2p} M_0^2 / a_{\min}$. Using (12.69) and (12.70), we get

$$\frac{\rho j}{2} \min\{\|\mathsf{B}_t \mathbf{x}_t^i\|^2 : i = 1, \ldots, j\} \leq \frac{(2+p)}{2a_{\min}\varepsilon^2} \varepsilon^2 \|\mathbf{b}_t\|^2. \tag{12.71}$$

Let us now denote $a = (2+p)/(a_{\min}\rho)$ and take for j the least integer which satisfies $a/j \leq \varepsilon^2$, so that

$$a/\varepsilon^2 \leq j \leq a/\varepsilon^2 + 1. \tag{12.72}$$

Denoting for any $t \in \mathscr{T}$

$$k_t = \arg\min\{\|\mathsf{B}_t \mathbf{x}_t^i\| : i = 1, \ldots, j\}, \tag{12.73}$$

we can use (12.72) with simple manipulations to obtain

$$\|\mathsf{B}_t \mathbf{x}_t^{k_t}\|^2 = \min\{\|\mathsf{B}_t \mathbf{x}_t^i\|^2 : i = 1, \ldots, j\} \leq \frac{a}{j\varepsilon^2} \varepsilon^2 \|\mathbf{b}_t\|^2 \leq \varepsilon^2 \|\mathbf{b}_t\|^2. \tag{12.74}$$

□

12.4.6 Optimality of the Inner Loop

We need the following simple lemma to prove the optimality of the inner loop (Step 1) implemented by the MPGP (or MPRGP) algorithm.

Lemma 12.4. *Let* $\{\mathbf{x}^k\}$ *and* $\{\boldsymbol{\lambda}^k\}$ *be generated by Algorithm 12.1 for the solution of (12.1) with* $\eta > 0$, $0 < \beta < 1$, $M > 0$, $\rho > 0$, *and* $\boldsymbol{\lambda}^0 \in \mathbb{R}^m$. *Let* $0 < a_{\min} \leq \lambda_{\min}(\mathsf{A})$, *where* $\lambda_{\min}(\mathsf{A})$ *denotes the least eigenvalue of* A.

Then for any $k \geq 0$

$$L(\mathbf{x}^k, \lambda^{k+1}, \rho) - L(\mathbf{x}^{k+1}, \boldsymbol{\lambda}^{k+1}, \rho) \leq \frac{\eta^2}{2a_{\min}}. \tag{12.75}$$

Proof. Let us denote

$$\hat{\mathbf{x}}^{k+1} = \arg\min_{\mathbf{x}\in\Omega_S} L(\mathbf{x}, \boldsymbol{\lambda}^{k+1}, \rho). \tag{12.76}$$

Using Lemma 11.2 and the definition of Step 1 of SMALSE-M, we get

$$\begin{aligned} L(\mathbf{x}^k, \boldsymbol{\lambda}^{k+1}, \rho) - L(\mathbf{x}^{k+1}, \boldsymbol{\lambda}^{k+1}, \rho) \leq & L(\mathbf{x}^k, \boldsymbol{\lambda}^{k+1}, \rho) - L(\widehat{\mathbf{x}}^{k+1}, \boldsymbol{\lambda}^{k+1}, \rho) \\ \leq & \frac{1}{2a_{\min}}\|\mathbf{g}^P\|^2 \leq \frac{\eta^2}{2a_{\min}}. \end{aligned}$$

□

Now, we are ready to prove the main result of this chapter: the optimality of Algorithm 12.1 (SMALSE-M) in terms of matrix-vector multiplications, provided Step 1 is implemented by Algorithm 11.2 (MPGP) or any other algorithm with an R-linear rate of convergence of the norm of the projected gradient, such as MPRGP for bound constraints described in Chap. 10.

Theorem 12.4. *Let* $0 < a_{\min} < a_{\max}$, $0 < c_{\max}$, *and* $\varepsilon > 0$ *be given constants and let the class of problems (12.66) satisfy*

$$a_{\min} \leq \lambda_{\min}(\mathsf{A}_t) \leq \lambda_{\max}(\mathsf{A}_t) \leq a_{\max} \ \text{ and } \ \|\mathsf{B}_t\| \leq c_{\max}. \tag{12.77}$$

Let $\{\mathbf{x}_t^k\}, \{\boldsymbol{\lambda}_t^k\}$, *and* $\{M_{t,k}\}$ *be generated by Algorithm 12.1 (SMALSE-M) for (12.66) with*

$$\|\mathbf{b}_t\| \geq \eta_t > 0,\ 0 < \beta < 1,\ M_{t,0} = M_0 > 0,\ \rho > 0,\ \text{and } \boldsymbol{\lambda}_t^0 = \mathbf{o}. \tag{12.78}$$

Let Step 1 of Algorithm 12.1 be implemented by Algorithm 11.2 (MPGP) with the parameters $\Gamma > 0$ *and* $\alpha \in (0, 2(a_{\max} + \rho c_{\max}^2)^{-1}]$ *to generate the iterates* $\mathbf{x}_t^{k,0}, \mathbf{x}_t^{k,1}, \ldots, \mathbf{x}_t^{k,l} = \mathbf{x}_t^k$ *for the solution of (12.66) starting from* $\mathbf{x}_t^{k,0} = \mathbf{x}_t^{k-1}$ *with* $\mathbf{x}_t^{-1} = \mathbf{o}$, *where* $l = l_{t,k}$ *is the first index satisfying*

$$\|g^P(\mathbf{x}_t^{k,l}, \boldsymbol{\lambda}_t^k, \rho)\| \leq M_{t,\ell}\|\mathsf{B}_t\mathbf{x}_t^{k,\ell}\|. \tag{12.79}$$

Then Algorithm 12.1 generates an approximate solution $\mathbf{x}_t^{k_t}$ *of any problem (12.66) which satisfies*

$$\|\mathbf{g}^P(\mathbf{x}_t^{k_t}, \boldsymbol{\lambda}_t^{k_t}, \rho)\| \le \varepsilon\|\mathbf{b}_t\| \quad \textit{and} \quad \|\mathsf{B}_t\mathbf{x}_t^{k_t}\| \le \varepsilon\|\mathbf{b}_t\| \tag{12.80}$$

at $O(1)$ *matrix-vector multiplications by the Hessian of the augmented Lagrangian for (12.66).*

Proof. Let $t \in \mathscr{T}$ be fixed and let $L_t(\mathbf{x}, \boldsymbol{\lambda}, \rho)$ denotes the augmented Lagrangian for problem (12.66). Then by (12.75) and the assumption $\eta_t \le \|\mathbf{b}_t\|$

$$L_t(\mathbf{x}_t^{k-1}, \boldsymbol{\lambda}_t^k, \rho) - L_t(\mathbf{x}_t^k, \boldsymbol{\lambda}_t^k, \rho) \le \frac{\eta_t^2}{2a_{\min}} \le \frac{\|\mathbf{b}_t\|^2}{2a_{\min}}. \tag{12.81}$$

Since the minimizer $\overline{\mathbf{x}}_t^k$ of $L_t(\mathbf{x}, \boldsymbol{\lambda}_t^k, \rho)$ subject to $\mathbf{x} \in \Omega_S^t$ satisfies (12.11) and is a possible choice for $\mathbf{x}_t^k$, it follows that

$$L_t(\mathbf{x}_t^{k-1}, \boldsymbol{\lambda}_t^k, \rho) - L_t(\overline{\mathbf{x}}_t^k, \boldsymbol{\lambda}_t^k, \rho) \le \frac{\|\mathbf{b}_t\|^2}{2a_{\min}}. \tag{12.82}$$

Using Theorem 11.2, we get that Algorithm 11.2 (MPGP) used to implement Step 1 of SMALSE-M starting from $\mathbf{x}_t^{k,0} = \mathbf{x}_t^{k-1}$ generates $\mathbf{x}_t^{k,l}$ satisfying

$$\|g_t^P(\mathbf{x}_t^{k,l}, \boldsymbol{\lambda}_t^k, \rho)\|^2 \le a_1\eta_\Gamma^l\big(L_t(\mathbf{x}_t^{k-1}, \boldsymbol{\lambda}_t^k, \rho) - L_t(\overline{\mathbf{x}}_t^k, \boldsymbol{\lambda}_t^k, \rho)\big) \le a_1\frac{\|\mathbf{b}_t\|^2}{2a_{\min}}\eta_\Gamma^l,$$

where a_1 and $\eta = \eta(\delta, \alpha) < 1$ are the constants specified for the class of problems (12.66) in Theorem 11.2. It simply follows by the inner stop rule (12.79) that the number l of the inner iterations in Step 1 is uniformly bounded by an index $l_{\max}$ which satisfies

$$a_1\frac{\|\mathbf{b}_t\|^2}{2a_{\min}}\eta_\Gamma^{l_{\max}} \le M_{t,\ell_{\max}}^2\varepsilon^2\|\mathbf{b}_t\|^2. \tag{12.83}$$

Thus, we have proved that MPGP completes Step 1 in a uniformly bounded number of iterations of MPGP. Since each iteration of MPRGP requires at most two matrix-vector multiplications, it follows that it can carry out Step 1 for any instance of the class of problems 12.1) in a uniformly bounded number of matrix-vector multiplications.

To finish the proof, notice that by Theorem 12.3, there is k_t such that

$$M_{t,k_t}\|\mathsf{B}\mathbf{x}^{k_t}\| \le \varepsilon\|\mathbf{b}\| \quad \text{and} \quad \|\mathsf{B}\mathbf{x}^{k_t}\| \le \varepsilon\|\mathbf{b}\|. \tag{12.84}$$

□

12.5 SMALSE-Mw for Strong Curvature

The SMALSE-M can face problems with a high sensitivity of the projected gradient when the solution is on a strongly curved part of the feasible set boundary, as illustrated in Fig. 11.1. In this case, the projected gradient $\mathbf{g}^P$ is very sensitive and can be large near the solution, so it isn't possible to control the precision of the inner loop of SMALSE-M effectively. The problem is especially with the chopped gradient $\boldsymbol{\beta}(\mathbf{x})$ and its component-wise projection onto Ω_S.

12.5.1 Reduced Projected Gradient (Quadratic Constraints)

Though the reduced gradient provides a nice and robust estimate of the solution error, closer analysis reveals that it provides little information about the local behavior of f in Ω_S. For example, the inequality

$$\mathbf{g}^T(\mathbf{y}-\mathbf{x}) \geq (\mathbf{g}^P)^T(\mathbf{y}-\mathbf{x}), \tag{12.85}$$

which is essential in the convergence analysis of SMALSE-M, need not hold when we replace the projected gradient $\mathbf{g}^P$ by the reduced gradient $\widetilde{\mathbf{g}}_\alpha$. Moreover, there may be no constant $C > 0$ such that for $\mathbf{x} \in \Omega_S$

$$C\|\mathbf{g}^P(\mathbf{x})\| \leq \|\widetilde{\mathbf{g}}_\alpha(\mathbf{x})\|.$$

The reason is that the free components of $\widetilde{\mathbf{g}}_\alpha(\mathbf{x})$ are different from the corresponding components of $\mathbf{g}(\mathbf{x})$, which defines the best linear approximation of the change of $f(\mathbf{x})$ near $\mathbf{x}$ in terms of the variation of free variables.

The remedy is the ($\alpha-$) *reduced projected gradient*

$$\widetilde{\mathbf{g}}^P_\alpha(\mathbf{x}) = \boldsymbol{\varphi}(\mathbf{x}) + \widetilde{\boldsymbol{\beta}}^\alpha(\mathbf{x}), \tag{12.86}$$

which is defined for each $\mathbf{x} \in \Omega$ by the free gradient $\boldsymbol{\varphi}(\mathbf{x})$ and the ($\alpha-$) *reduced chopped gradient* $\widetilde{\boldsymbol{\beta}}^\alpha(\mathbf{x})$ with the components

$$\boldsymbol{\varphi}_i(\mathbf{x}) = \mathbf{g}_i(\mathbf{x}) \text{ for } i \in \mathscr{F}(\mathbf{x}), \quad \boldsymbol{\varphi}_i(\mathbf{x}) = \mathbf{o} \text{ for } i \in \mathscr{A}(\mathbf{x}), \tag{12.87}$$

$$\widetilde{\boldsymbol{\beta}}^\alpha_i(\mathbf{x}) = \begin{cases} \mathbf{o} \text{ for } i \in \mathscr{F}(\mathbf{x}), \\ \mathbf{g}_i(\mathbf{x}) \text{ for } i \in \mathscr{A}(\mathbf{x}) \text{ and } \mathbf{n}_i^T\mathbf{g}_i > 0, \\ \frac{1}{\alpha}(\mathbf{x}_i - P_{\Omega_i}(\mathbf{x}_i - \alpha\mathbf{g}_i)) \text{ for } i \in \mathscr{A}(\mathbf{x}), \ \mathbf{n}_i^T\mathbf{g}_i \leq 0. \end{cases} \tag{12.88}$$

Observe that $\widetilde{\mathbf{g}}_{\alpha}^{P}$ is not a continuous function of $\mathbf{x}$ and the KKT conditions (11.2) are satisfied if and only if the reduced projected gradient is equal to zero.

Comparing the definitions of $\widetilde{\mathbf{g}}_{\alpha}$ and $\widetilde{\mathbf{g}}_{\alpha}^{P}$ and using (11.22), we get

$$\|\widetilde{\mathbf{g}}_{\alpha}\| \leq \|\widetilde{\mathbf{g}}_{\alpha}^{P}\| \leq \|\mathbf{g}^{P}\|. \tag{12.89}$$

We can also formulate the following lemma.

Lemma 12.5. *Let $\widehat{\mathbf{x}}$ solves (11.1), let $\widetilde{\mathbf{g}}_{\alpha}^{P} = \widetilde{\mathbf{g}}_{\alpha}^{P}(\mathbf{x})$ denote the reduced projected gradient at $\mathbf{x} \in \Omega_S$, and let $\alpha \in \left(0, 2(\lambda_{\min}(\mathsf{A}) + \|\mathsf{A}\|)^{-1}\right)$. Then*

$$\|\mathbf{x} - \widehat{\mathbf{x}}\| \leq \lambda_{\min}(\mathsf{A})^{-1}\|\widetilde{\mathbf{g}}_{\alpha}^{P}\| \leq \lambda_{\min}(\mathsf{A})^{-1}\|\mathbf{g}^{P}\|. \tag{12.90}$$

Proof. The inequalities are easy corollaries of Lemma 11.3 and (12.89).

Lemma 12.6. *Let the set Ω_S of problem (11.1) be defined by the convex functions h_i with continuous second derivatives and let $\widehat{\mathbf{x}}$ denote the solution of (11.1). Then there are constants $\varepsilon > 0$ and $C > 0$ such that if $\mathbf{x} \in \Omega$, $\|\mathbf{x} - \widehat{\mathbf{x}}\| \leq \varepsilon$, and $i \in \mathscr{S}$, then*

$$\|\mathbf{g}_i^{P}(\mathbf{x})\| \leq C\|\mathbf{x} - \widehat{\mathbf{x}}\|. \tag{12.91}$$

Proof. It is enough to prove (12.91) for C dependent on i. Let us first consider the components associated with the free or weakly binding set of the solution

$$\mathscr{Z} = \{i \in \mathscr{S} : \ \mathbf{g}_i(\widehat{\mathbf{x}}) = \mathbf{o}\}. \tag{12.92}$$

Then

$$\|\mathbf{g}_i(\mathbf{x})\| = \|\mathbf{g}_i(\mathbf{x}) - \mathbf{g}_i(\widehat{\mathbf{x}})\| = \|\mathsf{A}_{i*}(\mathbf{x} - \widehat{\mathbf{x}})\| \leq \|\mathsf{A}\|\|\mathbf{x} - \widehat{\mathbf{x}}\|, \tag{12.93}$$

where A_{i*} denotes the block row of A which corresponds to $\mathbf{g}_i$, $i \in \mathscr{Z}$. Observing that $\|\mathbf{g}_i^{P}(\mathbf{x})\| \leq \|\mathbf{g}_i(\mathbf{x})\|$ by (11.12), we get

$$\|\mathbf{g}_i^{P}(\mathbf{x})\| \leq C\|\mathbf{x} - \widehat{\mathbf{x}}\|, \quad C = \|\mathsf{A}\|. \tag{12.94}$$

Let us now consider the components of $\mathbf{g}(\mathbf{x})$ with indices in

$$\mathscr{B} = \{i \in \mathscr{A}(\mathbf{x}) : \mathbf{g}_i^{T}(\widehat{\mathbf{x}})\,\mathbf{n}_i(\widehat{\mathbf{x}}) < 0\}, \tag{12.95}$$

so there is $\varepsilon_0 > 0$ such that $\mathbf{g}_i(\mathbf{x}) \neq \mathbf{o}$ for $\|\mathbf{x} - \widehat{\mathbf{x}}\| \leq \varepsilon_0$, and notice that

$$\mathbf{g}_i^{T}(\widehat{\mathbf{x}})\nabla h_i(\widehat{\mathbf{x}}_i) = -\|\mathbf{g}_i^{T}(\widehat{\mathbf{x}})\|\|\nabla h_i(\widehat{\mathbf{x}}_i)\| < 0. \tag{12.96}$$

It follows that there is $\varepsilon \in (0, \varepsilon_0)$ such that if $\|\mathbf{x} - \widehat{\mathbf{x}}\| \leq \varepsilon$, then

$$\mathbf{g}_i^{T}(\mathbf{x})\nabla h_i(\mathbf{x}_i) < 0, \quad i \in \mathscr{B}. \tag{12.97}$$

Moreover, the mapping $\psi_i : \mathbb{R}^n \to \mathbb{R}^{\ell_i}$ defined for $\|\mathbf{x} - \widehat{\mathbf{x}}\| \leq \varepsilon$ by

$$\psi_i(\mathbf{x}) = \mathbf{g}_i(\mathbf{x}) - \frac{\mathbf{g}_i^T(\mathbf{x})\nabla h_i(\mathbf{x}_i)}{\|\nabla h_i(\mathbf{x}_i)\|^2}\nabla h_i(\mathbf{x}_i) \tag{12.98}$$

is differentiable at $\mathbf{x}$ and for $i \in \mathscr{B}$

$$\psi_i(\mathbf{x}) = \mathbf{g}_i(\mathbf{x}) - \frac{(\mathbf{g}_i^T(\mathbf{x})\nabla h_i(\mathbf{x}_i))^-}{\|\nabla h_i(\mathbf{x}_i)\|^2}\nabla h_i(\mathbf{x}_i) = \mathbf{g}_i^P(\mathbf{x}). \tag{12.99}$$

It follows that we can use the classical results of calculus to get that for each component $i \in \mathscr{B}$, there is $C > 0$ such $\|\mathbf{x} - \widehat{\mathbf{x}}\| \leq \varepsilon$ implies

$$\|\psi_i(\mathbf{x}) - \psi_i(\widehat{\mathbf{x}})\| \leq C\|\mathbf{x} - \widehat{\mathbf{x}}\|. \tag{12.100}$$

To finish the proof, recall that $\psi(\widehat{\mathbf{x}}) = \mathbf{o}$ and that for $i \in \mathscr{B}$ and $\|\mathbf{x} - \widehat{\mathbf{x}}\| \leq \varepsilon$

$$\mathbf{g}_i^P(\mathbf{x}) = \psi_i(\mathbf{x}). \tag{12.101}$$

□

Theorem 12.5. *Let the feasible set $\Omega_S = \Omega_1 \times \cdots \times \Omega_s$ of (11.1) be defined by the convex functions $h_i : \mathbb{R}^{\ell_i} \to \mathbb{R}$, each with a continuous second derivative. Then there is $C > 0$ and $\varepsilon > 0$ such that for each $\mathbf{x} \in \Omega_S$ and $\alpha \in \left(0, 2(\lambda_{\min}(\mathsf{A})^{-1} + \|\mathsf{A}\|)^{-1}\right)$*

$$\|\widetilde{\mathbf{g}}_\alpha^P(\mathbf{x})\| \leq \|\mathbf{g}^P(\mathbf{x})\| \leq C\|\widetilde{\mathbf{g}}_\alpha^P(\mathbf{x})\|. \tag{12.102}$$

Proof. Let $\alpha \in (0, 2\|\mathsf{A}\|^{-1})$ and $\mathbf{x} \in \Omega_S$ be fixed, so we can simplify the notation by

$$\mathbf{g} = \mathbf{g}(\mathbf{x}), \quad \widetilde{\mathbf{g}}^P = \widetilde{\mathbf{g}}_\alpha^P(\mathbf{x}), \quad \text{and} \quad \mathbf{g}^P = \mathbf{g}^P(\mathbf{x}). \tag{12.103}$$

The left inequality of (12.102) has already been established—see (12.89). We shall prove the right inequality of (12.102), observing that by the definitions

$$\|\mathbf{g}_i^P\| = \|\widetilde{\mathbf{g}}_i^P\| \quad \text{for} \quad i \in \mathscr{F}(\mathbf{x}). \tag{12.104}$$

If $i \in \mathscr{A}(\mathbf{x})$ and $\varepsilon > 0$, $C > 0$ are those of Lemma 12.6, then $\|\mathbf{x} - \widehat{\mathbf{x}}\| \leq \varepsilon$ and by Lemmas 12.5 and 12.6

$$\|\mathbf{g}_i^P\| \leq C\|\mathbf{x} - \widehat{\mathbf{x}}\| \leq C\lambda_{\min}(\mathsf{A})^{-1}\|\widetilde{\mathbf{g}}^P\|, \tag{12.105}$$

where $\nu(\alpha)$ is defined in Lemma 11.3. This proves the right inequality of (12.102) for $\mathbf{x}$ near the solution.

To prove (12.102) for $i \in \mathscr{A}(\mathbf{x})$ and $\|\mathbf{x} - \widehat{\mathbf{x}}\| \geq \varepsilon$, observe that

$$\|\mathbf{g}_i^P\| \leq \|\mathbf{g}_i\| = \|\left(\mathbf{g}_i - \mathbf{g}_i(\widehat{\mathbf{x}})\right) + \mathbf{g}_i(\widehat{\mathbf{x}})\| \leq \|\mathsf{A}\|\|\mathbf{x} - \widehat{\mathbf{x}}\| + C_1. \tag{12.106}$$

Moreover, by Lemma 12.5

$$\|\mathbf{x} - \widehat{\mathbf{x}}\| \le \lambda_{\min}(\mathsf{A})^{-1}\|\widetilde{\mathbf{g}}^P\|. \tag{12.107}$$

Thus for $\|\mathbf{x} - \widehat{\mathbf{x}}\| \ge \varepsilon$

$$\begin{aligned}\|\mathbf{g}_i^P\| \le & \|\mathsf{A}\|\|\mathbf{x} - \widehat{\mathbf{x}}\| + (C_1/\varepsilon)\varepsilon \le (\|\mathsf{A}\| + (C_1/\varepsilon))\,\|\mathbf{x} - \widehat{\mathbf{x}}\| \\ \le & (\|\mathsf{A}\| + (C_1/\varepsilon))\,\nu(\alpha)\|\widetilde{\mathbf{g}}^P\|. \end{aligned} \tag{12.108}$$

The rest follows by the finite dimension argument. □

We shall now consider a special case of problem (12.1), the minimization of a strictly convex quadratic function subject to *bound and equality constraints*

$$\min_{\mathbf{x}\in\Omega_{BE}} f(\mathbf{x}), \quad \Omega_{BE} = \{\mathbf{x} \in \mathbb{R}^n : \mathsf{B}\mathbf{x} = \mathbf{o} \quad \text{and} \quad \mathbf{x} \ge \ell\}, \tag{12.109}$$

where $f(\mathbf{x}) = \frac{1}{2}\mathbf{x}^T\mathsf{A}\mathbf{x} - \mathbf{x}^T\mathbf{b}$, $\mathbf{b} \in \mathbb{R}^n$, A is an $n \times n$ SPD matrix, and $\mathsf{B} \in \mathbb{R}^{m\times n}$. We consider the same assumptions as above, in particular, $\Omega_{BE} \ne \emptyset$, $\mathsf{B} \ne \mathsf{O}$, and $\operatorname{Ker}\mathsf{B} \ne \{\mathbf{o}\}$. We admit dependent rows of B and $\ell_i = -\infty$.

Algorithm 12.2. Semimonotonic augmented Lagrangians for QPQC problems with equality and separable inequality constraints with strong curvature (SMALSE-Mw)

Given an SPD matrix $\mathsf{A} \in \mathbb{R}^{n\times n}$, $\mathsf{B} \in \mathbb{R}^{m\times n}$, $\mathbf{b} \in R^n$, *constraints* $h_i(\mathbf{x}_i)$, $i = 1, \dots s$.

Step 0. *{Initialization.}*

Choose $0 < \beta < 1, M_{-1} = 0, 0 < M_0, \eta, \varrho; \boldsymbol{\lambda}^0 \in \mathbb{R}^m$, $\alpha \in (0, 2(\lambda_{\min}(\mathsf{A}) + \|\mathsf{A}\|))^{-1}$

for $k = 0, 1, 2, \dots$

Step 1. *{Inner iteration with adaptive precision control.}*

Find $\mathbf{x}^k \in \Omega_S$ *such that*

$$\|\widetilde{\mathbf{g}}_\alpha^P(\mathbf{x}^k, \boldsymbol{\lambda}^k, \varrho)\| \le \min\{M_k\|\mathsf{B}\mathbf{x}^k\|, \eta\} \tag{12.110}$$

Step 2. *{Updating the Lagrange multipliers.}*

$$\boldsymbol{\lambda}^{k+1} = \boldsymbol{\lambda}^k + \varrho\mathsf{B}\mathbf{x}^k \tag{12.111}$$

Step 3. *{Update M provided the increase of the Lagrangian is insufficient.}*

if $M_k = M_{k-1}$ *and*

$$L(\mathbf{x}^k, \boldsymbol{\lambda}^k, \varrho) < L(\mathbf{x}^{k-1}, \boldsymbol{\lambda}^{k-1}, \varrho) + \frac{\varrho}{2}\|\mathsf{B}\mathbf{x}^k\|^2, \tag{12.112}$$

$M_{k+1} = \beta M_k$

else

$M_{k+1} = M_k$

end if

end for

12.5.2 SMALSE-Mw Algorithm

The complete algorithm that we call SMALSE-Mw (w refers to widetilde recalling the reduced projected gradient) reads as follows.

In Step 1, we can use any algorithm for minimizing strictly convex quadratic functions subject to separable constraints provided it guarantees the convergence of the projected gradient to zero, such as MPGP of Chap. 11.

12.6 R-Linear Convergence of SMALBE-M

Theorem 12.4 guarantees that the number of iterations $k_t(\varepsilon)$ that are necessary to get an approximate solution of any problem from the class of problems (12.66) to the prescribed relative precision ε is uniformly bounded by

$$k(\varepsilon) = \max\{k_t(\varepsilon) : t \in \mathscr{T}\}. \tag{12.113}$$

Notice that $k(\varepsilon)$ does not depend on the matrices B_t which define the feasible sets Ω_t. This feature is not obvious from the standard analysis of the Uzawa-type methods, even for linear problems. However, Theorem 12.3 yields only

$$k(\varepsilon) \lesssim \varepsilon^{-2}, \tag{12.114}$$

i.e., we proved that there is $C > 0$ independent of B_t such that for any $\varepsilon > 0$

$$k(\varepsilon) \leq C\varepsilon^{-2}, \tag{12.115}$$

which is very pessimistic and has never been observed in practice.

Here we report a stronger result, namely that SMALBE-M enjoys the R-linear convergence of feasibility errors in a later stage of computations when the solution's free and strongly active variables' indices are identified and show that there are $k_t(\varepsilon)$ and $\overline{k}_t$ such that for sufficiently small ε

$$k_t(\varepsilon) - \overline{k}_t \lesssim |\log(\varepsilon)|. \tag{12.116}$$

The iterates can remain nonlinear at any stage of the solution procedure. The result does not assume independent equality constraints and remains valid with zero multipliers for active bound constraints. The following lemma shows that the binding and free sets are identified in a finite number of steps.

Lemma 12.7. *Let $\mathbf{x}^k$ and $\boldsymbol{\lambda}^k$ be generated by the SMALBE-M algorithm for the solution of (12.109) with $\rho > 0$. Then there is k_1 such that for $k \geq k_1$*

$$\widehat{\mathscr{F}} = \mathscr{F}(\widehat{\mathbf{x}}) \subseteq \mathscr{F}(\mathbf{x}^k) \quad \text{and} \quad \widehat{\mathscr{B}} = \mathscr{B}(\widehat{\mathbf{x}}) \subseteq \mathscr{B}(\mathbf{x}^k). \tag{12.117}$$

Proof. First observe that $\mathbf{g}^P(\mathbf{x}^k, \boldsymbol{\lambda}^k, \rho)$ converges to zero by Theorem 12.3. If $\widehat{\mathscr{B}} = \emptyset$ or $\widehat{\mathscr{F}} = \emptyset$, then the statements of our lemma concerning these sets are valid trivially. Hence, we assume that they are both nonempty and denote

$$\delta_{\widehat{\mathscr{F}}} = \min\{\widehat{x}_i - \ell_i : i \in \widehat{\mathscr{F}}\}, \quad \delta_{\widehat{\mathscr{B}}} = \min\{\widehat{g}_i : i \in \widehat{\mathscr{B}}\}. \tag{12.118}$$

Since $\mathbf{x}^k$ converges to $\widehat{\mathbf{x}}$, there is k' such that for $k \geq k'$ and $i \in \widehat{\mathscr{F}}$

$$\ell_i < \widehat{x}_i - \frac{1}{2}\delta_{\widehat{\mathscr{F}}} \leq x_i^k, \tag{12.119}$$

i.e., $i \in \mathscr{F}(\mathbf{x}^k)$.

To prove the second relation of (12.117), observe that $\mathbf{g}$ is a continuous function, so $\mathbf{g}^k = \mathbf{g}(\mathbf{x}^k, \boldsymbol{\lambda}^k, \rho)$ converges to $\widehat{\mathbf{g}} = \mathbf{g}(\widehat{\mathbf{x}}, \overline{\boldsymbol{\lambda}}, \rho)$. It follows that if $i \in \widehat{\mathscr{B}}$, then there is k'' such that for $k \geq k''$

$$g_i(\mathbf{x}^k, \boldsymbol{\lambda}^k, \rho) = g_i^k \geq \frac{1}{2}\delta_{\widehat{\mathscr{K}}}. \tag{12.120}$$

Moreover, since $\mathbf{g}^P(\mathbf{x}^k, \boldsymbol{\lambda}^k, \rho)$ converges to zero by the assumption, it follows that there is $k''' \geq k''$ such that for $k \geq k'''$ holds $x_i^k = \ell_i$, i.e., $i \in \mathscr{B}(x^k)$. We have thus proved the second relation of (12.117) with $k_1 = \max\{k', k'''\}$. □

Now, we are ready to show that the convergence is R-linear after the free and strong active constraints of the solution are identified.

Theorem 12.6. *Let $\{\mathbf{x}^k\}$, $\{\boldsymbol{\lambda}^k\}$, and $\{M_k\}$ be generated by SMALBE-M for the solution to (12.109) with $\eta > 0$, $0 < \beta < 1$, $M_0 > 0$, $\rho > 0$, and $\boldsymbol{\lambda}^0 \in \mathbb{R}^m$. Let the solution $\widehat{\mathbf{x}}$ of (12.109) be range regular and let k_0 and k_1 be those of Theorem 12.1 and Lemma 12.7, so that for $\overline{k} = \max\{k_0, k_1\}$*

$$M_{\overline{k}} = M_{\overline{k}+1} = \cdots \quad \textit{and} \quad \mathscr{F}(\widehat{\mathbf{x}}) = \mathscr{F}(\mathbf{x}^{\overline{k}}) = \mathscr{F}(\mathbf{x}^{\overline{k}+1}) = \cdots . \tag{12.121}$$

Let $\widehat{\sigma}_{\min}$ denote the smallest nonzero singular value of $\mathsf{B}_{*\widehat{\mathscr{F}}}$, let $\mathsf{H} = \mathsf{A} + \rho\mathsf{B}^T\mathsf{B}$, and denote

$$C_1 = M_{k_0}\frac{\kappa(\mathsf{H}) + 1}{\lambda_{\min}(\mathsf{H})} + \frac{\kappa(A^\rho)}{\widehat{\sigma}_{\min}}, \quad C_2 = \frac{M_{k_0}\kappa(\mathsf{A}^\rho) + \widehat{\sigma}_{\min}^{-1}\|\mathsf{H}\|}{\widehat{\sigma}_{\min}}, \quad C = \frac{2C_2}{\rho}.$$

Then the following relations hold:

(i) For any $k \geq \overline{k}$

$$\|\mathsf{B}\mathbf{x}^k\|^2 \leq (C+1)\left(\frac{C}{C+1}\right)^{k-\overline{k}}\|\mathsf{B}\mathbf{x}^{\overline{k}}\|^2. \tag{12.122}$$

(ii) For any $k \geq \overline{k}$

$$\|\mathbf{x}^k - \widehat{\mathbf{x}}\|^2 \le C_1(C+1)\left(\frac{C}{C+1}\right)^{k-\overline{k}} \|\mathsf{B}\mathbf{x}^{\overline{k}}\|^2. \tag{12.123}$$

(iii) If $\boldsymbol{\lambda}^0 \in \mathrm{Im}\mathsf{B}$*, then for any* $k \ge \overline{k}$

$$\|\boldsymbol{\lambda}^k - \boldsymbol{\lambda}_{LS}\|^2 \le C_2(C+1)\left(\frac{C}{C+1}\right)^{k-\overline{k}} \|\mathsf{B}\mathbf{x}^{\overline{k}}\|^2. \tag{12.124}$$

Proof. See [108, Theorem 4.4].

12.7 Adaptive Constraint Re-orthogonalization

The performance of the SMALBE can be improved by enhancing the information on the free set of current iterates into the reorthogonalization of equality constraints. The presentation follows Dostál et al. [109], see also Dostál and Vlach [86].

12.7.1 Algorithm

Let us first recall that the cost function of the problem (12.1) reads

$$f(\mathbf{x}) = \frac{1}{2}\mathbf{x}^T\mathsf{A}\mathbf{x} - \mathbf{b}^T\mathbf{x},$$

so that the augmented Lagrangian L for (12.1) reads

$$L(\mathbf{x}, \boldsymbol{\lambda}, \varrho) = \frac{1}{2}\mathbf{x}^T\left(\mathsf{A} + \varrho\mathsf{B}^T\mathsf{B}\right)\mathbf{x} - \mathbf{b}^T\mathbf{x} + \mathbf{x}^T\mathsf{B}^T\boldsymbol{\lambda}.$$

A little analysis of the proof of convergence of SMALBE reveals that the augmentation affects directly only free variables, i.e., the effective augmentation in k-th step is carried out by means of $\mathsf{B}_{*\mathscr{F}(\mathbf{x}^k)}$. Indeed, if $\mathscr{F} = \mathscr{F}(\mathbf{x}^k)$ denotes the set of indices of inactive constraints, i.e., $\nu_i > \ell_i$ for $i \in \mathscr{F}$, then the quadratic term of the Lagrangian reduces to

$$\frac{1}{2}\mathbf{x}_{\mathscr{F}}^T\left(\mathsf{A}_{\mathscr{F}\mathscr{F}} + \varrho\mathsf{B}_{*\mathscr{F}}^T\mathsf{B}_{*\mathscr{F}}\right)\mathbf{x}_{\mathscr{F}},$$

which can be too small and ill-conditioned to enforce effectively the equality constraints, especially if the number of active constraints is large.

A remedy is to carry out the iterations with orthonormal constraints as long as the active set of the iterates varies until we get a reasonable approximation $\mathscr{F}$ of the free set $\widehat{\mathscr{F}}$ of the solution $\widehat{\mathbf{x}}$. Assuming the full rank of $\mathsf{B}_{*\mathcal{F}}$

and denoting

$$\mathsf{B} = [\mathsf{B}_{*\mathscr{A}}, \mathsf{B}_{*\mathscr{F}}], \qquad \mathsf{B}_{*\mathscr{A}} \in \mathbb{R}^{m\times n_{\mathscr{A}}}, \quad \mathsf{B}_{*\mathscr{F}} \in \mathbb{R}^{m\times n_{\mathscr{F}}}, \quad n = n_{\mathscr{A}} + n_{\mathscr{F}},$$

we can introduce the Cholesky factor L of $\mathsf{B}_{*\mathscr{F}}\mathsf{B}_{*\mathscr{F}}^T$ and define

$$\widetilde{\mathsf{B}} = \mathsf{L}^{-1}\mathsf{B}, \quad \mathsf{L}\mathsf{L}^T = \mathsf{B}_{*\mathscr{F}}\mathsf{B}_{*\mathscr{F}}^T,$$

so that

$$\widetilde{\mathsf{B}}_{*\mathscr{F}}\widetilde{\mathsf{B}}_{*\mathscr{F}}^T = \mathsf{I}.$$

Thus, the quadratic term of the Lagrangian for modified equalities reads

$$\frac{1}{2}\mathbf{x}_{\mathscr{F}}^T\left(\mathsf{A}_{\mathcal{F}\mathcal{F}} + \varrho\,\mathsf{I}_{\mathscr{F}\mathscr{F}}\right)\mathbf{x}_{\mathscr{F}}.$$

The regularization term with $\widetilde{\mathsf{B}}^T\widetilde{\mathsf{B}}$ is more effective than that with $\mathsf{B}^T\mathsf{B}$ as long as the free set of $\mathbf{x}$ is near the free set $\mathscr{F}$ which defines $\widetilde{\mathsf{B}}$. However, it provides effective regularization for any $\mathbf{x} \in \mathbb{R}^n$ as

$$\mathbf{x}^T\mathsf{B}_{*\mathscr{F}}\mathsf{B}_{*\mathscr{F}}^T\mathbf{x} \le \mathbf{x}^T\mathsf{B}\mathsf{B}^T\mathbf{x} = \|\mathbf{x}\|^2, \quad \mathbf{x} \in \mathbb{R}^m,$$

implies

$$\|\widetilde{\mathsf{B}}\mathbf{x}\|^2 = \mathbf{x}^T\mathsf{B}^T\left(\mathsf{B}_{*\mathscr{F}}\mathsf{B}_{*\mathscr{F}}^T\right)^{-1}\mathsf{B}\mathbf{x} \ge \mathbf{x}^T\mathsf{B}^T\left(\mathsf{B}\mathsf{B}^T\right)^{-1}\mathsf{B}\mathbf{x} = \|\mathsf{B}\mathbf{x}\|^2. \quad (12.125)$$

To enhance these observations into SMALBE-M, recall that the solution of the problem (12.1) does not depend on the representation of Ω_{BE}, so we can improve the convergence by switching to $\widetilde{\mathsf{B}}$ associated with a current approximation $\mathscr{F}_k = \mathscr{F}(\mathbf{x}^k)$ of the free set $\widehat{\mathscr{F}}$ of the solution $\widehat{\mathbf{x}}$. Assuming that the orthogonalization of $\mathsf{B}_{*\mathscr{F}_k}$ is carried out by the Cholesky factor

$$\mathsf{L} = \mathit{Chol}\left(\mathsf{B}_{*\mathscr{F}_k}\mathsf{B}_{*\mathscr{F}_k}^T\right)$$

of the decomposition of $\mathsf{B}_{*\mathscr{F}_k}\mathsf{B}_{*\mathscr{F}_k}^T$, so that

$$\mathsf{L}\mathsf{L}^T = \mathsf{B}_{*\mathscr{F}_k}\mathsf{B}_{*\mathscr{F}_k}^T,$$

we get the orthogonalized equality constraints in the form

$$\mathsf{L}^{-1}\mathsf{B}\mathbf{x} = \mathbf{o}.$$

The corresponding Lagrangian $L_{\mathsf{L}^{-1}}$ reads

$$L_{\mathsf{L}^{-1}}(\mathbf{x}, \boldsymbol{\mu}, \varrho) = \frac{1}{2}\mathbf{x}^T\mathsf{PFP}\mathbf{x} - \mathbf{x}^T\mathsf{P}\mathbf{d} + \mathbf{x}^T\mathsf{B}^T\mathsf{L}^{-T}\boldsymbol{\mu} + \frac{\rho}{2}\|\mathsf{L}^{-1}\mathsf{B}\mathbf{x}\|^2,$$

and the gradient $\mathsf{B}_{\mathsf{L}^{-1}}$ of $L_{\mathsf{L}^{-1}}$ reads

$$\mathsf{B}_{\mathsf{L}^{-1}}(\mathbf{x}, \boldsymbol{\mu}, \varrho) = \nabla_{\mathbf{x}} L_{\mathsf{L}^{-1}}(\mathbf{x}, \boldsymbol{\mu}, \varrho) = \mathsf{PFPx} - \mathsf{Pd} + \mathsf{B}^T\mathsf{L}^{-T}\boldsymbol{\mu} + \varrho\mathsf{B}^T\mathsf{L}^{-T}\mathsf{L}^{-1}\mathsf{Bx}.$$

Notice that

$$\mathsf{B}_{\mathsf{L}^{-1}}(\mathbf{x}, \mathsf{L}^T\boldsymbol{\lambda}, 0) = \mathsf{B}(\mathbf{x}, \boldsymbol{\lambda}, 0),$$

so the multiplier $\boldsymbol{\mu}$ associated with $\mathsf{B}_{\mathsf{L}^{-1}}$ is related to $\boldsymbol{\lambda}$ by $\boldsymbol{\mu} = \mathsf{L}^T\boldsymbol{\lambda}$. Thus, to switch the representation of Ω, evaluate

$$\boldsymbol{\lambda}^k = \mathsf{L}^{-T}\boldsymbol{\mu}^k$$

with the actual L, evaluate new $\mathsf{L} = \mathit{Chol}\left(\mathsf{B}_{*,\mathscr{F}_k}\mathsf{B}^T_{*,\mathscr{F}_k}\right)$, set $\boldsymbol{\mu}^k = \mathsf{L}^T\boldsymbol{\lambda}^k$ in *Step 1*, and continue the iterations with a new Lagrangian until the active set of the iterates stabilizes. Since the transformations between $\boldsymbol{\mu}$ and $\boldsymbol{\lambda}$ are simple, only one set of multipliers is possible. The procedure is invoked whenever the free set of an iterate differs significantly from that approximated at the beginning of the cycle, and the active set is stabilized so that it changes only mildly. To describe the procedure explicitly, let $\#\mathscr{F}$ denote a number of elements of a free set $\mathscr{F}$. The modified SMALBE-M algorithms with an adaptive augmentation is described in detail as Algorithm 12.3.

In our experiments, we got excellent results even with one update of L (see Sect. 12.9.2). The specific implementation of the update rule depends on many factors, including the cost of orthogonalization (obtaining L). Observe that we do not define L by the exact free set of the solution.

Remark 12.3. We assumed that $\mathsf{B}_{*\mathscr{F}}$ is a full-row rank matrix. The algorithm can be modified to accept dependent rows as follows:

- If a zero pivot appears during the factorization, remember the index and skip the corresponding row and column.
- Reorder the vectors and matrices so the indices of skipped columns are at the end.
- Use the factor in the form

$$\mathsf{L} = \begin{bmatrix} \widetilde{\mathsf{L}} & \mathsf{O} \\ \mathsf{O} & \mathsf{I} \end{bmatrix},$$

  where $\widetilde{\mathsf{L}}$ is a factor of decomposed $\mathsf{B}_{*\mathscr{F}}\mathsf{B}^T_{*\mathscr{B}}$ without the rows and columns corresponding to approximately zero pivots.

Algorithm 12.3. Semi-monotonic augmented Lagrangians with adaptive augmentation for problem (12.1) **(Adaptive SMALBE-M)**

Step 0. {*Initialization*}
Choose $\eta > 0,\ 0 < \beta < 1,\ M_0 > 0,\ \varrho > \rho_0 > 0,\ \boldsymbol{\lambda}^0 \in \mathbb{R}^m,\ \mathsf{L} = \mathsf{I},\ \overline{k} = 0$
for $k = 0, 1, 2, \ldots$
Step 1. {*Inner iteration with adaptive precision control*}
Find $\mathbf{x}_I^k \geq \ell_I$ *such that*

$$\|\mathsf{B}^P_{\mathsf{L}^{-1}}(\mathbf{x}^k, \mathsf{L}^T\boldsymbol{\lambda}^k, \varrho)\| \leq \min\{M_k\|\mathsf{B}\mathbf{x}^k\|, \eta\} \tag{12.126}$$

Step 2. {*Updating the Lagrange multipliers*}

$$\boldsymbol{\lambda}^{k+1} = \boldsymbol{\lambda}^k + (\varrho - \varrho_0)\mathsf{L}^{-T}\mathsf{L}^{-1}\mathsf{B}\mathbf{x}^k$$

Step 3. {*Update M provided the increase of the Lagrangian is not sufficient*}
if $k > \overline{k}$ *and*

$$L(\mathbf{x}^k, \boldsymbol{\lambda}^k, \varrho) < L(\mathbf{x}^{k-1}, \boldsymbol{\lambda}^{k-1}, \varrho) + \frac{\varrho - \varrho_0}{2}\|\mathsf{B}\mathbf{x}^k\|^2 \tag{12.127}$$

$M_{k+1} = \beta M_k$
else
$M_{k+1} = M_k$
end else if
Step 4. {*Update* L}
if $|\#\mathscr{F}(\mathbf{x}^k) - \#\mathscr{F}(\mathbf{x}^k)|$ *is large*
and $Chol(\mathsf{B}_{*\mathscr{F}(\mathbf{x}^k)}\mathsf{B}^T_{*\mathscr{F}(\mathbf{x}^k)})$ *is nonsingular*
then $\mathsf{L} = Chol(\mathsf{B}_{*\mathscr{F}(\mathbf{x}^k)}\mathsf{B}^T_{*\mathscr{F}(\mathbf{x}^k)})$
$k = \overline{k} + 1$
end for

12.7.2 Convergence of Adaptive SMALBE-M

The theoretical results concerning adaptive SMALBE-M are very similar to those concerning original SMALBE-M.

Theorem 12.7. *Let* M_k, $\{\mathbf{x}^k\}$, *and* $\{\boldsymbol{\lambda}^k\}$ *be generated by SMALBE-M for the solution of* (12.1) *with*

$$M_0,\ \eta,\ \varrho > 0,\ 0 < \beta < 1,\ \text{and}\ \boldsymbol{\lambda}^0 \in \mathbb{R}^m.$$

Let the solution be regular. Then the following statements hold:

(i) If $\lambda_{\min}$ *denote the smallest eigenvalue of* A, *then*

$$M_k \geq \min\{M_0, \beta\sqrt{\varrho\lambda_{\min}}\}.$$

(ii) If $k > 0$ *and* $M_k = M_{k+1}$, *then*

$$L(\mathbf{x}^k, \boldsymbol{\lambda}^k, \varrho) \geq L(\mathbf{x}^{k-1}, \boldsymbol{\lambda}^{k-1}, \varrho) + \frac{\varrho}{2}\|\mathsf{B}\mathbf{x}\|^2. \tag{12.128}$$

(iii) The sequence $\{\boldsymbol{x}^k\}$ *converges to the solution* $\widehat{\mathbf{x}}$ *of* (12.1).

(iv) The series $\{\boldsymbol{\lambda}^k\}$ *converges to the uniquely determined vector* $\widehat{\boldsymbol{\lambda}}$ *of Lagrange multipliers of* (12.1).

(v) If ϱ *is sufficiently large, then the convergence of both the Lagrange multipliers and the feasibility error is linear.*

Proof. First notice that by (12.125) and (12.126)

$$\|\boldsymbol{g}^P_{\mathsf{L}^{-1}}(\mathbf{x}^k, \mathsf{L}^T\boldsymbol{\lambda}^k, \varrho)\| \leq \min\{M_k\|\mathsf{B}\mathbf{x}^k\|, \eta\} \leq \min\{M_k\|\mathsf{L}^{-1}\mathsf{B}\mathbf{x}^k\|, \eta\},$$

so the iterates $\mathbf{x}^k$, $\boldsymbol{\lambda}^k = \mathsf{L}^T\boldsymbol{\lambda}^k$ can be considered as iterates of the standard SMALSE algorithm for the problem with equality constraints in the form

$$\mathsf{L}^{-1}\mathsf{B}\mathbf{x} = \mathbf{o}.$$

Thus, we can apply Lemma 12.3(i) to get (i).

To prove (ii), notice that it follows by Lemma 12.3(i) and (12.125) that

$$\begin{aligned} L_{\mathsf{L}^{-1}}(\mathbf{x}^k, \boldsymbol{\lambda}^k, 0) &\geq L_{\mathsf{L}^{-1}}(\mathbf{x}^{k-1}, \boldsymbol{\lambda}^{k-1}, \varrho) = L_{\mathsf{L}^{-1}}(\mathbf{x}^{k-1}, \boldsymbol{\lambda}^{k-1}, 0) + \frac{\varrho}{2}\|\mathsf{L}^{-1}\mathsf{B}\mathbf{x}^k\|^2 \\ &\geq L_{\mathsf{L}^{-1}}(\mathbf{x}^{k-1}, \boldsymbol{\lambda}^{k-1}, 0) + \frac{\varrho}{2}\|\mathsf{B}\mathbf{x}^k\|^2. \end{aligned}$$

Using

$$L_{\mathsf{L}^{-1}}(\mathbf{x}^k, \boldsymbol{\lambda}^k, 0) = L_{\mathsf{L}^{-1}}(\mathbf{x}^k, \mathsf{L}^T\boldsymbol{\lambda}^k, 0) = L(\mathbf{x}^k, \boldsymbol{\lambda}^k, 0),$$

we get

$$L(\mathbf{x}^k, \boldsymbol{\lambda}^k, 0) \geq L(\mathbf{x}^{k-1}, \boldsymbol{\lambda}^{k-1}, 0) + \frac{\varrho}{2}\|\mathsf{B}\mathbf{x}^k\|^2. \tag{12.129}$$

This proves (ii). The remaining statements can be deduced from (ii) in the same way as in the proof of Theorem 12.2.

12.8 Implementation

Let us give you a few hints that can be helpful for the effective implementation of SMALSE-M (SMALBE-M) and SMALSE-Mw with the inner loop implemented by MPGP (MPRGP).

Before applying the algorithms presented to the problems with a well-conditioned Hessian A, we strongly recommend rescaling the equality constraints so that $\|\mathsf{A}\| \approx \|\mathsf{B}\|$. Considering the convergence rate estimate in Theorem 12.6, it is also helpful to orthonormalize or at least normalize the constraints. We strongly recommend using adaptive reorthogonalization.

We can add the stopping criterion not only after Step 1 but also into the procedure, which generates $\mathbf{x}^k$ in Step 1. In our experiments, we use

$$\|\nabla L(\mathbf{x}^k, \boldsymbol{\lambda}^k, \rho)\| \leq \varepsilon_g \|\mathbf{b}\| \quad \text{and} \quad \|\mathsf{B}\mathbf{x}^k\| \leq \varepsilon_f \|\mathbf{b}\|. \tag{12.130}$$

The relative precisions ε_f and ε_g should be judiciously determined. In our experiments, we often use for simplicity

$$\varepsilon = \varepsilon_g = \varepsilon_f.$$

Our stopping criterion in the inner loop reads

$$\|\mathbf{g}^P(\mathbf{y}^i, \boldsymbol{\lambda}^k, \rho)\| \leq \min\{M_k \|\mathsf{B}\mathbf{y}^i - \mathbf{c}\|, \eta\} \quad \text{or} \quad (9.57). \tag{12.131}$$

Thus, the inner loop is interrupted when the algorithm finds the solution or generates its new approximation $\mathbf{x}^k = \mathbf{y}^i$.

The SMALSE-M algorithm requires four parameters M_0, β, η, and ρ. The algorithm is not very sensitive to the choice of β and η and properly adjusts the balancing parameter M_k. A larger value of ρ increases the rate of convergence of SMALSE at the cost of slowing down the convergence in the inner loop. We give some hints concerning the initialization of the parameters and the implementation of SMALSE-M in Sect. 12.8.

The parameter η defines a fixed bound on the feasibility error. The algorithm is not sensitive concerning η; we use

$$\eta = 0.1\|\mathbf{b}\|.$$

The parameter β controls the increase of the precision control. Our experience indicates that $\beta = 0.2$ is a reasonable choice.

The regularization parameter ρ should compromise the fast speed of the outer loop with large ρ and slow convergence of the algorithm in the inner loop. For the problems arising from the dual formulation of the conditions of equilibrium of contact problems, the choice

$$\rho \approx \|\mathsf{A}\|$$

does not increase the upper bound on the spectrum of the Hessian of the augmented Lagrangian and seems to balance reasonably the speed of the inner and outer iterations. A little larger value of ϱ such as $\varrho = 2\|\mathsf{A}\|$ can speed up the convergence of the outer loop.

The basic strategy for initialization of M_0 should comply with the relation

$$M_k^2 < \rho \lambda_{\min}(\mathsf{A}), \tag{12.132}$$

which guarantees sufficient increase of the augmented Lagrangian. We use

$$M_0^2 \approx 100\rho\lambda_{\min}(\mathsf{A}), \tag{12.133}$$

which allows fast early updates of the Lagrange multipliers. Since the algorithm adjusts M_k automatically, the algorithm's performance is not sensitive to the choice of M_0.

We strongly recommend using adaptive reorthogonalization.

12.9 Numerical Experiments

We first illustrate the performance of SMALBE-M on minimization of the functions $f_{\mathrm{L},h}$ and $f_{\mathrm{LW},h}$ introduced in Sect. 8.6 subject to the multipoint constraints and the bound constraints used in our previous numerical tests. The experiments were carried out with the values of basic parameters equal to

$$M = 1, \quad \Gamma = 1, \quad \text{and} \quad \varrho = 10.$$

The norm of feasibility error is the norm of violation of the equality constraints. The other constraints are satisfied exactly in each iteration.

12.9.1 Balanced Reduction of Feasibility and Gradient Errors

Let us first show how SMALBE balances the norm of the feasibility error with the norm of the projected gradient on minimization of the quadratic function $f_{\mathrm{L},h}$ defined by the discretization parameter h (see Sect. 9.5.2) subject to the multipoint equality constraints introduced in Sect. 9.5.1 and the bound constraints defined in Sect. 10.8.1. The solution given in Fig. 12.2 illustrates also the solution of the benchmarks in Sects. 9.5.1 and 10.8.1.

The graph of the norms of the projected gradient and the feasibility error (vertical axis) in inner iterations for SMALBE is in Fig. 12.3. The results were obtained with $h = 1/33$, corresponding to $n = 1156$ unknowns and 131 equality constraints. We can see that the decrease in the projected gradient's norm is linear and proportional to the feasibility error norm. Let us recall that the Hessian $\mathsf{A}_{\mathrm{L},h}$ of $f_{\mathrm{L},h}$ is ill-conditioned with the spectral condition number $\kappa(\mathsf{A}_{\mathrm{L},h}) \approx h^{-2}$.

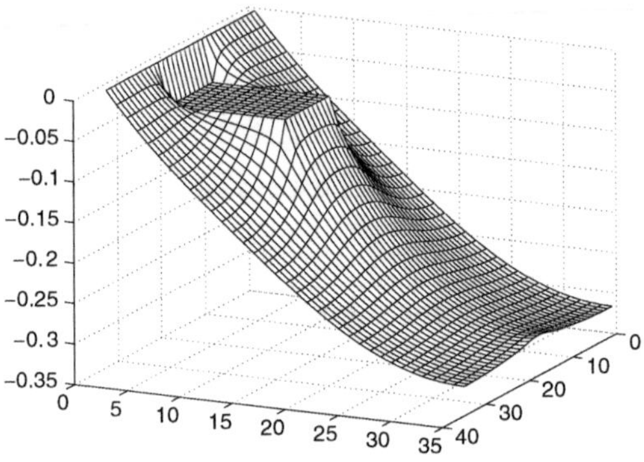

Fig. 12.2. Development of the norms of the projected gradient and feasibility error

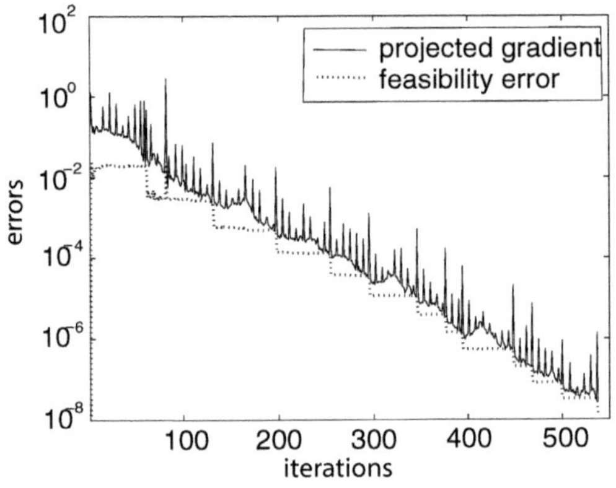

Fig. 12.3. Development of the norms of the projected gradient and feasibility error

12.9.2 Adaptive Augmentation—3D Hertz Problem

We tested the performance of the adaptive augmentation on the solution of the Hertz problem depicted in Fig. 12.4 by the TFETI domain decomposition method [104]. The benchmark appeared in Dostál et al. [109] and Dostál and Vlach [86]. See also the book [114]. The bodies were decomposed into 2×1000 subdomains, each discretized by 6^3 brick trilinear elements. The discretized problem was defined by 2,058,000 nodal variables, 3721 inequality constraints describing the nonpenetration, and 64,679 equality constraints describing the "gluing" of subdomains and discretized boundary conditions. The dimension of the dual bound and equality constrained problem was 718, 440 with 12,000 equality constraints and 3721 bound constraints, 3225 of which were active at the solution.

To check the effect of adaptive augmentation, we resolved our benchmark with the reorthogonalization invoked after each N_{reo} step, always starting after the first iteration, $N_{reo} \in \{1, 3, 5, \infty\}$, where ∞ stands for the original

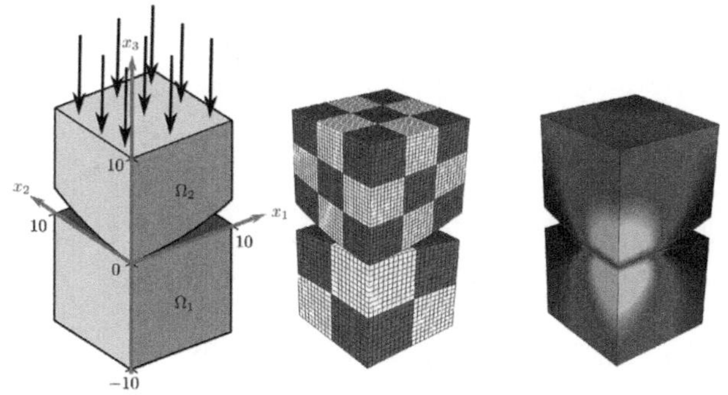

Fig. 12.4. 3d Herz example—setting (*left*); decomposition case (*middle*); von Mises stress (*right*)

SMALBE-M algorithm without reorthogonalization. Thus, the computations with step $N_{reo} = 5$ invoked the reorthogonalization after the 1st, 6th, etc. iterations. However, we skipped the orthogonalization when there was no change in the active set. We performed the computations to the relative precision $\varepsilon = 10^{-8}$.

	$N_{reo} = 1$	$N_{reo} = 3$	$N_{reo} = 5$	$N_{reo} = \infty$
Reorthogonalization steps	5/ 5	3/4	2/3	0/0
Outer iterations	8/10	10/13	10/13	64/121
Matrix × vector	91/121	117/161	119/158	333/557

Table 12.1. Effect of reorthogonalization ($\varepsilon = 10^{-5}/10^{-8}$)

The effects of reorthogonalization on the number of outer and inner iterations that are necessary to achieve the relative precision

$$\varepsilon_{out} = 10^{-5}, \quad \varepsilon_{in} = 10^{-5}/10^{-8}$$

iare in Table 12.1 and Figs. 12.5, 12.6, 12.7, and 12.8.

The results confirm that even straightforward implementation of the reorthogonalization strategy can considerably improve the algorithm's performance. An important class of problems can be solved much faster using the proposed reorthogonalization strategy. Some additional improvement can be achieved by implementing a more sophisticated update strategy depending on the cost of reorthogonalization. To illustrate the performance of SMALSE-M with reorthogonalization, let us point out that after identifying the active set of the solution, the algorithm required 34 conjugate gradient iterations to reduce the free gradient by 10^{-3}. We deduce that solving the related linear problem to the relative precision 10^{-8} would require some 90 conjugate iterations – nonlinearity was resolved at a cost smaller than two Newton steps. Notice that one nonlinear step of MPRGP requires two matrix-vector multiplications, and no FETI-type preconditioner was applied.

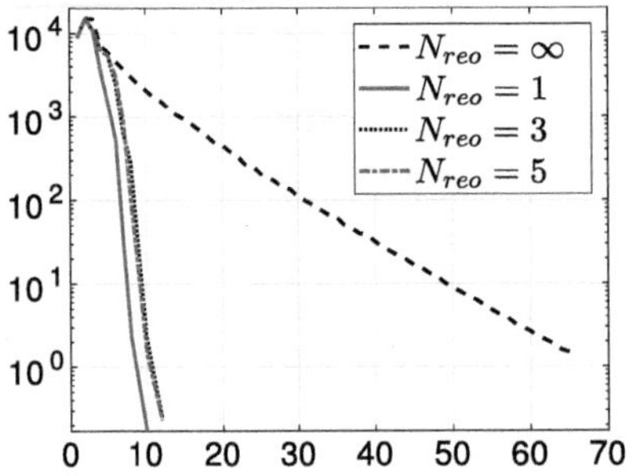

Fig. 12.5. $\|\mathsf{B}\mathbf{x}\|$ vs outer iterations

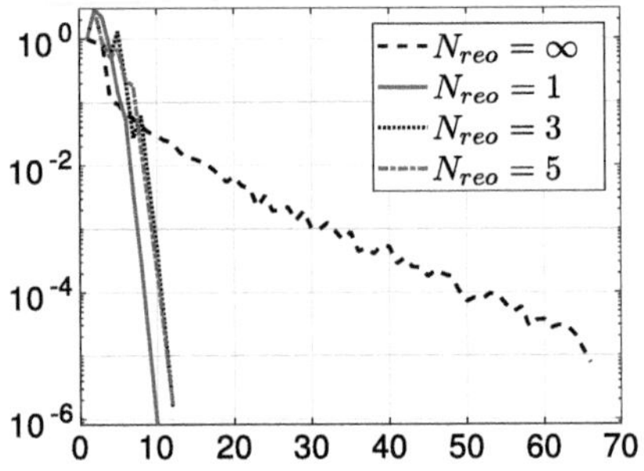

Fig. 12.6. $\|\mathbf{g}^P\|/\|\mathbf{g}_0^P\|$ vs outer iterations

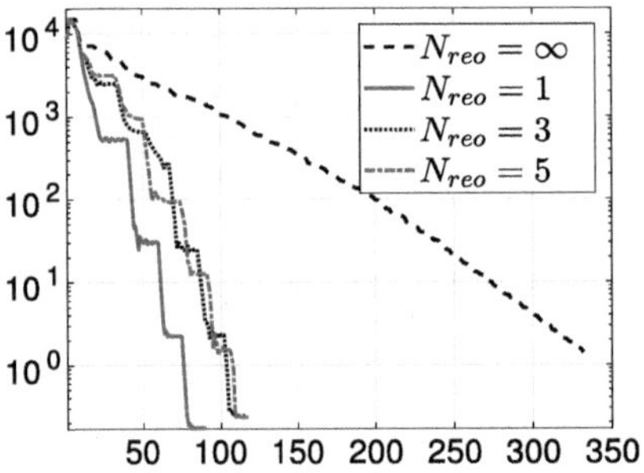

Fig. 12.7. $\|\mathsf{B}\mathbf{x}\|$ vs Hessian multiplications

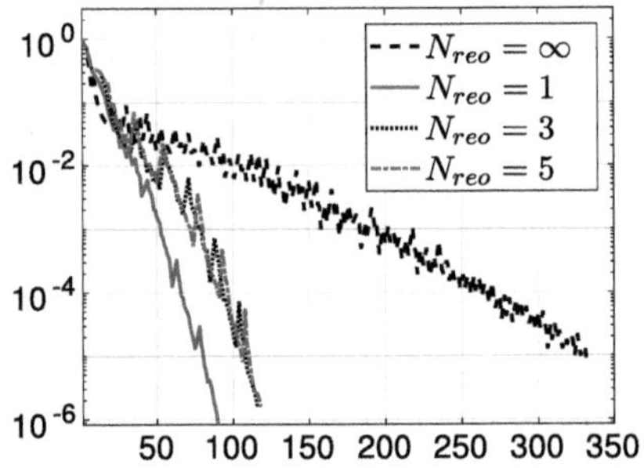

Fig. 12.8. $\|\mathbf{g}^P\|/\|\mathbf{g}_0^P\|$ vs Hessian multiplications

12.9.3 Numerical Demonstration of Optimality

To illustrate the optimality of SMALBE, we consider the class of well-conditioned problems to minimize the quadratic function $f_{\mathrm{LW},h}$ (see page 99) defined by the discretization parameter h subject to the orthonormal multipoint and bound constraints which describe the same feasible set as in Sect. 12.9.1.

The orthogonalization complies with the assumptions of Theorem 12.4. The matrices B of the equality constraints were obtained by the Gramm-Schmidt process applied to the matrices of the multipoint constraints of Sect. 9.5.1. The class of problems can be given a mechanical interpretation associated with the expanding and partly stiff spring systems on Winkler's foundation and an obstacle. The spectrum of the Hessian $\mathsf{A}_{\mathrm{LW},h}$ of $f_{\mathrm{LW},h}$ is located in the interval [2, 10]. Moreover

$$\|\mathsf{B}\| \leq 1 \quad \text{and} \quad \boldsymbol{\ell} \leq \mathbf{o}$$

(see Sect. 10.8.2), so that the assumptions of Theorem 12.4 are satisfied.

In Fig. 12.9, we can see the numbers of the CG iterations k_n (vertical axis) that were necessary to reduce the norm of the projected gradient and of the feasibility error to

$$\varepsilon = 10^{-6}\|\nabla f_{\mathrm{LW},h}(\mathbf{o})\|$$

for the problems with the dimension n ranging from 49 to 9409. The dimension n is on the horizontal axis. We can see that k_n varies mildly with varying n, in agreement with Theorem 12.4. The number of outer iterations decreased from 11 for $n = 49$ to 7 for $n = 9409$.

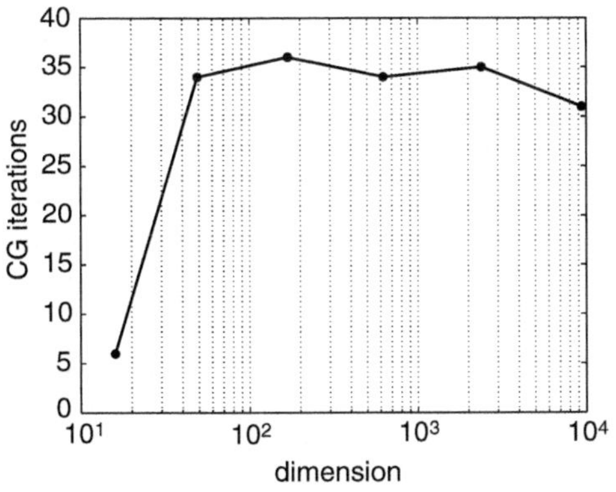

Fig. 12.9. Optimality of SMALBE for a class of well-conditioned problems

12.10 Comments

The development of SMALSE-M was inspired by the algorithm introduced by Conn et al. [39–41]; they adapted the augmented Lagrangian method of Powell [233] and Hestenes [163] to the solution of problems with a nonlinear cost function subject to nonlinear equality constraints and bound constraints. Conn, Gould, and Toint worked with the increasing penalty parameter and used a forcing series to control the precision of the auxiliary inequality-constrained problems. They proved that their algorithm converges and the potentially troublesome penalty parameter ρ_k is bounded [39]. Moreover, they used their algorithm to develop the LANCELOT [40] package for solving more general nonlinear optimization problems. More references are in their comprehensive book on trust region methods [41]. See also Bertsekas [15] or Glowinski and Le Tallec [144]. An excellent reference for applying augmented Lagrangian to more general problems is Birgin and Martínez [19]. For the augmented Lagrangian method with filter, see Friedlander and Leyfer [137].

The next improvement of augmented Lagrangian method algorithms included the adaptive precision control of the inner loop by the feasibility error, as introduced by Hager [159] and Dostál et al. [90]. These authors also proved basic convergence results for problems with a regular solution, including the linear convergence of Lagrange multipliers and feasibility error for a large initial penalty parameter ρ_0. The algorithm presented in [99] generated a forcing sequence for the increase of penalty parameter, which was sufficient to get the convergence results but spoiled any attempts to control the augmented Lagrangian Hessian's conditioning.

The subsequent modification, the update rule for the parameter ρ_k of the original SMALBE, which enforced a sufficient monotonic increase of $L(\mathbf{x}^k, \boldsymbol{\lambda}^k, \rho_k)$, was first published by Dostál [68]. The convergence analysis included the optimality of the outer loop and the bound on the penalty parameter.

We mentioned the possibility of keeping the regularization parameter fixed in the book [75, Chap. 6]. The algorithm for the bound and equality constraints, which keeps the regularization parameter fixed and uses the balancing parameter M for the precision control of auxiliary problems, was coined SMALBE-M. This innovation was necessary for the control of conditioning of the Hessian of the augmented Lagrangian and experimental demonstration of the numerical scalability of the algorithms for multibody contact problems. The analysis shows that it is possible to combine both strategies.

The performance of both SMALBE-M and SMALSE-M can be essentially improved by adaptive reorthogonalization [109].

The first optimality results for the bound and equality-constrained problems were proved by Dostál and Horák for the penalty method [79, 80]. The proof of the optimality of SMALBE with the auxiliary problems solved by MPRGP is in Dostál [70]; the generalization of the earlier results proved for the penalty method relied on a well-known observation that the basic augmented Lagrangian algorithm can be considered as a variant of the penalty method (see, e.g., Bertsekas [15, Sect. 4.4]). The presentation of SMALSE-M given here is based on Dostál and Kučera [82], and Dostál and Kozubek [81]. The SMALSE-Mw algorithm adapted for the elliptic constraints with strong excentricity appeared in Bouchala et al. [27]. The proof of linear convergence for SMALBE-M is in Dostál et al. [108]. The adaptive reorthogonalization was proposed recently by Dostál et al. [109], see also Dostál and Vlach [86].

Part IV
Case Studies

13
Elliptic Variational Inequalities

Variational inequalities describe a wide range of problems where one seeks a solution that minimizes some functional subject to certain constraints [117]. They provide a big source of QP and QCQP problems. Here, we consider their application to the formulation and solution of frictionless contact problems in mechanics.

We use the *FETI* (finite element tearing and interconnecting) methods introduced by Farhat and Roux [126] for unconstrained problems. The basic idea is to decompose the domain into subdomains connected by Lagrange multipliers and then eliminate the primal variables to get a small coarse problem and local problems, which we can solve in parallel. The procedure simultaneously transforms the general inequality constraints into bound constraints and improves the conditioning of the cost function's Hessian.

The bottleneck of the original FETI is caused by the coarse problem, which has a dimension proportional to the number of subdomains. The coarse problem is typically solved by a direct solver—its cost is negligible for a small number of subdomains, but it starts to dominate when the number of subdomains is large, currently some tens of thousands. The remedy offers the *hybrid FETI* (H-FETI-DP) method, which reduces the dimension of the coarse grid by interconnecting some adjacent subdomains by primal constraints.

Here, we introduce a model problem, the semi-coercive scalar variational inequality. We describe its discretization and decomposition into subdomains and clusters, reduce the problem by duality to the bound and equality-constrained QP problem, and report some theoretical and numerical results. The results show the unique efficiency of the SMALBE/MPRGP algorithms combined with the FETI methods. The numerical experiments include the bound and/or equality QP problems discretized by billions of nodal variables and a challenging roller-bearing problem.

Z. Dostál, *Optimal Quadratic Programming and QCQP Algorithms with Case Studies*, Springer Optimization and Its Applications 23,
https://doi.org/10.1007/978-3-031-95167-1_13

13.1 Model Problem

To explain the basic principles of domain decomposition methods for solving variational inequalities, we shall use a simple model problem, but our reasoning is also valid for more general cases. Let $\Omega = \Omega^1 \cup \Omega^2$, where $\Omega^1 = (0,1) \times (0,1)$ and $\Omega^2 = (1,2) \times (0,1)$ denote square domains with the boundaries Γ^1, Γ^2; their parts Γ_u^i, Γ_f^i, Γ_c^i are formed by the sides of Ω^i, $i = 1,2$, as in Fig. 13.1.

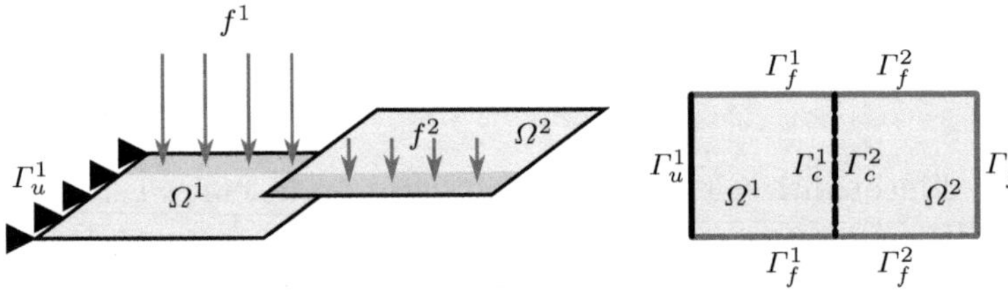

Fig. 13.1. Coercive model problem (left) and boundary conditions (right)

Let $H^1(\Omega^i)$, $i = 1,2$, denote the subspace of $L^2(\Omega^i)$ of elements with the first derivatives in $L^2(\Omega^i)$. Let

$$V^i = \{v^i \in H^1(\Omega^i) : v^i = 0 \quad \text{on} \quad \Gamma_u^i\}$$

denote closed subspaces of $H^1(\Omega^i)$, let $\mathcal{H} = H^1(\Omega^1) \times H^1(\Omega^2)$, and let

$$V = V^1 \times V^2 \qquad \text{and} \qquad \mathcal{K} = \{(v^1, v^2) \in V : v^2 - v^1 \geq 0 \quad \text{on} \quad \Gamma_c\}$$

denote a closed subspace and a closed convex subset of $\mathcal{H}$, respectively. The relations on the boundaries are in terms of traces. On $\mathcal{H}$, we define a symmetric bilinear form

$$a(u,v) = \sum_{i=1}^{2} \int_{\Omega^i} \left(\frac{\partial u^i}{\partial x} \frac{\partial v^i}{\partial x} + \frac{\partial u^i}{\partial y} \frac{\partial v^i}{\partial y} \right) \mathrm{d}\Omega$$

and a linear form

$$\ell(v) = \sum_{i=1}^{2} \int_{\Omega^i} f^i v^i \mathrm{d}\Omega,$$

where $f^i \in L^2(\Omega^i)$, $i = 1,2$, are nonzero and nonpositive. Thus we can define a problem to find

$$\min \ q(u) = \frac{1}{2} a(u,u) - \ell(u) \ \text{ subject to } \ u \in \mathcal{K}. \tag{13.1}$$

The solution of the model problem can be interpreted as the displacement of two membranes under the traction f. The left edge of the right membrane cannot penetrate below the right edge of the left membrane.

13.2 Domain Decomposition and Discretization

To enable efficient application of domain decomposition methods, we optionally decompose each Ω^i into $p = 1/H_s \times 1/H_s$, $i = 1, 2$, square subdomains. Misusing a little notation, we assign to each subdomain of Ω^1 a unique number $i \in \{1, \ldots, p\}$ and to each subdomain of Ω^2 a unique number $i \in \{p+1, \ldots, s\}$, $s = 2p$, as in Fig. 13.2. We call H_s a *decomposition parameter.*

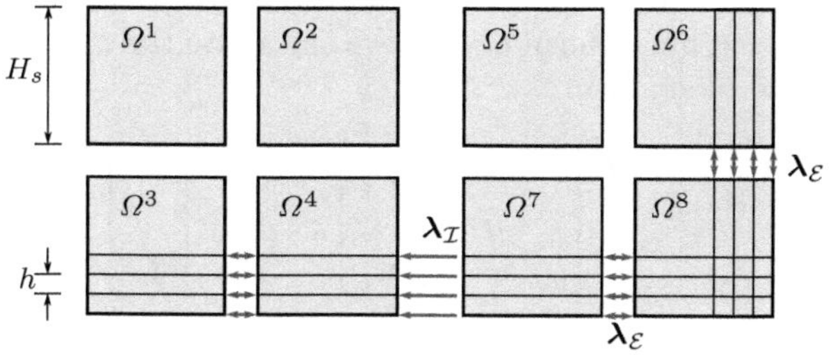

Fig. 13.2. Domain decomposition and discretization

To get a variational formulation of the decomposed problem, let

$$V_D^i = \{v^i \in H^1(\Omega^i) : v^i = 0 \quad \text{on} \quad \Gamma_U \cap \Gamma^i\}, \quad i = 1, \ldots s,$$

denote the closed subspaces of $H^1(\Omega^i)$, and let

$$V_D = V_D^1 \times \cdots \times V_D^s,$$
$$\mathcal{K}_D^C = \{v \in V_D : v^j - v^i \geq 0 \ \text{ on } \ \Gamma_C^1 \cap \Gamma_C^2, \ \ i \leq p < j\},$$
$$\mathcal{K}_D = \{v \in \mathcal{K}_D^C : v^i = v^j \ \text{ on } \ \Gamma^{ij}\}, \quad \Gamma^{ij} = \overline{\Gamma}^i \cap \overline{\Gamma}^j, \quad i,j \leq p \text{ or } \quad i,j > p.$$

On V_D, we define the scalar product

$$(u, v)_D = \sum_{i=1}^{s} \int_{\Omega^i} u^i v^i \mathrm{d}\Omega,$$

and the forms

$$a_D(u, v) = \sum_{i=1}^{s} \int_{\Omega^i} \left(\frac{\partial u^i}{\partial x_1} \frac{\partial v^i}{\partial x_1} + \frac{\partial u^i}{\partial x_2} \frac{\partial v^i}{\partial x_2} \right) \mathrm{d}\Omega \quad \text{and} \quad \ell_D(v) = (f, v)_D.$$

Using the above notation, it is a standard exercise [114, Sect. 10.2] to prove that (13.1) is equivalent to the problem to find $u \in \mathcal{K}_D$ such that

$$q_D(u) \leq q_D(v), \quad q_D(v) = \frac{1}{2} a_D(v, v) - \ell_D(v), \quad v \in \mathcal{K}_D. \tag{13.2}$$

After introducing regular grids with the discretization parameter h in the subdomains Ω^i (see Fig. 13.2), and using Lagrangian finite elements for the discretization, we get the discretized version of problem (13.2) with auxiliary domain decomposition

$$\min \frac{1}{2} \mathbf{u}^T \mathsf{K} \mathbf{u} - \mathbf{f}^T \mathbf{u} \quad \text{s.t.} \quad \mathsf{B}_I \mathbf{u} \leq \mathbf{o} \quad \text{and} \quad \mathsf{B}_E \mathbf{u} = \mathbf{o}. \tag{13.3}$$

In (13.3), $\mathsf{K} \in \mathbb{R}^{n \times n}$ denotes a block diagonal SPS (symmetric positive semidefinite) stiffness matrix, the full rank matrices B_I and B_E describe the non-penetration and interconnecting conditions, respectively, and $\mathbf{f}$ represents the discretized linear form $\ell_D(u)$. We can write the stiffness matrix and the vectors in the block form

$$\mathsf{K} = \begin{bmatrix} \mathsf{K}_1 & \mathsf{O} & \dots & \mathsf{O} \\ \mathsf{O} & \mathsf{K}_2 & \dots & \mathsf{O} \\ \dots & \dots & \dots & \dots \\ \mathsf{O} & \mathsf{O} & \dots & \mathsf{K}_s \end{bmatrix}, \quad \mathbf{u} = \begin{bmatrix} \mathbf{u}_1 \\ \dots \\ \mathbf{u}_s \end{bmatrix}, \quad \mathbf{f} = \begin{bmatrix} \mathbf{f}_1 \\ \dots \\ \mathbf{f}_s \end{bmatrix}, \quad s = 2p.$$

After a suitable scaling of the rows of $\mathsf{B} = [\mathsf{B}_E^T, \mathsf{B}_I^T]^T$, we can achieve $\mathsf{B}\mathsf{B}^T = \mathsf{I}$.

13.3 TFETI Quadratic Programming Problem

To reduce the problem to the subdomain boundaries using duality theory, let us introduce the Lagrangian associated with the problem (13.3) by

$$L(\mathbf{u}, \boldsymbol{\lambda}_I, \boldsymbol{\lambda}_E) = \frac{1}{2} \mathbf{u}^T \mathsf{K} \mathbf{u} - \mathbf{f}^T \mathbf{u} + \boldsymbol{\lambda}_I^T \mathsf{B}_I \mathbf{u} + \boldsymbol{\lambda}_E^T \mathsf{B}_E \mathbf{u}, \tag{13.4}$$

where $\boldsymbol{\lambda}_I$ and $\boldsymbol{\lambda}_E$ are the Lagrange multipliers associated with the inequalities and equalities, respectively. Introducing the notation

$$\boldsymbol{\lambda} = \begin{bmatrix} \boldsymbol{\lambda}_I \\ \boldsymbol{\lambda}_E \end{bmatrix} \quad \text{and} \quad \mathsf{B} = \begin{bmatrix} \mathsf{B}_I \\ \mathsf{B}_E \end{bmatrix},$$

we can observe that $\mathsf{B} \in \mathbb{R}^{m \times n}$ is a full rank matrix and write the Lagrangian as

$$L(\mathbf{u}, \boldsymbol{\lambda}) = \frac{1}{2} \mathbf{u}^T \mathsf{K} \mathbf{u} - \mathbf{f}^T \mathbf{u} + \boldsymbol{\lambda}^T \mathsf{B} \mathbf{u}.$$

The solution satisfies the KKT conditions, including

$$\mathsf{K}\mathbf{u} - \mathbf{f} + \mathsf{B}^T\boldsymbol{\lambda} = \mathbf{o}. \tag{13.5}$$

Equation (13.5) has a solution if and only if $\mathbf{f} - \mathsf{B}^T\boldsymbol{\lambda} \in \mathrm{Im}\mathsf{K}$, which can be expressed by means of a matrix R the columns of which span the null space of K as

$$\mathsf{R}^T(\mathbf{f} - \mathsf{B}^T\boldsymbol{\lambda}) = \mathbf{o}. \tag{13.6}$$

The matrix R can be formed directly, and $\mathsf{R}^T\mathsf{B}^T$ is non-singular.

Now assume that $\boldsymbol{\lambda}$ satisfies (13.6), so that we can evaluate $\boldsymbol{\lambda}$ from (13.5) by means of left generalized inverse matrix K^+ which satisfies $\mathsf{K}\mathsf{K}^+\mathsf{K} = \mathsf{K}$. We can verify directly that if $\mathbf{u}$ solves (13.5), then there is a vector $\boldsymbol{\alpha}$ such that

$$\mathbf{u} = \mathsf{K}^+(\mathbf{f} - \mathsf{B}^T\boldsymbol{\lambda}) + \mathsf{R}\boldsymbol{\alpha}. \tag{13.7}$$

After eliminating the primal variables $\mathbf{u}$, we can find $\boldsymbol{\lambda}$ by solving

$$\min \theta(\boldsymbol{\lambda}) \quad \text{s.t.} \quad \boldsymbol{\lambda}_{\mathcal{I}} \geq \mathbf{o} \quad \text{and} \quad \mathsf{R}^T(\mathbf{f} - \mathsf{B}^T\boldsymbol{\lambda}) = \mathbf{o}, \tag{13.8}$$

where

$$\theta(\boldsymbol{\lambda}) = \frac{1}{2}\boldsymbol{\lambda}^T\mathsf{B}\mathsf{K}^+\mathsf{B}^T\boldsymbol{\lambda} - \boldsymbol{\lambda}^T\mathsf{B}\mathsf{K}^+\mathbf{f}. \tag{13.9}$$

Once the solution $\widehat{\boldsymbol{\lambda}}$ of (13.8) is known, $\widehat{\mathbf{u}}$ (13.3) can be evaluated by (13.7) and

$$\boldsymbol{\alpha} = -(\mathsf{R}^T\widehat{\mathsf{B}}^T\widehat{\mathsf{B}}\mathsf{R})^{-1}\mathsf{R}^T\widehat{\mathsf{B}}^T\widehat{\mathsf{B}}\mathsf{K}^+(\mathbf{f} - \mathsf{B}^T\widehat{\boldsymbol{\lambda}}), \tag{13.10}$$

where $\widehat{\mathsf{B}} = [\widehat{\mathsf{B}}_I^T, \mathsf{B}_E^T]^T$, and the matrix $\widehat{\mathsf{B}}_I$ is formed by the rows $\mathbf{b}_i$ of B_I that correspond to the positive components of the solution $\widehat{\boldsymbol{\lambda}}_I$ characterized by $\widehat{\lambda}_i > 0$. A more effective procedure avoiding manipulation with $\widehat{\mathsf{B}}$ can be found in [170].

To proceed further, let us denote

$$\begin{aligned} \mathsf{F} &= \mathsf{B}\mathsf{K}^+\mathsf{B}^T = \widetilde{\mathsf{B}}\mathsf{S}^+\widetilde{\mathsf{B}}^T, & \widetilde{\mathbf{d}} &= \mathsf{B}\mathsf{K}^+\mathbf{f}, \\ \widetilde{\mathsf{G}} &= \mathsf{R}^T\mathsf{B}^T, & \widetilde{\mathbf{e}} &= \mathsf{R}^T\mathbf{f}, \end{aligned}$$

and let T defines orthonormalization of the rows of $\widetilde{\mathsf{G}}$ so that the matrix $\mathsf{G} = \mathsf{T}\widetilde{\mathsf{G}}$ has orthonormal rows. After introducing $\mathbf{e} = \mathsf{T}\widetilde{\mathbf{e}}$, problem (13.8) reads

$$\min \quad \frac{1}{2}\boldsymbol{\lambda}^T\mathsf{F}\boldsymbol{\lambda} - \boldsymbol{\lambda}^T\widetilde{\mathbf{d}} \quad \text{s.t.} \quad \boldsymbol{\lambda}_I \geq \mathbf{o} \quad \text{and} \quad \mathsf{G}\boldsymbol{\lambda} = \mathbf{e}. \tag{13.11}$$

After homogenization of the equality constraints and introducing orthogonal projectors, problem (13.11) turns into

$$\min \quad \overline{\theta}_\varrho(\boldsymbol{\lambda}) \quad \text{s.t.} \quad \mathsf{G}\boldsymbol{\lambda} = \mathbf{o} \quad \text{and} \quad \boldsymbol{\lambda}_I \geq -\widetilde{\boldsymbol{\lambda}}_I, \tag{13.12}$$

where ϱ is a positive constant, $\mathsf{G}\boldsymbol{\lambda}_I = \mathbf{e}$, and

$$\overline{\theta}_\varrho(\boldsymbol{\lambda}) = \frac{1}{2}\boldsymbol{\lambda}^T\mathsf{H}_\varrho\boldsymbol{\lambda} - \boldsymbol{\lambda}^T\mathsf{P}\mathbf{d}, \quad \mathsf{H}_\varrho = \mathsf{PFP} + \varrho\mathsf{Q}, \quad \mathsf{Q} = \mathsf{G}^T\mathsf{G}, \quad \mathsf{P} = \mathsf{I} - \mathsf{Q}.$$

The matrices P and Q are the orthogonal projectors onto $\mathrm{Ker}\mathsf{G}$ and $\mathrm{Im}\mathsf{G}^T$, respectively. It has been proved (see, e.g., Brenner [29] or Pechstein [226]) that there are constants $0 < c < C$ that depend neither on h nor H such that

$$c \le \lambda_{\min}(\mathsf{H}_\varrho) \le \max\{CH/h, \varrho\}.$$

13.4 Hybrid TFETI Problem and Optimality

The bottleneck of classical FETI methods is the rank d of the projector Q which is equal to the defect of stiffness matrix K, in our case $d = s$. To reduce the rank of Q, we use the idea of Klawonn and Rheinbach [183] to interconnect some subdomains on the primal level into clusters so that the defect of the stiffness matrix of the cluster is equal to the defect of one of the subdomain stiffness matrices.

For example, to couple subdomains with common corners $\mathbf{x}, \mathbf{y} \in \overline{\Omega}^i \cap \overline{\Omega}^j$, we can transform the nodal variables associated with $\widetilde{\Omega}^q = \overline{\Omega}^i \times \overline{\Omega}^j$ by the expansion matrix L^q obtained by replacing two columns of the identity matrix associated with $\mathbf{x}, \mathbf{y}$ by one column obtained as a normalized sum of the columns associated with the displacements of nodes $\mathbf{x}$ and $\mathbf{y}$. Feasible variables $\mathbf{u}^q$ of the cluster are related to global variables $\widetilde{\mathbf{u}}^q$ by $\mathbf{u}^q = \mathsf{L}^q\widetilde{\mathbf{u}}^q$ and the stiffness matrix $\widetilde{\mathsf{K}}^q$ of such cluster in global variables can be obtained by

$$\widetilde{\mathsf{K}}^q = (\mathsf{L}^q)^T \mathrm{diag}(\mathsf{K}^i, \mathsf{K}^j)\mathsf{L}^q.$$

Let us denote by $\mathbf{e}$ and $\overline{\mathbf{e}}$ the vectors with all components equal to 1 and $1/\|\mathbf{e}\|$, respectively. To describe the coupling by averages, we use the transformation of bases proposed by Klawonn and Widlund [185], see also Klawonn and Rheinbach [184]. The basic idea is a rather trivial observation that if

$$[\mathsf{C}\ \overline{\mathbf{e}}] = [\mathbf{c}_1, \ldots, \mathbf{c}_{p-1}, \overline{\mathbf{e}}], \quad \overline{\mathbf{e}} = \frac{1}{\sqrt{p}}\mathbf{e},$$

denote an orthonormal basis of $\mathbb{R}^p$, then the last coordinate of a vector $\mathbf{x} \in \mathbb{R}^p$ in this basis is given by $x_p = \overline{\mathbf{e}}^T\mathbf{x}$. If we apply the transformation to the variables associated with the interiors of adjacent edges, we can join them by the expansion mapping L as above.

The procedure can be generalized to specify the feasible vectors of any cluster connected by the edge averages of adjacent edges. Using a proper numbering of variables by subdomains, in each subdomain setting first the variables that are not affected by the interconnecting, then the variables associated with the averages ordered by edges, we get the matrix Z with orthonormal columns the range of which represents the feasible displacements of the cluster,

$$\mathsf{Z} = \begin{bmatrix} \mathsf{C} \; \mathsf{E} \end{bmatrix}, \; \mathsf{C} = \mathrm{diag}(\mathsf{C}^1, \dots \mathsf{C}^s), \; \mathsf{E} = 1/\sqrt{2} \begin{bmatrix} \cdot & & \cdot & \cdot \\ \cdot & \overline{\mathbf{e}}^{ij} & & \cdot \\ \cdot & & \cdot & \cdot \\ \cdot & \overline{\mathbf{e}}^{ji} & & \cdot \\ \cdot & & \cdot & \cdot \end{bmatrix}, \; (i,j) \in \mathcal{C}. \quad (13.13)$$

In (13.13), $\mathcal{C}$ denotes a set of ordered couples of indices (i,j), $i < j$, s here denotes the number of subdomains in the cluster, and $\overline{\mathbf{e}}^{ij}$ denotes the basis vectors associated with the edge averages. Each couple $(i,j) \in \mathcal{C}$ defines the connection of the adjacent edges of Ω^i and Ω^j by averages. The procedure is very similar to that described in the introduction of this section; the only difference is that we replace the expansion matrix L^q by the basis of feasible displacements of the cluster Z^q. The feasible variables of the cluster are related to global variables $\widetilde{\mathbf{u}}^q$ by

$$\mathbf{u}^q = \mathsf{Z}^q \widetilde{\mathbf{u}}^q$$

and the Schur complement $\widetilde{\mathsf{S}}^q$ of such a cluster in global variables can be obtained by

$$\widetilde{\mathsf{S}}^q = (\mathsf{Z}^q)^T \mathrm{diag}(\mathsf{S}^i, \mathsf{S}^j, \dots, \mathsf{S}^\ell) \mathsf{Z}^q.$$

Assuming that the set of all subdomains is decomposed into c clusters interconnected by the edge averages, we can use the global transformation matrix with orthonormal columns

$$\mathsf{Z} = \mathrm{diag}(\mathsf{Z}^1, \cdots, \mathsf{Z}^c)$$

to connect the groups of $m \times m$ subdomains into clusters to get the stiffness matrix

$$\widetilde{\mathsf{S}} = \mathsf{Z}^T \mathsf{S} \mathsf{Z} = \mathrm{diag}(\widetilde{\mathsf{S}}^1, \cdots \widetilde{\mathsf{S}}^c)$$

and the matrices

$$\widetilde{\mathsf{B}}, \quad \widetilde{\mathsf{R}} = \mathrm{diag}(\widetilde{\mathbf{e}}^1, \dots, \widetilde{\mathbf{e}}^c), \quad \widetilde{\mathsf{G}} = \mathsf{T} \widetilde{\mathsf{R}}^T \widetilde{\mathsf{B}}^T,$$

where $\widetilde{\mathsf{B}}$ denotes a matrix that enforces interconnecting constraints that are not enhanced on the primal level and T denotes an orthogonalization matrix so that $\widetilde{\mathsf{G}}\widetilde{\mathsf{G}}^T = \mathsf{I}$. It is easy to achieve that

$$\widetilde{\mathsf{B}}\widetilde{\mathsf{B}}^T = \mathsf{I}. \quad (13.14)$$

Notice that $\widetilde{\mathsf{B}}$ enforces both constraints that connect subdomains into clusters and those connecting the clusters. Moreover, Ker$\widetilde{\mathsf{B}}$ =KerBZ, but BZ need not have orthonormal rows. Using the above transformation, we reduced problem (13.12) to

$$\min \quad \widetilde{\theta}_{\varrho}(\boldsymbol{\lambda}) \quad \text{s.t.} \quad \widetilde{\mathsf{G}}\boldsymbol{\lambda} = \mathbf{o} \quad \text{and} \quad \boldsymbol{\lambda}_I \geq -\widetilde{\boldsymbol{\lambda}}_I, \tag{13.15}$$

where ϱ is a positive constant and

$$\widetilde{\theta}_{\varrho}(\boldsymbol{\lambda}) = \frac{1}{2}\boldsymbol{\lambda}^T \widetilde{\mathsf{H}}_{\varrho}\boldsymbol{\lambda} - \boldsymbol{\lambda}^T \widetilde{\mathsf{P}}\widetilde{\mathbf{d}}, \tag{13.16}$$
$$\widetilde{\mathsf{H}}_{\varrho} = \widetilde{\mathsf{P}}\widetilde{\mathsf{F}}\widetilde{\mathsf{P}} + \varrho\widetilde{\mathsf{Q}}, \quad \widetilde{\mathsf{Q}} = \widetilde{\mathsf{G}}^T\widetilde{\mathsf{G}}, \quad \widetilde{\mathsf{P}} = \mathsf{I} - \widetilde{\mathsf{Q}} \quad \widetilde{\mathsf{F}} = \widetilde{\mathsf{B}}\widetilde{\mathsf{S}}^{+}\widetilde{\mathsf{B}}^T.$$

$\widetilde{\mathsf{P}}$ and $\widetilde{\mathsf{Q}}$ are the orthogonal projectors onto Ker$\widetilde{\mathsf{G}}$ and Im$\widetilde{\mathsf{G}}^T$, respectively. Notice that the number of the rows of G is m^2 times larger that that of $\widetilde{\mathsf{G}}$, so that the cost of $(\mathsf{G}^T\mathsf{G})^{-1}$ is about m^4 times larger than that of $(\widetilde{\mathsf{G}}\widetilde{\mathsf{G}})^{-1}$.

The following theorem provides the bounds on the spectrum of $\widetilde{\mathsf{H}}_{\varrho}$.

Theorem 13.1. *Let $\varrho \approx \|\widetilde{\mathsf{F}}\|$ and let the parameters $H_s, m,$ and h specify problem* (13.15). *Then there are constants $c, C > 0$ independent of H_s, m, h such that*

$$c \leq \lambda_{\min}(\widetilde{\mathsf{H}}_{\varrho}) \leq \|\widetilde{\mathsf{H}}_{\varrho}\| \leq CmH_s/h. \tag{13.17}$$

Proof. See Dostál et al. [110].

Let $C_1 > 2$ denote an arbitrary constant and consider a class of HTHETI problems (13.15) generated with the parameters H_s, m, h, such that

$$mH_s/h \leq C_1.$$

Then we can use Theorem 12.4 to get that there is a constant $M_{\max}$ such that the SMALBE algorithm with the inner loop implemented by MPRGP can find an approximate solution of any problem (13.15) in at most $M_{\max}$ matrix–vector multiplications. See also Dostál et al. [111].

13.5 Numerical Experiments

We performed numerical experiments to check the bounds and compare H-TFETI-DP with TFETI. Including a related linear problem enables assessing the effect of inequality constraints. If not specified, we use the default relative precision stopping criterion with $\varepsilon = 10^{-4}$.

13.5.1 Billions Equality Constrained QP Benchmark

Here, we give an example of the extreme efficiency of the dual formulation of the related equality-constrained QP problem. We present the results achieved with H-TFETI-DP and standard TFETI on solving the unit square Poisson benchmark discretized by Q1 finite elements on a regular grid with parameters h and H_s, $H_s/h = 100$ [110]. The domain was decomposed into $n_c \times n_c$ clusters, $n_c = 6, 18, 54$, each cluster comprising 15×15 square subdomains partly joined by edge averages. The equality constrains interconnect both the clusters into domains and the subdomains into clusters.

We used the ESPRESO (ExaScale PaRallel FETI SOlver) package [121] developed at the Czech National Supercomputing Center in Ostrava. Notice that H-TFETI-DP outperforms TFETI due to the small coarse problem and cheap iterations. The computation time is almost constant for the H-TFETI-DP, unlike TFETI, because of a bottleneck in the solution of TFETI coarse problems for a larger number of domains.

Table 13.1. Billion variables Poisson—unpreconditioned H-TFETI-DP and TFETI, m=15, see [110]

Clusters	Subdomains	Cores	Unknowns	H-TFETI-DP (iter/sec)	TFETI (iter/sec)
36	8100	108	81,018,001	117/26.0	45/14.5
324	72,900	972	729,054,001	118/27.7	42/40.2
2916	656,100	8.748	6,561,162,001	116/28.0	41/61.0

Remark 13.1. The original formulation of the benchmark is the unconstrained minimization of the energy function associated with the discrete SPD Laplacian. Both TFETI and H-TFETI-DP transform the problem to the minimization of the block-diagonal energy function subject to the equality constraints that interconnects the subdomains. The latter problem is transformed by duality to an unconstrained problem in Lagrange multipliers that is solved by the conjugate gradient method.

13.5.2 Model Variational Inequality

We used the above procedure to get the discretized H-TFETI-DP QP problem (13.1) that we solved by a combination of the SMALBE-M algorithm with the inner loop resolved by MPRGP. We implemented both algorithms in the PETSc based package PERMON [227] developed at the Department of Applied Mathematics of the VSB-Technical University of Ostrava and the Institute of Geonics of the Czech Academy of Science.

We can see that the number of the outer SMALBE iterations decreases with the coarse problem dimension, i.e., with the increasing number of subdomains in clusters. Since the outer iterations are cheap, the number of inner MPRGP iterations is more critical. The number of inner iterations is rather stable. Comparing Tables 13.1 and 13.2, we see that the non-penetration inequality constraints increased the cost of computation about two times. Let

Table 13.2. Semicoercive variational inequality, primal dimension 20,480,000, inequalities 3169, see [111]

m=	1	2	4	8
Outer iter	52	25	16	12
Matrix × vector	243	252	186	218
Coarse problem dimension	2048	512	128	32

us recall that the gradient iterations of MPRGP require two matrix-vector iterations. For details, see [111].

13.5.3 Elastic Cube on Rigid Obstacle

The variational inequalities that describe frictionless contact problems have the same structure as our model problem 13.1. Here, we consider a clamped cube over a sinus-shaped obstacle depicted in Fig. 13.3. It is loaded by its weight, decomposed into $4 \times 4 \times 4$ clusters, $H_s/h = 14$, using the ESPRESO [121] implementation of H-TFETI-DP for contact problems. We can see in Table 13.3 that TFETI needs fewer iterations, but H-TFETI-DP is still faster due to 64 times smaller coarse space and better exploitation of the node-core memory organization. In general, if we use $m \times m \times m$ clusters, the hybrid strategy reduces the dimension and cost of the coarse problem by m^6. For details, see [113].

Table 13.3. Clamped elastic cube over the sinus-shaped obstacle, $m = 4$, $H_s/h = 14$, see [112]

Clusters	Subdomains	Cores	Unknowns (10^6)	H-TFETI-DP (iter/sec)	TFETI (iter/sec)
64	4096	192	13	169/23.9	117/24.9
512	72,900	1536	99	208/30.2	152/115.1
1000	656,100	3000	193	206/42.6	173/279.9

We can see that the number of inner iterations is rather stable. Since the cost of each iteration is proportional to the dimension of the problem (in our case from 13,000,000 to 193,000,000 nodal variables), we conclude that the cost of the solution of the cube problem depends linearly on the dimension of the problem. This shows that SMALBE/MPRGP is in a sense optimal for solving such problems.

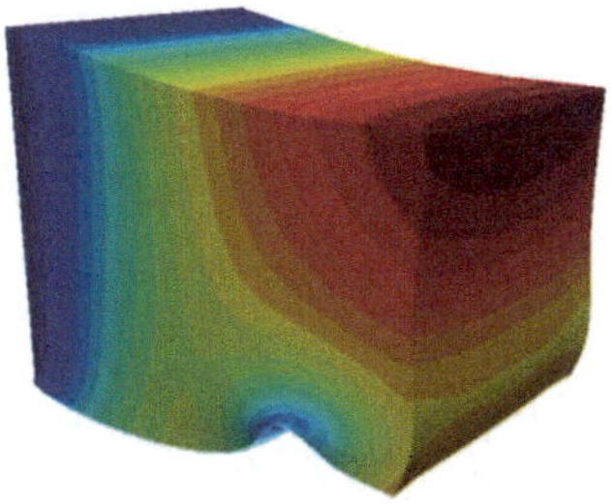

Fig. 13.3. Displacements of a clamped elastic cube over the sinus-shaped obstacle

13.5.4 Roller Bearings of Wind Generator

We have also tested the above algorithms on real-world problems, including the stress analysis in the roller bearings of a wind generator depicted in Fig. 13.4 (see [104]). The problem comprises 73 bodies in mutual contact; only one is fixed in space. The problem is complex because of balancing many "floating" bodies and the transfer of information through many highly nonlinear interfaces. Moreover, the solution is dual degenerate due to smooth contact interfaces.

Fig. 13.4. Roller bearings of wind generator (real bearing, decomposed model, and solution)

The solution to the problem discretized by 2,730,000/459,800 primal/dual variables and decomposed into 700 subdomains required 837 matrix–vector multiplications, including outer iterations for exact non-penetration. Though the number of iterations is not small, the parallel scalability of the algorithm enables to obtain the solution in a reasonable time.

13.6 Comments

The chapter shows how to enhance specific information about the problem in the solution algorithm.

The first observation is that the ill-conditioning and large dimension of the original QP problem can be reduced by decomposing the domain into smaller subdomains interconnected by equality constraints. The kernels of subdomain stiffness matrices can improve the conditioning of the modified system, as observed by Farhat and Roux [125]. The second observation concerns the feasible region that is sufficiently complicated to make the gradient projection expensive. The remedy is to switch to a dual problem. The last observation concerns the possibility of effectively inverting the stiffness matrices of partly interconnected subdomains that form clusters as proposed by Klawonn and Rheinbach [184]. Putting everything together results in H-TFETI-DP for variational inequalities [111].

14
Contact Problem with Friction

Contact problems with friction are much more complicated than the frictionless problems treated in the previous chapter. The most popular friction law, the Coulomb law, makes the problem intrinsically non-convex. As a result, we face new challenges, particularly the non-uniqueness of a solution, nonlinear inequality constraints in dual formulation, and intractable minimization formulation—moreover, the problems with friction share with the frictionless problems large dimension and ill-conditioned discretized operators.

Our first step toward the practical solution of the contact problem with Coulomb friction is the definition of a simplified friction law known as given (Tresca) friction. This friction model assumes that the normal pressure is a priori known on the contact interface. Such an assumption is unrealistic and violates the laws of physics, e.g., it admits positive contact pressure on the part of the contact interface that is not active in the solution, but it makes the problem well-posed. Moreover, solving the problem with Tresca's friction minimizes a convex non-differentiable cost function subject to a convex set defined by separable quadratic constraints.

Then we switch to the dual formulation to get a convex QCQP (Quadratically Constrained Quadratic Programming) problem with linear equality constraints and separable quadratic inequality constraints defined by the slip bounds. The structure of the problem complies well with TFETI, so we can achieve the final QCQP problem being well-conditioned and solvable with asymptotically linear complexity. Finally, we plugged the Tresca solver into the fix-point iterations to solve the original problem with Coulomb friction. We demonstrate the linear complexity of the Tresca solver by solving a two-block problem discretized by millions of nodal variables, recomputing the same problem with Coulomb friction, and presenting the realistic yielding clamp connection analysis.

Z. Dostál, *Optimal Quadratic Programming and QCQP Algorithms with Case Studies*, Springer Optimization and Its Applications 23,
https://doi.org/10.1007/978-3-031-95167-1_14

14.1 Coulomb Friction

Let us consider a system of elastic bodies $\Omega^1, \ldots, \Omega^s$ with the boundaries $\Gamma^1, \ldots \Gamma^s$, and let us examine the effect of friction on the equilibrium in a point on the tangential plane of a smooth contact interface, assuming it complies with the Coulomb friction law.

Let us denote by $\mathscr{S}$ the set of the ordered couples of indices (p, q), $p < q$, of the bodies that can come into contact. If $(p, q) \in \mathcal{S}$, then we define the mapping

$$\chi^{pq} : \Gamma^{pq} \to \Gamma^{qp}$$

which assigns to each $\mathbf{x} \in \Gamma^{pq}$ an approximate orthogonal projection $\chi^{pq}(\mathbf{x})$ of $\mathbf{x}$ onto Γ^{qp}. Assuming that the displacement

$$\mathbf{u} = (\mathbf{u}^1, \ldots, \mathbf{u}^s), \quad \mathbf{u}^p : \overline{\Omega^p} \to \mathbb{R}^3,$$

is sufficiently smooth, we can write the non-penetration contact conditions and the conditions of equilibrium in the normal direction $\mathbf{n}$ briefly as

$$[u_n] \le g, \quad \lambda_n \ge 0, \quad \lambda_n([u_n] - g) = 0, \quad \mathbf{x} \in \Gamma_C^{pq}. \tag{14.1}$$

Here λ_n denotes the normal traction on the slave face (i.e., $\Gamma^{pq} \subseteq \Gamma^p$) of the contact interface, $[u_n]$ denotes the jump of the normal components of boundary displacements, and g denote the initial gap

$$g = g(\mathbf{x}) \approx \|\chi^{pq}(\mathbf{x}) - \chi^{pq}(\mathbf{x})\|.$$

To formulate the conditions of equilibrium in the tangential plane, let us introduce the notation

$$\boldsymbol{\lambda}_T = \boldsymbol{\lambda} - \boldsymbol{\lambda}_N, \quad \mathbf{u}_N = u_n \mathbf{n}^p, \quad \mathbf{u}_T = \mathbf{u} - \mathbf{u}_N, \quad \mathbf{x} \in \Gamma_C^{pq},$$

and

$$[\mathbf{u}_T] = \mathbf{u}_T^p - \mathbf{u}_T^q \circ \chi^{pq}, \quad \mathbf{x} \in \Gamma_C^{pq}, \quad (p, q) \in \mathscr{S}.$$

Let $\Phi > 0$ denote the *friction coefficient*, which can depend on $\mathbf{x}$. In this case, we assume it is bounded away from zero. The *Coulomb friction law* at $\mathbf{x} \in \Gamma_C^{pq}$, $(p, q) \in \mathscr{S}$, can be written in the form

$$\text{if } [u_n] = g, \quad \text{then} \quad \lambda_n \ge 0, \tag{14.2}$$
$$\text{if } [u_n] = g \text{ and } \|\boldsymbol{\lambda}_T\| < \Phi\lambda_n, \quad \text{then} \quad \mathbf{u}_T = \mathbf{o}, \tag{14.3}$$
$$\text{if } \|\boldsymbol{\lambda}_T\| = \Phi\lambda_n, \text{ then there is } \mu > 0 \text{ such that } [\mathbf{u}_T] = \mu[\boldsymbol{\lambda}_T], \tag{14.4}$$

where

$$g = \left(\chi^{pq} - \mathsf{Id}\right) \cdot \mathbf{n}^p.$$

The relations (14.1)–(14.4) with the classical Dirichlet and Neumann boundary conditions and the condition of equilibrium describe completely the kinematics and equilibrium of a system of elastic bodies in contact obeying the Coulomb friction law.

14.2 Variational Formulation

The classical formulation of contact problems with friction law (14.1)–(14.4) by partial differential equations makes sense only when the solutions satisfy strong regularity assumptions similar to those required by the classical solutions of frictionless problems. This problem is addressed by a variational formulation that characterizes the solutions by local averages. We shall use tools similar to those in Sect. 13.1, but the symmetric positive semidefinite bilinear energy form $a(\mathbf{u}, \mathbf{v})$ is defined by the elasticity operator (see, e.g., [114, Chap. 12]). We shall use the linear functional $\ell(\mathbf{v})$, the spaces

$$\begin{aligned} V^p &= \left\{\mathbf{v} \in \left(H^1(\Omega^p)\right)^3 : \mathbf{v} = \mathbf{o} \ \text{ on } \ \Gamma_U^p\right\}, \ p = 1, \dots, s, \\ V &= V^1 \times \dots \times V^s, \end{aligned}$$

and the convex set

$$\mathscr{K} = \{\mathbf{v} \in V : [v_n] \le g \ \text{ on } \ \Gamma_C^{pq}, \ \ (p, q) \in \mathscr{S}\}.$$

The effect of friction will be enhanced by a functional j, which represents the virtual work of frictional forces

$$j(\mathbf{u}, \mathbf{v}) = \sum_{(p,q)\in\mathscr{S}} \int_{\Gamma_C^{pq}} \Phi|\lambda_n(\mathbf{u})| \, \|[\mathbf{v}_T]\| \, \mathrm{d}\Gamma, \quad \mathbf{u}, \mathbf{v} \in V. \tag{14.5}$$

The functional j is nonlinear and non-differentiable, which is a source of complications. It is possible to prove that the classical solution $\mathbf{u} \in \mathscr{K}$ of the contact problem with Coulomb's friction (14.1)–(14.4) satisfies

$$a(\mathbf{u}, \mathbf{v} - \mathbf{u}) - \ell(\mathbf{v} - \mathbf{u}) + j(\mathbf{u}, \mathbf{v}) - j(\mathbf{u}, \mathbf{u}) \ge 0, \quad \mathbf{v} \in \mathscr{K}. \tag{14.6}$$

The problem (14.6) is well defined, but the theoretical results concerning the existence and uniqueness of the solution require additional assumptions. Moreover, no true potential energy functional exists for the problem with friction. For more detailed discussion, see the books by Hlaváček et al. [168], Kikuchi and Oden [180], or Wohlmuth [263]. See also Dostál et al. [114, Chap. 12].

14.3 Tresca Friction and Dualization

Let us consider a simplification of the variational inequality (14.6) that can be reduced to a minimization problem with separable constraints, which can be solved effectively by the methods described in Chaps. 11 and 12.

The basic idea assumes that the normal traction λ_n on the contact interface is known. Though it is possible to contrive a physical situation that such a model can capture, the main motivation is the possibility of using the simplified model in the fixed-point iterations for solving contact problems with Coulomb's friction. If we assume that λ_n is known, we get the problem to find a sufficiently smooth displacement field

$$\mathbf{u} = (\mathbf{u}^1, \dots, \mathbf{u}^s), \quad \mathbf{u}^p : \overline{\Omega^p} \rightarrow \mathbb{R}^3,$$

such that for any $(p,q) \in \mathscr{S}$ and $\mathbf{x} \in \Gamma^{pq}$

$$\text{if } [u_n] = g, \quad \text{then} \quad \lambda_n \geq 0, \tag{14.7}$$

$$\text{if } [u_n] = g \quad \text{and} \quad \|\boldsymbol{\lambda}_T\| < \Psi, \quad \text{then} \quad \mathbf{u}_T = \mathbf{o}, \tag{14.8}$$

$$\text{if } \|\boldsymbol{\lambda}_T\| = \Psi, \quad \text{then there is } \mu > 0 \quad \text{such that} \quad [\mathbf{u}_T] = \mu[\boldsymbol{\lambda}_T], \tag{14.9}$$

where $\Psi \geq 0$ denotes a sufficiently smooth prescribed slip bound. After substituting into the friction functional j (14.5) and simplifying the notation, we get

$$j(\mathbf{v}) = \sum_{(p,q)\in\mathscr{S}} \int_{\Gamma_C^{pq}} \Psi \|[\mathbf{v}_T]\| \, d\Gamma, \tag{14.10}$$

and the variational problem (14.6) reduces to the problem to find $\mathbf{u} \in \mathscr{K}$ which satisfies

$$a(\mathbf{u}, \mathbf{v} - \mathbf{u}) - \ell(\mathbf{v} - \mathbf{u}) + j(\mathbf{v}) - j(\mathbf{u}) \geq 0, \quad \mathbf{v} \in \mathscr{K}. \tag{14.11}$$

Observing that

$$\Psi(\mathbf{x}) \|[\mathbf{v}_T(\mathbf{x})]\| = \max_{\|\boldsymbol{\tau}\| \leq \Psi(\mathbf{x})} \boldsymbol{\tau} \cdot [\mathbf{v}_T(\mathbf{x})],$$

we can write the non-differentiable term in (14.10) as

$$j(\mathbf{v}) = \sum_{(p,q)\in\mathscr{S}} \int_{\Gamma_C^{pq}} \Psi \|[\mathbf{v}_T]\| \, d\Gamma = \sum_{(p,q)\in\mathscr{S}} \int_{\Gamma_C^{pq}} \max_{\|\boldsymbol{\tau}\| \leq \Psi(\mathbf{x})} \boldsymbol{\tau} \cdot [\mathbf{v}_T(\mathbf{x})] \, d\Gamma. \tag{14.12}$$

For the development of scalable algorithms for solving contact problems with Tresca friction, it is important that any solution of the variational boundary value inequality (14.11) solves

$$\begin{aligned} &\min f(\mathbf{v}) \quad \text{subject to} \quad \mathbf{v} \in \mathscr{K}, \\ &f(\mathbf{v}) = \frac{1}{2}a(\mathbf{v},\mathbf{v}) - \ell(\mathbf{v}) + j(\mathbf{v}). \end{aligned} \tag{14.13}$$

Denoting by $\mathbf{u}(\Psi)$ a solution of (14.13), we can try to find a solution to the problem with Coulomb's friction by the fixed-point iterations

$$\begin{aligned} \mathbf{u}^0 &= \mathbf{u}(\Psi^0), \\ \mathbf{u}^{k+1} &= \mathbf{u}\left(\Phi\lambda_n(\mathbf{u}^k)\right). \end{aligned} \tag{14.14}$$

The procedure was proposed by Panagiotopoulos [224].

14.4 Domain Decomposition and Discretization

To simplify the solution, let us decompose each domain Ω^p into subdomains with sufficiently smooth boundaries as in Chap. 13, assign a unique number to each subdomain, decompose appropriately the boundaries of subdomain into the parts of contact boundaries Γ_C^{ij} and the boundaries of adjacent boundaries $\Gamma_{DD}^{k\ell}$, and introduce the notation

$$\begin{aligned} V_{DD}^p &= \left\{\mathbf{v} \in H^1(\Omega^p)^3 : \mathbf{v} = \mathbf{o} \text{ on } \Gamma_U^p\right\}, \quad p = 1, \ldots s_D, \\ V_{DD} &= V^1 \times \cdots \times V^{s_D}, \\ \mathscr{K} &= \{\mathbf{v} \in V_{DD} : [v_n] \le g \text{ on } \Gamma_C^{pq},\ (p,q) \in \mathscr{S}_D\}. \\ \mathscr{K}_{DD} &= \{\mathbf{v} \in \mathcal{K} : \mathbf{v}^p = \mathbf{v}^q \text{ on } \Gamma_{DD}^{pq},\ p,q = 1,\ldots,s_D\}. \end{aligned}$$

The procedure is similar to that used in Sect. 13.2, but more general; s_D and $\mathscr{S}_D$ denote the number of subdomains and the coupling set defined by the indices of the subdomains that share the contact interface, respectively.

The variational decomposed problem with the Tresca (given) friction defined by a slip bound Ψ reads

$$\text{find } \mathbf{u} \in \mathscr{K}_{DD} \text{ such that } f(\mathbf{u}) \le f(\mathbf{v}), \quad \mathbf{v} \in \mathscr{K}_{DD}, \tag{14.15}$$

where

$$f(\mathbf{v}) = \frac{1}{2}a(\mathbf{v},\mathbf{v}) - \ell(\mathbf{v}) + j(\mathbf{v}).$$

After the finite element discretization, we get the block diagonal stiffness matrix $\mathsf{K} \in \mathbb{R}^{n \times n}$, the vector of nodal forces $\mathbf{f} \in \mathbb{R}^n$, the constraint matrix $\mathsf{B}_I \in \mathbb{R}^{m_I}$, the discretized gap function $\mathbf{c}_I \in \mathbb{R}^{m_I}$, and the matrix $\mathsf{B}_E \in \mathbb{R}^{m_I}$ that describes the gluing and Dirichlet conditions.

To define the discretized non-differentiable term j (14.10), we introduce $2m_I \times n$ matrix T such that m_I blocks of which are $\mathsf{T}_i = \mathsf{T}_i(\mathbf{x}_i) \in \mathbb{R}^{2\times n}$ are formed by appropriately placed orthonormal tangential vectors $\mathbf{t}_1(\mathbf{x}_i)$ and $\mathbf{t}_2(\mathbf{x}_i)$ so that the tangential component of the displacement $\mathbf{u}$ is given by $\mathsf{T}_i\mathbf{u}$. After applying numerical integration to (14.10), we get the discretized dissipative term in the form

$$j(\mathbf{u}) = \sum_{i=1}^{m_c} \Psi_i \|\mathsf{T}_i\mathbf{u}\|, \tag{14.16}$$

where Ψ_i is the slip bound associated with $\mathbf{x}_i$. Using (14.12), we get

$$j(\mathbf{u}) = \sum_{i=1}^{m_c} \Psi_i \|\mathsf{T}_i\mathbf{u}\| = \sum_{i=1}^{m_c} \max_{\|\boldsymbol{\tau}_i\| \le \Psi_i} \boldsymbol{\tau}_i^T \mathsf{T}_i\mathbf{u}, \tag{14.17}$$

where $\boldsymbol{\tau}_i \in \mathbb{R}^2$ can be interpreted as a Lagrange multiplier. The right-hand side of (14.17) is differentiable.

The discretized problem (14.15) reads

$$\text{find } \mathbf{u} \in \mathscr{K} \text{ such that } f(\mathbf{u}) \le f(\mathbf{v}), \quad \mathbf{v} \in \mathscr{K}, \tag{14.18}$$

where

$$\begin{aligned} f(\mathbf{v}) &= \frac{1}{2}\mathbf{v}^T\mathsf{K}\mathbf{v} - \mathbf{f}^T\mathbf{v} + j(\mathbf{v}), \\ \mathscr{K} &= \{\mathbf{v} \in \mathbb{R}^n : \mathsf{B}_I\mathbf{u} \le \mathbf{c}_I \text{ and } \mathsf{B}_E\mathbf{u} = \mathbf{o}\}. \end{aligned} \tag{14.19}$$

We assume that B_I, B_E, and T are full rank matrices. Notice that

$$\mathsf{T}^T\mathsf{B}_I = \mathsf{O}.$$

If each node is involved in at most one inequality, then it is possible to achieve that

$$\mathsf{B}\mathsf{B}^T = \mathsf{I}, \quad \mathsf{B} = \begin{bmatrix} \mathsf{B}_E \\ \mathsf{B}_I \\ \mathsf{T} \end{bmatrix}. \tag{14.20}$$

14.5 Dual Formulation

The problem (14.18) is not suitable for numerical solution, even if we replace j in the cost function by (14.17). The reasons are that the stiffness matrix K is typically large, ill-conditioned, and singular, and the feasible set is, in general, so complex that the projections onto it can hardly be effectively computed.

Under the circumstances, it would be complicated to solve auxiliary problems efficiently and to identify the solution's active set effectively.

As in Chap. 13, the complications can be essentially reduced by applying the duality theory of convex programming. In the dual formulation of problem (14.18), we use three types of Lagrange multipliers, namely $\boldsymbol{\lambda}_I \in \mathbb{R}^{m_I}$ associated with the non-interpenetration condition, $\boldsymbol{\lambda}_E \in \mathbb{R}^{m_E}$ associated with the "gluing" and prescribed displacements, and

$$\boldsymbol{\tau} = [\boldsymbol{\tau}_1^T, \boldsymbol{\tau}_2^T, \dots, \boldsymbol{\tau}_{m_T}^T]^T \in \mathbb{R}^{2m_T}, \quad \boldsymbol{\tau}_i \in \mathbb{R}^2, \ \ i = 1, \dots, m_T,$$

which are used to smooth the non-differentiability. The Lagrangian associated with problem (14.18) reads

$$L(\mathbf{u}, \boldsymbol{\lambda}_I, \boldsymbol{\lambda}_E, \boldsymbol{\tau}) = f(\mathbf{u}) + \boldsymbol{\tau}^T \mathsf{T}\mathbf{u} + \boldsymbol{\lambda}_I^T(\mathsf{B}_I\mathbf{u} - \mathbf{c}_I) + \boldsymbol{\lambda}_E^T(\mathsf{B}_E\mathbf{u} - \mathbf{c}_E).$$

To simplify the notation, we denote $m = m_E + m_I + 2m_T = m_E + 3m_I$,

$$\boldsymbol{\lambda} = \begin{bmatrix} \boldsymbol{\lambda}_E \\ \boldsymbol{\lambda}_I \\ \boldsymbol{\tau} \end{bmatrix}, \quad \mathsf{B} = \begin{bmatrix} \mathsf{B}_E \\ \mathsf{B}_I \\ \mathsf{T} \end{bmatrix}, \quad \mathbf{c} = \begin{bmatrix} \mathbf{o} \\ \mathbf{c}_I \\ \mathbf{o} \end{bmatrix},$$

and

$$\Lambda(\Psi) = \{(\boldsymbol{\lambda}_E^T, \boldsymbol{\lambda}_I^T, \boldsymbol{\tau}^T)^T \in \mathbb{R}^m : \ \boldsymbol{\lambda}_I \geq \mathbf{o}, \|\tau_i\| \leq \Psi_i, \ i = 1, \dots, m_T\},$$

so that we can write the Lagrangian briefly as

$$L(\mathbf{u}, \boldsymbol{\lambda}) = \frac{1}{2}\mathbf{u}^T \mathsf{K}\mathbf{u} - \mathbf{f}^T\mathbf{u} + \boldsymbol{\lambda}^T(\mathsf{B}\mathbf{u} - \mathbf{c}).$$

Using the convexity of the cost function and constraints, we can reformulate problem (14.18) by duality to get

$$\min_{\mathbf{u}} \sup_{\boldsymbol{\lambda} \in \Lambda(\Psi)} L(\mathbf{u}, \boldsymbol{\lambda}_I, \boldsymbol{\lambda}_E, \boldsymbol{\tau}) = \max_{\boldsymbol{\lambda} \in \Lambda(\Psi)} \min_{\mathbf{u}} L(\mathbf{u}, \boldsymbol{\lambda}_I, \boldsymbol{\lambda}_E, \boldsymbol{\tau}).$$

We conclude that problem (14.18) is equivalent to the saddle point problem

$$L(\widehat{\mathbf{u}}, \widehat{\boldsymbol{\lambda}}) = \max_{\boldsymbol{\lambda} \in \Lambda(\Psi)} \min_{\mathbf{u}} L(\mathbf{u}, \boldsymbol{\lambda}). \tag{14.21}$$

Recall that B is a full rank matrix. For a fixed $\boldsymbol{\lambda}$, the Lagrange function $L(\cdot, \boldsymbol{\lambda})$ is convex in the first variable and the minimizer $\mathbf{u}$ of $L(\cdot, \boldsymbol{\lambda})$ satisfies

$$\mathsf{K}\mathbf{u} - \mathbf{f} + \mathsf{B}^T\boldsymbol{\lambda} = \mathbf{o}. \tag{14.22}$$

Equation (14.22) has a solution if and only if

$$\mathbf{f} - \mathsf{B}^T\boldsymbol{\lambda} \in \mathrm{Im}\mathsf{K}, \tag{14.23}$$

which can be expressed more conveniently using a matrix $\mathsf{R} \in \mathbb{R}^{n\times 6s}$ the columns of which span the null space of K as

$$\mathsf{R}^T(\mathbf{f} - \mathsf{B}^T\boldsymbol{\lambda}) = \mathbf{o}.$$

The matrix R can be formed directly.

Now assume that $\boldsymbol{\lambda}$ satisfies (14.23) and denote by K^+ any matrix which satisfies

$$\mathsf{K}\mathsf{K}^+\mathsf{K} = \mathsf{K}. \tag{14.24}$$

Let us note that the action of a generalized inverse which satisfies (14.24) can be evaluated at the cost comparable with that of Cholesky's decomposition applied to the regularized K (see Sect. 1.5). It can be verified directly that if $\mathbf{u}$ solves (14.22), then there is a vector $\boldsymbol{\alpha} \in \mathbb{R}^{6s}$ such that

$$\mathbf{u} = \mathsf{K}^+(\mathbf{f} - \mathsf{B}^T\boldsymbol{\lambda}) + \mathsf{R}\boldsymbol{\alpha}. \tag{14.25}$$

After substituting expression (14.25) into a problem (14.21), changing the signs, and omitting the constant term, we get that $\boldsymbol{\lambda}$ solves the minimization problem

$$\min \Theta(\boldsymbol{\lambda}) \quad \text{s.t.} \quad \boldsymbol{\lambda} \in \Lambda(\Psi) \quad \text{and} \quad \mathsf{R}^T(\mathbf{f} - \mathsf{B}^T\boldsymbol{\lambda}) = \mathbf{o}, \tag{14.26}$$

$$\Theta(\boldsymbol{\lambda}) = \frac{1}{2}\boldsymbol{\lambda}^T\mathsf{B}\mathsf{K}^+\mathsf{B}^T\boldsymbol{\lambda} - \boldsymbol{\lambda}^T(\mathsf{B}\mathsf{K}^+\mathbf{f} - \mathbf{c}).$$

Once the solution $\widehat{\boldsymbol{\lambda}}$ of (14.26) is known, the solution $\widehat{\mathbf{u}}$ of (14.18) can be evaluated by (14.25) with

$$\boldsymbol{\alpha} = (\mathsf{R}^T\widetilde{\mathsf{B}}^T\widetilde{\mathsf{B}}\mathsf{R})^{-1}\mathsf{R}^T\widetilde{\mathsf{B}}^T(\widetilde{\mathbf{c}} - \widetilde{\mathsf{B}}\mathsf{K}^+(\mathbf{f} - \mathsf{B}^T\widehat{\boldsymbol{\lambda}})),$$

where $\widetilde{\mathsf{B}}$ and $\widetilde{\mathbf{c}}$ are formed by the rows of B and $\mathbf{c}$ corresponding to all equality constraints and all free inequality constraints.

14.6 Preconditioning by Projectors to Rigid Body Modes

Even though problem (14.26) is more suitable for computations than (14.18), further improvement can be achieved by adapting the observations from the previous chapters. Let us denote

$$\widetilde{\mathsf{F}} = \mathsf{B}\mathsf{K}^+\mathsf{B}^T, \qquad F = \|\widetilde{\mathsf{F}}\|,$$
$$\mathsf{F} = F^{-1}\widetilde{\mathsf{F}}, \qquad \widetilde{\mathbf{d}} = F^{-1}(\mathsf{B}\mathsf{K}^+\mathbf{f} - \mathbf{c}),$$
$$\widetilde{\mathsf{G}} = \mathsf{R}^T\mathsf{B}^T, \qquad \widetilde{\mathbf{e}} = \mathsf{R}^T\mathbf{f},$$

and let U denote a regular matrix that defines the orthonormalization of the rows of $\widetilde{\mathbf{G}}$ so that the matrix

$$\mathsf{G} = \mathsf{U}\widetilde{\mathsf{G}}$$

has orthonormal rows. After denoting

$$\mathbf{e} = \mathsf{U}\widetilde{\mathbf{e}},$$

problem (14.26) reads

$$\min \frac{1}{2}\boldsymbol{\lambda}^T\mathsf{F}\boldsymbol{\lambda} - \boldsymbol{\lambda}^T\widetilde{\mathbf{d}} \quad \text{s.t.} \quad \boldsymbol{\lambda} \in \Lambda(\Psi) \quad \text{and} \quad \mathsf{G}\boldsymbol{\lambda} = \mathbf{e}. \tag{14.27}$$

Next, we transform the problem of minimization on a subset of the affine space to that on a subset of the vector space using $\widetilde{\boldsymbol{\lambda}}$ which satisfies

$$\mathsf{G}\widetilde{\boldsymbol{\lambda}} = \mathbf{e}. \tag{14.28}$$

We can use $\widetilde{\boldsymbol{\lambda}}$ which solves

$$\min \frac{1}{2}\|\boldsymbol{\lambda}\|^2 \quad \text{s.t.} \quad \boldsymbol{\lambda} \in \Lambda(\Psi) \quad \text{and} \quad \mathsf{G}\boldsymbol{\lambda} = \mathbf{e}, \tag{14.29}$$

or, if the problem is coercive, then we can use [114, Lemma 11.2] to get

$$\widetilde{\boldsymbol{\lambda}} = \begin{bmatrix} \mathbf{o}_{\mathscr{I}} \\ \mathsf{G}_{\mathscr{E}}^T(\mathsf{G}_{\mathscr{E}}\mathsf{G}_{\mathscr{E}}^T)^{-1}\mathbf{e} \end{bmatrix}.$$

The above choices of $\widetilde{\boldsymbol{\lambda}}$ are important for the proof of optimality as they guarantee that $\mathbf{o}$ is a feasible vector of the modified problem. For practical computations, we can use the least squares solution of

$$\mathsf{G}\boldsymbol{\lambda} = \mathbf{e}$$

given by

$$\widetilde{\boldsymbol{\lambda}} = \mathsf{G}^T\mathbf{e}.$$

Having $\widetilde{\boldsymbol{\lambda}}$, we can look for the solution of (14.27) in the form

$$\boldsymbol{\lambda} = \boldsymbol{\mu} + \widetilde{\boldsymbol{\lambda}}.$$

To carry out the transformation, denote $\boldsymbol{\lambda} = \boldsymbol{\mu} + \widetilde{\boldsymbol{\lambda}}$ and $\widetilde{\Lambda}(\Psi) = \Lambda(\Psi) - \widetilde{\boldsymbol{\lambda}}$, so that

$$\frac{1}{2}\boldsymbol{\lambda}^T \mathsf{F}\boldsymbol{\lambda} - \boldsymbol{\lambda}^T\widetilde{\mathbf{d}} = \frac{1}{2}\boldsymbol{\mu}^T\mathsf{F}\boldsymbol{\mu} - \boldsymbol{\mu}^T(\widetilde{\mathbf{d}} - \mathsf{F}\widetilde{\boldsymbol{\lambda}}) + \frac{1}{2}\widetilde{\boldsymbol{\lambda}}^T\mathbf{F}\widetilde{\boldsymbol{\lambda}} - \widetilde{\boldsymbol{\lambda}}^T\widetilde{\mathbf{d}}$$

and problem (14.27) turns, after returning to the old notation, into

$$\min \frac{1}{2}\boldsymbol{\lambda}^T\mathsf{F}\boldsymbol{\lambda} - \boldsymbol{\lambda}^T\mathbf{d} \text{ s.t. } \mathsf{G}\boldsymbol{\lambda} = \mathbf{o} \text{ and } \boldsymbol{\lambda} \in \widetilde{\Lambda}(\Psi), \quad \mathbf{d} = \widetilde{\mathbf{d}} - \mathsf{F}\widetilde{\boldsymbol{\lambda}}. \tag{14.30}$$

We can achieve that $\mathbf{o} \in \widetilde{\Lambda}(\Psi)$.

Our final step is based on the observation that problem (14.30) is equivalent to

$$\min \overline{\theta}_\rho(\boldsymbol{\lambda}) \quad \text{s.t.} \quad \mathsf{G}\boldsymbol{\lambda} = \mathbf{o} \quad \text{and} \quad \boldsymbol{\lambda} \in \widetilde{\Lambda}(\Psi), \tag{14.31}$$

where ρ is an arbitrary positive constant,

$$\mathsf{Q} = \mathsf{G}^T\mathsf{G} \qquad \text{and} \qquad \mathsf{P} = \mathsf{I} - \mathsf{Q}$$

denote the orthogonal projectors on the image space of G^T and on the kernel of G, respectively, and

$$\overline{\theta}_\rho(\boldsymbol{\lambda}) = \frac{1}{2}\boldsymbol{\lambda}^T(\mathsf{PFP} + \rho\mathsf{Q})\boldsymbol{\lambda} - \boldsymbol{\lambda}^T\mathsf{P}\mathbf{d}.$$

The following theorem provides the bounds on the spectrum of

$$\mathsf{H}_\varrho = \mathsf{PFP} + \rho\mathsf{Q}.$$

Theorem 14.1. *Let $\varrho \approx \|\mathsf{F}\|$ and let the parameters H_s and h specify problem (14.31). Then there are constants $c, C > 0$ independent of H_s, h such that*

$$c \le \lambda_{\min}(\mathsf{H}_\varrho) \le \|\mathsf{H}_\varrho\| \le CH_s/h. \tag{14.32}$$

Proof. See Dostál et al. [106].

Let $C_1 > 2$ denote an arbitrary constant and consider a class of problems (14.31) generated with the parameters H_s, h, such that

$$H_s/h \le C_1.$$

Then we can use Theorem 12.4 to get that there is a constant $M_{\max}$ such that the SMALBE algorithm with the inner loop implemented by MPRGP can find an approximate solution of any problem (13.15) in at most $M_{\max}$ matrix–vector multiplications. See also [106].

14.7 Numerical Experiments

Here, we give some results of numerical experiments resolved in [106] that illustrate the performance of the TFETI-based algorithms implemented for the solution of contact problems with friction using their implementation in MatSol [190]. All the computations were carried out with the parameters recommended in Chaps. 11 and 12, i.e.,

$$M_0 = 1,\ \eta = 0.1\|\mathsf{Pd}\|,\ \rho = 1 \approx \|\mathsf{PFP}\|, \Gamma = 1,\ \beta = 0.2,\ \boldsymbol{\lambda}^0 = \mathbf{o},\ \text{and}\ \boldsymbol{\mu}^0 = \mathbf{o}.$$

Our exposition is based on Dostál et al. [106]. The relative precision of the computations was $\varepsilon = 10^{-4}$ (see (12.80)).

14.7.1 Cantilever Beams with Isotropic Friction

The first benchmark is a 3D semicoercive contact problem of two cantilever beams of sizes $2 \times 1 \times 1$ [m] in mutual contact without friction. The beams are depicted in Fig. 14.1. Zero horizontal displacements were prescribed on the right face of the upper beam. The lower beam (shifted by 1 [m]) was fixed on its left face. The vertical traction $\mathbf{f} = 20$ [MPa] was prescribed on the upper and left faces of the upper beam.

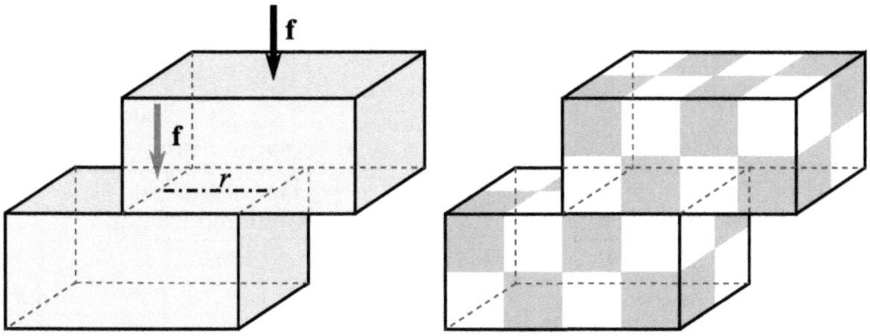

Fig. 14.1. Two beams benchmark and its domain decomposition

The discretizations and the decompositions were defined by the discretization parameter h and the decomposition parameter H, respectively. For each h and H, the bodies were decomposed into cubic subdomains and discretized by the regular mesh with hexahedral elements. We kept $H/h = 8$, so that the assumptions of Theorem 14.1 were satisfied. The normal and tangential tractions along the axis of the contact interface are in Fig. 14.2 (see [106]). This figure also shows that in the solution of the largest problem, most of the linear and quadratic inequality constraints were active.

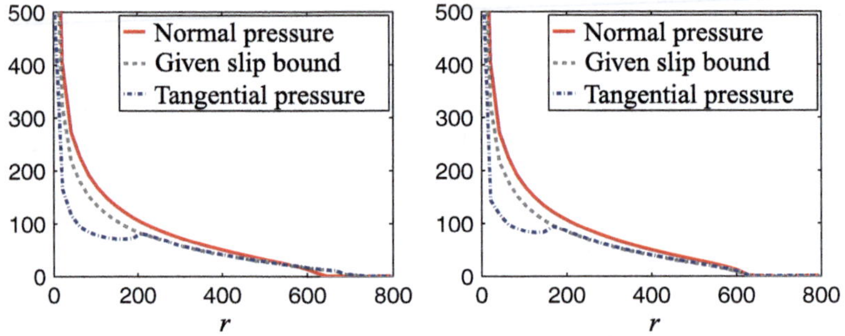

Fig. 14.2. Tresca (*left*) and Coulomb (*right*) friction—normal and contact pressures along r (length of r is given in milimeters)

The performance of the algorithms is documented in the above graphs. The numbers of outer iterations of SMALSE-M and multiplications by the Hessian F of the dual function depending on the primal dimension n for the problem with Tresca friction are depicted in Fig. 14.3. We can see stable numbers of outer iterations for n ranging from 431,244 to 11,643,588. The dual dimension of the problems ranged from 88,601 to 2,728,955. We conclude that the algorithm's performance is in agreement with the theory. The problem is complex because many nodes on the contact interface nearly touch each other, making identifying the contact interface difficult.

Similar results were obtained for the solution of the problem with the Coulomb friction defined by the friction coefficient $\Phi = 0.1$. The results are also given in the above graphs. The number of iterations is naturally grater.

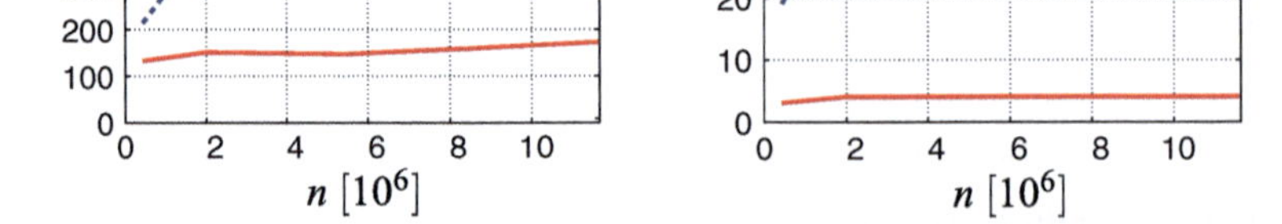

Fig. 14.3. Tresca and Coulomb friction: numerical scalability of cantilever beams—matrix–vector multiplications by F (*left*) and outer (SMALSE-M) iterations (*right*)

14.7.2 Yielding Clamp Connection

Yielding clamp connections are parts of mining support that provide ductile behavior and dissipate energy during seismic events, reducing the risk of collapse. Here, we model the yielding clamp connection as a contact with the Coulomb friction, where the coefficient of friction was $\Phi = 0.5$. The problem was decomposed into 250 subdomains using METIS (see Fig. 14.4) and discretized by 1,592,853 and 216,604 primal and dual variables, respectively. The total displacements for the Coulomb friction are depicted in Fig. 14.5 (see [106]). It required 1922 matrix-vector multiplications to find the solution.

Fig. 14.4. Yielding clamp connection—domain decomposition

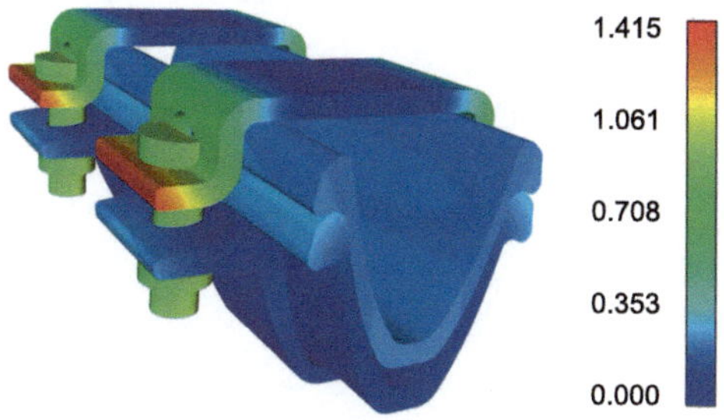

Fig. 14.5. Coulomb friction: total displacement

15
Model Predictive Control

Optimal control is a source of many interesting quadratic programming problems. The QP problems in optimal control have a smaller dimension than those discussed in the remaining case studies, typically some hundreds of unknown variables. However, their solution poses some new challenges:

- there is a limit on the solution time corresponding to the small change of the state of the real problem,
- the solution of the control problem typically requires the solution of many related QP problems,
- lower precision solutions can be acceptable so that alternative problem formulations can be considered.

Another challenge is related to practical implementation. The matrices that define the QP problems are typically structured. The QP problems can be formulated with parameters, so the conditioning of the Hessians can vary from problem to problem. Some computations can be carried out offline.

Here, we consider bound-constrained QP problems that arise when applying the Model Predictive Control (MPC). We limit our attention to the *oscillating masses* regulator control problem from Wang and Boyd [261]. We briefly formulate the problems and discuss the effect of parameters on the conditioning of constraints. We compare the impact of using the Newton direction and conjugate gradient direction in MPRGP and PRGPN and discuss the effective exploitation of the projected gradient to identify the active set of the solution. Numerical experiments with variants of MPRGP and PRGPN illustrate the discussion. Our exposition is based on Šantin et al. [243].

Z. Dostál, *Optimal Quadratic Programming and QCQP Algorithms with Case Studies*, Springer Optimization and Its Applications 23,
https://doi.org/10.1007/978-3-031-95167-1_15

15.1 Linear Model Predictive Control (MPC)

Let us consider a discrete-time linear time-invariant system described by

$$\mathbf{x}(k+1) = \mathsf{A}\mathbf{x}(k) + \mathsf{B}\mathbf{u}(k), \quad k = 0, 1, 2, \ldots, \tag{15.1}$$

where $\mathbf{x}(k) \in \mathbb{R}^{n_x}$ and $\mathbf{u}(k) \in \mathbb{R}^{n_u}$ denote the *state* and *input (control) vectors*, respectively, $\mathsf{A} \in \mathbb{R}^{n_x \times n_x}$, $\mathsf{B} \in \mathbb{R}^{n_x \times n_u}$. If we are given the state $\mathbf{x}(k)$ and control vectors

$$\mathbf{u}(k), \mathbf{u}(k+1), \ldots, \mathbf{u}(k+N-1),$$

we can predict the states over the prediction horizon as

$$\begin{aligned}
\mathbf{x}(k+1) &= \mathsf{A}\mathbf{x}(k) + \mathsf{B}\mathbf{u}(k), \\
\mathbf{x}(k+2) &= \mathsf{A}\mathbf{x}(k+1) + \mathsf{B}\mathbf{u}(k+1) = \mathsf{A}^2\mathbf{x}(k) + \mathsf{AB}\mathbf{u}(k), \\
\mathbf{x}(k+3) &= \mathsf{A}^3\mathbf{x}(k) + \mathsf{A}^2\mathsf{B}\mathbf{u}(k), \\
&\;\;\vdots \\
\mathbf{x}(k+N) &= \mathsf{A}^{N-1}\mathbf{x}(k) + \mathsf{A}^{N-2}\mathsf{B}\mathbf{u}(k).
\end{aligned}$$

Introducing the notation

$$\mathbf{X}(k) = \begin{bmatrix} \mathbf{x}^k \\ \mathbf{x}^{k+1} \\ \vdots \\ \mathbf{x}^{k+N-1} \end{bmatrix} \in R^{N.n_x}, \quad \mathbf{U}(k) = \begin{bmatrix} \mathbf{u}^k \\ \mathbf{u}^{k+1} \\ \vdots \\ \mathbf{u}^{k+N-1} \end{bmatrix} \in \mathbb{R}^{N.n_u},$$

and

$$\mathbf{A} = \begin{bmatrix} \mathsf{A} \\ \mathsf{A}^2 \\ \vdots \\ \mathsf{A}^{N-1} \end{bmatrix}, \quad \mathbf{B} = \begin{bmatrix} \mathsf{B} & \mathsf{O} \cdots & \mathsf{O} & \mathsf{O} \\ \mathsf{AB} & \mathsf{B} & \mathsf{O} & \mathsf{O} \\ \vdots & \vdots \ddots & & \vdots \\ \mathsf{A}^{N-2}\mathsf{B} & \mathsf{A}^{N-3}\mathsf{B} \cdots & \mathsf{B} & \mathsf{O} \\ \mathsf{A}^{N-1}\mathsf{B} & \mathsf{A}^{N-2}\mathsf{B} \cdots & \mathsf{AB} & \mathsf{B} \end{bmatrix},$$

we can rewrite the state predictions over a finite prediction horizon

$$k+1, \ldots, k+N$$

as a function of the initial state $\mathbf{x}(k)$ and control $\mathbf{U}(k)$ in the compact form

$$\mathbf{X}(k+1) = \mathbf{A}\mathbf{x}(k) + \mathbf{B}\mathbf{U}(k). \tag{15.2}$$

To guarantee that we can get the prescribed final state $\mathbf{x}_N$, we assume that the block row

$$\mathbf{B}_N = [\mathsf{A}^{N-1}\mathsf{B},\ \mathsf{A}^{N-2}\mathsf{B},\ \cdots,\mathsf{AB},\ \mathsf{B}]$$

has a full row rank. The initial state $\mathbf{x}(k)$ can be replaced by its estimate $\widetilde{\mathbf{x}}(k) \approx \mathbf{x}(k)$. More can be found, e.g., in Maciejowski [209].

15.2 Regulator Problem

We shall restrict our presentation to the *regulator problem*; its goal is to drive the system state $\mathbf{x}(k+N)$ most effectively to the origin with the minimum control cost measured by a suitable norm of $\mathbf{U}(k)$. We shall use the so-called dense formulation, which has only $n_U = n_u \times N$ box constraints on the components of the control input variables of $\mathbf{U}(k)$.

The effective formulation of the regulator problem reduces to the minimization of suitable norms of U and $\mathbf{x}(N)$ subject to (15.2). Moreover, it seems reasonable to control also the norm of the state variables $\mathbf{x}(k+i)$, $i = 1,\ldots,N-1$. Taking into account the above reasoning, we can consider the cost function in the most general form

$$\begin{aligned} f\left(\mathbf{X}(k),\mathbf{U}(k),\mathbf{x}(k+N)\right) &= \frac{1}{2}\mathbf{x}(k+N)^T\mathsf{Q}_N\mathbf{x}(k+N) \\ &+ \frac{1}{2}\left(\mathbf{X}^T(k)\mathsf{Q}_X\mathbf{X}(k) + \mathbf{U}(k)^T\mathsf{Q}_U\mathbf{U}(k)\right), \end{aligned} \tag{15.3}$$

where $\mathsf{Q}_N \in \mathbb{R}^{n_x \times n_x}$ and $\mathsf{Q}_X \in \mathbb{R}^{n_X \times n_X}$ are SPD matrices, $Q_U \in \mathbb{R}^{n_U \times n_U}$ is an SPD matrix. In (15.3), k and N denote the current time step and the prediction horizon.

Noticing that all arguments of f depend on k, we can skip k in the description of algorithms in what follows. In particular, we get the cost function in the form

$$f(\mathbf{X},\mathbf{U},\mathbf{x}_N) = \frac{1}{2}\mathbf{x}_N^T\mathsf{Q}_N\mathbf{x}_N + \frac{1}{2}\left(\mathbf{X}^T\mathsf{Q}_X\mathbf{X} + \mathbf{U}^T\mathsf{Q}_U\mathbf{U}\right).$$

Given $\mathbf{x}_0$ and the set of admissible control variables Ω_U, the 'most precise solution' of the linear regulator MPC problem amounts to finding

$$\min f(\mathbf{X},\mathbf{U}) \text{ subject to } \mathbf{U} \in \Omega_U,\ \mathbf{X} = \mathbf{Ax} + \mathbf{BU},\ \mathbf{x}_N = \mathbf{o},\ \mathbf{x}_0 = \mathbf{x}. \tag{15.4}$$

We shall assume that Ω_U is defined by the box constraints on the entries of the control vector $\mathbf{U}$. We can also assume that the state variables only modestly change so that we can solve (15.4) with prescribed $\mathbf{X}$ to get the simplified problem

$$\min f\big(\mathbf{U},\mathbf{x}_N(\mathbf{U})\big) \text{ subject to } \mathbf{U} \in \Omega_U, \quad \mathbf{x}_N(\mathbf{U}) = \mathsf{A}\mathbf{x} + \mathbf{B}_N\mathbf{U}. \tag{15.5}$$

Remark 15.1. The Receding Horizon Control assumes that the problem (15.4) is recomputed at each sampling instant, which provides a new

$$\mathbf{x}_0 = \widetilde{\mathbf{x}}.$$

The algorithm uses the initial state $\mathbf{x}_0$ to obtain new control inputs $\mathbf{U}$, of which only the first control $\mathbf{u}_0$ is applied to the system. The new control input $\mathbf{u}_0$ should be available by the next sampling time. The latter requirement strongly motivates the development of fast QP algorithms for these relatively small problems.

15.3 Oscillating Masses Regulator Problem

We shall illustrate the performance of the QP algorithms on the oscillating masses model problem from Wang and Boyd [261] (see also Šantin et al. [243]). It consists of six masses interconnected by spring dampers, with the first and the last masses fixed to the walls, as in Fig. 15.1.

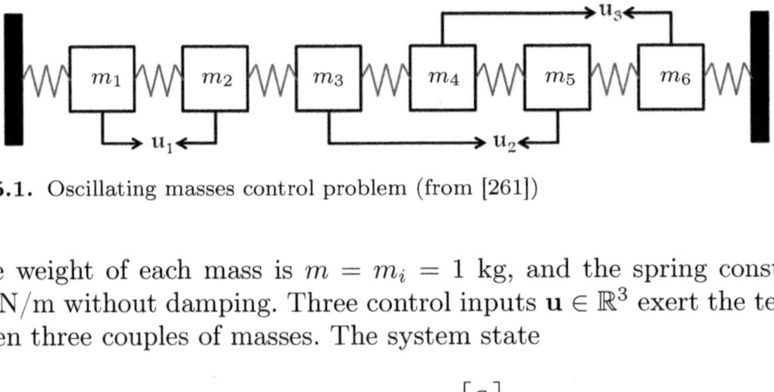

Fig. 15.1. Oscillating masses control problem (from [261])

The weight of each mass is $m = m_i = 1$ kg, and the spring constant is $k = 1$ N/m without damping. Three control inputs $\mathbf{u} \in \mathbb{R}^3$ exert the tensions between three couples of masses. The system state

$$\mathbf{x} \in \mathbb{R}^{12}, \quad \mathbf{x} = \begin{bmatrix} \mathbf{s} \\ \dot{\mathbf{s}} \end{bmatrix}$$

represents the displacement from the steady state and the velocities of particular masses. The continuous time-invariant system is described by the initial conditions $\mathbf{s}(0) = \mathbf{o}$, $\dot{\mathbf{s}}(0) = \mathbf{o}$ and

$$\begin{aligned} m\ddot{s}_1 &= -ks_1 + k(s_2 - s_1) + u_1 \\ m\ddot{s}_2 &= -k(s_2 - s_1) + k(s_3 - s_2) - u_1 \\ m\ddot{s}_3 &= -k(s_3 - s_2) + k(s_4 - s_3) + u_2 \\ m\ddot{s}_4 &= -k(s_4 - s_3) + k(s_5 - s_4) + u_3 \\ m\ddot{s}_5 &= -k(s_5 - s_4) + k(s_6 - s_5) - u_2 \\ m\ddot{s}_6 &= -k(s_6 - s_5) - ks_6 - u_3. \end{aligned} \tag{15.6}$$

We assume the control variables

$$-0.5\,\mathbf{e} \le \mathbf{u} \le 0.5\,\mathbf{e}, \quad \mathbf{e} = [1, 1, \dots, 1]^T,$$

and a random external disturbance $\mathbf{w} \in \mathbb{R}^6$ with a uniform distribution

$$-0.5\,\mathbf{e} \le \mathbf{w} \le 0.5\,\mathbf{e}, \quad \mathbf{e} = [1, 1, \dots, 1]^T.$$

The inequalities should be interpreted component-wise. The distribution acts additionally on the displacement state of each mass; see [261].

To make our numerical experiments comparable with the published ones [243], we used the same discretization of (15.6) with the time-step $\Delta = 0.5$. We assembled the matrices **A**,**B** that define the relations between the state and input vectors introduced in Sect. 15.1. We used the cost function (15.5) with

$$\mathsf{Q}_U = \mathsf{I}, \; \mathsf{Q}_N = \mathsf{Q}_X = \mu\mathsf{I}, \; \mu \in \{1, 1000\}.$$

We carried out the numerical experiments with the prediction horizon $N \in \{10, 50, 100, 200\}$, so that the dimension of the related QP problem was

$$n_U = N.n_u \in \{30, 150, 300, 600\}.$$

Since the dimension of the discretized QP problems is not very large and admits direct solvers of auxiliary linear problems, PRGPN from Sect. 10.5.4 is a reasonable choice from the algorithms described in this book. We also consider the original MPRGP algorithm, an interesting combination of proportioning and the Newton projected line search coined NPP proposed by O. Šantin [242]. We performed the tests with recommended parameters, including $\Gamma = 1$ and the precision $\varepsilon = 1e - 6\|\mathsf{A}\|$. More details are in Šantin et al. [243] or Šantin [242].

We compared the performance of these methods with three very different state-of-the-art solvers. In particular, we considered the active set-based code qpOASES [129] by Ferreau et al., the FiOrdOs code using the first-order methods by Ullmann [257], and the interior point-based code FORCES by Domahidi et al. [53]. The solution from the previous sample time was used as a warm start in all the tests to speed up the convergence of the tested algorithms. The results of numerical experiments are in Figs. 15.2 and 15.3.

We can see that both MPRGP and PRGPN are competitive. Still, the best results were obtained by combining the proportioning with the Newton projected line search NPP algorithm. The latter algorithm can exploit some specific features of optimal control problems, particularly a good initial approximation of the active set and relatively small dimensions of these problems.

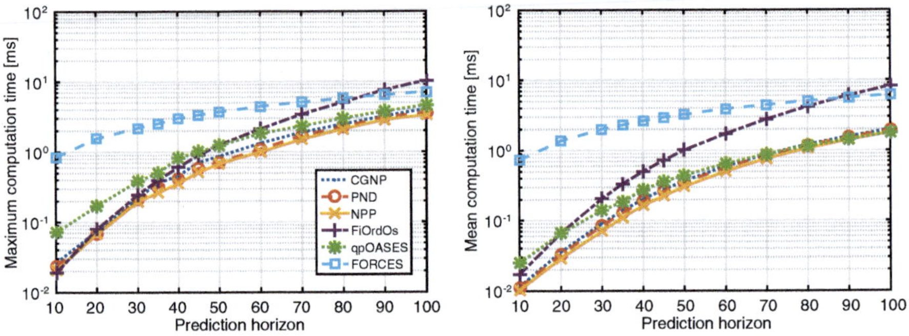

Fig. 15.2. Max/min computation time for well-conditioned ($\mu = 1$) oscillating masses MPRGP and PRGPN are labeled as CGNP and PND (courtesy of O. Šantin [242]). Part of the results appeared first in Šantin et al. [243]

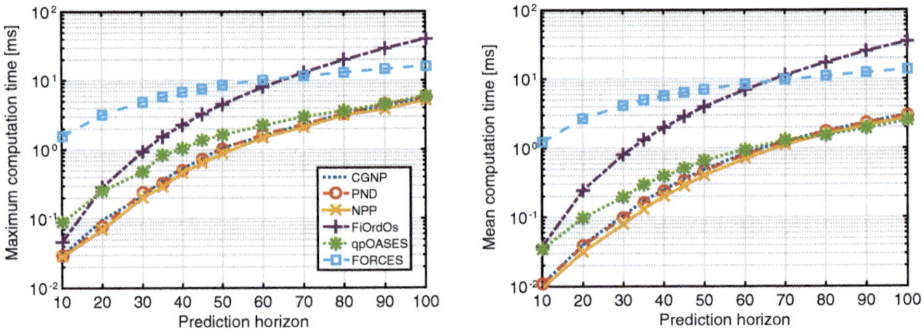

Fig. 15.3. Max/min computation time for ill-conditioned ($\mu = 1000$) oscillating masses. MPRGP and PRGPN are labeled as CGNP and PND (courtesy of O. Šantin [242]). Part of the results appeared first in Šantin et al. [243]

16
Support Vector Machines

Practical quadratic programming algorithms provide a valuable tool for solving optimization problems in artificial intelligence. Here, we consider one such problem: binary classification with Support Vector Machines (SVM). To illustrate the binary classification problem, consider a training dataset of vectors $\mathbf{x}_i \in \mathbb{R}^d$ with patients' information, including *features* such as age, gender, blood pressure, and other relevant medical history. We call the vectors $\mathbf{x}_i$ *samples.* Let each patient in the dataset be labeled as either having diabetes (positive class) or not having diabetes (negative class). The goal is to classify new samples, i.e., to predict which class it belongs.

The QP problem, the heart of SVM, looks for the hyperplane in the feature's space that separates different classes of samples while maximizing the margin. The SVM have some specific features:

- the Hessian of the cost function is SPS,
- the rank of the primal cost function Hessian is typically very small,
- the number of linear inequality constraints can be large,
- constraints are typically defined by very sparse matrices,
- only a few inequality constraints are active in the primal solution,
- the requirements on the solution precision are relaxed, so a low precision solution is acceptable,
- the formulation can include constants enhancing additional information or preference.

Here, we describe the primal and dual soft margin SVM problems and give the results of numerical experiments with solving simplified dual SVM problems with soft margin [47, Chap. 12]. We solved the resulting QP problems with SPS or SPD Hessian and bound constraints on four classification datasets from LIBSVM dataset web page [204]. For basic information on SVM, see Diesenroth, Faisal, and Ong [47, Chap. 12].

Z. Dostál, *Optimal Quadratic Programming and QCQP Algorithms with Case Studies*, Springer Optimization and Its Applications 23,
https://doi.org/10.1007/978-3-031-95167-1_16

16.1 Primal SVM Quadratic Programming Problems

Let

$$\mathcal{X} = \{(\mathbf{x}_i, y_i) : \mathbf{x}_i \in \mathbb{R}^d,\ y_i \in \{-1, 1\},\ i = 1, \ldots n\}$$

denote a training set of samples labeled by -1 or 1. We aim to split $\mathbb{R}^d$ into two parts corresponding to the labels. We shall use a linear function

$$\ell : \mathbb{R}^d \ni \mathbf{z} \to \mathbf{z}^T\mathbf{w} + b \in \mathbb{R}$$

parametrized by $\mathbf{w}$ and b. Our goal is to find the parameters $\mathbf{w}$ and b so that the hyperplane

$$\mathcal{H} = \{\mathbf{z} \in \mathbb{R}^d : \ell(\mathbf{z}) = 0\}$$

separates the two classes. However, the solution to this problem is typically not unique, as illustrated in Fig. 16.1 left, and there may be no separating hyperplane as in Fig. 16.1 right.

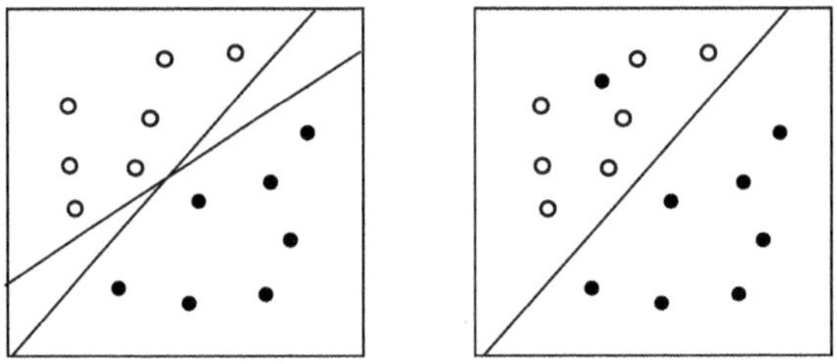

Fig. 16.1. Separating hyperplanes (left) and linearly non-separable data (right)

If the classes can be separated, then we can attempt to find the parameters $\mathbf{w}, b$ so that the distance of $\mathcal{H}$ from the nearest members of the classes is as large as possible as in Fig. 16.2. Notice that the hyperplane $\mathcal{H}$ is typically defined by a few samples called support vectors. This observation gives the name of SVM. We can even achieve that

$$|\ell(\mathbf{x}_i)| \geq 1,\ i = 1, \ldots, n.$$

The parameters $\mathbf{w}$ and b solve the *SVM hard margin problem* to find

$$\min_{\mathbf{w},b} \frac{1}{2}\|\mathbf{w}\|^2 \text{ subject to } \quad y_i(\mathbf{w}^T\mathbf{x}_i + b) \geq 1, \quad i = 1, \ldots, n. \tag{16.1}$$

For more details, see, e.g., [47, Chap. 12].

If the two classes cannot be separated as in Fig. 16.1 right, we introduce a fixed positive constant $C > 0$ and the non-negative *slack variables* $\boldsymbol{\xi} = [\xi_i] \in \mathbb{R}^n$. We shall use them to formulate the *SVM soft margin problem*

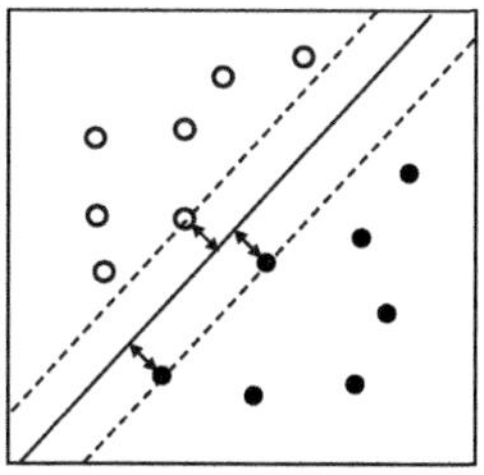

Fig. 16.2. Support vectors and maximal margin

$$\min_{\mathbf{w},b,\boldsymbol{\xi}} \frac{1}{2}\|\mathbf{w}\|^2 + C\sum_{i=1}^{n} \xi_i \tag{16.2}$$

subject to

$$y_i(\mathbf{w}^T\mathbf{x}_i + b) \geq 1 - \xi_i, \quad \xi_i \geq 0, \quad i = 1, \ldots, n. \tag{16.3}$$

Notice that $\xi_i > 0$ allows the placing of the sample $\mathbf{x}_i$ within the margin or on the wrong side of the separating hyperplane, so we call this modification *soft margin SVM.* Thus, large values of C give more weight to the samples that do not lie on the correct side of the margin. The rank of the Hessian of the cost functions 16.1 and 16.2 is equal to d. The dimensions of the slacks and Hessian are n and $d+1+n$, respectively. Denoting

$$\mathbf{x} = \begin{bmatrix} \mathbf{w} \\ b \\ \boldsymbol{\xi} \end{bmatrix}, \quad \mathsf{A} = \begin{bmatrix} \mathsf{I}_d & \mathbf{o} & \mathsf{O} \\ \mathbf{o}^T & 0 & \mathbf{o}^T \\ \mathsf{O} & \mathbf{o} & \mathsf{O}_n \end{bmatrix}, \quad \text{and} \quad \mathbf{b} = -C\begin{bmatrix} \mathbf{o}_d \\ 0 \\ \mathbf{e}_n \end{bmatrix},$$

where

$$\mathbf{e}_n = [1 \ldots 1]^T \in \mathbb{R}^n,$$

we can write the cost function (16.2) in the form used in Chap. 2 as

$$f(\mathbf{x}) = \frac{1}{2}\mathbf{x}^T\mathsf{A}\mathbf{x} - \mathbf{b}^T\mathbf{x}. \tag{16.4}$$

Recall that we omit the dimensions of blocks when they can be deduced from the assumption that the block matrix is well-defined.

Similarly, to rewrite the constraints (16.3) in the standard form, let us denote

$$\mathsf{X} = \begin{bmatrix} y_1\mathbf{x}_1^T \\ \vdots \\ y_n\mathbf{x}_n^T \end{bmatrix}, \quad \mathbf{y} = \begin{bmatrix} y_1 \\ \vdots \\ y_n \end{bmatrix}, \quad \text{and} \quad \mathbf{c} = -\begin{bmatrix} \mathbf{e}_n \\ \mathbf{o}_n \end{bmatrix}.$$

Using this notation and

$$\mathsf{B} = -\begin{bmatrix} \mathsf{X} & \mathbf{y} & \mathsf{I}_n \\ \mathsf{O} & \mathbf{o} & \mathsf{I}_n \end{bmatrix} \in \mathbb{R}^{2n\times(d+1+n)}, \tag{16.5}$$

we can rewrite the soft margin SVM problem in the standard form as

$$\min \frac{1}{2}\mathbf{x}^T\mathsf{A}\mathbf{x} - \mathbf{b}^T\mathbf{x} \quad \text{subject to} \quad \mathsf{B}\mathbf{x} \le \mathbf{c}. \tag{16.6}$$

16.2 Dual SVM Quadratic Programming Problems

General inequality constraints and a low-rank Hessian matrix in (16.6) prevent the direct application of SMALBE/MPRGP algorithms. Our cure is switching to the dual problem and effective regularization. A unique benefit of this approach is the dimension of the dual problem independent of the number of features.

Here, we shall use the dual formulation of the soft margin SVM problem to allow the application of the MPRGP algorithm. Recall that (16.6) is a QP problem with SPS Hessian A and inequality constraints, so the corresponding dual problem introduced in Sect. 2.7 reads

$$\max_{\boldsymbol{\lambda}\in\Omega_{BE}} \Theta(\boldsymbol{\lambda}), \quad \Omega_{BE} = \{\boldsymbol{\lambda} \in \mathbb{R}^m : \boldsymbol{\lambda} \ge \mathbf{o},\ \mathsf{R}^T\mathsf{B}^T\boldsymbol{\lambda} = \mathsf{R}^T\mathbf{b}\}, \tag{16.7}$$

where

$$\Theta(\boldsymbol{\lambda}) = -\frac{1}{2}\boldsymbol{\lambda}^T\mathsf{B}\mathsf{A}^+\mathsf{B}^T\boldsymbol{\lambda} + \boldsymbol{\lambda}^T(\mathsf{B}\mathsf{A}^+\mathbf{b} - \mathbf{c}) - \frac{1}{2}\mathbf{b}^T\mathsf{A}^+\mathbf{b} \tag{16.8}$$

and

$$\mathsf{R} = \begin{bmatrix} \mathsf{O}_{d,n+1} \\ \mathsf{I}_{n+1} \end{bmatrix}$$

denotes the kernel of A. Moreover, there is $\overline{\boldsymbol{\alpha}} \in \mathbb{R}^d$ such that $(\overline{\boldsymbol{\lambda}}, \overline{\boldsymbol{\alpha}})$ is a KKT pair for problem (16.7) and by (2.104)

$$\overline{\mathbf{x}} = \mathsf{A}^+(\mathbf{b} - \mathsf{B}^T\overline{\boldsymbol{\lambda}}) + \mathsf{R}\overline{\boldsymbol{\alpha}}. \tag{16.9}$$

Recall that A^+ denotes any left generalized inverse that satisfies

$$\mathsf{A}\mathsf{A}^+\mathsf{A} = \mathsf{A},$$

so that we can take

$$\mathsf{A}^+ \overset{(1.7)}{=} \mathsf{A}^\#, \tag{16.10}$$

and $\overline{\alpha}$ is the Lagrange multiplier for the dual problems that can be evaluated by (2.119). Let us rewrite the components of the dual problem (16.7) and (16.8) in terms of input data:

$$\mathsf{BA}^{+} = -\begin{bmatrix} \mathsf{X} & \mathbf{y} & \mathsf{I}_n \\ \mathsf{O} & \mathbf{o} & \mathsf{I}_n \end{bmatrix} \begin{bmatrix} \mathsf{I}_d & \mathbf{o} & \mathsf{O} \\ \mathbf{o}^T & 0 & \mathbf{o}^T \\ \mathsf{O} & \mathbf{o} & \mathsf{O}_n \end{bmatrix} = -\begin{bmatrix} \mathsf{X} & \mathbf{o} & \mathsf{O} \\ \mathsf{O} & \mathbf{o} & \mathsf{O}_n \end{bmatrix},$$

$$\mathsf{BA}^{+}\mathsf{B}^T = \begin{bmatrix} \mathsf{X} & \mathbf{o} & \mathsf{O} \\ \mathsf{O} & \mathbf{o} & \mathsf{O}_n \end{bmatrix} \begin{bmatrix} \mathsf{X}^T & \mathsf{O}_n \\ \mathbf{y}^T & \mathbf{o}^T \\ \mathsf{I}_n & \mathsf{I}_n \end{bmatrix} = \begin{bmatrix} \mathsf{X}\mathsf{X}^T & \mathsf{O} \\ \mathsf{O} & \mathsf{O}_n \end{bmatrix},$$

$$\mathsf{BA}^{+}\mathbf{b} = -C \begin{bmatrix} \mathsf{X} & \mathbf{o} & \mathsf{O} \\ \mathsf{O} & \mathbf{o} & \mathsf{O}_n \end{bmatrix} \begin{bmatrix} \mathbf{o}_d \\ 0 \\ \mathbf{e}_n \end{bmatrix} = \mathbf{o}_{2n},$$

$$\mathsf{R}^T\mathsf{B}^T = -\left[\mathsf{O}_{n+1,d}\ \mathsf{I}_{n+1}\right] \begin{bmatrix} \mathsf{X}^T & \mathsf{O}_{d,n} \\ \begin{bmatrix} \mathbf{y}^T & \mathbf{o}^T \\ \mathsf{I}_n & \mathsf{I}_n \end{bmatrix} \end{bmatrix} = -\begin{bmatrix} \mathbf{y}^T & \mathbf{o}_n^T \\ \mathsf{I}_n & \mathsf{I}_n \end{bmatrix},$$

$$\mathsf{R}^T\mathbf{b} = -C\left[\mathsf{O}_{n+1,d}\ \mathsf{I}_{n+1}\right] \begin{bmatrix} \mathbf{o}_d \\ \begin{bmatrix} 0 \\ \mathbf{e}_n \end{bmatrix} \end{bmatrix} = -C \begin{bmatrix} 0 \\ \mathbf{e}_n \end{bmatrix},$$

$$\mathsf{A}^{+}\mathbf{b} = -C \begin{bmatrix} \mathsf{I}_d & \mathbf{o} & \mathsf{O} \\ \mathbf{o}^T & 0 & \mathbf{o}^T \\ \mathsf{O} & \mathbf{o} & \mathsf{O}_n \end{bmatrix} \begin{bmatrix} \mathbf{o}_d \\ 0 \\ \mathbf{e}_n \end{bmatrix} = \mathbf{o}_{d+1+n}.$$

Denoting

$$\boldsymbol{\lambda} = \begin{bmatrix} \boldsymbol{\lambda}_1 \\ \boldsymbol{\lambda}_2 \end{bmatrix}, \quad \boldsymbol{\lambda}_1, \boldsymbol{\lambda}_2 \in \mathbb{R}^n,$$

we can rewrite the dual problem (16.7) as

$$\max -\frac{1}{2}\boldsymbol{\lambda}_1^T\mathsf{X}\mathsf{X}^T\boldsymbol{\lambda}_1 + \boldsymbol{\lambda}_1^T\mathbf{e}_n \text{ subject to } \boldsymbol{\lambda} \geq \mathbf{o},\ \mathbf{y}^T\boldsymbol{\lambda}_1 = 0, \text{ and } \boldsymbol{\lambda}_1 + \boldsymbol{\lambda}_2 = C\mathbf{e}_n,$$

or equivalently (without $\boldsymbol{\lambda}_2$ and the second set of the equality constraints) as

$$\min \theta(\boldsymbol{\lambda}_1) \text{ subject to } \mathbf{y}^T\boldsymbol{\lambda}_1 = 0,\ \mathbf{o} \leq \boldsymbol{\lambda}_1 \leq C\mathbf{e}_n,\ \boldsymbol{\lambda}_1 \in \mathbb{R}^n, \tag{16.11}$$

$$\theta(\boldsymbol{\lambda}) = \frac{1}{2}\boldsymbol{\lambda}^T\mathsf{X}\mathsf{X}^T\boldsymbol{\lambda} - \boldsymbol{\lambda}^T\mathbf{e}_n. \tag{16.12}$$

Substituting into (16.9) and (2.104) and denoting $\boldsymbol{\lambda} = \boldsymbol{\lambda}_1$, we get

$$\mathbf{w} = \mathsf{X}^T\boldsymbol{\lambda} \quad \text{and} \quad b = \overline{\alpha}_{d+1}. \tag{16.13}$$

Examining (16.11), we get that the constant C in (16.13) defines bounds on the Lagrange multipliers with a large C for relaxed constraints. We can achieve a similar effect, i.e., reducing the Lagrange multipliers, by adding a regularization term to the cost function θ. We shall get the modified dual cost function

$$\theta_2(\boldsymbol{\lambda}) = \frac{1}{2}\boldsymbol{\lambda}^T\left(\mathsf{X}\mathsf{X}^T + C^{-1}\mathsf{I}\right)\boldsymbol{\lambda} - \boldsymbol{\lambda}^T\mathbf{e}_n$$

and the *dual soft margin SVM problem with ℓ_2-loss function*

$$\min \theta_2(\boldsymbol{\lambda}) \text{ subject to } \boldsymbol{\lambda} \geq \mathbf{o}. \tag{16.14}$$

Notice that large C makes the regularization small, as in (16.11).

16.3 Numerical Experiments

Before embarking on numerical experiments, let us recall that our goal here is just to illustrate the performance of the MPRGP algorithm on SVM problems. The performance of the algorithm can be considerably improved by specific modifications based on the experience and particular features of the problems, see, e.g., successive 'shrinking' of the optimization problem in Joachims [176]. Here, we only use the observation that the number of support vectors is very small so that most dual constraints are active at the solution. The experiments were carried out with the PERMON software [227].

16.3.1 Mushroom and Phishing Classification Problem

Here we report the results of numerical experiments with the *mushrooms* and *pissing* classification datasets [34] carried out by Crisci et al. [43]. We use the *mushrooms* and *pissing* datasets to train classifiers to recognize edible mushrooms and malicious websites, respectively. The main characteristics of the datasets are summarized in Table 16.1.

Table 16.1. Characteristics of the classification datasets

Dataset	Samples+	Samples−	Features	Training data	Testing data
mashrooms	2830	2613	112	5443	2681
phising	4073	3333	68	7406	3649

Tables 16.2 and 16.3 report the results on the datasets by using the MPRGP algorithm with recommended parameters and a variant of the spectral gradient algorithm PBBF of Sect. 11.7.1 with

$$K = 10 \quad \text{and } \overline{\alpha} = 1.95 \quad \|\mathsf{A}\|^{-1}.$$

The computations were carried out with the relative precision of the projected gradient

$$\|\mathbf{g}^P(\mathbf{x}^k)\| \leq 0.1\|\mathbf{g}^P(\mathbf{x}^0)\|.$$

We provide for each method the number of Hessian multiplications, the computational time, and the accuracy

$$A = 100\frac{TP + TN}{TP + FP + TN + FN} \quad (\%) \tag{16.15}$$

of the classifier expressed in terms of true positive (TP), true negative (TN), false positive (FP), and false negative (FN) samples.

Table 16.2. Dataset *mashrooms* ℓ_2-loss function

C	Algorithm	Iterations	Time	Accuracy
1	MPRGP	962	23.20	99.89
1	PBBF	57	1.16	99.89
10	MPRGP	998	22.22	99.89
10	PBBF	72	1.53	100.00
100	MPRGP	994	14.17	99.89
100	PBBF	66	0.82	99.89

Table 16.3. Dataset *phishing* ℓ_2-loss function

C	Algorithm	Matrix× vector	Time	Accuracy
1	MPRGP	2503	59.50	93.92
1	PBBF	224	5.27	93.61
10	MPRGP	3231	76.97	94.08
10	PBBF	72	1.53	93.75
100	MPRGP	10,480	250.23	93.81
100	PBBF	4670	119.54	93.70

Looking at the tables, we find that PBBF outperforms MPRGP. The explanation is low requirements on the solution precision and a large active set identified much faster by the long steps of PBBF.

16.3.2 ExCAPE Data Set

The second benchmark reports the analysis results of a dataset from the ExCAPE project [122]. The project aims to predict compound bioactivity for the pharmaceutical industry. The tested dataset is related to the Pham protein database. It contains over 226100 samples with 2048 attributes. The goal of the experiments was to examine the efficiency of the MPRGP algorithm and to establish the relationship between the precision of the solution to the dual problem. We also estimated the training rate given by the above accuracy (16.15) and the value of the regularization constant C. The numerical experiments were carried out using PERMON software [227] and appeared first in Kružík et al. [196].

In each iteration, the approximate value of w was used to evaluate the margin $2/\|w\|$ and the training rate. The initial guess was set to zero vector. The relative precision $\varepsilon = 0.1$ of the projected gradient norm was used in all experiments. The results are in Table 16.4 and Figs. 16.3 and 16.4.

Table 16.4. Dataset *phishing* ℓ_2-loss function

C	Best accuracy	Converged accuracy	Best accuracy iter.	Converged acc. iter.
1e−5	76.87	73.15	4	24
1e−4	82.41	81.89	58	277
1e−3	84.34	83.93	756	2868
1e−2	84.52	84.13	8415	22,240
1e−1	84.69	84.06	9976	80,306

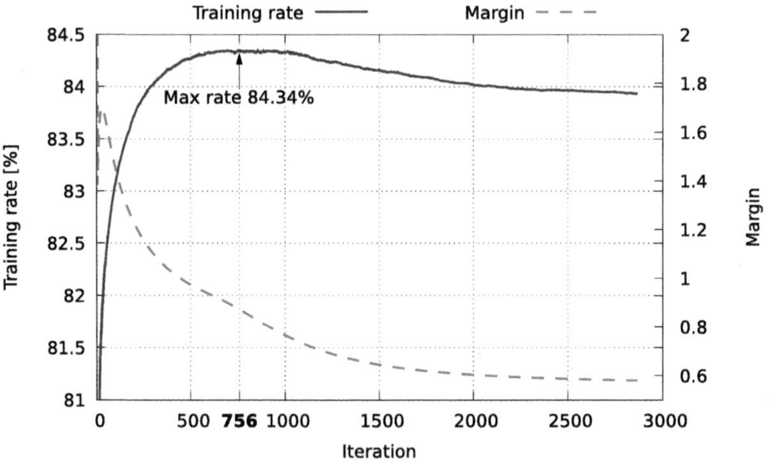

Fig. 16.3. ExCAPE dataset, $C = 1e-3$: The training rate and margin relation. The bold-printed iteration number corresponds to the maximum training rate

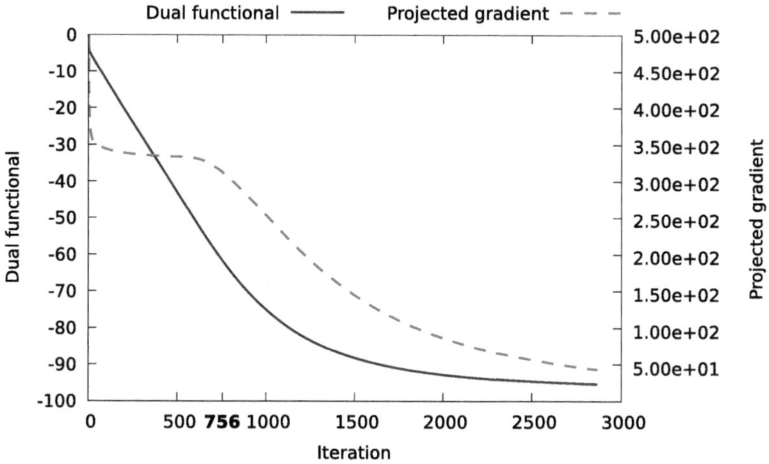

Fig. 16.4. ExCAPE dataset, $C = 1e - 3$: Dual function value and the norm of projected gradient. The maximum training rate was reached in bold-printed iteration number

17
PERMON and ESPRESO Software

Many algorithms presented in this book were implemented in PERMON and ESPRESO software packages.

17.1 PERMON

PERMON (Parallel Efficient Robust Modular Object Numerical) [227] is a collection of libraries for solution of QP problems, FETI domain decomposition methods, and support vector machines. Its development has started at the Department of Applied Mathematics of FEECS VSB-Technical University of Ostrava in 2011, since 2018 with participation of the Institute of Geonics of the Czech Academy of Sciences.

The main PERMON module is the PermonQP library, which consists of the object for QP problem definition, QP transformations, QP solvers, and supporting functions. PermonQP provides a number of QP transformations that are used to modify a given QP problem so that the formulation can be solved by the provided solvers and/or improved so that the solution is cheaper. PermonQP contains many solvers described in this book.

The second PERMON module is called PermonSVM. It provides a library and computer program to train linear support vector machine classifiers. PermonSVM uses PermonQP for the QP problems solution.

The PERMON libraries are written in C and utilize the PETSc library as the linear algebra backend. Moreover, the TAO optimization solvers and linear systems KSP solvers, including their preconditioners, can be utilized directly to solve QP problems. The PERMON libraries have the same object design as PETSc, making them easy to use for users familiar with PETSc. Both modules are published under the permissive BSD-2-Clause open-source license.

Z. Dostál, *Optimal Quadratic Programming and QCQP Algorithms with Case Studies*, Springer Optimization and Its Applications 23,
https://doi.org/10.1007/978-3-031-95167-1_17

17.2 ESPRESO

ESPRESO is an acronym for ExaScale PaRallel FETI SOlver developed at the IT4Innovations National Supercomputing Center in Ostrava to provide a highly efficient parallel solver. It enhances the H-TFETI method, which is designed to run on massively parallel machines with thousands of compute nodes and hundreds of thousands of CPU cores. The algorithms can be seen as a multilevel FETI method designed to overcome the main bottleneck of the standard FETI methods, a large coarse problem, which arises when solving large problems decomposed into a large number of subdomains.

The ESPRESO solver can use modern HPC accelerators to speed up the solver runtime. To achieve this, it uses a dense representation of sparse system matrices in the form of Schur complements. The main asset of using this implementation of the FETI solvers is the essential reduction of the iteration time. Instead of calling a solve routine of the sparse direct solver in every iteration, which is a sequential operation, the solver can use the dense matrix-vector multiplication (GEMV) routine. GEMV offers the parallelism required by modern accelerators and delivers up to a three-to-five-times speed-up, depending on the hardware configuration. More information on ESPRESO can be found on its website: https://numbox.it4i.cz.

Acronyms and Notations

Acronyms

ACQ	Abadie constraint qualification
BB	Barzilai and Borwein step-length, gradient method
CG	Conjugate Gradient method, algorithm, step, etc.
CGPP	CG with projector preconditioning
cgSLS	CG for semidefinite least square problems
FETI	Finite Element Tearing and Interconnecting
KKT	Karusch-Kuhn-Tucker conditions, pair
LCQ	Linear Constraint Qualification
MPGP	Modified Proportioning with Gradient Projections
MPRGP	Modified Proportioning with Reduced Gradient Projections
MPRGPS	MPRGP for problems with SPS Hessian
PCG	Preconditioned Conjugate Gradient method, step, etc.
PRGPN	Proportioning with Reduced Gradient Projection and Newton step
PBBF	Projected Barzilai-Borwein with Fallback
QCQP	Quadratically Constrained Quadratic Programming
QP	Quadratic Programming
RSVD	Reduced SVD
SMALBE-M	Semi-Monotonic Augmented Lagrangian algorithm for Bound and Equality constraints
SMALE	Semi-Monotonic Augmented Lagrangian for Equalities
SMALSE-M	Semi-Monotonic Augmented Lagrangian algorithm

Z. Dostál, *Optimal Quadratic Programming and QCQP Algorithms with Case Studies*, Springer Optimization and Its Applications 23,
https://doi.org/10.1007/978-3-031-95167-1

SMALSE-Mw	SMALSE-M for strong curvature of the feasible set boundary
SPD	Symmetric Positive Definite (matrix)
SPS	Symmetric Positive Semidefinite (matrix)
SVD	Singular Value Decomposition
SVM	Support Vector Machines
TFETI	Total FETI

Notations

$\mathbf{a}, \mathbf{b}, \mathbf{c}, \ldots$	vectors
$\mathbf{o}, \mathbf{o}_n$	zero vector
$[\mathbf{a}]_i = \mathbf{a}_i = a_i$	component of $\mathbf{a}$, $\mathbf{a} = [a_1, \ldots, a_n]^T$
$\mathbf{a}^T, \mathbf{b}^T, \mathbf{c}^T, \ldots$	transposed vectors
$(\mathbf{a}, \mathbf{b})$, $\mathbf{a}^T\mathbf{b}$	scalar product
$\mathrm{Span}\{\mathbf{a}_1, \ldots, \mathbf{a}_k\}$	span $\mathbf{a}_1, \ldots, \mathbf{a}_k$
$\mathcal{U}, \mathcal{V}, \ldots$	vector spaces
$\mathcal{U}^\perp$	orthogonal complement of $\mathcal{U}$
$\mathcal{U} \oplus \mathcal{V}$	direct sum
$\mathsf{A}, \mathsf{B}, \mathsf{C}, \ldots$	matrices
O, $\mathsf{O}_{m,n}$, $\mathsf{I}, \mathsf{I}_{m,n}$	zero and identity matrices
$[\mathsf{A}_{ij}] = a_{ij}$	(i,j)-th component of A
$\mathrm{Ker}\mathsf{A}$, $\mathrm{Im}\mathsf{A}$	null and image spaces of A
$\mathsf{A}\vert\mathcal{U}$	mapping $\mathbf{u} \to \mathsf{A}\mathbf{u}$ restricted to $\mathcal{U}$
A^T	transpose of A
A^{-1}	inverse of A (Sect. 1.4)
$\mathsf{A}^+, \mathsf{A}^\#$	left generalized inverse to A (Sect. 1.4)
$\mathsf{A}^\dagger$	Moore–Penrose generalized inverse of A Sect. (1.4)
λ_i,$\lambda_i(\mathsf{A})$	eigenvalue of A
$\sigma(\mathsf{A}) = \{\lambda_1, \ldots, \lambda_n\}$	spectrum of A
$\kappa(\mathsf{A})$	spectral condition number of A
$\overline{\kappa}(\mathsf{A})$	regular spectral condition number of A
σ_i, $\sigma_i(\mathsf{A})$	singular value of A
$\Vert \cdot \Vert_1$	ℓ_1 norm
$\Vert \cdot \Vert_2$	ℓ_2 (Euclidean) norm
$\Vert \cdot \Vert_\infty$	ℓ_∞ norm
$\Vert \cdot \Vert_\mathsf{A}$	A-norm (energy norm)
f	cost function
$\mathbf{g}, \nabla f, \nabla_\mathbf{x} L$	gradient of the cost function or Lagrangian
$\mathbf{g}^P = \boldsymbol{\varphi} + \boldsymbol{\beta}$	projected gradient (11.10)
$\widetilde{\mathbf{g}}_\alpha$	$(\alpha-)$ reduced gradient (11.20)
$\widetilde{\mathbf{g}}_\alpha^P = \boldsymbol{\varphi} + \boldsymbol{\beta}^\alpha$	$(\alpha-)$ reduced projected gradient (12.86)
$L(\mathbf{x}, \boldsymbol{\lambda})$, $L_0(\mathbf{x}, \boldsymbol{\lambda})$	Lagrangian (2.19)
$L(\mathbf{x}, \boldsymbol{\lambda}, \varrho)$	augmented Lagrangian (6.5)
$\mathsf{A}_\varrho = \mathsf{A} + \varrho\mathsf{B}^T\mathsf{B}$,	augmented matrix A
ϱ	regularization (penalization) parameter
Γ	balancing parameter

References

1. Antonelli, L., De Simone, V., di Serafino, D.: Spatially adaptive regularization in image segmentation, Algorithms **13**, 226–235 (2020)
2. Arrow, K.J., Hurwitz, L., Uzawa, H.: Studies in Nonlinear Programming. Stanford University Press, Stanford (1958)
3. Asmundis, R., Di Serafino, D., Riccio, F., Toraldo, G.: On spectral properties of steepest descent methods. IMA J. Numer. Anal. **33**, 1416–1435 (2013)
4. Avery, P., Rebel, G., Lesoinne, M., Farhat, C.: A numerically scalable dual–primal substructuring method for the solution of contact problems – part I: the frictionless case. Comput. Methods Appl. Mech. Eng. **193**, 2403–2426 (2004)
5. Axelsson, O.: A class of iterative methods for finite element equations. Comput. Methods Appl. Mech. Eng. **9**, 127–137 (1976)
6. Axelsson, O.: Iterative Solution Methods. Cambridge University Press, Cambridge (1994)
7. Axelsson, O.: Unified analysis of preconditioning methods for saddle point matrices. Num. Linear Algebra Appl. **22**, 233–253 (2015)
8. Axelsson, O., Lindskøg, G.: On the rate of convergence of the preconditioned conjugate gradient method. Numer. Math. **48**, 499–523 (1986)
9. Badea, L., Tai, X.-C., Wang, J.: Convergence rate analysis of a multiplicative Schwarz method for variational inequalities. SIAM J. Numer. Anal. **41**(3), 1052–1073 (2003)
10. Balay, S., Gropp, W., McInnes, L.C., Smith, B.: PETSc 2.0 Users Manual. Argonne National Laboratory. http://www.mcs.anl.gov/petsc/
11. Barzilai, J., Borwein, J.: Two-point stepsize gradient methods. IMA J. Numer. Anal. **8**, 141–148 (1988)
12. Bazaraa, M.S., Shetty, C.M., Sherali, H.D.: Nonlinear Programming, Theory and Algorithms, 2nd edn. Wiley, New York (1993)

Z. Dostál, *Optimal Quadratic Programming and QCQP Algorithms with Case Studies*, Springer Optimization and Its Applications 23,
https://doi.org/10.1007/978-3-031-95167-1

13. Belytschko, T., Neal, K.O.: Contact–impact by the pinball algorithm with penalty and Lagrangian methods. Int. J. Num. Methods Eng. **31**(3), 547–572 (1991)
14. Benzi, M., Golub, G.H., Liesen, J.: Numerical solution of saddle point problems. Acta Numer. 1–137 (2005)
15. Bertsekas, D.P.: Constrained Optimization and Lagrange Multiplier Methods. Academic Press, London (1982)
16. Bertsekas, D.P.: Nonlinear Optimization. Athena Scientific, Belmont (1999)
17. Bertsekas, D.P., Tsitsiklis, J.N.: Introduction to Linear Optimization. Athena Scientific, Belmont (1997)
18. Bielschowski, R.H., Friedlander, A., Gomes, F.A.M., Martínez, J.M., Raydan, M.: An adaptive algorithm for bound constrained quadratic minimization. Invest. Oper. **7**, 67–102 (1997)
19. Birgin, E.M., Martínez, J.M.: Practical Augmented Lagrangian Method. SIAM, Philadelphia (2014)
20. Birgin, E., Martínez, J.M., Ryadan, M.M.: Monotone spectral projected gradient method on convex sets. SIAM J. Optim. **10**, 1196–1211 (2000)
21. Blum, E., Oettli, W.: Direct proof of the existence theorem in quadratic programming. Oper. Res. **20**, 165–167 (1971)
22. Blum, H., Braess, D., Suttmeier, F.T.: A cascadic multigrid algorithm for variational inequalities. Comput. Vis. Sci. **7**(3–4), 153–157 (2004)
23. Bomze, I.M., Rinaldi, F., Zeffiro, D.: Active set complexity of the away-step Frank–Wolfe algorithm. SIAM J. Optim. **30**, 2470–2500 (2020)
24. Bouchala, J., Dostál, Z., Sadowská, M.: Scalable BETI for Variational Inequalities. In: Langer, U., et al. (eds.) Domain Methods in Science and Engineering XVII. Lecture Notes in Computational Science and Engineering (LNCSE), vol. 60, pp. 167–174. Springer, Berlin (2008)
25. Bouchala, J., Dostál, Z., Sadowská, M.: Theoretically supported scalable BETI Method for variational inequalities. Computing **82**, 53–75 (2008)
26. Bouchala, J., Dostál, Z., Vodstrčil, P: Separable spherical constraints and the decrease of a quadratic function in the gradient projection step. JOTA **157**, 132–140 (2013)
27. Bouchala, J., Dostál, Kozubek, T., Pospíšil, L., Vodstrčil, P.: On the solution of convex QPQC problems with elliptic and other separable constraints with strong curvature. Appl. Math. Comput. **247**, 848–864 (2014)
28. Bouchala, J., Dostál, Z., Kozubek, T., Pospíšil, L., Vodstrčil, P.: On the solution of convex QCQP problems with elliptic and other separable constraints. Appl. Math. Comput. **247**, 848–874 (2014)
29. Brenner, C.S.: The condition number of the Schur complement in domain decomposition. Numer. Math. **83**, 187–203 (1999)
30. Calamai, P.H., Moré, J.J.: Projected Gradient Methods for Linearly Constrained Problems. Math. Program. **39**, 93–116 (1987)

31. Calvetti, D., Reichel, L., Zhang, Q.: Conjugate gradient algorithms for symmetric inconsistent linear systems. In: Chu, M.T., Plemmons, R.J., Brown, J.D., Ellison, D.C. (eds.) Proceedings of the Lanczos Centenary Conference, pp. 267–272. SIAM, Philadelphia (1994)
32. Cauchy, A.: Méthode générale pour la résolution des systems d'équations simultanées. Comp. Rend. Sci. Paris **25**, 536–538 (1847)
33. Chen, Ke: Matrix Preconditioning Techniques and Applications. Cambridge University Press, Cambridge (2005)
34. Chih-Chung, C., Chih-Jen, L.: LIBSVM – A Library for Support Vector Machines. https://www.csie.ntu.edu.tw/~cjlin/libsvm/
35. Chvátal, V.: Linear Programming. W.H. Freeman and Company, New York (1983)
36. Coleman, T.F., Hulbert, L.A.: A globally and superlinearly convergent algorithm for convex quadratic programs with simple bounds. SIAM J. Optim. **3**, 298–321 (1993)
37. Coleman, T.F., Lin, Y.: An interior trust region approach for nonlinear minimization subject to bounds. SIAM J. Optim. **6**, 418–445 (1996)
38. Coleman, T.F., Liu, J.G.: An interior Newton method for quadratic programming. Math. Program. **85**, 491–523 (1999)
39. Conn, A.R., Gould, N.I.M., Toint, Ph.L.: A globally convergent augmented Lagrangian algorithm for optimization with general constraints and simple bounds. SIAM J. Numer. Anal. **28**, 545–572 (1991)
40. Conn, A.R., Gould, N.I.M., Toint, Ph.L.: LANCELOT: A FORTRAN Package for Large Scale Nonlinear Optimization (Release A), No. 17. Springer Series in Computational Mathematics. Springer, New York (1992)
41. Conn, A.R., Gould, N.I.M., Toint, Ph.L.: Trust Region Methods. SIAM, Philadelphia (2000)
42. Cottle, R., Pang, J., Stone, R.: The Linear Complementarity Problems. Academic Press, New York (1992)
43. Comparison of active-set and gradient projection-based algorithms for box-constrained quadratic programming. Soft Comput. **24**, 17761–17770 (2020)
44. Dai, Y.-H., Fletcher, R.: Projected Batzilai-Borwain gradient methods for large-scale box-constraine quadratic programming. Numer. Math. **100**, 21–47 (2005)
45. Dai,Y.-H., Liao L.-Z.: R-linear convergence of the Barzilai and Borwein gradient method. IMA J. Numer. Anal. **22**, 1–10 (2002)
46. Daniel, J. W.: The conjugate gradient method for linear and nonlinear operator equations. SIAM J. Numer. Anal. **4**, 10–26 (1967)
47. Deisenroth, M.P., Faisal, A.A., Ong, C.S.: Mathematics for Machine Learning. Cambridge University Press, Cambridge (2020)
48. Dembo, R.S., Tulowitzski, U.: On minimization of a quadratic function subject to box constraints. Working Paper No. 71, Series B, School of Organization and Management. Yale University, New Haven (1983)

49. Demmel, J.W.: Applied Numerical Linear Algebra. SIAM, Philadelphia (1997)
50. Diamond, M.A.: The solution of a quadratic programming problem using fast methods to solve systems of linear equations. Int. J. Syst. Sci. **5**, 131–136 (1974)
51. Diniz-Ehrhardt, M.A., Dostál, Z., Gomes-Ruggiero, M.A., Martínez, J.M., Santos, S.A.: Nonmonotone strategy for minimization of quadratics with simple constraints. Appl. Math. **46**(5), 321–338 (2001)
52. Diniz-Ehrhardt, M.A., Gomes-Ruggiero, M.A., Santos, S.A.: Numerical analysis of the leaving-face criterion in bound-constrained quadratic minimization. Optim. Methods Softw. **15**, 45–66 (2001)
53. Domahidi, A., Zraggen, U., Zeilinger, M.N., Morari, M., Jones, C.N.: Efficient interior methods for multistage problems arising in receding horizon control. In: Proceedings of 51st IEEE Conference on Decision and Control. IEEE Maui, Hawai (2012)
54. Domorádová, M.: Projector Preconditioning for the Solution of Large Scale Bound Constrained Problems. M.Sc. Thesis, FEECS VŠB—Technical University of Ostrava, Ostrava (2006)
55. Domorádová, M., Dostál, Z.: Projector preconditioning for partially bound constrained quadratic optimization. Numer. Linear Algebra Appl. **14**(10), 791–806 (2007)
56. Dorn, W.S.: Duality in quadratic programming. Quart. Appl. Math. **18**, 155–162 (1960)
57. Dorn, W.S.: Self-dual quadratic programs. J. Soc. Ind. Appl. Math. **9**, 51–54 (1961)
58. Dostál, Z.: The critical exponent of operators with constrained spectral radius Comment. Math. Univ. Carolin. **19**(2), 315–318 (1978)
59. Dostál, Z.: ℓ_∞-norm of iterates and the spectral radius of matrices. Math. Bohem. (Časopis pro pěstování matematiky) **105**(3), 256–260 (1980)
60. Dostál, Z.: Conjugate gradient method with preconditioning by projector. Int. J. Comput. Math. **23**, 315–323 (1988)
61. Dostál, Z.: On the penalty approximation of quadratic programming problem. Kybernetika **27**(2), 151–154 (1991)
62. Dostál, Z.: Directions of Large Decrease and Quadratic Programming. In: Marek, I. (ed.) Software and Algorithms of Numerical Mathematics 1993. University of West Bohemia, Charles University, and Union of Czech Mathematicians and Physicists, pp. 9–22 (1993)
63. Dostál, Z.: Box constrained quadratic programming with proportioning and projections. SIAM J. Optim. **7**(3), 871–887 (1997)
64. Dostál, Z.: Inexact solution of auxiliary problems in Polyak type algorithms. Acta Univ. Palacki. Olomuc. Fac. Rerum Nat. Math. **38**, 25–30 (1999)
65. Dostál, Z.: On preconditioning and penalized matrices. Numer. Linear Algebra Appl. **6**, 109–114 (1999)

66. Dostál, Z.: A proportioning based algorithm for bound constrained quadratic programming with the rate of convergence. Numer. Algorithms **34**(2–4), 293-302 (2003)
67. Dostál, Z.: Semi-monotonic inexact augmented Lagrangians for quadratic programming with equality constraints. Optim. Methods Softw. **20**(6), 715–727 (2005)
68. Dostál, Z.: Inexact semi-monotonic augmented Lagrangians with optimal feasibility convergence for quadratic programming with simple bounds and equality constraints. SIAM J. Numer. Anal. **43**(1), 96–115 (2005)
69. Dostál, Z.: Optimal algorithms for large sparse quadratic programming problems with uniformly bounded spectrum. In: Di Pillo, G., Roma, M. (eds.) Large-Scale Nonlinear Optimization. Nonconvex Optimization and Its Applications, vol. 83, pp. 83–93. Springer, Berlin (2006)
70. Dostál, Z.: An optimal algorithm for bound and equality constrained quadratic programming problems with bounded spectrum. Computing **78**, 311–328 (2006)
71. Dostál, Z.: Inexact semi-monotonic augmented Lagrangians with optimal feasibility convergence for quadratic programming with simple bounds and equality constraints. SIAM J. Numer. Anal. **45**(2), 500–513 (2007)
72. Dostál, Z.: An optimal algorithm for a class of equality constrained quadratic programming problems with bounded spectrum. Comput. Optim. Appl. **38**(1), 47–59 (2007)
73. Dostál, Z.: On the decrease of a quadratic function along the projected–gradient path. ETNA **31**, 25–29 (2008)
74. Dostál, Z.: On solvability of convex non-coercive quadratic programming problems. JOTA **143**, 413-416 (2009)
75. Dostál, Z.: Optimal Quadratic Programming Algorithms, with Applications to Variational Inequalities, 1st edn. Springer, New York (2009)
76. Dostál, Z., Beremlijski, P.: On convergence of inexact Augmented Lagrangians for separable and equality convex QCQP problems without constraint qualification. Adv. Electr. Electron. Eng. **15**(2), 242–250 (2017)
77. Dostál, Z., Horák, D.: Scalability and FETI based algorithm for large discretized variational inequalities. Math. Comput. Simul. **61**(3–6), 347–357 (2003)
78. Dostál, Z., Horák, D.: Scalable FETI with optimal dual penalty for semicoercive variational inequalities. Contemp. Math. **329**, 79–88 (2003)
79. Dostál, Z., Horák, D.: Scalable FETI with optimal dual penalty for a variational inequality. Numer. Linear Algebra Appl. **11**(5–6), 455–472 (2004)
80. Dostál, Z., Horák, D.: Theoretically supported scalable FETI for numerical solution of variational inequalities. SIAM J. Numer. Anal. **45**, 500–513 (2007)

81. Dostál, Z., Kozubek, T.: An optimal algorithm with superrelaxation for minimization of a quadratic function subject to separable constraints with applications. Math. Program. Ser. A **135**, 195–220 (2012)
82. Dostál, Z., Kučera, R.: An optimal algorithm for minimization of quadratic functions with bounded spectrum subject to separable convex inequality and linear equality constraints. SIAM J. Optim. **20** 2913–2938 (2010)
83. Dostál, Z., Pospíšil, L.: Minimizing quadratic functions with semidefinite Hessian subject to bound constraints. Comput. Math. Appl. **70**, 2014–2028 (2015)
84. Dostál, Z., Pospíšil, L.: Conjugate gradients for symmetric positive semidefinite least-squares problems. Int. J. Comput. Math. **95**(11), 2229–2239 (2018)
85. Dostál, Z., Schöberl, J.: Minimizing quadratic functions over non-negative cone with the rate of convergence and finite termination. Comput. Optim. Appl. **30**(1), 23–44 (2005)
86. Dostál, Z., Vlach, O.: An accelerated augmented Lagrangian algorithm with adaptive orthogonalization strategy for bound and equality constrained quadratic programming and its application to large-scale contact problems of elasticity. J. Comput. Appl. Math. **394**, 113565 (2021)
87. Dostál, Z., Vondrák, V.: Duality based solution of contact problems with Coulomb friction. Arch. Mech. **49**(3), 453–460 (1997)
88. Dostál, Z., Friedlander, A., Santos, S.A.: Solution of contact problems of elasticity by FETI domain decomposition. In: Mandel, J., Farhat, C., Cai, X.-C. (eds.) Domain Decomposition Methods 10. Contemporary Mathematics, vol. 218, pp. 82–93. AMS, Providence (1998)
89. Dostál, Z., Friedlander, A., Santos, S.A.: Adaptive precision control in quadratic programming with simple bounds and/or equalities. In: De Leone, R., Murli, A., Pardalos P.M., Toraldo, G. (eds.) High Performance Software for Non-linear Optimization. Applied Optimization, vol. 24, pp. 161–173. Kluwer, Dordrecht (1998)
90. Dostál, Z., Friedlander, A., Santos, S.A.: Augmented Lagrangians with adaptive precision control for quadratic programming with equality constraints. Comput. Optim. Appl. **14**, 37–53 (1999)
91. Dostál, Z., Gomes, F.A.M., Santos, S.A.: Duality based domain decomposition with adaptive natural coarse grid projectors for contact problems. In: Whiteman, J. (ed.) Proceedings of MAFELAP, pp. 259–270. Elsevier, Amsterdam (2000)
92. Dostál, Z., Gomes, F.A.M., Santos, S.A.: Solution of contact problems by FETI domain decomposition with natural coarse space projection. Comput. Methods Appl. Mech. Eng. **190**(13–14), 1611–1627 (2000)
93. Dostál, Z., Gomes, F.A.M., Santos, S.A.: Duality based domain decomposition with natural coarse space for variational inequalities. J. Comput. Appl. Math. **126**(1–2), 397–415 (2000)

94. Dostál, Z., Gomes, F.A.M., Santos, S.A.: FETI domain decomposition for contact 3D problems. In: Neittaanmäki, P., Křížek, M. (eds.) Proceedings of International Conference on Finite Element Method in Three-Dimensional Problems, GAKUTO International Series, Tokyo, Math. Sci. Appl., vol. 15, pp. 54–65 (2001)
95. Dostál, Z., Vondrák, V., Rasmussen, J.: FETI based semianalytic sensitivity analysis in contact shape optimization. In: Hoffmann, K.H., Hoppe, R.H.W., Schulz, V. (eds.) Fast Solution of Discretized Optimization Problems. International Series of Numerical Mathematics, vol. 138, pp. 98–106. Birkhäuser, Basel (2001)
96. Dostál, Z., Friedlander, A., Santos S.A., Alesawi, K.: Augmented Lagrangians with adaptive precision control for quadratic programming with equality constraints: corrigendum and addendum. Comput. Optim. Appl. **23**(1), 127–133 (2002)
97. Dostál, Z., Haslinger, J., Kučera, R.: Implementation of fixed point method for duality based solution of contact problems with friction. J. Comput. Appl. Math. **140**(1–2), 245–256 (2002)
98. Dostál, Z., Friedlander, A.,Gomes, F.A.M., Santos, S.A.: Preconditioning by projectors in the solution of contact problems: a parallel implementation. Ann. Oper. Res. **117**, 117–129 (2002)
99. Dostál, Z., Friedlander, A., Santos, S.A.: Augmented Lagrangians with adaptive precision control for quadratic programming with simple bounds and equality constraints. SIAM J. Optim. **13**, 1120–1140 (2003)
100. Dostál, Z., Horák, D., Kučera, R., Vondrák, V., Haslinger, J., Dobiáš, J., Pták, S.: FETI based algorithms for contact problems: scalability, large displacements, and 3D Coulomb friction. Comput. Methods Appl. Mech. Eng. **194**(2–5), 395–409 (2005)
101. Dostál, Z., Horák, D., Kučera, R.: Total FETI - an easier implementable variant of the FETI method for numerical solution of elliptic PDE. Commun. Numer. Methods Eng. **22**, 1155–1162 (2006)
102. Dostál, Z., Vondrák, V., Horák, D., Farhat, C., Avery, P.: Scalable FETI algorithms for frictionless contact problems. In: Langer, U., et al. (eds.) Domain Methods in Science and Engineering XVII. Lecture Notes in Computational Science and Engineering (LNCSE), vol. 60, pp. 263–270. Springer, Berlin (2008)
103. Dostál, Z., Domorádová, M., Sadowská, M.: Superrelaxation in minimizing quadratic functions subject to bound constraints. Comput. Optim. Appl. **48**(1), 23–44 (2010)
104. Dostál, Z., Kozubek, T., Vondrák, V., Brzobohatý, T., Markopoulos, A.: Scalable TFETI algorithm for the solution of multibody contact problems of elasticity. Int. J. Numer. Methods Eng. **82**, 1384–1405 (2010)
105. Dostál, Z., Kozubek, T., Markopoulos, A., Menšík, M.: Cholesky factorization of a positive semidefinite matrix with known kernel. Appl. Math. Comput. **217** 6067–6077 (2011)

106. Dostál, Z., Kozubek, T., Markopoulos, A., Brzobohatý, T., Vondrák, V., Horyl, P.: Theoretically supported scalable TFETI algorithm for the solution of multibody 3D contact problems with friction. Comput. Methods Appl. Mech. Eng. **205**, 110–120 (2012)
107. Dostál, Z., Horák, D., Vodstrčil, P.: On R-linear convergence of semi-monotonic inexact augmented Lagrangians for saddle point problems. Comput. Opt. Appl. **58**(1), 87–103 (2014)
108. Dostál, Z., Brzobohatý, T., Horák, D., Kozubek, T., Vodstrčil, P.: On R-linear convergence of semi-monotonic inexact augmented Lagrangians for bound and equality constrained quadratic programming problems with application. Comput. Math. Appl. **67**(3), 515–526 (2014)
109. Dostál, Z., Vlach, O., Brzobohat, T.: Scalable TFETI based algorithm with adaptive augmentation for contact problems with variationally consistent discretization of contact conditions. Finite Elem. Anal. Des. **156**, 24–43 (2019)
110. Dostál, Z., Horák, Brzobohatý, Vodstrčil, Bounds on the spectra of Schur complements of large H-TFETI-DP clusters for 2D Laplacian. Numer. Linear Algebra Appl. **28** (2), article no. e2344 (2021)
111. Dostál, Z., Horák, Brzobohatý, Kružík, J., Vlach, O.: Highly scalable hybrid domain decomposition method for the solution of huge scalar variational inequalities. Numer. Algorithms **91**, 773–801 (2022)
112. Dostál, Z., Brzobohatý, T., Horák, D., Kružík, J., Vlach, O.: Scalable hybrid TFETI-DP methods for large boundary variational inequalities. In: Brenner, S., et al. (eds.) Proc. of Domain Decomposition Methods in Science and Engineering XXVI. LNCSE, vol. 145, pp. 29–40. Springer, Cham (2022)
113. Dostál, Z., Brzobohatý, T., Vlach, O., Meca, M., Sadowská, M.: Hybrid TFETI domain decomposition with the clusters joined by faces' rigid modes for solving huge 3D elastic problems. Comput. Mech. **71**(2), 333–347 (2023)
114. Dostál, Z., Kozubek, T., Sadowská, M., Vondrák, V.: Scalable Algorithms for Contact Problems, 2nd edn. Birkhäuser, Cham (2023)
115. Dostál, Z., Horák, D., Kružík, J.: Tight upper bound on the heavy ball iterations' error (submitted, 2025)
116. Dureisseix, D., Farhat, C.: A numerically scalable domain decomposition method for solution of frictionless contact problems. Int. J. Numer. Methods Eng. **50**(12), 2643–2666 (2001)
117. Duvaut, G., Lions, J.-L.: Les Inéquations an Mécanique et Physique. Dunod, Paris (1972)
118. Eaves, B.C.: On quadratic programming. Manage. Sci. **17**, 698–711 (1971)
119. Eck, C., Jarůšek, J., Krbec, M.: Unilateral Contact Problems. Variational Methods and Existence Theorems. Chapman & Hall/CRC, Boca Raton (2006)

120. Elman, H.C., Sylvester, D.J.,Wathen, A.J.: Finite Elements and Fast Iterative Solvers: With Applications in Incompressible Fluid Dynamics. Oxford University Press, Oxford (2005)
121. ESPRESO - Highly Parallel Framework for Engineering Applications. http://numbox.it4i.cz
122. ExCAPE-Exascale Compound activity prediction. http://www.excape-h2020.eu
123. Farhat, C., Gérardin, M.: On the general solution by a direct method of a large scale singular system of linear equations: application to the analysis of floating structures. Int. J. Numer. Methods Eng. **41**, 675–696 (1998)
124. Farhat, C., Mandel J.: The two-level FETI method for static and dynamic plate problems – Part I: an optimal iterative solver for biharmonic systems. Comput. Methods Appl. Mech. Eng. **155**, 129–152 (1994)
125. Farhat, C., Mandel, J., Roux, F.-X.: Optimal convergence properties of the FETI domain decomposition method. Comput. Methods Appl. Mech. Eng. **115**, 365–385 (1994)
126. Farhat, C., Roux, F.-X.: A method of finite element tearing and interconnecting and its parallel solution algorithm. Int. J. Numer. Methods Eng. **32**, 1205–1227 (1991)
127. Farhat, C., Roux, F.-X.: An unconventional domain decomposition method for an efficient parallel solution of large-scale finite element systems. SIAM J. Sci. Comput. **13**, 379–396 (1992)
128. Farhat, C., Lesoinne, M., LeTallec, P., Pierson, K., Rixen, D.: FETI-DP: a dual-primal unified FETI method – a faster alternative to the two-level FETI method. Int. J. Numer. Methods Eng. **50**, 1523–1544 (2001)
129. Ferreau, H.J., Kozma, A., Potschka, A., Bock, H.G., Diehl, M.: qpOASES: a parametric active-set algorithm for quadratic programming. Math. Program. Comput. **6**(4), 1–37 (2014)
130. Fiedler, M., Pták, V.: On matrices with non-positive off-diagonal elements and positive principal minors. Czech. Math. J. **12**, 123–128 (1962)
131. Fletcher, R.: Practical Methods of Optimization. Wiley, Chichester (1997)
132. Fletcher, R.: On the Barzilai–Borwein Method. In: Qi, L., Teo, K., Yang, X. (eds.) Optimization and Control with Applications. Applied Optimization, vol. 96, pp. 235–256. Springer, Boston (2005)
133. Forsgren, A, Gill, Ph.E., Wright, M.H.: Interior methods for augmented nonlinear optimization. SIAM Rev. **44**, 525–597 (2002)
134. Fortin, M., Glowinski, R.: Augmented Lagrangian Methods: Applications to the Numerical Solution of Boundary Value Problems. North-Holland, Amsterdam (1983)
135. Fox, L., Huskey, H.D.,Wilkinson, J.H.: Notes on the solution of algebraic linear simultaneous equations. Quart. J. Mech. Appl. Math. **1**, 149–173 (1948)

136. Frank, M., Wolfe, P.: An algorithm for quadratic programming. Nav. Res. Logist. Q. **3**, 95–110 (1956)
137. Friedlander, M.P., Leyfer, S.: Global and finite termination of a two-phase augmented Lagrangian filter method for general quadratic programs. SIAM J. Sci. Comput. **30**(4), 1706–1729 (2008)
138. Friedlander, A., Martínez, J.M.: On the maximization of a concave quadratic function with box constraints. SIAM J. Optim. **4**, 177–192 (1994)
139. Friedlander, A., Martínez, J.M., Santos, S.A.: A new trust region algorithm for bound constrained minimization. Appl. Math. Optim. **30**, 235–266 (1994)
140. Friedlander, A., Martínez, J.M., Raydan, M.: A new method for large scale box constrained quadratic minimization problems. Optim. Methods Softw. **5**, 57–74 (1995)
141. Gass, S.I.: Linear Programming: Methods and Applications. McGraw Hill, New York (1958)
142. Giraud, L., Langou, J., Rozložník, M., van den Eshof, J.: Rounding error analysis of the classical Gram–Schmidt orthogonalization process. Numer. Math. **101**, 87–100 (2005)
143. Glowinski, R.: Variational Inequalities. Springer, Berlin (1980)
144. Glowinski, R., Le Tallec, P.: Augmented Lagrangian and Operator-Splitting Methods in Nonlinear Mechanics. SIAM, Philadelphia (1989)
145. Glowinski, R., Lions, J.-L., Trèmoliéres, R.: Numerical Analysis of Variational Inequalities. North-Holland, Amsterdam (1981)
146. Golub, G.H., O'Leary, D.P.: Some history of the conjugate gradient and Lanczos methods. SIAM Rev. **31**, 50–102 (1989)
147. Golub, G.H., Van Loan, C.F.: Matrix Computations, 4th edn. Johns Hopkins University Press, Baltimore (2013)
148. Gould, N.I.M., Toint, Ph.L.: A Quadratic Programming Bibliography. Rutherford Appleton Laboratory Numerical Analysis Group Internal Report 2000-1, Chilton Oxfordshire (2003) (see also ftp://ftp.numerical.rl.ac.uk/pub/qpbook/qpbook/bib)
149. Greenbaum, A.: Behaviour of slightly perturbed Lanczos and conjugate gradient algorithms. Linear Algebra Appl. **113**, 7–63 (1989)
150. Greenbaum, A.: Iterative Methods for Solving Linear Systems. SIAM, Philaledelphia (1997)
151. Greenbaum, A., Strakoš, Z.: Behaviour of slightly perturbed Lanczos and conjugate gradient algorithms. SIAM J. Matrix Anal. Appl. **22**, 121–137 (1992)
152. Grippo, L., Lampariello, F., Lucidi, S.: A nonmonotone line search technique for Newton's method. SIAM J. Numer. Anal. **23**(4), 707–716 (1986)
153. Griva, I., Nash, S.G., Sofer, A.: Linear and Nonlinear Optimization. SIAM, Philadelphia (2009)

154. Gwinner, J.: A penalty approximation for a unilateral contact problem in nonlinear elasticity. Math. Methods Appl. Sci. **11**(4), 447–458 (1989)
155. Hackbusch, W.: Multigrid Methods and Applications. Springer, Berlin (1985)
156. Hackbusch, W.: Iterative Solution of Large Sparse Systems. Springer, Berlin (1994)
157. Hager, W.W.: Dual techniques for constraint optimization. J. Optim. Theory Appl. **55**, 37–71 (1987)
158. Hager, W.W.: Applied Numerical Linear Algebra. Prentice Hall, Englewood Cliffs (1988)
159. Hager, W.W.: Analysis and implementation of a dual algorithm for constraint optimization. J. Optim. Theory Appl. **79**, 37–71 (1993)
160. Hager, W.W.: Iterative methods for nearly singular linear systems. SIAM J. Sci. Comput. **22**(2), 747–766 (2000)
161. Hager, W.W., Zhang, H.: A new active set algorithm for box constrained optimization. SIAM J. Optim. **17**, 526–557 (2006)
162. Haslinger, J., Kučera, R. Kozubek, T.: Convex programming with separable ellipsoidal constraints: application in contact problems with orthotropic friction. In: LCNSE, vol. 101, pp. 221–242 (2014)
163. Hestenes, C.: Multiplier and gradient methods. J. Optim. Theory Appl. **4**, 303–320 (1969)
164. Hestenes, M.R., Stiefel, E.: Methods of conjugate gradients for solving linear systems. J. Res. Natl. Bur. Stand. **49**, 409–436 (1952)
165. Heyn, T.D., Anitescu, M., Tasora, A., Negrut, D.: Using subspace and spectral methods for solving complementarity problems in many-body contact dynamic simulations. Int. J. Numer. Methods Eng. **95**, 541–561 (2013)
166. Hintermüller,M., Ito, K., Kunisch, K.: The primal–dual active set strategy as a semismooth Newton method. SIAM J. Optim. **13**, 865–888 (2003)
167. Hintermüller, M., Kovtumenko, V., Kunisch, K.: The primal–dual active set strategy as a semismooth Newton method. Adv. Math. Sci. Appl. **14**, 513–535 (2004)
168. Hlaváček, I., Haslinger, J., Nečas, J., Lovíšek, J.: Solution of Variational Inequalities in Mechanics. Springer, Berlin (1988)
169. Horák, D.: FETI based domain decomposition for variational inequalities. Ph.D. Thesis, FEECS VŠB–Technical University of Ostrava, Ostrava (2007)
170. Horák, D., Dostál, Z., Sojka, R.: On the efficient reconstruction of displacements in FETI methods for contact problems. Adv. Electr. Electron. Eng. **15**, 237–241 (2017)
171. Hrušková, P., Dostál, Z., Vlach, O., Vodstrčil, P.: On multipoint constraints in FETI methods. Appl. Math. **70**(1), 47–64 (2025)

172. Hüeber, S., Stadler, G., Wohlmuth, B.I.: A primal-dual active set algorithm for three-dimensional contact problems with Coulomb friction. SIAM J. Sci. Comput. **30**, 572–596 (2008)
173. Hunek, I.: On a penalty formulation for contact–impact problems. Comput. Struct. **48**(2), 193–203 (1993)
174. Iontcheva, A.H., Vassilevski, P.S.: Monotone multigrid methods based on element agglomeration coarsening away from the contact boundary for the Signorini's problem. Numer. Linear Algebra Appl. **11**(2–3), 189–204 (2004)
175. Ipsen, I.C., Mayer, C.D.: The angle between complementary subspaces. Amer. Math. Mon. **102**, 904–911 (1995)
176. Joachims, T.: Making large-scale SVM practical. In: Schölkopf, B., Burges, J.C., Smola, A.J. (eds.) Advances in Kernel Methods - Support Vector Learning. MIT Press, Cambridge (1998)
177. Kanto, Y., Yagawa, G.: A dynamic contact buckling analysis by the penalty finite element method. Int. J. Numer. Methods Eng. **29**(4), 755–774 (1990)
178. Kantorovich, L.V.: Functional analysis and applied mathematics. Uspekhy Mat. Nauk **3**, 89–185 (1948)
179. Keller, G., Gould, N.I.M., Wathen, A.J.: Constraint preconditioning for indefinite linear systems. SIAM J. Matrix Anal. Appl. **21**, 1300–1317 (2000)
180. Kikuchi, N., Oden, J.T.: Contact problems in elasticity. SIAM, Philadelphia (1988)
181. Kinderlehrer, D., Stampaccia, G.: An introduction to variational inequalities and their applications. Academic Press, London (1980)
182. Klawonn, A., Rheinbach, O.: A parallel implementation of dual–primal FETI methods for three dimensional linear elasticity using a transformation of basis. SIAM J. Sci. Comput. **28**(5), 1886–1906 (2007)
183. Klawonn, A.,. Rheinbach, O.: A hybrid approach to 3-level FETI. Proc. Appl. Math. Mech. **8**, 10841–10843 (2010)
184. Klawonn, A., Rheinbach, O.: Highly scalable parallel domain decomposition methods with an application to biomechanics. Z. Angew. Math. Mech. **90**, 5–32 (2010)
185. Klawonn, A., Widlund, O.: Dual–primal FETI method for linear elasticity. Commun. Pure Appl. Math. **LIX**, 1523–1572 (2006)
186. Klawonn, A., Widlund, O.B., Dryja, M.: Dual–primal FETI methods for three-dimensional elliptic problems with heterogeneous coefficients. SIAM J. Numer. Anal. **40**, 159–179 (2002)
187. Kornhuber, R.: Monotone multigrid methods for elliptic variational inequalities. II. Numer. Math. **72**(4), 481–499 (1996)
188. Kornhuber, R.: Adaptive Monotone Multigrid Methods for Nonlinear Variational Problems. Teubner–Verlag, Stuttgart (1997)
189. Kornhuber, R., Krause, R.: Adaptive multigrid methods for Signorini's problem in linear elasticity. Comput. Vis. Sci. **4**(1), 9–20 (2001)

190. Kozubek, T., Markopoulos, A., Brzobohatý, T., Kučera, R., Vondrák, V., Dostál, Z.: MatSol–MATLAB efficient solvers for problems in engineering. The library is no longer updated and is replaced by the ESPRESO framework. http://numbox.it4i.cz
191. Kraus, J., Margenov, S.: Robust Algebraic Multilevel Methods and Algorithms. Radon Series on Computational and Applied Mathematics, vol. 5. De Gruyter, Berlin (2009)
192. Krause, R.H.: On the Multiscale Solution of Constrained Minimization Problems. In: Langer, U., et al. (eds.) Domain Methods in Science and Engineering XVII. Lecture Notes in Computational Science and Engineering (LNCSE), vol. 60, pp. 93–104. Springer, Berlin (2008)
193. Krause, R.H., Wohlmuth, B.I.: A Dirichlet-Neumann type algorithm for contact problems with friction. Comput. Vis. Sci. **5**(3), 139–148 (2002)
194. Kruis, J.: Domain Decomposition Methods for Distributed Computing. Saxe-Coburg Publications, Stirling (2006)
195. Kruis, J., Matouš, K., Dostál, Z.: Solving laminated plates by domain decomposition. Adv. Eng. Softw. **33**(7–10), 445–452 (2002)
196. Kružík, J., Pecha, M., Hapla, V., Horák, D., Čermák, M.: Investigating convergence of linear SVM implemented in PermonSVM employing MPRGP algorithm. In: Kozubek, T., et al. (eds.) High Performance Computing in Science and Engineering. LCNS, vol. 11087. Springer, Cham (2017)
197. Kučera, R.: Convergence rate of an optimization algorithm for minimizaing quadratic functions with separable convex constraints. SIAM J. Optim. **19**(2), 846–862 (2008)
198. Kunisch, K., Rendl, F.: An infeasible active set method for quadratic problems with simple bounds. SIAM J. Optim. **14**(1), 35–52 (2002)
199. Lanczos, C.: Solution of systems of linear equations by minimized iterations. J. Res. Natl. Bur. Stand. **49**, 33–53 (1950)
200. Langer, U., Steinbach, O.: Boundary element tearing and interconnecting methods. Computing **71**, 205–228 (2003)
201. Laursen, T.: Computational Contact and Impact Mechanics. Springer, Berlin (2002)
202. Laursen, T., Simo, C.: An augmented Lagrangian treatment of contact problems involving friction. Comput. Struct. **42**(1), 97–116 (1992)
203. Li, D., Sun, R.: On a faster R-linear convergence rate of the Barzilai-Borwein method. arXiv (2021). https://doi.org/https://arxiv.org/abs/2101.00205v2
204. LIBSVM Data: Classification (Binary Class). https://www.csie.ntu.edu.tw/~cjlin/libsvmtools/datasets/binary.html
205. Lions, P.-L., Duvaut, G.: Les Inéquations en Méchanique et en Physique. Dunod, Paris (1972)
206. Lukáš, D., Dostál, Z.: Optimal multigrid preconditioned semi–monotonic augmented lagrangians applied to the Stokes problem. Numer. Linear Algebra Appl. **14**, 741–750 (2007)

207. Lukšan, L., Vlček, J.: Indefinitely preconditioned inexact Newton method for large sparse non-linear programming problems. Numer. Linear Algebra Appl. **5**, 157–187 (1999)
208. Luo, Z.Q., Tseng, P.: Error bounds and convergence analysis of feasible descent methods: a general approach. Ann. Oper. Res. **46**, 157–178 (1993)
209. Maciejowski, J.: Predictive Control with Constraints. Rearson Education, Harlow (2001)
210. Mandel, J.: Étude algébrique d'une méthode multigrille pour quelques problèmes de frontière libre (French). C. R. Acad. Sci. Ser. I **298**, 469–472 (1984)
211. Mandel, J., Tezaur, R.: On the convergence of a dual-primal substructuring method. Numer. Math. **88**, 543–558 (2001)
212. Mandel, J., Tezaur, R., Farhat, C.: A scalable substructuring method by Lagrange multipliers for plate bending problems. SIAM J. Numer. Anal. **36**, 1370–1391 (1999)
213. Marchuk, G.I., Kuznetsov, Yu.A.: Theory and applications of the generalized conjugate gradient method. Adv. Math. Suppl. Stud. **10**, 153–167 (1986)
214. Menšík, M.: Cholesky decomposition of positive semidefinite matrices with known kernel. Bc. Thesis, FEECS VŠB–Technical University of Ostrava, Ostrava (2007; in Czech)
215. Meurant, G., Strakoš, Z.: The Lanczos and conjugate gradient algorithms in finite precision arithmetic. Acta Numer. **15**, 471–542 (2006)
216. Mohy-ud-Din, H., Robinson, D.P.: A solver for nonconvex bound-constrained quadratic optimization. SIAM J. Optim. **25**(4), 2385–2407 (2015)
217. Moré, J.J., Toraldo, G.: On the solution of large quadratic programming problems with bound constraints. SIAM J. Optim. **1**, 93–113 (1991)
218. Nesterov, Yu.E.: A method of solving convex programming problem with convergence rate $O(1/k^2)$. Soviet. Math. Dokl. **27**, 372–376 (1983)
219. Nicolaides, R.A.: Deflation of conjugate gradients with applications to boundary value problems. SIAM J. Numer. Anal. **24**, 355–365 (1987)
220. Nocedal, J., Wright, S.F.: Numerical Optimization. Springer, New York (2000)
221. Of, G.: BETI - Gebietszerlegungsmethoden mit schnellen Randelementverfahren und Anwendungen. Ph.D. Thesis, University of Stuttgart (2006)
222. O'Leary, D.P.: A generalised conjugate gradient algorithm for solving a class of quadratic programming problems. Linear Algebra Appl. **34**, 371–399 (1980)
223. Oviedo, H., Dalmau, O. , Herrera, R.: An accelerated minimal gradient method with momentum for strictly convex quadratic optimization. BIT Numer. Math. **62**(2), 591–606 (2022)

224. Panagiotopoulos, P.D.: A nonlinear programming approach to the unilateral contact and friction boundary value problem. Ing.-Archiv **44**, 421–432 (1975)
225. Parlett, B.N., Scott, D.S.: The Lanczos algorithm with selective orthogonalization. Math. Comput. **33**, 217–238 (1979)
226. Pechstein, C.: Finite and Boundary Element Tearing and Interconnecting Solvers for Multiscale Problems. LNCSE 90. Springer, Heidelberg (2013)
227. PERMON – Parallel, Efficient, Robust, Modular, Object-oriented Numerical software toolbox. http://permon.vsb.cz/
228. Polyak, B.T.: Some methods of speeding up the convergence of iteration methods. USSR Comput. Math. Math. Phys. **4**(5), 791–803 (1964)
229. Polyak, B.T.: The conjugate gradient method in extremal problems. USSR Comput. Math. Math. Phys. **9**, 94–112 (1969)
230. Poole, G., Boullion, T.: A survey on M-Matrices. SIAM Rev. **16**(4), 419–427 (1974)
231. Pospíšil, L.: Development of Algorithms for Solving Minimizing Problems with Convex Quadratic Function on Special Convex Sets and Applications. Ph.D. Thesis, VSB-Technical University of Ostava (2015)
232. Pospíšil, L., Dostál, Z,: The projected Barzilai–Borwein method with fall-back for strictly convex QCQP problems with separable constraints. Math. Comput. Simul. **145**, 79–89 (2018)
233. Powell, M.J.D.: A method for nonlinear constraints in minimization problems. In: Fletcher, A. (ed.) Optimization, pp. 283–298. Academic Press, New York (1969)
234. Pshenichny, B.N., Danilin, Yu.M.: Numerical Methods in Extremal Problems. Mir, Moscow (1982)
235. Pták, V.: Spectral radius, norms of iterates, and the critical Exponent. Linear Algebra Appl. **1**, 245–260 (1968)
236. Raydan, M.: The Barzilai and Borwein gradient method for the large unconstrained minimization problem. SIAM J. Optim. **7**, 26–33 (1997)
237. Rennie, B.C.: An inequality which includes that of Kantorovich. Am. Math. Mon. **70**, 982 (1963)
238. Ronovský, A., Vašatova, A.: Elastic image registration based on domain decomposition with mesh adaptation. Adv. Electr. Electron. Eng. **15**(2), 322–330 (2017); Číslo: 2/2017 Periodikum: Advances in Electrical and Electronic Engineering
239. Rusten, T., Winther, R.: A preconditioned iterative method for saddle point problems. SIAM J. Matrix Anal. Appl. **13**, 887–904 (1992)
240. Saad, Y.: Iterative Methods for Large Linear Systems. SIAM, Philadelphia (2002)
241. Sadowská, M.: Scalable Total BETI for 2D and 3D Contact Problems. Ph.D. Thesis, FEECS VŠB-Technical University of Ostrava (2008)

242. Šantin, O.: Numerical Algorithms of Quadratic Programming for Model Predictive Control. Ph.D. Thesis, Czech Technical University, Prague (2016)
243. Šantin, O., Jarosová, M., Havlena, V., Dostál, Z.: Proportioning with second-order information for model predictive control. Optim. Methods Softw. **32**(3), 436–454 (2017)
244. Schöberl, J.: Solving the Signorini problem on the basis of domain decomposition techniques. Computing **60**(4), 323–344 (1998)
245. Schöberl, J.: Efficient contact solvers based on domain decomposition techniques. Int. J. Comput. Math. Appl. **42**, 1217–1228 (2001)
246. Simo, J.C., Laursen, T.A.: An augmented Lagrangian treatment in contact problems involving friction. Comput. Struct. **42**, 97–116 (1992)
247. Sluis, A. van der, Vorst, H.A. van der: The rate of convergence of the conjugate gradients. Numer. Math. **48**, 543–560 (1986)
248. Smith, R.L.: Some interlacing properties of the Schur complement of a Hermitian matrix. Linear Algebra Appl. **177**, 137–144 (1992)
249. Stewart, G.W.: Conjugate direction methods for solving systems of linear equations. Numer. Math. **21**, 285–297 (1973)
250. Stewart, G.W.: Error and perturbation bounds for subspace associated with certain eigenvalue problems. SIAM Rev. **15**(4), 727–763 (1973)
251. Strang, G.: Linear Algebra and Its Applications. Academic Press, New York (1976)
252. Szegö, G.: Orthogonal Polynomials. AMS, Providence (1967)
253. Tai, X.-C., Tseng, P.: Convergence rate analysis of an asynchronous space decomposition method for convex minimization. Math. Comput. **71**, 1105–1135 (2001)
254. Tarvainen, P.: Two-level Schwarz method for unilateral variational inequalities. IMA J. Numer. Anal. **14**, 1–18 (1998)
255. Tasora, A., Serban, R., Mazhar, A., Pazouki, H., Melanz, D., Fleischmann et al.: Chrono: an open source multi-physics dynamics engine. In Kozubek, T., et al. (eds.) Proceedings of High Performance Computing in Science and Engineering Conference. LNCS 9611, pp. 19–49. Springer, Berlin (2016)
256. Toselli, A., Widlund, O.B.: Domain Decomposition Methods – Algorithms and Theory. Springer Series on Computational Mathematics, vol. 34. Springer, Berlin (2005)
257. Ullmann, F.: FiOrdOs: A MATLAB Toolbox for C-code Generation for First Order Methods. M.Sc. Thesis, Swiss Federal Institute of Technology Zurich. ETH, Zurich (2011)
258. Vanderbei, R.J.: Linear Programming. International Series in Operations Research & Management Science, vol. 37. Springer, New York (2001)
259. Vorst, H. van der: Iterative Krylov Methods for Large Linear Systems. Cambridge University Press, Cambridge (2003)

260. Wohlmuth, B.I.: Discretization Methods and Iterative Solvers Based on Domain Decomposition. Springer, Berlin (2001)
261. Wang, Y., Boyd, S.: Fast model predictive control using online optimization. IEEE Trans. Control Syst. Technol. **18**(2), 267–278 (2010)
262. Watt, J., Borhani, R., Katsaggelos, A.K.: Machine Learning Refined. Cambridge University Press, Cambridge (2020)
263. Wohlmuth, B.I.: Variationally consistent discretization scheme and numerical algorithms for contact problems. Acta Numer. **20**, 569–734 (2011)
264. Wohlmuth, B., Krause, R.: Monotone methods on nonmatching grids for nonlinear contact problems. SIAM J. Sci. Comput. **25**, 324–347 (2003)
265. Wriggers, P.: Contact Mechanics. Springer, Berlin (2005)
266. Wright, S.J.: Primal-Dual Interior Point Methods. SIAM, Philadelphia (1997)
267. Yang, E.K., Tolle, J.W.: A class of methods for solving large, convex quadratic programs subject to box constraints. Math. Program. **51**, 223–228 (1991)
268. Yuan, Y.-X.: A new stepsize for the steepist descent method. J. Comput. Math. **24**, 149–156 (2006)
269. Zeidler, E.: Nonlinear Functional Analysis and Its Applications III: Variational Methods and Optimization. Springer, New York (1984)
270. Zeng, J., Zhou, S.: On monotone and geometric convergence of Schwarz methods for two-sided obstacle problem. SIAM J. Numer. Anal. **35**, 600–616 (1998)
271. Zou, Q., Magulés, F.: Fast gradient method with alignment for symmetric linear system without Cauchy step. J. Comput. Appl. Math. **381**, 113–133 (2021)

Index

Z. Dostál, *Optimal Quadratic Programming and QCQP Algorithms with Case Studies*, Springer Optimization and Its Applications 23,
https://doi.org/10.1007/978-3-031-95167-1

T

W

GPSR Compliance

The European Union's (EU) General Product Safety Regulation (GPSR) is a set of rules that requires consumer products to be safe and our obligations to ensure this.

If you have any concerns about our products, you can contact us on ProductSafety@springernature.com

In case Publisher is established outside the EU, the EU authorized representative is:

Springer Nature Customer Service Center GmbH
Europaplatz 3
69115 Heidelberg, Germany

Batch number: 09177073

Printed by Printforce, the Netherlands